Lecture Notes in
Mathematics

Edited by A. Dold and B. Eckmann

605

Leo Sario Mitsuru Nakai
Cecilia Wang Lung Ock Chung

Classification Theory of Riemannian Manifolds

Harmonic, quasiharmonic and biharmonic functions

Springer-Verlag
Berlin Heidelberg New York 1977

Authors

Leo Sario
Department of Mathematics
University of California
Los Angeles, CA 90024
USA

Cecilia Wang
Department of Mathematics
Arizona State University
Tempe, AZ 85281
USA

Mitsuru Nakai
Department of Mathematics
Nagoya Institute of Technology
Gokiso, Showa, Nagoya 466
Japan

Lung Ock Chung
Department of Mathematics
North Carolina State University
Raleigh, NC 27607
USA

Library of Congress Cataloging in Publication Data
Main entry under title:

Classification theory of Riemannian manifolds.

(Lecture notes in mathematics ; 605)
Bibliography: p.
Includes indexes.
1. Harmonic functions. 2. Riemannian manifolds.
I. Sario, Leo. II. Series: Lecture notes in mathematics (Berlin) ; 605.
QA3.L28 no. 605 [QA405] 510'.8s [515'.53] 77-22197

AMS Subject Classifications (1970): 31BXX

ISBN 3-540-08358-8 Springer-Verlag Berlin Heidelberg New York
ISBN 0-387-08358-8 Springer-Verlag New York Heidelberg Berlin

Printing and binding: Beltz Offsetdruck, Hemsbach/Bergstr.
2141/3140-543210

To

Angus E. Taylor

TABLE OF CONTENTS

CHAPTER II

QUASIHARMONIC FUNCTIONS

CHAPTER III

BOUNDED BIHARMONIC FUNCTIONS

CHAPTER V

BOUNDED DIRICHLET FINITE BIHARMONIC FUNCTIONS

CHAPTER VI

HARMONIC, QUASIHARMONIC, AND BIHARMONIC DEGENERACIES

CHAPTER VII

RIESZ REPRESENTATION OF BIHARMONIC FUNCTIONS

CHAPTER VIII

BIHARMONIC GREEN'S FUNCTION γ

CHAPTER IX

BIHARMONIC GREEN'S FUNCTION β: DEFINITION AND EXISTENCE

CHAPTER X

RELATION OF O^N_β TO OTHER NULL CLASSES

CHAPTER XI

HADAMARD'S CONJECTURE ON THE GREEN'S FUNCTION OF A CLAMPED PLATE

PREFACE AND HISTORICAL NOTE

The purpose of the present research monograph is to systematically develop a classification theory of Riemannian manifolds based on the existence or nonexistence of harmonic, quasiharmonic, and biharmonic functions with various boundedness properties. By definition, a function u is harmonic, quasiharmonic, or biharmonic, according as $\Delta u = 0$, $\Delta u = 1$, or $\Delta\Delta u = 0$, where Δ signifies the Laplace-Beltrami operator $-\mathrm{div\ grad} = d\delta + \delta d$.

With two exceptions, all results presented herein are new in that they have not appeared in any other book. The exceptions are the classical definitions of Δ in Chapter 0 and the basic harmonic inclusion relations I.1.11 and I.2.1 (see Table of Contents).

This monograph, the result of an eight year research project, is offered to the mathematical public with some serious claims to being a new theory, the object of which we have found exceptionally fascinating.

Six phases can be distinguished in the historical development of classification theory:

(1) Riemann's mapping theorem.

(2) The classical type problem.

(3) Harmonic and analytic classification of Riemann surfaces.

(4) Harmonic classification of Riemannian manifolds.

(5) Biharmonic classification of Riemannian manifolds.

(6) The biharmonic type problem.

We shall briefly discuss each of these six phases, the last three of which are the topic of the present book.

First phase

Riemann's mapping theorem classifies all simply connected plane regions into two types, according as the region is conformally equivalent to the disk or the plane. Riemann's use of the then unproved Dirichlet principle was criticized by Weierstrass, who exhibited a functional similar to the Dirichlet integral, the infimum of which was not furnished by any competing function. Schwarz and Neumann then developed their alternating methods which put Riemann's results on a sound basis. A rigorous proof of the Dirichlet principle was subsequently furnished by Hilbert.

These early results were subject to certain smoothness conditions imposed upon the boundary. We owe to Osgood the proof of the Riemann mapping theorem for arbitrary simply connected plane regions.

We have received Riemann's mapping theorem in mother's milk, as it were, and we do not always stop to think of the depth of the theorem. For an illuminating well-known example, take the square $\{0 < x < 1,\ 0 < y < 1\}$, with the segments $\{x = n^{-1},\ 0 < y \leq 2^{-1}\}$, $n = 1,2,\ldots$, deleted. That this region can be conformally mapped onto the disk despite the unimaginable distortions that take place near the origin is a result that amply satisfies Hardy's criterion for a deep theorem: it transcends intuition.

Second phase

The second phase in the history of classification theory started from the observation that the conformal equivalence of a simply connected region to the disk amounted to the existence of the harmonic Green's function. This existence problem was extended to arbitrary regions and gained momentum when Weyl, in the wake of Klein's idea of a "dachziegelartig" covering, introduced in his 1913 classic what we now call a manifold. In particular, he defined an abstract Riemann surface as a 2-manifold with conformal structure; his additional requirement of a countable basis was later shown redundant by Radó. The existence of the harmonic Green's function on Weyl's abstract Riemann surfaces, whether or not presented as covering surfaces of the complex plane, was the classical type problem. It was of primary interest to several leading analysts in the thirties and fourties. Kakutani related the problem to Brownian motion, Kobayashi introduced his nets, Nevanlinna employed the harmonic measure, Ahlfors used conformal metrics, Teichmüller Euclidean and Lobatschewsky metrics, and Myrberg symmetries. Despite striking results so obtained, the classical type problem remains unsolved to-date. Conditions are known that are either necessary or sufficient, but none that are both.

Third phase

From the existence of the harmonic Green's function there was a natural step to the third phase of classification theory: the problem of existence of harmonic functions with various boundedness properties. Relations between the classes of Riemann surfaces without such functions and those without the Green's function turned into perhaps the most challenging problem in the history of Riemann surfaces. A breakthrough was obtained by Ahlfors in his invited address before the Annual Meeting of the American Mathematical Society in New York in 1949. He established the strictness of one of the inclusion relations and set the stage for that of the others, soon obtained primarily by Tôki. The by now classical string of strict inclusions $O_G < O_{HP} < O_{HB} < O_{HD} = O_{HC}$ between classes of Riemann surfaces not carrying harmonic Green's functions or harmonic functions which are positive, bounded, Dirichlet finite, or bounded Dirichlet finite, respectively, was the crowning achievement in the theory of harmonic functions on Riemann surfaces.

A corresponding classification scheme of Riemann surfaces based on the existence of analytic functions was obtained by Ahlfors, Beurling, Pfluger, and others.

For the convenience of the reader who is further interested in the above first three phases in the development of classification theory, we list here some books that include accounts of various aspects of these phases:

(a) Analytic Function Theory, by E. Hille, Vol. II, Ginn & Co., 1962, 496 pp. (see pp. 320-321).

(b) Theorie der Riemannschen Flächen, by A. Pfluger, Springer-Verlag, Grundlehren 89, 1957, 248 pp.

(c) Riemann Surfaces, by L. Ahlfors and L. Sario, Princeton University Press, 1960, 382 pp.

(d) Value Distribution Theory, by L. Sario and K. Noshiro, in collaboration with T. Kuroda, K. Matsumoto, and M. Nakai, Van Nostrand (now Springer-Verlag), University Series, 1966, 236 pp.

(e) Principal Functions, by B. Rodin and L. Sario, in collaboration with M. Nakai, Van Nostrand (now Springer-Verlag), University Series, 1968, 347 pp.

(f) Capacity Functions, by L. Sario and K. Oikawa, Springer-Verlag, Grundlehren 149, 1969, 361 pp.

(g) Classification Theory of Riemann Surfaces, by L. Sario and M. Nakai, Springer-Verlag, Grundlehren 164, 1970, 446 pp.

Fourth phase

Why had the dimension been restricted to 2 in the first three phases of the history of classification theory? This question gave rise to the fourth phase, the harmonic classification of Riemannian manifolds. On them, harmonicity is well defined by the Laplace-Beltrami operator $\Delta = d\delta + \delta d$, regardless of the dimension, even or odd. To be sure, some problems, such as the strictness of $O_G^N < O_{HP}^N < O_{HB}^N$, were rendered trivial for $N > 2$ by N-space and punctured N-space. But in other problems, such as the strictness of $O_{HB}^N < O_{HD}^N$, the higher dimensions brought in challenging difficulties that were only recently overcome. The main gain in the shift to Riemannian manifolds was the availability of new aspects that were not meaningful on abstract Riemann surfaces. The L^p finiteness of the function and the completeness of the manifold are typical of these. An account of this fourth phase of classification theory is given in Chapter I of the present monograph.

Fifth phase

To understand the inauguration of the fifth phase of classification theory, the biharmonic classification of Riemannian manifolds, we have to go back to the origin of biharmonic functions and to Airy, Astronomer Royal. In fact, at this point we intentionally go somewhat beyond the topic at hand, as we are here bordering on some of the most dramatic events in the history of science.

When Newton was preparing the greatest scientific book ever written, he needed observational data on the solar system. These were provided for him (too slowly, much to Newton's anger) by the first Astronomer Royal, Flamsteed. His successor in this position, after a few generations, was Airy. About half way between the times of Flamsteed and Airy, Hershel had discovered something unbelievable: the millennia old belief in the sacred number 7 of the moving heavenly bodies was broken; Hershel had found a new planet, Uranus. But Uranus was misbehaving, though not badly: the naked eye could not have discerned the difference between its observed and theoretical positions. The mystery was finally solved by Adams, then a graduate student, who had accomplished the remarkable feat of computing from those infinitesimal disturbances of Uranus the existence and orbit of yet another planet, Neptune. Adams' result was the most striking proof yet of what order Newton had brought into the apparently chaotic motions of heavenly bodies by his simple inverse square law.

But Airy's telescope was too busy with scheduled observations to have time to look at Adams' new planet. Thus the glory of Neptune's discovery slipped to Leverrier in France, whose computations led Galle in Berlin to find Neptune.

As much as history may criticize Airy for this act, or lack of act, as an astronomer, it owes him a debt for his excursion into another field, with which he came in contact in his studies of the earth's crust: elasticity. Here he introduced the fundamental stress function now bearing his name. This leads us to another immortal in the history of science: Maxwell discovered that Airy's stress function was biharmonic. The discovery was made several years before Maxwell's main achievement, the electromagnetic equations. Biharmonic functions soon turned out to play a fundamental role in elasticity and hydrodynamics, and Académie des Sciences posed as its annual prize question the biharmonic Dirichlet problem for a clamped elastic plate. The problem was solved by Hadamard in his monumental 1908 memoir, to which we shall return later in connection with Hadamard's conjecture. In the hands of several other leading analysts of our century, the theory of biharmonic functions has further evolved into an elegant branch of modern analysis.

There has been a "gap", however: virtually the entire theory has been restricted to Euclidean regions. But biharmonicity is a locally defined concept, and the full richness of the theory cannot be expected on an arbitrarily chosen global carrier, such as a Euclidean region. A more natural carrier is a locally defined one. We recall what impetus the shift from the plane and 3-space to abstract Riemann surfaces gave to the theory of harmonic functions. Similar gains are likely in the case of biharmonic functions. However, biharmonicity is not meaningful on an abstract Riemann surface, since the Laplacian is not invariant under a parametric change. In contrast, the Laplace-Beltrami operator $\Delta = d\delta + \delta d$ is well defined on a Riemannian manifold and so is, a fortiori, the biharmonic operator $\Delta^2 = \Delta\Delta$. Developing a systematic biharmonic classification theory of Riemannian manifolds of any dimension thus appears to be of compelling importance.

This task, which initiated the fifth phase of classification theory, was set to a UCLA seminar given by the first named author in 1969-1976. Professors C. Wang, Y. K. Kwon, J. Rader, M. Range, D. Hada, N. Mirsky, and L. Chung participated in the seminar and substantially contributed to the theory. Our colleague at UCLA Professor B. Walsh collaborated with us on our very first paper on biharmonic functions. Professor M. Nakai visited UCLA in 1970-71, a period of vigorous collaboration with him that brought new problems, methods and results into the theory. With Professors Wang and Chung, the collaboration has continued ever since they joined our seminar in 1969 and 1971, respectively. The result of the eight year work, which somewhat concurrently covered both the fourth and fifth phases of classification theory, is presented in Chapters I-VII of this monograph.

Is it not premature to attempt a biharmonic classification of Riemannian manifolds since not even a topological classification has been carried out? Fortunately, the methods we shall use have no relation to the topology of the manifold. These two directions can be pursued independently. We are interested in analysis only.

What geometric information about the Riemannian manifold do we gain by our analytic considerations? None. Again, the book is written by and for analysts, and the Riemannian manifold only acts as the most natural locally defined carrier.

Apart from some brief sections on completeness, there is no discussion whatever of geometric properties of the manifold. Here the field lies wide open for further research. For some striking results in this direction we refer to the pioneering work of Greene and Wu on curvature and analysis.

If analysis only is our topic, does not the entire classification problem degenerate by the shift from Riemann surfaces to Riemannian manifolds? While on Riemann surfaces the conformal structure provides rigidity and thus lends significance to existence problems, is it not always possible to find on a given manifold a Riemannian metric which simultaneously gives any existence or non-existence properties one wishes? That this is not so is best shown by the impressive array of inclusion relations, many of them strict, obtained in Chapters II-VII. Those in VI.2.4 are particularly illuminating of this point:

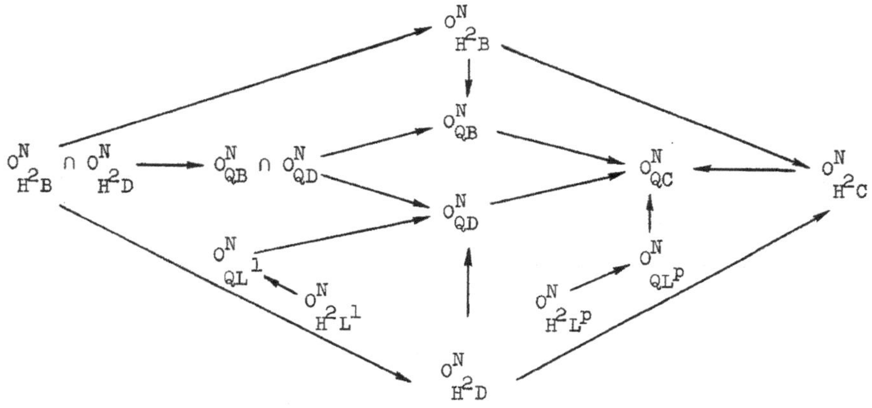

Here an arrow means strict inclusion.

<u>Sixth phase</u>

It is time to proceed to the sixth phase of the evolution of classification theory, the biharmonic type problem. By this we mean, in analogy with the classical harmonic type problem, the problem of existence of biharmonic Green's functions. It is noteworthy that, whereas the harmonic type problem chronologically preceded the general harmonic classification of Riemann surfaces, the order was reversed in the case of biharmonic functions on Riemannian manifolds. The reason is that phenomena related to Hadamard's conjecture, which we shall soon describe, entailed difficulties in defining the biharmonic Green's functions on a noncompact Riemannian manifold. The development thus struck the path of least resistance. First the theory was developed for regular biharmonic functions as presented in Chapters II-VII. Only after sufficient familiarity with biharmonic functions had been so gained, it was possible to successfully tackle the biharmonic Green's functions. The resulting theory of these functions, developed concurrently with the fifth phase of classification theory since 1974, in collaboration with Professor J. Ralston in 1975 and with Professor M. Nakai since 1975, is presented in Chapters VIII-XI.

A most fruitful contrast with harmonic Green's functions is that there are two biharmonic Green's functions, β and γ, according as the boundary conditions are $\beta = \partial\beta/\partial n = 0$ or $\gamma = \Delta\gamma = 0$. In the concrete case of a thin elastic plate under a point load, these functions are the deflections of a clamped or simply supported plate, respectively. For higher dimensions, we shall analogously speak of biharmonic Green's functions of clamped and simply supported bodies, although the physical meaning of deflection under a point load is lost. Of the two functions, γ is simpler and will be discussed first, in Chapter VIII. The rest of the book, Chapters IX-XI are devoted to the intriguing problem of the existence and properties of β.

There is another interesting contrast with the classical type problem. As so many other concepts that have richly endowed pure analysis, the harmonic Green's function was introduced by its initiator as a physical concept, the electrostatic potential of an electrical unit charge in a grounded system. From this concrete beginning the harmonic Green's function became a fruitful topic of pure analytic inquiry already quite early in this century. In striking contrast, the study of the biharmonic Green's functions has been largely restricted to their physical role as the deflection of a thin elastic plate under a point load. One of the main purposes of the present monograph is to discuss β and γ free from any physical connections, and on Riemannian manifolds of arbitrary dimension.

The aforementioned difficulty in defining a biharmonic Green's function on a noncompact Riemannian manifold does not concern γ, as the corresponding function γ_Ω on a regular subregion Ω increases with Ω. To describe the phenomenon related to Hadamard's conjecture that causes complications in the definition of β, we return to Hadamard's memoir of 1908. In addition to solving in it the biharmonic Dirichlet problem, he made the famous conjecture that the deflection of every thin elastic clamped plate under a point load is of constant sign. Some four decades later, the conjecture was disproved by various counterexamples constructed by Duffin, Garabedian, Loewner, and Szegö. It follows that β_Ω can be of nonconstant sign on even quite simple subregions Ω. It is this phenomenon that causes difficulty in defining β on a noncompact Riemannian manifold.

In Chapters IX-XI we first introduce a definition of β that is independent of the pole and the exhaustion, then compare the degeneracy of β with other degeneracies, present a new simple counterexample to Hadamard's conjecture, give a generalization to higher dimensions, and furnish a new unified noncomputational proof of some of the classical counterexamples.

About this book

The above historical survey of the fourth, fifth, and sixth phases tells what this book is about. Beyond that, we do not include in this Preface any description of the contents of the book, but relegate it to the beginning of each chapter and most sections. The reader is specifically referred to these "local" introductions which also contain essential information on the plan of the book. For a preliminary bird's-eye view of the book, we hope that the detailed Table of Contents is also useful.

The "Notes" at the end of each section give both bibliographical references and occasional suggestions for further research. Classification theory is far from being a closed book.

The Bibliography includes, we hope, all work on harmonic, quasiharmonic, and biharmonic classification theory on Riemannian manifolds.

The chapters can be read quite independently, as the cross-references pinpoint what results may be needed from an earlier chapter. Typically, a reader interested in Hadamard's conjecture can read directly Chapter XI, starting with its introduction.

The cross-reference system is simple: III.2.4 means Chapter III, Section 2, Subsection 2.4, and the theorem therein is referred to as Theorem III.2.4. Similarly, III.§2 stands for Chapter III, Section 2. In references within the same chapter, the chapter number is omitted.

In our presentation, no changes have been made for the sake of changes. Where the authors had no improvements to report, their original presentations have been closely followed.

The book is self-contained in that it only presupposes rudimentary knowledge of complex and real analysis, functional analysis, ordinary and partial differential equations, algebra, and differential geometry, routinely included in a normal Ph.D. curriculum. Beyond these "prerequisites", we occasionally use a well-known result from some standard monograph; an explicit reference, with or without page numbers, is then given.

The order of authors on the title page is that of seniority. The plan of the book was done by the senior author, and all four authors have done their full share in the preparation of the book.

The undersigned is fortunate to have again had, in 1970-71 and since 1975, the advantage of collaboration with his old friend Professor Mitsuru Nakai, with whom he has collaborated on three earlier monographs (see Bibliography). Professor Nakai's technical mastery has been a conditio sine qua non for the creation of much of the present theory.

Professor Cecilia Wang, whose doctoral dissertation it was my privilege to direct in 1969-70, has been my indispensable coauthor throughout these eight years. Without her unfailing devotion to the task and her rapid grasp of essentials, this large scale project of new work could never have been completed. Most of the development of the theory in 1971-75, in particular during the year 1973-74 when she was on the faculty of UCLA, was in collaboration with her.

Professor Lung Ock Chung, whose thesis advisor I also had the advantage of being in 1972-74, impressed us by rapidly shifting from his original field, logic, into mathematics, in which he never took an undergraduate course. In a short time, he not only acquainted himself with classical and modern analysis related to our field, but also, in a number of papers, settled some challenging problems we had encountered on biharmonic functions. He devoted the academic year 1974-75 to collaboration on this book.

Acknowledgments

Trying to create a new theory is a risky and thankless task: by definition, there is no previous knowledge and hence no a priori interest in it, and a non-expert is tempted to expect results serving established fields. We are, therefore, bound to gratitude to Professor M. Schiffer, who followed our work with interest through the years and who perused the entire manuscript. The exceptional combination of his expertise in classification theory, biharmonic functions, and Riemannian manifolds made his judgment particularly significant.

We are also indebted to Professor S. S. Chern for reading the manuscript and making his valued comments and to Professor B. Eckmann for including our book in this uniquely successful series.

We are deeply grateful to the U.S. Army Research Office for support throughout the seven years the preparation of this monograph has taken. Drs. A. S. Galbraith and J. Chandra cooperated in every way to make this book a reality.

The renowned efficiency of Springer-Verlag, who now also carries our monographs (d)-(g) listed on pages 3-4, is an axiom that needs no elaboration. The Mathematics Editor, Mr. R. Minio, was extremely helpful.

To Mrs. Elaine Barth, Mrs. Laurie Beerman, Ms. Charlotte Johnson, and Miss Julie Honig, this is the sixth book under our research projects for which they have typed, with speed and accuracy, numerous versions of the manuscript. If Olympics were held in the demanding art of mathematical typing, the UCLA team would probably walk off with all the medals.

Santa Monica, June, 1977

Leo Sario

CHAPTER 0

LAPLACE-BELTRAMI OPERATOR

A fundamental concept throughout this monograph is that of harmonicity on a Riemannian manifold. In the present preparatory chapter, we shall review the two main definitions of the Laplace-Beltrami operator.

We start ab ovo: in §1 we first briefly compile what will be needed of tensor analysis, and define a Riemannian manifold. This is followed by a detailed introduction of the operators grad, div, and, by means of these, the Laplace-Beltrami operator Δ.

Another approach to Δ is then presented in §2, where the elements of E. Cartan's exterior differential calculus are briefly reviewed, and Δ is defined in terms of the exterior derivative d and coderivative δ. Some basic formulas related to Δ are also deduced.

§1. RIEMANNIAN MANIFOLDS

1.1. Covariant and contravariant vectors. Let R be a connected, countable, oriented C^∞ manifold of dimension N, $2 \leq N < \infty$, with local coordinates $x = (x^1, x^2, \ldots, x^N)$. Under a change of coordinates from $x = (x^i)$ to $x' = (x'^p)$, the differential

$$dx'^p = \frac{\partial x'^p}{\partial x^i} dx^i$$

is the simplest contravariant vector, or contravariant tensor of the first order,

$$T'^p = \frac{\partial x'^p}{\partial x^i} T^i.$$

Here and later we use the Einstein convention: whenever an index, i, appears both in the upper and lower position, it is understood that summation for

$i = 1,\ldots,N$ is carried out.

The gradient of a function,

$$\frac{\partial f}{\partial x'^p} = \frac{\partial x^i}{\partial x'^p}\frac{\partial f}{\partial x^i},$$

exemplifies the simplest covariant vector, or covariant tensor of the first order,

$$T'_p = \frac{\partial x^i}{\partial x'^p} T_i.$$

The product of a covariant and a contravariant vector is invariant,

$$a'_p b'^p = a_i b^i.$$

In fact,

$$a'_p b'^p = \frac{\partial x^i}{\partial x'^p}\frac{\partial x'^p}{\partial x^j} a_i b^j = \delta^i_j a_i b^j = a_i b^i.$$

Here the Kronecker delta δ^i_j is the simplest example of a mixed tensor,

$$T'^p_q = \frac{\partial x'^p}{\partial x^i}\frac{\partial x^j}{\partial x'^q} T^i_j.$$

Indeed,

$$\frac{\partial x'^p}{\partial x^i}\frac{\partial x^j}{\partial x'^q}\delta^i_j = \frac{\partial x'^p}{\partial x^i}\frac{\partial x^i}{\partial x'^q} = \frac{\partial x'^p}{\partial x'^q} = \delta'^p_q.$$

1.2. Metric tensor. A contravariant tensor of the second order is, by definition,

$$T'^{pq} = \frac{\partial x'^p}{\partial x^i}\frac{\partial x'^q}{\partial x^j} T^{ij},$$

and a covariant tensor of the second order,

$$T'_{pq} = \frac{\partial x^i}{\partial x'^p}\frac{\partial x^j}{\partial x'^q} T_{ij}.$$

Suppose there is given on R a covariant tensor of the second order, g_{ij}, with the properties that each g_{ij} is C^∞ on the parametric ball of its definition, $g_{ij} = g_{ji}$, and $g = \det(g_{ij}) \neq 0$. The expression $g_{ij}dx^i dx^j$ is invariant,

$$g'_{pq}dx'^p dx'^q = \frac{\partial x^i}{\partial x'^p}\frac{\partial x^j}{\partial x'^q} g_{ij} \frac{\partial x'^p}{\partial x^n}\frac{\partial x'^q}{\partial x^m} dx^n dx^m$$

$$= \frac{\partial x^i}{\partial x^n}\frac{\partial x^j}{\partial x^m} g_{ij} dx^n dx^m = g_{ij}dx^i dx^j,$$

and defines an invariant arc element ds by

$$ds^2 = g_{ij}dx^i dx^j.$$

By definition, the tensor g_{ij} is the <u>Riemannian metric tensor</u>, or the <u>fundamental tensor</u>, and the manifold endowed with ds is a <u>Riemannian N-manifold</u> $R = \{R,ds\}$.

The arc element ds determines the geometry of R. In particular, since

$$g' = \det(g'_{pq}) = \left|\frac{\partial x^i}{\partial x'^p}\frac{\partial x^j}{\partial x'^q} g_{ij}\right|$$

$$= \left|\frac{\partial x^i}{\partial x'^p}\right|\left|\frac{\partial x^j}{\partial x'^q} g_{ij}\right|$$

$$= \left|\frac{\partial x^i}{\partial x'^p}\right|\left|\frac{\partial x^j}{\partial x'^q}\right| |g_{ij}| = \left|\frac{\partial x}{\partial x'}\right|^2 g$$

and

$$dx'^1 \cdots dx'^N = \left|\frac{\partial x'}{\partial x}\right| dx^1 \cdots dx^N,$$

we have the invariant <u>volume element</u>

$$dx = dV_x = g^{1/2}dx^1 \cdots dx^N.$$

The <u>conjugate metric tensor</u> g^{ij} is defined by the tensor equation

$$g^{ik}g_{kj} = \delta^i_j.$$

The fact that g^{ij} is a contravariant tensor is seen by first taking a tensor

$\bar{g}^{ij}$ which satisfies the above equation in a fixed parametric region. Since $\bar{g}^{ij}$ is a tensor, it satisfies the above tensor equation in every parametric region, and we have $g^{ij} = \bar{g}^{ij}$ on all of R.

By means of g^{ij}, we "raise the indices" of a covariant vector b_i,

$$b^i = g^{ij}b_j.$$

The resulting b^i is a contravariant vector:

$$b'^p = g'^{pq}b'_q = \frac{\partial x'^p}{\partial x^i}\frac{\partial x'^q}{\partial x^j}g^{ij}\frac{\partial x^k}{\partial x'^q}b_k$$

$$= \frac{\partial x'^p}{\partial x^i}\frac{\partial x^k}{\partial x^j}g^{ij}b_k = \frac{\partial x'^p}{\partial x^i}g^{ij}b_j = \frac{\partial x'^p}{\partial x^i}b^i.$$

Similarly, "lowering the indices" of a contravariant vector b^i by

$$b_i = g_{ij}b^j$$

gives a covariant vector b_i:

$$b'_p = g'_{pq}b'^q = \frac{\partial x^i}{\partial x'^p}\frac{\partial x^j}{\partial x'^q}g_{ij}\frac{\partial x'^q}{\partial x^k}b^k$$

$$= \frac{\partial x^i}{\partial x'^p}g_{ij}b^j = \frac{\partial x^i}{\partial x'^p}b_i.$$

Thus we can consider b_i and b^i as two component systems of the same vector b.

The product of two vectors a and b is defined as the <u>contraction</u>

$$a \cdot b = a^i b_i = a_i b^i,$$

and the length $|a|$ of a by

$$|a|^2 = a \cdot a.$$

An important vector in the present book is the gradient of a function φ, grad $\varphi = \nabla\varphi$, with covariant components

$$(\text{grad } \varphi)_i = \frac{\partial\varphi}{\partial x^i},$$

contravariant components

$$(\mathrm{grad}\ \varphi)^i = g^{ij} \frac{\partial\varphi}{\partial x^j},$$

length $|\nabla\varphi|$ with

$$|\nabla\varphi|^2 = (\nabla\varphi)_i \cdot (\nabla\varphi)^i = g^{ij} \frac{\partial\varphi}{\partial x^i} \frac{\partial\varphi}{\partial x^j},$$

and the Dirichlet integral

$$D(\varphi) = \int_R |\nabla\varphi|^2 dx = \int_R g^{ij} \frac{\partial\varphi}{\partial x^i} \frac{\partial\varphi}{\partial x^j} g^{1/2} dx^1 \dots dx^N.$$

Here and later, all functions to be considered are postulated sufficiently differentiable to justify the operations applied to them.

The unit vector n normal to the hypersurface $\varphi = c$ with $\nabla\varphi \neq 0$ is given by

$$n_i = \frac{\partial\varphi}{\partial x^i} |\nabla\varphi|^{-1},$$

or

$$n^i = g^{ij} \frac{\partial\varphi}{\partial x^j} |\nabla\varphi|^{-1},$$

and the directional derivative of a function f in the direction n by

$$\frac{\partial f}{\partial n} = \mathrm{grad}\ f \cdot n = g^{ij} \frac{\partial f}{\partial x^i} \frac{\partial\varphi}{\partial x^j} |\nabla\varphi|^{-1}.$$

1.3. Laplace-Beltrami operator. The divergence of a vector field b, $\mathrm{div}\ b = \partial b^i/\partial x^i$, generalizes from the Euclidean space to a Riemannian manifold as

$$\mathrm{div}\ b = g^{-1/2} \frac{\partial}{\partial x^i}(g^{1/2} b^i).$$

To see this, we have to prove that

$$g'^{-1/2} \frac{\partial}{\partial x'^p}(g'^{1/2} b'^p) = g^{-1/2} \frac{\partial}{\partial x^i}(g^{1/2} b^i),$$

where $g' = \det(g'_{pq})$. We shall use the formula

$$\frac{\partial}{\partial x^i} \log\left|\frac{\partial x'}{\partial x}\right| = \frac{\partial^2 x'^p}{\partial x^i\, \partial x^j} \frac{\partial x^j}{\partial x'^p}$$

to compute

$$g'^{-1/2} \frac{\partial}{\partial x'^p}(g'^{1/2} b'^p) = \frac{\partial b'^p}{\partial x'^p} + b'^p \frac{\partial}{\partial x'^p} \log g'^{1/2}.$$

Here

$$\frac{\partial b'^p}{\partial x'^p} = \frac{\partial}{\partial x'^p}\left(\frac{\partial x'^p}{\partial x^i} b^i\right) = b^i \frac{\partial^2 x'^p}{\partial x^j\, \partial x^i} \frac{\partial x^j}{\partial x'^p} + \frac{\partial x'^p}{\partial x^i} \frac{\partial x^k}{\partial x'^p} \frac{\partial b^i}{\partial x^k}$$

$$= b^i \frac{\partial}{\partial x^i} \log\left|\frac{\partial x'}{\partial x}\right| + \frac{\partial b^i}{\partial x^i}$$

and

$$b'^p \frac{\partial}{\partial x'^p} \log g'^{1/2} = b^i \frac{\partial}{\partial x^i}\left(\log g^{1/2} + \log\left|\frac{\partial x}{\partial x'}\right|\right).$$

Therefore,

$$\frac{\partial b'^p}{\partial x'^p} + b'^p \frac{\partial}{\partial x'^p} \log g'^{1/2} = \frac{\partial b^i}{\partial x^i} + b^i \frac{\partial}{\partial x^i} \log g^{1/2}$$

$$= g^{-1/2} \frac{\partial}{\partial x^i}(g^{1/2} b^i),$$

and we have established the invariance of div b.

The Laplace-Beltrami operator Δ acting on a function f is defined as

$$\Delta f = -\text{div grad } f.$$

A function f is called harmonic, quasiharmonic, or biharmonic according as $\Delta f = 0$, $\Delta f = 1$, or $\Delta^2 f = \Delta\Delta f = 0$, respectively.

In local coordinates, Δ has the invariant expression

$$\Delta f = -g^{-1/2} \frac{\partial}{\partial x^i}\left(g^{1/2} g^{ij} \frac{\partial f}{\partial x^j}\right),$$

which will be used throughout the book. In the Euclidean case is reduces to

$$\Delta f = -\sum_{i=1}^{N} \frac{\partial^2 f}{\partial x^{i2}} .$$

On Riemannian manifolds, we always choose the above minus sign to comply with the natural definition of Δ in terms of the exterior derivative and coderivative, to which we now proceed.

§2. HARMONIC FORMS

2.1. Differential p-forms. For further insight into the nature of harmonic functions, we shall now deduce the above expression of Δ in terms of the exterior differential calculus of E. Cartan. We start by briefly reviewing, with or without proofs, those rudiments of this calculus that we shall use in our later reasoning. For a comprehensive and rigorous presentation, we refer the reader to de Rham's classic [1].

The exterior, or wedge, product of two differentials satisfies, by definition, the conditions

$$dx^i \wedge dx^j = -dx^j \wedge dx^i, \qquad dx^i \wedge dx^i = 0,$$

$$f(dx^i \wedge dx^j) = (fdx^i) \wedge dx^j = dx^i \wedge (fdx^j).$$

As before, all functions to be considered will be postulated sufficiently differentiable to justify the operations applied to them. For $1 \le p \le N$, and identical sets of positive integers $(i_1,\dots,i_p)$ and $(j_1,\dots,j_p)$ in $[1,N]$, let $\varepsilon^{i_1\dots i_p}_{j_1\dots j_p}$ be the signature ± 1 of the permutation $\begin{pmatrix} i_1\dots i_p \\ j_1\dots j_p\end{pmatrix}$. By definition,

$$dx^{i_1} \wedge \dots \wedge dx^{i_p} = \varepsilon^{i_1\dots i_p}_{j_1\dots j_p}\, dx^{j_1} \wedge \dots \wedge dx^{j_p},$$

and the product vanishes if and only if at least two of the integers $i_1,\dots,i_p$

are the same. The definition of the signature is extended to any $i_1 < \cdots < i_p$ and $j_1 < \cdots < j_p$ by

$$\varepsilon^{i_1 \cdots i_p}_{j_1 \cdots j_p} = \begin{cases} 1 & \text{if each } i_k = j_k, \\ 0 & \text{otherwise.} \end{cases}$$

For a covariant tensor of the pth order, $\alpha_{i_1 \cdots i_p}$, $0 \leq p \leq N$,

$$\alpha_{i_1 \cdots i_p} dx^{i_1} \wedge \cdots \wedge dx^{i_p}$$

is invariant, with each index i running independently from 1 to N. If for any permutation $(j_1, \ldots, j_p)$ of $(i_1, \ldots, i_p)$, $\alpha_{i_1 \cdots i_p} = \varepsilon^{i_1 \cdots i_p}_{j_1 \cdots j_p} \alpha_{j_1 \cdots j_p}$, then we define a <u>differential p-form</u>

$$\alpha = \sum_{i_1 < \cdots < i_p} \alpha_{i_1 \cdots i_p} dx^{i_1} \wedge \cdots \wedge dx^{i_p},$$

$$= \frac{1}{p!} \alpha_{i_1 \cdots i_p} dx^{i_1} \wedge \cdots \wedge dx^{i_p}.$$

In view of the constant factor $1/p!$, we can henceforth use all tensor equations with $i_1 < \cdots < i_p$.

The vector space of differential p-forms, $0 \leq p \leq N$, is denoted by A^p, with A^0 the vector space of functions, and linear combinations of $\alpha, \beta \in A^p$ defined in an obvious manner. The exterior product of

$$\alpha = \sum_{i_1 < \cdots < i_p} \alpha_{i_1 \cdots i_p} dx^{i_1} \wedge \cdots \wedge dx^{i_p} \in A^p$$

and

$$\beta = \sum_{j_1 < \cdots < j_q} \beta_{j_1 \cdots j_q} dx^{j_1} \wedge \cdots \wedge dx^{j_q} \in A^q$$

is defined as

$$\alpha \wedge \beta = \sum_{i_1 < \cdots < i_p} \sum_{j_1 < \cdots < j_q} \alpha_{i_1 \cdots i_p} \beta_{j_1 \cdots j_q} dx^{i_1} \wedge \cdots \wedge dx^{i_p} \wedge dx^{j_1} \wedge \cdots \wedge dx^{j_q}.$$

__2.2.__ __Hodge operator.__ Associated with the covariant tensor $\alpha_{i_1 \cdots i_p}$ is the contravariant tensor

$$\alpha^{i_1 \cdots i_p} = g^{i_1 j_1} \cdots g^{i_p j_p} \alpha_{j_1 \cdots j_p}.$$

For $\alpha, \beta \in A^p$, we define the invariant __local inner product__

$$\alpha \cdot \beta = \sum_{i_1 < \cdots < i_p} \alpha^{i_1 \cdots i_p} \beta_{i_1 \cdots i_p} = \sum_{i_1 < \cdots < i_p} \alpha_{i_1 \cdots i_p} \beta^{i_1 \cdots i_p},$$

and the __global inner product__ over a Riemannian manifold R,

$$(\alpha, \beta) = \int_R \alpha \cdot \beta \, dx = \int_R \alpha \cdot \beta \, g^{1/2} dx^1 \wedge \cdots \wedge dx^N.$$

We shall seek a differential $(N - p)$-form, to be denoted by $*\beta$, such that (α,β) can be conveniently written as

$$(\alpha, \beta) = \int_R \alpha \wedge *\beta = \int_R \beta \wedge *\alpha.$$

This is the only purpose of introducing the Hodge star operator $*$. It is readily verified that the above condition is met by

$$*\beta = (*\beta)_{j_1 \cdots j_{N-p}} dx^{j_1} \wedge \cdots \wedge dx^{j_{N-p}},$$

with $j_1 < \cdots < j_{N-p}$ the indices complementary to $i_1 < \cdots < i_p$, and

$$(*\beta)_{j_1 \cdots j_{N-p}} = \varepsilon^{1 \cdots N}_{i_1 \cdots i_p \; j_1 \cdots j_{N-p}} \beta^{i_1 \cdots i_p} g^{1/2}.$$

In particular,

$$*1 = dx.$$

At any given point of R, we can find an orthonormal coordinate system $x^1, \ldots, x^N$, that is, such that $ds^2 = \Sigma_1^N dx^{i2}$ at that point. To this end, we simply take a suitable linear combination of coordinates in any given system in a neighborhood of that point. To prove any tensor equation, or its equivalent for $i_1 < \cdots < i_p$, at a point, it suffices to establish it in this special coordinate system.

Here we show that, for any $\alpha \in A^p$,

$$**\alpha = (-1)^{Np+p}\alpha.$$

Indeed, in an orthonormal coordinate system, $g_{ij} = g^{ij} = \delta^i_j$, $g^{1/2} = 1$, $\alpha^{i_1\cdots i_p} = \alpha_{i_1\cdots i_p}$, and therefore,

$$*\alpha = \varepsilon^{1\cdots N}_{i_1\cdots i_p\ j_1\cdots j_{N-p}}\ \alpha^{i_1\cdots i_p}\ dx^{j_1} \wedge \cdots \wedge dx^{j_{N-p}}.$$

A fortiori,

$$**\alpha = \varepsilon^{j_1\cdots j_{N-p}\ i_1\cdots i_p}_{i_1\cdots i_p\ j_1\cdots j_{N-p}}\ \alpha_{i_1\cdots i_p}\ dx^{i_1} \wedge \cdots \wedge dx^{i_p} = (-1)^{Np+p}\ \alpha.$$

In particular, $**1 = *dx = 1$, which gives

$$*(dx^1 \wedge \cdots \wedge dx^N) = g^{-1/2}.$$

2.3. Exterior derivative and coderivative. The exterior derivative d is an operator from A^p to A^{p+1} defined for $f \in A^0$ by

$$df = \frac{\partial f}{\partial x^i}\, dx^i,$$

and for $\alpha \in A^p$ by

$$d\alpha = \sum_{i_1<\cdots<i_p} \frac{\partial \alpha_{i_1\cdots i_p}}{\partial x^i}\, dx^i \wedge dx^{i_1} \wedge \cdots \wedge dx^{i_p}.$$

For $\alpha \in A^N$, we have $d\alpha = 0$, and for $\alpha \in A^p$, $\beta \in A^q$,

$$d(\alpha \wedge \beta) = d\alpha \wedge \beta + (-1)^p\, \alpha \wedge d\beta.$$

In view of

$$d^2\alpha = \frac{1}{2} \sum_{i_1<\cdots<i_p} \left(\frac{\partial^2\alpha_{i_1\cdots i_p}}{\partial x^i\,\partial x^j} - \frac{\partial^2\alpha_{i_1\cdots i_p}}{\partial x^j\,\partial x^i} \right) dx^i \wedge dx^j \wedge dx^{i_1} \wedge \cdots \wedge dx^{i_p},$$

we obtain

$$d^2\alpha = 0.$$

The <u>coderivative</u> δ is an operator from A^p to A^{p-1} defined for $\alpha \in A^p$ by

$$\delta\alpha = (-1)^{Np+N+1} {*d*\alpha}.$$

The motivation of this definition is a search for an operator δ such that, for $\alpha \in A^{p-1}$, $\beta \in A^p$, and either $\operatorname{supp} \alpha$ or $\operatorname{supp} \beta$ compact,

$$(d\alpha,\beta) = (\alpha,\delta\beta).$$

We merely sketch the proof. By virtue of $d(\alpha \wedge *\beta) = d\alpha \wedge *\beta + (-1)^{p-1}\alpha \wedge d*\beta$ and the compactness of $\operatorname{supp}(\alpha \wedge *\beta)$,

$$\begin{aligned}
(d\alpha,\beta) &= \int_R d\alpha \wedge *\beta = (-1)^p \int_R \alpha \wedge d*\beta \\
&= (-1)^{p+N(N-p+1)+N-p+1} \int_R \alpha \wedge **d*\beta \\
&= (-1)^{Np+N+1} \int_R \alpha \wedge **d*\beta \\
&= \int_R \alpha \wedge *\delta\beta = (\alpha,\delta\beta),
\end{aligned}$$

as claimed.

By $d\alpha = 0$ for $\alpha \in A^N$, we have $\delta f = 0$ for $f \in A^0$. In view of $\delta^2\alpha = \pm *d**d*\alpha = \pm *dd*\alpha$ and $d^2(*\alpha) = 0$,

$$\delta^2\alpha = 0$$

for all $\alpha \in A^p$.

<u>2.4.</u> <u>Laplace-Beltrami operator.</u> For $\alpha \in A^p$, we define the Laplace-Beltrami operator Δ by

$$\Delta\alpha = (d\delta + \delta d)\alpha.$$

A p-form α is <u>harmonic</u> if $\Delta\alpha = 0$.

We have $\Delta : A^p \to A^p$. For $\beta \in A^p$ and $\operatorname{supp}\alpha$ or $\operatorname{supp}\beta$ compact,

$$(\Delta\alpha,\beta) = (\alpha,\Delta\beta).$$

Indeed,

$$((d\delta + \delta d)\alpha,\beta) = (\delta\alpha,\delta\beta) + (d\alpha,d\beta) = (\alpha,(d\delta + \delta d)\beta).$$

For $f \in A^0$, $\Delta f = \delta df = (-1)^{N\cdot 1+N+1} *d*df$ so that

$$\Delta f = - *d*df.$$

Here,

$$*df = *\left(\frac{\partial f}{\partial x^i}\, dx^i\right) = \varepsilon^{1\cdots N}_{i1\cdots\hat{i}\cdots N}\, g^{1/2}\, g^{ij}\, \frac{\partial f}{\partial x^j}\, dx^1 \wedge \cdots \wedge \widehat{dx^i} \wedge \cdots \wedge dx^N,$$

with the roof symbol standing for omission. It follows that

$$d*df = (-1)^{i-1} \frac{\partial}{\partial x^i}\left(g^{1/2}\, g^{ij}\, \frac{\partial f}{\partial x^j}\right)(-1)^{i-1}\, dx^1 \wedge \cdots \wedge dx^N$$

and therefore,

$$\Delta f = -g^{-1/2} \frac{\partial}{\partial x^i}\left(g^{1/2}\, g^{ij}\, \frac{\partial f}{\partial x^j}\right).$$

Thus we have obtained the same local expression for Δf both as $-\operatorname{div}\operatorname{grad} f$ and $(d\delta + \delta d)f$.

For the product of two functions, the classical formula

$$\Delta(\varphi\psi) = \Delta\varphi \cdot \psi + \varphi \cdot \Delta\psi - 2\nabla\varphi \cdot \nabla\psi$$

remains valid, with the minus sign reflecting the convention on the sign of Δ. In fact,

$$\Delta(\varphi\psi) = -\left(\frac{\partial}{\partial x^i} \log g^{1/2}\right) \cdot g^{ij}\left(\frac{\partial\varphi}{\partial x^j} \cdot \psi + \varphi\, \frac{\partial\psi}{\partial x^j}\right) - \frac{\partial}{\partial x^i}\left[g^{ij}\left(\frac{\partial\varphi}{\partial x^j} \cdot \psi + \varphi\, \frac{\partial\psi}{\partial x^j}\right)\right].$$

For $\alpha \in A^{N-1}$ and a regular subregion Ω of R, Stokes' formula

$$\int_{\partial\Omega} \alpha = \int_{\Omega} d\alpha$$

give all the basic relations between volume and surface integrals. Again, we only sketch the proof and refer to the rigorous treatment in, e.g., de Rham [1]. For the Euclidean plane and $\alpha = a(x,y)dx + b(x,y)dy$, we have $d\alpha = (\partial b/\partial x - \partial a/\partial y)dxdy$, and are, therefore, dealing with the elementary formula, proved by integrating the two terms of $d\alpha$ with respect to x and y, respectively. For an arbitrary R, a suitable subdivision of Ω permits us to assume that $\bar{\Omega}$ lies in a parametric region. Then for

$$\alpha = \alpha_{1\ldots\hat{i}\ldots N} dx^1 \wedge \ldots \wedge \widehat{dx^i} \wedge \ldots \wedge dx^N,$$

we have

$$\int_{\Omega} d\alpha = \int_{\Omega} \frac{\partial\alpha_{1\ldots\hat{i}\ldots N}}{dx^i} dx^i \wedge dx^1 \wedge \ldots \wedge \widehat{dx^i} \wedge \ldots \wedge dx^N,$$

which after integrating with respect to each x^i, in analogy with the case $N = 2$, can be seen to take the form

$$\int_{\partial\Omega} \alpha_{1\ldots\hat{i}\ldots N} dx^1 \wedge \ldots \wedge \widehat{dx^i} \wedge \ldots \wedge dx^N = \int_{\partial\Omega} \alpha .$$

One of the simple consequences of Stokes' formula we shall often use is

$$\int_{\partial\Omega} u{*}dv - v{*}du = -\int_{\Omega} u{*}\Delta v - v{*}\Delta u$$

for $u,v \in A^0$. We have

$$\int_{\partial\Omega} u{*}dv - v{*}du = \int_{\Omega} du \wedge {*}dv - dv \wedge {*}du + \int_{\Omega} ud{*}dv - vd{*}du,$$

where the first integral vanishes by $(du,dv) = (dv,du)$, and the second integral is

$$(-1)^{N^2+N}\int_\Omega u{*}{*}d{*}dv - v{*}{*}d{*}du = -\int_\Omega u{*}\Delta v - v{*}\Delta u$$

as maintained.

CHAPTER I

HARMONIC FUNCTIONS

We designate the family of harmonic functions by H, and the subfamily of functions in a given class X by HX. We are interested in the classes $X = P, B, D, C$ of functions which are positive, bounded, Dirichlet finite, or bounded and Dirichlet finite, respectively. Let O^N_{HX} be the class of Riemannian N-manifolds R, $N \geq 2$, for which $HX(R)$ consists of constant, and denote by O^N_G the class of parabolic N-manifolds, characterized by the nonexistence of harmonic Green's functions g. In §1, we establish the strict inclusion relations $O^N_G < O^N_{HP} < O^N_{HB}$. For $N = 2$, we base the proof of the strictness on the original counterexamples establishing these relations for Riemann surfaces. For $N > 2$, these relations are trivial.

In §2, we show that $O^N_{HB} < O^N_{HD} = O^N_{HC}$. It is of interest that, although other inclusion relations carry over without difficulty from Riemann surfaces to Riemannian manifolds, the strictness of the inclusion $O^N_{HB} < O^N_{HD}$ for $N > 2$ was established only recently, over a decade after the other relations. To prove the strictness, we introduce a concept which will be a proving ground, and a recurring theme, throughout the present book: the Poincaré N-ball. It is the unit N-ball with the Poincaré-type metric $ds = (1 - r^2)^\alpha |dx|$, with α a constant. The Poincaré N-ball is ideally suited for testing the dependence, in a concrete setup, of the existence of functions with various boundedness properties on the metric of a fixed base manifold.

Using a Riemannian manifold, rather than a Riemann surface, as the carrier, has the additional advantage that the important class of L^p functions becomes meaningful. In §3, we give a systematic discussion of $O^N_{HL^p}$.

The chapter closes with a brief account, in §4, of the relationships of the harmonic null classes O^N_{HX} and completeness of a manifold.

§1. RELATIONS $O_{\omega}^N = O_G^N < O_{HP}^N < O_{HB}^N$

1.1. Definitions. By a Riemannian manifold we shall mean a noncompact, connected, oriented C^∞ manifold of dimension $N \geq 2$ with local parameters $x = (x^1,\ldots,x^N)$ and a C^∞ metric tensor g_{ij} such that the 2-form $g_{ij}x^ix^j$ is positive definite. Throughout this book, we use the Einstein summation convention. The metric tensor g_{ij}, its determinant g, and the conjugate metric tensor g^{ij} determine the arc element squared,

$$ds^2 = g_{ij}dx^idx^j,$$

the volume element

$$dx = *1 = g^{1/2}dx^1 \cdots dx^N,$$

the normal derivative of $f \in C^1$ on a hypersurface $\varphi = \text{const}$, $\varphi \in C^1$,

$$\frac{\partial f}{\partial n} = \frac{g^{ij}\dfrac{\partial f}{\partial x^i}\dfrac{\partial \varphi}{\partial x^j}}{\left(g^{ij}\dfrac{\partial \varphi}{\partial x^i}\dfrac{\partial \varphi}{\partial x^j}\right)^{1/2}},$$

the mixed Dirichlet integral of f_1 and f_2 in C^1,

$$D(f_1,f_2) = \int_R df_1 * df_2 = \int_R g^{ij}\frac{\partial f_1}{\partial x^i}\frac{\partial f_2}{\partial x^j}\,dx,$$

and the Dirichlet integral of f,

$$D(f) = D(f,f) = \int_R *|\text{grad } f|^2.$$

The Laplace-Beltrami operator $\Delta = d\delta + \delta d$ acting on $f \in C^2$ gives

$$\Delta f = -g^{-1/2}\frac{\partial}{\partial x^i}\left(g^{1/2}g^{ij}\frac{\partial f}{\partial x^j}\right)$$

(cf. Ch. 0). A function h is harmonic, by definition, if $\Delta h = 0$. The class of harmonic functions on an open set S of R is denoted by $H(S)$.

An open subset Ω of R will be called smooth if for every $x \in \partial\Omega$, there exists a parametric ball B about x such that $\partial\Omega \cap B$ is an open subset of a hyperplane in terms of some local coordinates. A smooth subset Ω will be called a <u>regular subregion</u> of R if the closure $\overline{\Omega}$ is compact.

Let Ω_0, Ω be regular subregions of R with $\overline{\Omega}_0 \subset \Omega$, and denote by ω_Ω the harmonic function on $\Omega - \overline{\Omega}_0$ with boundary values $\omega_\Omega | \partial\Omega_0 = 0$, $\omega_\Omega | \partial\Omega = 1$. As Ω increases, ω_Ω decreases, and the directed limit $\omega = \lim_{\Omega \to R} \omega_\Omega$ under an exhaustion of R by $\{\Omega\}$ is called the <u>harmonic measure</u> of the ideal boundary of R on $R - \Omega_0$. We denote by O_ω^N the class of Riemannian manifolds with $\omega \equiv 0$.

Given a point y of a Riemannian manifold R, and a regular subregion Ω of R with $y \in \Omega$, let $g_\Omega(x,y)$ be the harmonic Green's function on $\overline{\Omega}$ with pole y. The directed limit $\lim_{\Omega \to R} g_\Omega$ under an exhaustion of R by $\{\Omega\}$ is either identically ∞ or the harmonic Green's function $g(x,y)$ on R. The Riemannian N-manifolds with these properties are called parabolic or hyperbolic, respectively, and their classes are denoted by O_G^N and $\tilde{O}_G^N$. In general, we shall use the notation $\tilde{O}^N$ for the class of Riemannian N-manifolds which do not belong to a given null class O^N.

<u>1.2.</u> <u>Principal functions.</u> To prove the equality of O_ω^N and O_G^N, we recall some fundamentals of principal functions. For a regular subregion Ω_0 of a Riemannian manifold R, set $\alpha = \partial\Omega_0$, $A = R - \overline{\Omega}_0$. An operator L from $C(\alpha)$ to $H(A) \cap C(\overline{A})$ is called <u>normal</u> if

$$\begin{cases} L(c_1 f_1 + c_2 f_2) = c_1 L f_1 + c_2 L f_2, \\ Lf|\alpha = f, \\ \min f \le Lf \le \max f \quad \text{on } A, \\ \displaystyle\int_{\alpha_1} *dLf = 0, \end{cases}$$

where $\alpha_1 \subset A$, $\alpha_1 \sim \alpha$. Important normal operators are L_0 and L_1, defined by $L_i f = \lim_{\Omega \to R} L_{i\Omega} f$, $i = 0,1$, where $L_{i\Omega} : C(\alpha) \to H(\Omega - \overline{\Omega}_0) \cap C(\overline{\Omega} - \Omega_0)$ with

$$\frac{\partial L_{0\Omega}f}{\partial n} = 0 \quad \text{on} \quad \partial\Omega,$$

$$L_{1\Omega}f = \text{const} \quad \text{on} \quad \partial\Omega.$$

Let $s \in H(A) \cap C^1(\bar{A})$. The principal function problem for given R, A, L, and s consists in constructing a function $p \in H(R)$ which on A imitates the behavior of the "singularity function" s in the sense that

$$p = s + L(p - s) \quad \text{on} \quad \bar{A},$$

where $L(p - s)$ stands for $L((p - s)|\alpha)$. In view of $\int_{\alpha_1} {*dL(p - s)} = \int_{\alpha_1} {*dp} = 0$, it is necessary for the existence of p that $\int_{\alpha_1} {*ds} = 0$. That this condition is also sufficient is the content of the Main Existence Theorem of principal functions (e.g., Rodin-Sario [1]). The function p is unique up to an additive constant. The functions p corresponding to L_0 and L_1 are denoted by p_0 and p_1.

<u>1.3. Equality of O_ω^N and O_G^N.</u> We claim:

<u>THEOREM. The equality $O_\omega^N = O_G^N$ is valid for every $N \geq 2$.</u>

<u>Proof.</u> If $R \in \tilde{O}_G^N$, take a Green's function $g(x,y)$ on R and a regular subregion Ω_0 containing y. For $m = \min_{\partial\Omega_0} g(x,y)$ and a regular subregion Ω of R with $\bar{\Omega}_0 \subset \Omega$, $m(1 - \omega_\Omega) < g$ on $\Omega - \Omega_0$, hence $m(1 - \omega) \leq g$ on $R - \Omega_0$. A fortiori, $\omega \not\equiv 0$ and $R \in \tilde{O}_\omega^N$.

Conversely, if $R \in \tilde{O}_\omega^N$, with $\omega > 0$ defined on $R - \bar{\Omega}_0$, set again $\alpha = \partial\Omega_0$. Throughout this book, the boundary of a region will be oriented positively with respect to the region (e.g., de Rham [1]). Since $\omega > 0$ on $A = R - \bar{\Omega}_0$, we have $\int_\alpha {*d\omega} > 0$, for otherwise $\partial\omega/\partial n \equiv 0$ on α, and $\omega \equiv 0$. Here we have taken the liberty of integrating along α, as we may replace, if necessary, α by the boundary of a slightly larger region with boundary $\alpha_1 \sim \alpha$ as in 1.2. Set $\omega_1 = -\omega/\int_\alpha {*d\omega}$. For $y \in \Omega_0$, take a parametric ball B_y about y, $\bar{B}_y \subset \Omega_0$, and set $\alpha_y = \partial B_y$, $A_y = B_y - y$. Normalize the Green's function $g_y(x,y)$ on B_y by

$\int_{\alpha_y} {*dg_y} = -1$. Apply the Main Existence Theorem on principal functions in 1.2 to the Riemannian manifold $R_0 = R - y$ and the boundary neighborhood $A_0 = A \cup A_y$. For the normal operator choose L_1 on A and the Dirichlet operator L_D on A_y. Here L_D associates with every $f \in C(\alpha_y)$ the restriction $L_D f$ to $\bar{B}_y - y$ of the function in $H(B_y) \cap C(\bar{B}_y)$ with boundary values f on α_y. For the "singularity function" choose $s = \omega_1$ on $\bar{A}$, and $s = g_y$ on A_y. Then $\int_{\partial(\Omega_0 - B_y)} {*ds} = 0$, and there exists a principal function $p \in H(R_0)$ with the properties

$$p|A = \omega_1 + L_1(p - \omega_1), \quad p|A_y = g_y + L_D(p - g_y).$$

Since $p - \inf_R p > 0$, we have $g_\Omega < p - \inf_R p$, hence $g = \lim_\Omega g_\Omega$ exists, and $R \in \tilde{O}_G^N$.

1.4. Inclusions $O_G^N \subset O_{HP}^N \subset O_{HB}^N$. In the notation of the introduction to this chapter, we claim that

$$O_G^N \subset O_{HP}^N$$

for every $N \geq 2$. For the proof, suppose there exists a nonconstant $h \in HP$. Given any normal operator L, set $h_0 = -h + L(h|\alpha)$ on A. We have $h_0(x_0) > 0$ for some $x_0 \in A$, as otherwise $\partial h_0/\partial n \leq 0$ on α, hence by $\int_\alpha {*dh_0} = 0$, $h_0 = \text{const}$. Consequently, $\omega_\Omega \sup_A h_0 \geq h_0$ on $\Omega - \Omega_0$, $\omega \sup_A h_0 \geq h_0$ on $R - \Omega_0$, and therefore, $\omega(x_0) \geq h_0(x_0)/\sup_A h_0 > 0$. A fortiori, $\omega \not\equiv 0$ and $R \in \tilde{O}_G^N$.

Inclusion $O_{HP}^N \subset O_{HB}^N$ is trivial.

1.5. Strictness. We proceed to the main step:

THEOREM. The strict inclusions $O_G^N < O_{HP}^N < O_{HB}^N$ are valid for every $N \geq 2$.

The proof will be given in 1.6 - 1.11, with the case $N = 2$ occupying most of the reasoning.

1.6. Base manifold for $N = 2$. Let $\{q_m\}$ be the sequence of odd primes; then the quantities $\mu = \mu(m,n) = q_m 2^n$ are all different for $m,n = 1,2,\ldots$. Set

$r_i = 1 - 2^{-i}$ and consider the disk $|z| < 1$, $z = re^{i\theta}$, with radial slits

$$R^{\nu}_{mn} = \{r_{2\mu} \le r \le r_{2\mu+1}, \ \theta = \nu \cdot 2\pi/2^{m+\lambda}\},$$

where $\nu = 1,\ldots,2^{m+\lambda}$, and $\lambda = \lambda(\mu)$ is a positive integer to be specified later. For each $m = 1,2,\ldots$, we have an infinite sequence $(n = 1,2,\ldots)$ of collections of slits $(\nu = 1,\ldots,2^{m+\lambda})$.

For each m and each $k = 1,\ldots,2^{m-1}$, let S_{mk} denote the sector

$$(k - 1)\cdot 2\pi/2^{m-1} \le \theta \le k\cdot 2\pi/2^{m-1}.$$

Identify, by pairs, those edges of R^{ν}_{mn} that lie in the same sector S_{mk}, symmetrically located with respect to the bisecting ray d_{mk} of S_{mk}. The edges facing d_{mk} are here identified, and so are the edges away from d_{mk}. In particular, the edges of a slit on d_{mk} are mutually identified, and the left edge of a slit on $\theta = (k - 1)\cdot 2\pi/2^{m-1}$ is identified with the right edge of the slit on $\theta = k\cdot 2\pi/2^{m-1}$.

The points thus identified on the slits will be denoted by p and $p_m = p_m(p)$. For an end point p of a slit on the boundary of S_{mk}, there are 2^{m-1} identified points $p^i_m(p)$, $i = 1,\ldots,2^{m-1}$.

When this identification is carried out for each $m = 1,2,\ldots$, a surface W is obtained.

1.7. Conformal structure. W becomes a Riemann surface when it is endowed with a conformal structure consisting of a covering C of W by open sets O and their homeomorphic mappings $t = \varphi(p)$ onto parametric disks as follows.

Let $z = z(p)$ signify the projection of $p \in W$ into $|z| < 1$. For a point p not on a slit, let O be a disk about p not touching any slit, and let $t = \varphi(p) \equiv z(p)$. If p lies on the edge of a slit but is different from its end points, O is to consist of two half-disks on W with their (equal) diameters, one centered at p, the other at $p_m(p)$, neither reaching the end points nor touching other slits. The half-disks are then transferred, by proper rigid rotations $\tilde{p}(p)$ about $z = 0$, so as to form a connected full disk; the mapping φ is taken as $t = z(\tilde{p}(p))$.

Then let p be an end point of a slit R^{ν}_{mn} that does not lie on the boundary of S_{mk}. The neighborhood O of p shall consist of two slit disks of equal radius, one centered at p, the other at $p_m(p)$, neither disk reaching the other end points or touching other slits. By proper rigid rotations $\tilde{p}(p)$ about $z = 0$, the two slit disks are transferred so as to form a connected doubly covered disk. The mapping $t = (z(\tilde{p}(p)))^{1/2}$ now serves as φ.

Finally, if p is an end point of a slit on the boundary of S_{mk}, O shall consist of 2^{m-1} slit disks centered at the points $p^i_m(p)$, $i = 1,\ldots,2^{m-1}$. By suitable rigid rotations $\tilde{p}(p)$ about $z = 0$, the slit disks are again transferred so as to form a connected 2^{m-1}-fold disk. The mapping φ of O is now $t = (z(\tilde{p}(p))^{2^{-m+1}}$.

The collection of the sets O thus chosen is an open covering C of W, and the functions φ form a family Φ of homeomorphic mappings of O onto parametric disks. For any $O_1, O_2 \in C$ and corresponding $\varphi_1, \varphi_2 \in \Phi$, the change of parameter $t_2 = \varphi_2(\varphi_1^{-1}(t_1))$ is directly conformal on $\varphi_1(O_1 \cap O_2)$. Thus (C,Φ) is a conformal structure on W, and (W,C,Φ) is a Riemann surface. In the sequel, we shall speak of the Riemann surface W without explicit reference to its structure.

<u>1.8.</u> <u>Reflection function</u>. For $m = 1$, the sectors S_{mk} reduce to the single sector $S_{11} = \{0 \leq \theta \leq 2\pi\}$, and the bisecting ray d_{11} is the negative real axis. For every p on W, we define $p_1 = p_1(p)$ as the symmetric point with respect to the real axis. The uniqueness of $p_1(p)$ is established as follows.

If p lies on a slit, so does p_1, the edges corresponding in an obvious manner. In particular, for $p \in R^{\nu}_{1n}$, p_1 coincides with p_1 used in the definition of the identification. If $p \in R^{\nu}_{mn}$ with $m > 1$, we first exclude the case in which p is an end point of an R^{ν}_{mn} on the boundary of S_{mk}. Then p is identical with $p_m(p)$, and the operation p_1 leads to two points, $p_1(p)$ and $p_1(p_m(p))$. Since the S_{mk} are, by pairs, symmetrically located about the real axis, these two points are identified by $p_m(p)$, and the operation p_1 becomes unique.

Finally, if p is an end point of an R^{ν}_{mn} that lies on the boundary of S_{mk}, then p is one of the 2^{m-1} identified points $p^i_m(p)$, $i = 1,\ldots,2^{m-1}$. The points $p_1(p^i_m(p))$ are, in a different order, identical with the $p^i_m(p)$, and the operation p_1 reduces to $p = p_1(p)$.

Thus $p_1(p)$ is uniquely defined on all of W. In terms of the local parameters $t = \varphi(p)$, $t_1 = \varphi_1(p_1)$, the corresponding transformation $t_1(t) = \varphi_1(p_1(\varphi^{-1}(t)))$ is a sense-reversing conformal mapping of W onto itself.

1.9. Positive harmonic functions. Let $h(t)$ be a positive harmonic function on W. In order to prove that h reduces to a constant, we shall first show that h is symmetric with respect to the real axis.

We may normalize h so that $h(t_0) = 1$, where t_0 corresponds to the point of W that covers $z = 0$. By virtue of the indirect conformality of $t_1(t)$, the function $h(t_1(t))$ is harmonic on W. The same is true of the difference

$$h_1(t) = h(t) - h(t_1(t)),$$

with $h_1(t) = 0$ on R^{ν}_{1n}, and we infer that

$$\int_{K_r} |h_1(t(z))|\,d\theta \leq \int_{K_r} h(t(z))\,d\theta + \int_{K_r} h(t_1(t(z)))\,d\theta = 4\pi$$

for $K_r = \{r = \text{const} < 1\}$. In fact, each integral on the right is equal to $2\pi h(t_0) = 2\pi$.

Consider the annulus

$$A_\mu = \{r_{2\mu-1} \leq r \leq r_{2\mu+2}\}$$

for $m = 1$, that is, $\mu = 3\cdot 2^n$. The function $h_1(t(z))$ vanishes on the slits in A_μ, but it need not be harmonic on them. Let $v_\mu(z)$ be the harmonic function on A_μ with $v_\mu(z) = |h_1(t(z))|$ on the boundary β_μ of A_μ. Then

$$|h_1(t(z))| \leq v_\mu(z)$$

on A_μ. We denote by $g(\zeta,z)$ the Green's function on A_μ with pole at z, normalized by $\int_{\beta_\mu} *dg(\zeta,z) = -1$, and consider the annulus

$$B_\mu = \{r_{2\mu} \leq r \leq r_{2\mu+1}\}.$$

If C_i is the circle $r = r_i$, we have

$$\int v_\mu(z)d\theta = \int |h_1(t(z))|d\theta \leq 4\pi$$

along $C_{2\mu-1}$ and $C_{2\mu+2}$. It follows that, for $z \in B_\mu$,

$$v_\mu(z) = \int_{\beta_\mu} v_\mu(\zeta) \frac{\partial g(\zeta,z)}{\partial n} ds \leq 8\pi \max \frac{\partial g(\zeta,z)}{\partial n} = M_\mu,$$

where $\partial/\partial n$ is in the direction of the normal interior to A_μ and the maximum is taken for $\zeta \in \beta_\mu$ and $z \in B_\mu$. In view of the definition of v_μ, we conclude that $|h_1| \leq M_\mu$ on B_μ, with the bound M_μ independent of h and λ.

If we let the number $2^{1+\lambda}$ of the slits in B_μ tend to infinity, the width of the sectors T_{1n}^ν, bounded by R_{1n}^ν, $R_{1n}^{\nu+1}$, $C_{2\mu}$, and $C_{2\mu+1}$, tends to zero. Consider the harmonic function $s(z)$ on T_{1n}^ν with $s = 0$ on $R_{1n}^\nu \cup R_{1n}^{\nu+1}$, and $s = M_\mu$ on $C_{2\mu} \cup C_{2\mu+1}$. We choose for $\lambda = \lambda(\mu)$ the smallest positive integer for which $s(z) \leq 2^{-\mu}$, say, on the circle $D_\mu = \{r = \frac{1}{2}(r_{2\mu} + r_{2\mu+1})\}$. Since $h_1 = 0$ on $R_{1n}^\nu \cup R_{1n}^{\nu+1}$, and $|h_1| \leq M_\mu$ on $C_{2\mu} \cup C_{2\mu+1}$, we have $|h_1| \leq s$ on B_μ, and consequently, $|h_1| \leq 2^{-\mu}$ on D_μ. On letting $\mu \to \infty$, we obtain $h_1 \equiv 0$ on W.

Thus $h(t)$ is symmetric about the real axis and $\partial h/\partial\theta = 0$ on the slits R_{1n}^ν and at the points on the real axis that do not lie on any slit.

1.10. Symmetry about bisectors. We proceed to prove, by induction, the symmetry of $h(t)$ on S_{mk} about d_{mk} for any m. Suppose that the symmetry has been established for d_{ik} with $i = 1,\ldots,m-1$ and $k = 1,\ldots,2^{i-1}$; furthermore, $\partial h/\partial\theta = 0$ on R_{in}^ν and at the points on d_{ik} that do not lie on any slit.

For p in S_{mk}, let $p_m = p_m(p)$ be the symmetric point about d_{mk}. If p

is interior to S_{mk}, p_m lies on a slit R^{ν}_{hn} if and only if p does. In the affirmative case, the uniqueness of the operation p_m is established as in the case $m = 1$ by $p_h(p_m(p_h(p))) = p_m(p)$. The corresponding transformation $t_m(t)$ is sense-reversing and conformal, and the function $h(t_m(t))$ is harmonic in the interior of S_{mk}. The same remains true if p lies on (an edge of) R^{ν}_{jn}, $j \geq m$, or on the boundary of S_{mk}.

The (radial) boundary of S_{mk} consists of some d_{in} with $i < m$. If p lies on this boundary but not on any R^{ν}_{jn} with $j \geq m$, then a sufficiently small neighborhood U of p is transformed by $p_m(p)$ onto two half-disks. The diameters of the latter lie again on some d_{in} with $i < m$, and are located either on some slits R^{ν}_{in} with $i < m$ or not on any slit. But at such points, $\partial h/\partial\theta = 0$, and we conclude, by the symmetry of $h(t)$ about every d_{in} with $i < m$, that $h(t_m(t))$ is harmonic on U. Thus the function

$$h_m(t) = h(t) - h(t_m(t))$$

is harmonic on all of W.

In the same manner as for $m = 1$, we conclude that, with $\lambda = \lambda(\mu)$ properly chosen, $h_m(t) \equiv 0$ on W, and $h(t)$ on S_{mk} is symmetric about d_{mk}. Furthermore, $\partial h/\partial\theta = 0$ on the R^{ν}_{mn} and at the points on d_{mk} that do not lie on any slit.

1.11. Relations $O^N_G < O^N_{HP} < O^N_{HB}$. Let $C_\rho = \{|z| = \rho\}$ be a fixed circle without common points with any R^{ν}_{mn}. Since $\partial h/\partial\theta = 0$ at $C_\rho \cap d_{mk}$, we find, on letting $m \to \infty$, that $\partial h/\partial\theta = 0$ on all of C_ρ. Consequently, $h(t) = \text{const}$ and $W \in O_{HP}$. Here and below we use, for any null class O^N of Riemannian manifolds, the notation O for the corresponding null class of Riemann surfaces.

On the other hand, $g(z,0) = -(2\pi)^{-1} \log |z|$ is the Green's function on W with pole 0, and we have shown that the Riemann surface W belongs to $\tilde{O}_G \cap O_{HP}$.

We now map the universal covering surface W_∞ of W conformally onto the unit disk $D = \{|w| < 1\}$ and take on D the Poincaré metric $ds = (1 - |w|^2)^{-1}|dw|$. It induces on W a conformal metric $ds = \lambda(t)|dt|$ which makes W into a

Riemannian 2-manifold R. We indicate by the subscript E operations and quantities with respect to the (Euclidean) parametric disks of W, and by the subindex R those with respect to the same disks of R. For $f \in C^2(R)$,

$$\Delta_R f = -g_R^{-1/2} \frac{\partial}{\partial x^i}(g_R^{1/2} g_R^{ii} \frac{\partial f}{\partial x^i})$$
$$= -\frac{1}{\lambda^2} \sum_{i=1}^{2} \frac{\partial}{\partial x^i}(\lambda^2 \lambda^{-2} \frac{\partial f}{\partial x^i}) = \frac{1}{\lambda^2} \Delta_E f.$$

Therefore, a function is harmonic on the Riemann surface W if and only if it is harmonic on the Riemannian manifold R. A fortiori, we have established the strict inclusion $O_G^2 < O_{HP}^2$ for Riemannian 2-manifolds.

For $N > 2$, the strictness is trivial. In fact, the Euclidean N-space E^N has the Green's function $g(x,0) = cr^{-N+2}$, $x = (r,\theta^1,\ldots,\theta^{N-1})$. For $h \in HP$, the Poisson integral gives the Harnack inequality on a ball of radius ρ,

$$\left(\frac{\rho}{\rho + r}\right)^{N-2} \frac{\rho - r}{\rho + r} h(0) \le h(x) \le \left(\frac{\rho}{\rho - r}\right)^{N-2} \frac{\rho + r}{\rho - r} h(0)$$

(e.g., Courant-Hilbert [1]). As $\rho \to \infty$, we obtain $h = \text{const}$, and conclude that $O_G^N < O_{HP}^N$ for all $N \ge 2$.

The inclusion $O_{HP}^N \subset O_{HB}^N$ is immediate, every $h \in HB$ giving rise to an $h + c \in HP$. To prove the strictness for $N = 2$, consider the Riemann surface W_0 and the Riemannian 2-manifold R_0, obtained by puncturing the above W and R, respectively, at the origin. The function $h(z) = -\log |z|$ belongs to $HP(W_0)$, hence also to $HP(R_0)$. On the other hand, the origin is a removable singularity for every $h \in HB(W_0)$, hence for every $h \in HB(R_0)$, so that the latter extends to an $h \in HB(R)$. By $O_{HP}^N \subset O_{HB}^N$, $h = \text{const}$, and we have $O_{HP}^2 < O_{HB}^2$.

For $N > 2$, consider the space E_0^N obtained by puncturing the Euclidean N-space E^N at the origin. Poisson's formula shows at once that every $h \in HB(E^N)$ is constant. The same is true of every $h \in HB(E_0^N)$, the origin being a removable singularity. On the other hand, $r^{-N+2} \in HP(E_0^N)$, and we have shown that $O_{HP}^N < O_{HB}^N$.

The proof of Theorem 1.5 is complete.

NOTES TO §1. Principal functions were introduced in Sario [1], [3], [4], and a systematic presentation of their theory and applications given in Rodin-Sario [1]. In the present book, we only need the Main Existence Theorem, and the Main Extremum Theorem for $p_0 - p_1$ to be used in §2.

The proof of $O_\omega = O_G$ for Riemann surfaces of arbitrary finite or infinite genus was first given in Sario [2], and in fact motivated the introduction of the theory of principal functions.

The first proof of $O_G < O_{HP} < O_{HB}$ for Riemann surfaces, reproduced above, utilizes ideas of Ahlfors [1] and Tôki [1] and was given in Sario [5]. An independent proof was given by Tôki [2]. The relation $O_G < O_{HB}$ had been earlier proved by Tôki [1]. For Riemannian manifolds of dimension $N > 2$, the relations $O_G^N < O_{HP}^N < O_{HB}^N$ were included in Sario-Nakai [1].

§2. RELATIONS $O_{HB}^N < O_{HD}^N = O_{HC}^N$

2.1. Inclusion $O_{HB}^N \subset O_{HD}^N = O_{HC}^N$. We retain the notation in the introduction to Chapter I. To prove that $O_{HB}^N \subset O_{HD}^N$, we make use of principal functions. Given a Riemannian manifold R, take $a, b \in R$, let B_a, B_b with $\bar{B}_a \cap \bar{B}_b = \emptyset$ be parametric balls about a, b, and set $\alpha_a = \partial B_a$, $\alpha_b = \partial B_b$. Consider the Green's functions $g_a = g(x,a)$ and $g_b = g(x,b)$ on $\bar{B}_a$, $\bar{B}_b$ respectively, normalized by $\int_{\alpha_a} {*dg_a} = -1$, $\int_{\alpha_b} {*dg_b} = -1$. Let Ω_0 be a regular subregion of R with $\bar{B}_a \cup \bar{B}_b \subset \Omega_0$, and set $A_a = B_a - a$, $A_b = B_b - b$, $A = R - \bar{\Omega}_0$. For the Riemannian manifold of the Main Existence Theorem of principal functions take $R_0 = R - a - b$; for the boundary neighborhood, $A_0 = A_a \cup A_b \cup A$; for the singularity function, $s|A_a = g_a$, $s|A_b = -g_b$, $s|A \equiv 0$; for the normal operator, $L = L_D$ for A_a and A_b, $L = L_i$, $i = 0,1$, for A. The corresponding principal functions p_i then have the properties $p_i|A_a = g_a + h_i$, $p_i|A_b = -g_b + k_i$, where $h_i \in H(B_a)$, $k_i \in H(B_b)$. We normalize the additive constants of p_i by $k_i(b) = 0$, and set $p_2 = p_0 - p_1$.

The Main Extremum Theorem of principal functions (e.g., Rodin-Sario [1, p. 243 ff.]) states that p_2 minimizes the functional $D(h) - 2h(a)$ among all

$h \in HD(R)$ with $h(b) = 0$. Explicitly,

$$D(h) - 2h(a) = -p_2(a) + D(h - p_2).$$

The choice $h = 0$ (or $h = p_2$) gives $D(p_2) = p_2(a)$, hence $p_2 \in HD$. Suppose there exists a nonconstant $h \in HD$, $h(b) = 0$. Since a can be arbitrarily chosen, we may assume that $h(a) \neq 0$. Then $p_2 \neq$ const, for otherwise $D(p_2) = p_2(a) = 0$, which gives $-2h(a) = 0$, a contradiction. In view of the third property defining a normal operator in 1.2, each p_i is bounded, hence $p_2 \in HC$ and $R \in \tilde{O}^N_{HC}$. This proves both relations $O^N_{HB} \subset O^N_{HD} = O^N_{HC}$.

2.2. Strictness. We include the above identity in the following Theorem, where the essence is the strict inequality to be proved:

THEOREM. The relations $O^N_{HB} < O^N_{HD} = O^N_{HC}$ are valid for every $N \geq 2$.

The proof will be given in 2.3 - 2.9. For $N = 2$, the counterexample is again based on the corresponding Riemann surface. For $N > 2$, we introduce a Riemannian manifold, the unit N-ball with a Poincaré-type metric, which will render us great services throughout the development of the harmonic, quasiharmonic and biharmonic classifications.

2.3. Case $N = 2$. We retain the notation of 1.6 and consider the annulus $1 < |z| < 3$ with radial slits

$$Q^{\nu}_{mn} = \{2 + r_{2\mu} \leq r \leq 2 + r_{2\mu+1};\ 2 - r_{2\mu+1} \leq r \leq 2 - r_{2\mu};\ \theta = \nu \cdot 2\pi/2^{m+\lambda}\},$$

where $\mu = q_m \cdot 2^n$, $m,n = 1,2,\ldots$, and $\nu = 1,\ldots,2^{m+\lambda}$. Take an infinite collection $\{F(k)\}$, $k = 1,2,\ldots$, of copies of these slit annuli F and, successively for each fixed $m = 1,2,\ldots$ and subsequently fixed $j = 0,1,\ldots$, join $F(i + 2^m j)$ with $F(i + 2^{m-1} + 2^m j)$, $i = 1,\ldots,2^{m-1}$, along the edges of $E_m = \bigcup_{n,\nu} Q^{\nu}_{mn}$, with a folding at each edge. Upon the surface thus obtained, a conformal structure is imposed in a manner analogous to that of W in 1.7 so as to form a Riemann surface V.

With each point $p \in V$, we associate a point $p_m = p_m(p)$ as follows. Let p on the copy $F(k)$ be denoted by $p(k)$. The operation $p_m(p)$ shall assign the point $p_m = p(k \pm 2^{m-1})$ to $p(k)$. Here p_m lies, in an identical location, on the copy $F(k \pm 2^{m-1})$, which was joined with $F(k)$ along E_m; the choice of the minus or plus sign depends on the sheet on which p lies.

An ambiguity in the operation p_m seems to arise if $p \in F(k)$ belongs to some E_h with $h \neq m$. In fact, then $p(k)$ is identified with $p(k \pm 2^{h-1})$, and p_m carries the former into $p(k \pm 2^{m-1})$, the latter into $p(k \pm 2^{h-1} \pm 2^{m-1})$; here the signs of 2^{m-1}, 2^{h-1} are individually the same as above. But $p(k \pm 2^{h-1} \pm 2^{m-1})$, lying on an E_h, is identified with $p(k \pm 2^{h-1} \pm 2^{m-1} \mp 2^{h-1})$ $= p(k \pm 2^{m-1})$, and the two locations of $p_m(p)$ coincide. Thus the operation $p_m(p)$ is uniquely determined.

The corresponding transformation of the local parameter will be denoted by $t_m = t_m(t)$.

If $h(t) \in HC$ on V, the function $h_m(t) = h(t) - h(t_m(t))$ vanishes on E_m, and we conclude again that, for properly chosen λ, $h_m \equiv 0$. For $m \to \infty$, it follows that $h(t)$ takes on identical values on all copies of F. This gives, the Dirichlet integral being finite, the desired relation $h = \text{const}$. The proof of $V \in \tilde{O}_{HB} \cap O_{HD}$ is completed by the fact that $\log|z| \in HB$ on V.

To show that $\tilde{O}^2_{HB} \cap O^2_{HD} \neq \emptyset$, we endow V with a conformal metric in the same manner as in 1.11. The resulting Riemannian 2-manifold R continues to carry nonconstant HB functions. That it does not carry nonconstant HD functions follows from the fact that endowing V with a conformal metric does not alter the Dirichlet integral. Indeed, if operations and quantities with respect to R and V are again indicated by subscripts, we have for $h \in H$, the Dirichlet integral over a parametric disk O

$$D_R(h) = \int_O \sum_{i=1}^{2} g^{ii}\left(\frac{\partial h}{\partial x^i}\right)^2 g^{1/2} dx^1 dx^2$$

$$= \int_O \sum_{i=1}^{2} \lambda^{-2}\left(\frac{\partial h}{\partial x^i}\right)^2 \lambda^2 dx^1 dx^2$$

$$= \int_O \sum_{i=1}^{2} \left(\frac{\partial h}{\partial x^i}\right)^2 dx^1 dx^2 = D_V(h).$$

<u>2.4. Poincaré N-ball</u> B_α^N. To prove $O_{HB}^N < O_{HD}^N$ for $N > 2$, we consider what we call the Poincaré N-ball B_α^N. It is the unit N-ball $\{x = (x^1,\ldots,x^N) \mid |x| = r < 1\}$, endowed with the Poincaré-type metric

$$ds_\alpha = \lambda(r)|dx|, \quad \lambda(r) = (1 - r^2)^\alpha, \quad \alpha \text{ a real constant.}$$

We know from the reasoning in 1.11 that for $N = 2$ the harmonicity of a function is independent of a conformal metric, hence of α on the Poincaré disk B_α^2, and the reasoning reduces to the case of the Euclidean disk B_0^2, which trivially carries all functions under consideration. For this reason, we shall, here and later, consider the Poincaré N-ball of dimension $N > 2$ only.

Let $(r,\theta) = (r,\theta^1,\ldots,\theta^{N-1})$ be the Euclidean coordinates. Then

$$ds_\alpha^2 = \lambda^2 dr^2 + \lambda^2 r^2 \sum_{i=1}^{N-1} \gamma_i(\theta) d\theta^{i2},$$

where the γ_i are certain trigonometric functions of $\theta = (\theta^1,\ldots,\theta^{N-1})$. The volume element is

$$dx = g^{1/2}\, dr d\theta = \lambda^N r^{N-1}\, \gamma(\theta) dr d\theta,$$

with $d\theta = d\theta^1 \cdots d\theta^{N-1}$, $\gamma(\theta) = (\gamma_1(\theta) \cdots \gamma_{N-1}(\theta))^{1/2}$.

A function $S_n(\theta)$ is, by definition, a spherical harmonic of degree $n = 0,1,2,\ldots,$ if $r^n S_n(\theta)$ is harmonic in the Euclidean metric. Every $S_n(\theta)$ is a unique linear combination of linearly independent fundamental spherical harmonics $S_{nm}(\theta)$, $m = 1,\ldots,m_n$, where m_n is given by the power series

$$\frac{1 + x}{(1 - x)^{N-1}} = \sum_{n=0}^{\infty} m_n x^n$$

(Müller [1, p. 4]). We also recall that $\{S_{nm}\}$ forms an orthogonal system with respect to the inner product $(f,g) = \int_\omega fg\,d\omega$, with ω the $(N-1)$-dimensional unit sphere, and $d\omega = \gamma(\theta)d\theta^1 \dots d\theta^{N-1}$, the Euclidean volume of the hypersurface element of ω. A sufficiently smooth function on ω can be expanded in an absolutely and uniformly convergent series in terms of the spherical harmonics (Courant-Hilbert [1, p. 314 ff.], Chevalley [1, p. 213]).

2.5. Representation of harmonic functions on B^N_α. By 2.4, a harmonic function h on B^N_α has an expansion on an r-sphere,

$$h(r,\theta) = \sum_{n=0}^{\infty} \sum_{m=1}^{m_n} d_{nm}(r)S_{nm}(\theta).$$

We examine the $d_{nm}(r)$. Suppose $f(r)S_{nm}(\theta)$ is nonconstant harmonic on B^N_α. Then

$$0 = \Delta_\alpha(f(r)S_{nm}(\theta)) = \Delta_\alpha f(r)\cdot S_{nm}(\theta) + f(r)\Delta_\alpha S_{nm}(\theta),$$

where Δ_α is the Laplace-Beltrami operator corresponding to the metric ds_α. The Euclidean Laplace operator Δ gives $\Delta(r^n S_{nm}(\theta)) = 0$, hence

$$\Delta S_{nm}(\theta) = n(n + N - 2)r^{-2}S_{nm}(\theta).$$

By direct computation we see that, for a smooth function $S(\theta)$, $\Delta_\alpha S(\theta) = \lambda^{-2}\Delta S(\theta)$ and therefore,

$$\Delta_\alpha S_{nm}(\theta) = \lambda^{-2}n(n + N - 2)r^{-2}S_{nm}(\theta).$$

It follows that

$$0 = -\lambda^{-2}\left[f'' + \left(\frac{N-1}{r} + \frac{(N-2)\lambda'}{\lambda}\right)f' - n(n + N - 2)r^{-2}f\right]S_{nm}.$$

Hence fS_{nm} is harmonic on B^N_α if and only if f satisfies

$$r^2(1 - r^2)f'' + r\{(N-1) - [(N-1) + 2(N-2)\alpha]r^2\}f' - n(n + N - 2)(1 - r^2)f = 0.$$

We solve this equation for each n. Since all coefficients can be expanded into power series of r, the origin is a regular singular point of the equation.

Thus there exists at least one solution of the equation in the form

$$f_n(r) = r^{p_n} \sum_{i=0}^{\infty} c_{ni} r^i, \qquad c_{n0} = 1.$$

On substituting $f_n(r)$ into the equation, we obtain

$$\sum_{i=0}^{\infty} [(p_n + i - 1)(p_n + i) + (N - 1)(p_n + i) - n(n + N - 2)] c_{ni} r^{p_n+i}$$

$$- \sum_{i=2}^{\infty} \{(p_n + i - 3)(p_n + i - 2) + [(N-1) + 2(N-2)\alpha](p_n + i - 2) - n(n+N-2)\} c_{n,i-2} r^{p_n+i} = 0.$$

To determine p_n, we equate to 0 the coefficient of r^{p_n} and obtain the indicial equation

$$(p_n - 1)p_n + (N - 1)p_n - n(n + N - 2) = 0.$$

The roots are $p_n = n$ and $p_n = -(n + N - 2)$. Since the origin is in B_α^N, p_n cannot be negative, and we have $p_n = n$.

We then equate to 0 the coefficient of r^{p_n+1} and obtain $c_{n1} = 0$. The coefficient c_{ni} for $i > 1$ is similarly deduced from that of r^{p_n+i}:

$$c_{n,2i} = \prod_{j=1}^{i} \frac{(n + 2j - 2)[n + 2j + N - 4 + 2(N - 2)\alpha] - n(n + N - 2)}{(n + 2j)(n + 2j + N - 2) - n(n + N - 2)}$$

and $c_{n,2i+1} = 0$.

The limit of

$$f_n(r) = r^n + \sum_{i=1}^{\infty} c_{n,2i}\, r^{2i+n}$$

as $r \to 1$ exists, since the $c_{n,2i}$ are of constant sign for all sufficiently large i. Furthermore, this limit cannot be zero, for otherwise $\lim_{r\to 1} f_n(r) S_{nm}(\theta) \equiv 0$, which would imply $f_n(r) \equiv 0$, in violation of $c_{n0} = 1$. Similarly, $f_n(r) \neq 0$ for $0 < r < 1$. Hence for an arbitrary but fixed $r_0 \in (0,1)$, there exist constants a_{nm} such that $a_{nm} f_n(r_0) = d_{nm}(r_0)$. Consequently,

$$\sum_{n=0}^{\infty} \sum_{m=1}^{m_n} a_{nm} f_n(r) S_{nm}(\theta)$$

is a series of functions harmonic on B_α^N which converges absolutely and uniformly to $h(r,\theta)$ on the $(N - 1)$-sphere $\{r = r_0\}$, hence also on the ball $\{r < r_0\}$.

Now choose $r' \in (r_0,1)$. The same argument as above provides us with constants a'_{nm} such that

$$\sum_{n=0}^{\infty} \sum_{m=1}^{m_n} a'_{nm} f_n(r) S_{nm}(\theta)$$

converges to h on the $(N - 1)$-sphere $\{r = r'\}$, hence also on the ball $\{r < r'\}$. The sums of these two series of harmonic functions are identical on the ball $\{r < r_0\}$, so that $a_{nm} = a'_{nm}$ for all (n,m). Thus the expansion is unique.

We have proved:

LEMMA. *Every harmonic function* $h(r,\theta)$ *on the Poincaré* N-*ball* B_α^N *has the expansion in fundamental spherical harmonics* S_{nm},

$$h(r,\theta) = \sum_{n=0}^{\infty} f_n(r) \sum_{m=1}^{m_n} a_{nm} S_{nm}(\theta),$$

where

$$f_n(r) = r^n + \sum_{i=1}^{\infty} c_{n,2i}\, r^{2i+n}$$

and the $c_{n,2i}$ *have the product expressions given above.*

2.6. Parabolicity. We shall need the following characterization:

LEMMA. $B_\alpha^N \in O_G^N$ *if and only if* $\alpha \geq 1/(N - 2)$.

For the proof, note that if $h(r)$ is harmonic, then

$$\Delta h = -\lambda^{-N} r^{-(N-1)}(\lambda^{N-2} r^{N-1} h')' = 0.$$

Therefore, the Green's function with pole at the origin is of the form

$$g(x,0) = c\int_r^1 (1 - r^2)^{-(N-2)\alpha} r^{-(N-1)} dr.$$

It exists if and only if $\alpha < 1/(N - 2)$.

2.7. Asymptotic behavior of harmonic functions on B_α^N. To study the order of growth of $f_n(r)$ for a nonconstant $f_n S_n \in H$ as $r \to 1$, we change the variable to $\rho = 1 - r$. For economy of notation, we write $f_n(\rho)$ for $f_n(r) = f_n(1 - \rho)$. Then the differential equation in 2.5 to be satisfied by f_n is transformed into

$$\rho^2 f_n''(\rho) + \rho a(\rho) f_n'(\rho) + b(\rho) f_n(\rho) = 0,$$

where

$$\begin{cases} a(\rho) = \dfrac{2(N - 2)\alpha(1 - \rho)}{2 - \rho} - \dfrac{(N - 1)\rho}{1 - \rho}, \\ b(\rho) = -n(n + N - 2)\left(\dfrac{\rho}{1 - \rho}\right)^2. \end{cases}$$

This is also a linear equation with $\rho = 0$ a regular singular point. The roots of the indicial equation

$$p_n(p_n - 1) + a(0)p_n + b(0) = p_n(p_n - 1) + (N - 2)\alpha p_n = 0$$

are $p_n = 0$ and $p_n = 1 - (N - 2)\alpha$.

LEMMA. *For* $N > 2$, $f_n S_n \in H(B_\alpha^N)$, *and* $n > 0$,

$$f_n(\rho) \sim \begin{cases} c\rho^{1-(N-2)\alpha}, & \alpha > 1/(N - 2), \\ -c \log \rho, & \alpha = 1/(N - 2), \\ c, & \alpha < 1/(N - 2), \end{cases}$$

and

$$f_n'(\rho) \sim \begin{cases} c\rho^{-(N-2)\alpha}, & \alpha > 0, \\ c, & \alpha \leq 0, \end{cases}$$

as $\rho \to 0$, with $c = c(n,\alpha,N)$ a positive constant independent of ρ.

Proof. By the maximum principle, $f_n(r) \neq 0$ for $r \neq 0$. Since $f_n(r)/r^n \to 1$ as $r \to 0$, we have $f_n(r) > 0$ for all r. Therefore, if c exists, it must be ≥ 0. That $c > 0$ is again a consequence of the maximum principle.

For $\alpha > 1/(N-2)$, two linearly independent solutions are of the form

$$\begin{cases} f_{n1} = \sigma_1, \\ f_{n2} = \rho^{1-(N-2)\alpha}\ \sigma_2 + d \log \rho \cdot \sigma_1, \end{cases}$$

where σ_1, σ_2 are certain power series in ρ with $\sigma_1(0) \neq 0$, $\sigma_2(0) \neq 0$, and the constant d vanishes if $(N-2)\alpha$ is not an integer (see, e.g., Golomb-Shanks [1, pp. 351-357, 365-368]). There exist constants a,b such that $f_n = af_{n1} + bf_{n2}$.

The function f_{n1} cannot be bounded, for otherwise $f_{n1}S_n \in HB$, in violation of $B_\alpha^N \in O_G^N$. Since f_{n1} is bounded near $\rho = 0$, it must have a singularity at $\rho = 1$, that is, $r = 0$. Thus $b \neq 0$, for otherwise $f_n = af_{n1}$, contrary to the fact that f_n does not have a singularity at $r = 0$. Hence $f_n \sim c\rho^{1-(N-2)\alpha}$.

In the case $\alpha = 1/(N-2)$,

$$\begin{cases} f_{n1} = \sigma_1, \\ f_{n2} = \rho\sigma_2 + \log \rho \cdot \sigma_1 \end{cases}$$

are linearly independent solutions, and the reasoning is the same as above.

If $\alpha < 1/(N-2)$,

$$\begin{cases} f_{n1} = \rho^{1-(N-2)\alpha}\ \sigma_1, \\ f_{n2} = \sigma_2 + d \log \rho \cdot \rho^{1-(N-2)\alpha}\ \sigma_1, \end{cases}$$

where $d = 0$ if $(N - 2)\alpha$ is not an integer. Thus $f_n = af_{n1} + bf_{n2} \sim c$.

It remains to prove the estimates for f_n'. First suppose $\alpha > 1/(N - 2)$. Then

$$f_n' = a\sigma_1' + b\rho^{1-(N-2)\alpha}\sigma_2' + b_1\rho^{-(N-2)\alpha}\sigma_2 + bd\rho^{-1}\sigma_1 + bd \log \rho \cdot \sigma_1',$$

where $b \neq 0$. If $(N - 2)\alpha$ is not an integer, that is, if $d = 0$, then

$$f_n' \sim c\rho^{-(N-2)\alpha}.$$

If $(N - 2)\alpha$ is an integer, then $-(N - 2)\alpha \leq -2$, and the above estimate still holds.

In the case $\alpha = 1/(N - 2)$, a straightforward computation yields $f_n' \sim c\rho^{-1}$.

If $0 < \alpha < 1/(N - 2)$, then $(N - 2)\alpha$ is not an integer, so that

$$f_n = a\rho^{1-(N-2)\alpha}\sigma_1 + b\sigma_2.$$

Therefore,

$$f_n' = a[1 - (N - 2)\alpha]\rho^{-(N-2)\alpha}\sigma_1 + a\rho^{1-(N-2)\alpha}\sigma_1' + b\sigma_2' \sim c\rho^{-(N-2)\alpha}.$$

In the Euclidean case $\alpha = 0$, we know that $f_n = c_n(1 - \rho)^n$, hence $f_n' \sim c$.

In the remaining case $\alpha < 0$,

$$f_n = a\rho^{1-(N-2)\alpha}\sigma_1 + b(\sigma_2 + d \log \rho \cdot \rho^{1-(N-2)\alpha}\sigma_1),$$

where $d = 0$ if $(N - 2)\alpha$ is not an integer. We infer that $f_n' \sim c$.

<u>2.8. Characterization of O_{HP}^N and O_{HB}^N.</u> The following characterizations are now readily established.

<u>LEMMA</u>. $B_\alpha^N \in O_{HP}^N \Leftrightarrow \alpha \geq 1/(N - 2)$.

$B_\alpha^N \in O_{HB}^N \Leftrightarrow \alpha \geq 1/(N - 2)$.

In fact, if $\alpha < 1/(N - 2)$, Lemma 2.7 gives $f_n \sim c$, hence $f_n S_n \in HB$, and $B_\alpha^N \in \tilde{O}_{HB}^N \subset \tilde{O}_{HP}^N$.

If $\alpha \geq 1/(N-2)$, then by 2.6 and 1.4, $B_\alpha^N \in O_G^N \subset O_{HP}^N \subset O_{HB}^N$.

2.9. Characterization of O_{HD}^N and completion of proof. We claim:

LEMMA. $B_\alpha^N \in O_{HD}^N$ if and only if $|\alpha| \geq 1/(N-2)$.

Proof. The Dirichlet integral of $f_n S_n$ over B_α^N is

$$D(f_n S_n) = \int_0^1 \int_\omega \left[\lambda^{-2}(f_n' S_n)^2 + \lambda^{-2} r^{-2} \sum_{i=1}^{N-1} \gamma_i^{-1} \left(\frac{\partial(f_n S_n)}{\partial\theta^i}\right)^2\right] \gamma(\theta) \lambda^N r^{N-1} d\theta dr,$$

where ω is the $(N-1)$-dimensional unit sphere, $\gamma(\theta) = (\gamma_1(\theta) \cdots \gamma_{N-1}(\theta))^{1/2}$, and $d\theta = d\theta^1 \cdots d\theta^{N-1}$. Therefore,

$$D(f_n S_n) = \int_0^1 \left[\lambda^{-2} f_n'^2 \int_\omega S_n^2 d\omega + \lambda^{-2} r^{-2} f_n^2 D_\omega(S_n)\right] \lambda^N r^{N-1} dr$$

$$= \int_0^1 (c_1 f_n'^2 + c_2 r^{-2} f_n^2)(1 - r^2)^{(N-2)\alpha} r^{N-1} dr,$$

where $d\omega$ is the Euclidean volume of the hypersurface element of ω, D_ω is the Euclidean Dirichlet integral over ω, and c_1, c_2 are positive constants. By Lemma 2.7, we have for $\alpha \in (0, 1/(N-2))$,

$$D(f_n S_n) \leq \int_0^1 [c_1(1 - r)^{-2(N-2)\alpha} + c_2 r^{-2}](1 - r^2)^{(N-2)\alpha} r^{N-1} dr < \infty,$$

and for $\alpha \in (-1/(N-2), 0]$,

$$D(f_n S_n) \leq \int_0^1 [c_1 + c_2 r^{-2}](1 - r^2)^{(N-2)\alpha} r^{N-1} dr < \infty.$$

This proves the necessity of the condition.

To prove the sufficiency, we first recall that, for $\alpha \geq 1/(N-2)$, $B_\alpha^N \in O_G^N \subset O_{HD}^N$. It, therefore, remains to consider the case $\alpha \leq -1/(N-2)$. Take a nonconstant $h(r,\theta) \in H(B_\alpha^N)$. It has an expansion in spherical harmonics $h(r,\theta) = \Sigma_0^\infty f_n(r) S_n(\theta)$. By Stokes' formula,

$$D_{B^N_\alpha}(f_nS_n, f_pS_p) = \lim_{r\to 1}\int_{|x|<r} d(f_nS_n)*d(f_pS_p) = \lim_{r\to 1} g\int_\omega S_nS_p d\omega,$$

where g is a function of r. For $n \neq p$, we therefore have the Dirichlet orthogonality

$$D_{B^N_\alpha}(f_nS_n, f_pS_p) = 0.$$

As a consequence,

$$D(h) = \sum_{n=0}^{\infty} D(f_nS_n) \geq D(f_nS_n)$$

$$= \int_0^1 (c_1 f_n'^2 + c_2 r^{-2} f_n^2)\lambda^{N-2} r^{N-1} dr$$

$$> c_2 \int_0^1 f_n^2 \lambda^{N-2} r^{N-3} dr$$

for some positive constants c_1 and c_2. For $\alpha \leq -1/(N-2)$, we have by Lemma 2.7

$$D(h) > c\int_{1/2}^1 (1-r^2)^{(N-2)\alpha} r^{N-3} dr = \infty.$$

We conclude that $B^N_\alpha \in O^N_{HD}$.

From Lemma 2.9 thus established and Lemma 2.8, we see that $B^N_\alpha \in \tilde{O}^N_{HB} \cap O^N_{HD}$ for $\alpha \leq -1/(N-2)$ and $N > 2$. This shows the strictness of the inclusion $O^N_{HB} \subset O^N_{HD}$ and completes the proof of Theorem 2.2.

2.10. Summary on harmonic functions on the Poincaré N-ball. For later reference, we compile here Lemmas 2.6, 2.8, and 2.9 on the existence of harmonic functions on the Poincaré N-ball into the following comprehensive

THEOREM. For $N \geq 2$,

$$B^N_\alpha \in \begin{cases} O^N_G \Leftrightarrow \alpha \geq 1/(N-2), \\ O^N_{HP} \Leftrightarrow \alpha \geq 1/(N-2), \\ O^N_{HB} \Leftrightarrow \alpha \geq 1/(N-2), \\ O^N_{HD} \Leftrightarrow |\alpha| \geq 1/(N-2), \\ O^N_{HC} \Leftrightarrow |\alpha| \geq 1/(N-2). \end{cases}$$

The last statement is a consequence of Theorem 2.2.

2.11. Generalization. We shall now give another proof of the strictness of the important inclusion $O^N_{HB} \subset O^N_{HD}$ for $N > 2$. Actually, we shall establish a general criterion containing, as a special case, the above result $B^N_\alpha \in \tilde{O}^N_{HB} \cap O^N_{HD}$ if and only if $\alpha \leq -1/(N-2)$.

In terms of polar coordinates $(r,\theta) = (r,\theta^1,\ldots,\theta^{N-1})$, $r = |x|$, of $x = (x^1,\ldots,x^N)$ in the unit N-ball B^N, $N \geq 2$, the Euclidean line element $|dx|$ is given by

$$|dx|^2 = dr^2 + r^2 \sum_{i=1}^{N-1} \gamma_i(\theta)d\theta^{i2},$$

with the γ_i trigonometric functions of θ as in 2.4. Endow B^N with the metric

$$ds^2 = \varphi(r)^2dr^2 + \psi(r)^2r^2 \sum_{i=1}^{N-1} \gamma_i(\theta)d\theta^{i2},$$

where φ, ψ are strictly positive C^∞ functions on $[0,1)$, with $\varphi(0) = \psi(0) = 1$, $\varphi'(0) = \psi'(0) = 0$. Denote by $B^N_{\varphi\psi}$ the resulting Riemannian manifold. We shall prove:

THEOREM. *The manifold $B^N_{\varphi\psi}$ belongs to $\tilde{O}^N_{HB}$ if and only if (φ,ψ) satisfies*

(a) $$\int_0^1 \left(\int_0^r \varphi(\rho)\psi(\rho)^{N-3}d\rho\right) \frac{\varphi(r)}{\psi(r)^{N-1}}\, dr < \infty.$$

<u>The manifold</u> $B^N_{\varphi\psi}$ <u>belongs to</u> $\tilde{O}^N_{HD}$ <u>if and only if</u> (φ,ψ) <u>satisfies</u>

(b)
$$\int_0^1 \left(\int_0^1 \varphi(\rho)\psi(\rho)^{N-3} d\rho \right) \frac{\varphi(r)}{\psi(r)^{N-1}}\, dr < \infty .$$

Suppose $\varphi(r) = \psi(r)$ on $[0,1)$, that is, $ds = \varphi(|x|)\,|dx|$ on B^N. For $B^N_\varphi = B^N_{\varphi\varphi}$, we have $B^N_\varphi \in \tilde{O}^N_{HB}$ if and only if

(a')
$$\int_0^1 \left(\int_0^r \varphi(\rho)^{N-2} d\rho \right) \varphi(r)^{-(N-2)} dr < \infty ,$$

and $B^N_\varphi \in \tilde{O}^N_{HD}$ if and only if

(b')
$$\int_0^1 \left(\int_0^1 \varphi(\rho)^{N-2} d\rho \right) \varphi(r)^{-(N-2)} dr < \infty .$$

The method in 2.4 - 2.10 is only applicable to quite simple φ such as $(1 - r^2)^\alpha$, for which the computations can be carried out. In our present approach, it suffices to assume φ to be C^∞, or even C^1, on $[0,1)$. If $\varphi(r) = (1 - r^2)^\alpha$, then (a') is equivalent to $\alpha < 1/(N - 2)$, and (b') to $|\alpha| < 1/(N - 2)$, so that the characterization $B^N_\alpha \in \tilde{O}^N_{HB} \cap O^N_{HD}$ if and only if $\alpha \leq -1/(N - 2)$ follows at once.

The proof of the above theorem will be given in 2.12 - 2.18.

<u>2.12.</u> <u>Radial harmonic functions.</u> The Laplace-Beltrami operator $\Delta_{\varphi\psi}$ on $B^N_{\varphi\psi}$ takes the form

$$\Delta_{\varphi\psi} u = - \frac{1}{\varphi\psi^{N-1} r^{N-1}} \frac{\partial}{\partial r} \left(\frac{\psi^{N-1} r^{N-1}}{\varphi} \frac{\partial u}{\partial r} \right) - \frac{1}{\psi^2 r^2 \gamma} \sum_{i=1}^{N-1} \frac{\partial}{\partial \theta^i} \left(\gamma \gamma_i^{-1} \frac{\partial u}{\partial \theta^i} \right) ,$$

where $\gamma(\theta) = \Pi_1^{N-1} \gamma_i(\theta)^{1/2}$. Note that $d\omega = \gamma(\theta) d\theta^1 \cdots d\theta^{N-1}$ is the Euclidean surface element on the unit sphere $\omega = \{|x| = 1\}$, and

$$\int_\omega d\omega = 2\pi^{N/2}/\Gamma(N/2) = A_N$$

is the Euclidean area of ω. Fix an arbitrary $a \in (0,1)$, let Ω^N be the annulus $a < |x| < 1$ in E^N, and Γ^N the inner boundary $|x| = a$. We use the notation $\Omega^N_{\varphi\psi}$ for Ω^N viewed as a subregion of $B^N_{\varphi\psi}$. The harmonic measure $e_0(x)$ of the ideal boundary $|x| = 1$ of $B^N_{\varphi\psi}$ on $\Omega^N_{\varphi\psi}$ is a harmonic function on $\Omega^N_{\varphi\psi}$ with boundary values 0 on Γ^N and 1 "on the ideal boundary". We know from 1.3 that $e_0(x) \equiv 0$ if and only if $B^N_{\varphi\psi} \in O^N_G$. Clearly, $e_0(x)$ is radial, that is, $e_0(x) = e_0(|x|)$ on Ω^N. Therefore, $e_0(r)$ is a solution of

$$L_0 u = \left(\frac{\psi^{N-1} r^{N-1}}{\varphi} u'\right)' = 0$$

on $[a,1)$ with $e_0(a) = 0$ and "$e_0(1) = 1$". Hence,

$$e_0(r) = c \int_a^r \frac{\varphi(\rho)}{\psi(\rho)^{N-1} \rho^{N-1}} d\rho$$

with a suitable constant $c \geq 0$, and $B^N_{\varphi\psi} \in \tilde{O}^N_G$ if and only if $c > 0$, or, equivalently,

(c) $$\int_0^1 \frac{\varphi(r)}{\psi(r)^{N-1}} dr < \infty.$$

Either one of conditions (a) and (b) implies this inequality. On the other hand, either one of the memberships $B^N_{\varphi\psi} \in \tilde{O}^N_{HB}$ and $B^N_{\varphi\psi} \in \tilde{O}^N_{HD}$ entails $B^N_{\varphi\psi} \in \tilde{O}^N_G$. In the proof of our theorem we thus may, and henceforth will, assume that (c) is valid, or, equivalently, that $e_0(x) > 0$ on $\Omega^N_{\varphi\psi}$. In this case, $e_0(r)$ is an increasing function and

$$e_0(1) = \lim_{r \to 1} e_0(r) = 1.$$

2.13. Reduction of the problem. Denote by $HX(B^N_{\varphi\psi})$ the classes of harmonic functions on $B^N_{\varphi\psi}$ with $X = B, D$, or C. Let $HX(\Omega^N_{\varphi\psi})_0$ be the class of functions $u \in HX(\Omega^N_{\varphi\psi})$ with boundary values 0 on Γ^N. Suppose $u \in HX(\Omega^N_{\varphi\psi})_0$ is radial, that is, $u(x) = u(|x|)$, for $X = B$ or C. Then $u(r)$ is a bounded solution of $L_0 u = 0$ on $[a,1)$ with $u(a) = 0$ and has a representation of the above form for $e_0(r)$. Therefore, there exists a constant k such that $u = ke_0$. On the other hand, it is well known (e.g., Sario-Nakai [1]) that

$$HX(B^N_{\varphi\psi}) \simeq HX(\Omega^N_{\varphi\psi})_0$$

in the sense of linear isomorphisms for X = B, D, and C. These two remarks together with the identity $O^N_{HD} = O^N_{HC}$ yield:

The manifold $B^N_{\varphi\psi}$ belongs to O^N_{HB} (O^N_{HD}, resp.) if and only if the class $HB(\Omega^N_{\varphi\psi})_0$ ($HC(\Omega^N_{\varphi\psi})_0$, resp.) contains a nonradial harmonic function u, that is, $u(x) \neq u(|x|)$.

To prove Theorem 2.11 we thus have to show that (a) ((b), resp.) is equivalent to the property of $HB(\Omega^N_{\varphi\psi})_0$ ($HC(\Omega^N_{\varphi\psi})_0$, resp.) containing a nonradial function.

2.14. Arbitrary harmonic functions. Take a complete orthonormal system $\{A_N^{-1/2}, S_{nm}(\theta)\}$, $n = 1,2,\ldots$, $m = 1,\ldots,m_n$, in $L^2(\omega, d\omega)$, where $\{S_{nm}(\theta)\}$, $m = 1,\ldots,m_n$, spans the linear space of spherical harmonics of degree $n \geq 1$. Since $\Delta_{11}(r^n S_{nm}) = 0$ implies $\Delta_{11} S_{nm} = n(n + N - 2)r^{-2} S_{nm}$ at (r,θ), we have

$$\Delta_{\varphi\psi} S_{nm} = \frac{n(n + N - 2)}{\psi^2 r^2} S_{nm}.$$

For $f, g \in C^1$, multiplication of the covariant components of grad f by the contravariant components of grad g gives

$$\nabla f \cdot \nabla g = \varphi^{-2} \frac{\partial f}{\partial r} \frac{\partial g}{\partial r} + \sum_{i=1}^{N-1} \psi^{-2} r^{-2} \gamma_i^{-1} \frac{\partial f}{\partial \theta^i} \frac{\partial g}{\partial \theta^i},$$

and integration by parts yields

$$\int_\omega \nabla S_{nm} \cdot \nabla S_{pq}\, d\omega = 0$$

for $(n,m) \neq (p,q)$.

Take an arbitrary harmonic function h on $\Omega^N_{\varphi\psi}$ with boundary values 0 on Γ^N. For a fixed $r \in [a,1)$, expand h into the Fourier series

$$h(r,\theta) = A_N^{-1/2}\, h_0(r) + \sum_{n=1}^{\infty} \sum_{m=1}^{m_n} h_{nm}(r) S_{nm}(\theta)$$

on ω, with

$$\begin{cases} h_0(r) = A_N^{-1/2} \int_\omega h(r,\theta)\,d\omega, \\ h_{nm}(r) = \int_\omega h(r,\theta) S_{nm}(\theta)\,d\omega. \end{cases}$$

Here h_0 is a solution of $L_0 u = 0$ on $[a,1)$ with $h_0(a) = 0$, and, in view of $\Delta_{\varphi\psi}(h_{nm}S_{nm}) = 0$, h_{nm} is a solution of the equation

$$L_n u = \left(\frac{\psi^{N-1} r^{N-1}}{\varphi}\, u' \right)' - n(n + N - 2)\varphi\psi^{N-3} r^{N-3} u = 0$$

on $[a,1)$ with $h_{nm}(a) = 0$, $n \geq 1$, $m = 1,\ldots,m_n$.

2.15. Reduction to solution types. Consider a solution u_t of $L_n u = 0$ on $[a,t)$, $a < t < 1$, with $u_t(a) = 0$ and $u_t(t) = 1$. By the maximum principle, $\{u_t\}$ is a decreasing net, and by Harnack's principle, $e_n = \lim_{t\to 1} u_t$ exists and is a solution of $L_n u = 0$ on $[a,1)$ with $e_n(a) = 0$ and $0 \leq e_n < 1$ on $(a,1)$. We know that either $e_n > 0$ on $(a,1)$ with $\lim_{r\to 1} e_n(r) = e_n(1) = 1$, or else $e_n \equiv 0$ on $(a,1)$. We denote by (B) the class of operators L_n yielding the former case. It is readily seen that u_t for L_n satisfies $L_{n+1}\, u_t < 0$ and $L_{n+1}\, u_t^\beta > 0$ for $\beta = (n + 1)(n + N - 1)/n(n + N - 2)$. A fortiori, by the comparison principle (Nakai [8]),

$$e_n^\beta \leq e_{n+1} \leq e_n$$

on $[a,1)$. We conclude that either $L_n \in$ (B) for every $n \geq 1$ or else $L_n \notin$ (B) for every $n \geq 1$.

Suppose $L_1 \in$ (B). On integrating $L_1 e_1 = 0$ from a to r, multiplying by $\varphi(r)\psi(r)^{-N+1} r^{-N+1}$, and integrating from a to 1, we obtain in view of (c),

$$1 = \frac{\psi(a)^{N-1}a^{N-1}e_1'(a)}{\varphi(a)} \int_a^1 \frac{\varphi(r)}{\psi(r)^{N-1}r^{N-1}}\,dr$$

$$+ (N-1)\int_a^1 \left(\int_a^r \varphi(\rho)\psi(\rho)^{N-3}\rho^{N-3}e_1(\rho)d\rho\right) \frac{\varphi(r)}{\psi(r)^{N-1}r^{N-1}}\,dr.$$

Since e_1 has a positive infimum on, e.g., $[a/2,1)$, relation (a) follows.

Conversely, suppose (a) is valid. Let u_t be as above for L_1. By exactly the same reasoning we obtain

$$1 = u_t(t) = \frac{\psi(a)^{N-1}a^{N-1}u_t'(a)}{\varphi(a)} \int_a^t \frac{\varphi(r)}{\psi(r)^{N-1}r^{N-1}}\,dr$$

$$+ (N-1)\int_a^t \left(\int_a^r \varphi(\rho)\psi(\rho)^{N-3}\rho^{N-3}u_t(\rho)d\rho\right) \frac{\varphi(r)}{\psi(r)^{N-1}r^{N-1}}\,dr.$$

Since $u_t'(a) \to e_1'(a)$ and $u_t \to e_1$ on $[a,t)$, both decreasingly as $t \to 1$, the Lebesgue convergence theorem applied to this equality yields the equality preceding it. We infer that $e_1 \not\equiv 0$, that is, $L_1 \in (B)$.

We summarize:

<u>The operator</u> L_n <u>belongs to</u> (B), <u>for some and hence for every</u> $n \geq 1$, <u>if and only if</u> (φ,ψ) <u>satisfies condition</u> (a).

<u>2.16.</u> <u>Existence of</u> HB <u>functions.</u> We are ready to prove the first part of Theorem 2.11. Suppose $B^N_{\varphi\psi} \in \tilde{O}^N_{HB}$. Then there exists a nonradial h in $HB(\Omega^N_{\varphi\psi})_0$, and by its expansion in 2.14, there exists an $n \geq 1$ such that $h_{nm} \not\equiv 0$ for some $m = 1,\ldots,m_n$. By the expression for $h_{nm}(r)$ in 2.14, $k = \sup_{a\leq r<1}|h_{nm}(r)| < \infty$. Let u_t be as in 2.15 for L_n. Then $|h_{nm}(r)| \leq ku_t(r)$ on $[a,t]$ and a fortiori $|h_{nm}| \leq ke_n$, that is, $L_n \in (B)$. Hence (a) is valid.

Conversely, suppose (a) holds. Then $L_1 \in (B)$, and e_1S_{11}, say, is a nonradial member of $HB(\Omega^N_{\varphi\psi})_0$, hence $B^N_{\varphi\psi} \in \tilde{O}^N_{HB}$.

Observe that (b) implies (a), and $B^N_{\varphi\psi} \in \tilde{O}^N_{HD}$ entails $B^N_{\varphi\psi} \in \tilde{O}^N_{HB}$. In the proof of the second part of Theorem 2.11 we may and shall, therefore, assume that not only (c) but also (a) is satisfied.

2.17. Dirichlet integrals. Let f be of class C^1 on Ω^N. The Dirichlet integral $D(f)$ of f over $\Omega^N_{\varphi\psi}$ is

$$D(f) = \int_a^1 \left(\int_\omega \nabla f(r,\theta)\cdot\nabla f(r,\theta)d\omega \right) \varphi(r)\psi(r)^{N-1}r^{N-1}dr.$$

By virtue of (a) and (c), the functions h_{nm} in 2.14 for an $h \in HC(\Omega^N_{\varphi\psi})_0$ are multiples of $e_n > 0$, that is,

$$h(r,\theta) = A_N^{-1/2} c_0 e_0(r) + \sum_{n=1}^{\infty} \sum_{m=1}^{m_n} c_{nm} e_n(r) S_{nm}(\theta).$$

By 2.14 and the orthogonality of the basis in $L^2(\omega, d\omega)$,

$$D(h) = c_0^2 D(e_0) = \sum_{n=1}^{\infty} \sum_{m=1}^{m_n} c_{nm}^2 D(e_n S_{nm}).$$

Moreover,

$$D(e_n S_{nm}) = D(e_n) + D^*(e_n S_{nm}),$$

where we have set

$$D^*(e_n S_{nm}) = \int_a^1 \left(\int_\omega \nabla S_{nm}\cdot\nabla S_{nm} d\omega \right) e_n(r)^2 \varphi(r)\psi(r)^{N-1}r^{N-1}dr.$$

Since the harmonic measure always has a finite Dirichlet integral $D(e_0) < \infty$ (e.g., Sario-Nakai [1]), we are only interested in the finiteness of $D(e_n S_{nm})$. Observe that

$$\int_\omega \nabla S_{nm}\cdot\nabla S_{nm} d\omega = \psi(r)^{-2} r^{-2} \int_\omega \left(\sum_1^{N-1} \gamma_i(\theta)^{-1} \left(\frac{\partial}{\partial\theta^i} S_{nm}(\theta) \right)^2 \right) d\omega \sim \psi(r)^{-2}$$

as $r \to 1$. Therefore,

$$D^*(e_n S_{nm}) \approx \int_0^1 \varphi(r)\psi(r)^{N-3}dr,$$

where $A \approx B$ means that A and B are simultaneously finite or infinite.

It remains to estimate

$$D(e_n) = A_N \int_a^1 e_n'(r)^2 \varphi(r)^{-2} \varphi(r)\psi(r)^{N-1} r^{N-1} dr$$

$$\approx \int_a^1 e_n'(r)^2 \frac{\psi(r)^{N-1}}{\varphi(r)} dr.$$

Since $L_n e_n = 0$, we have

$$e_n'(r) = \frac{\varphi(r)}{\psi(r)^{N-1} r^{N-1}} \left(\frac{\psi(a)^{N-1} a^{N-1} e_n'(a)}{\varphi(a)} + n(n + N - 2) \int_a^r \varphi(\rho)\psi(\rho)^{N-3} \rho^{N-3} e_n(\rho) d\rho \right).$$

A fortiori, as $r \to 1$,

$$e_n'(r) \sim \frac{\varphi(r)}{\psi(r)^{N-1}} \left(1 + \int_0^r \varphi(\rho)\psi(\rho)^{N-3} d\rho \right)$$

and

$$e_n'(r)^2 \sim \left(\frac{\varphi(r)}{\psi(r)^{N-1}} \right)^2 \left(1 + \left(\int_0^r \varphi(\rho)\psi(\rho)^{N-3} d\rho \right)^2 \right).$$

On substituting this into the above expression for $D(e_n)$ and observing (c), we obtain

$$D(e_n) \approx \int_0^1 \left(\int_0^r \varphi(\rho)\psi(\rho)^{N-3} d\rho \right)^2 \frac{\varphi(r)}{\psi(r)^{N-1}} dr.$$

2.18. Existence of HD functions. We are ready to prove the latter half of Theorem 2.11. Suppose $B^N_{\varphi\psi} \in \tilde{O}^N_{HD}$. Then there exists a nonradial $h \in HC(\Omega^N_{\varphi\psi})_0$ and, by 2.17, a nonzero $c_{nm} e_n(r) S_{nm}(\theta)$. Again by 2.17, $D(e_n S_{nm}) < \infty$ and

$$\int_0^1 \varphi(r)\psi(r)^{N-3} dr < \infty.$$

This with (c) implies (b).

Conversely, suppose (b) is satisfied. By (c), we then have the above inequality, that is, $D^*(e_n S_{nm}) < \infty$ for every $n \geq 1$ and $m = 1,\dots,m_n$. In view of 2.17, we also obtain $D(e_n) < \infty$ for every $n \geq 1$. Thus,

$$D(e_n S_{nm}) < \infty, \qquad n = 1,2,\dots, \qquad m = 1,\dots,m_n,$$

that is, every $e_n S_{nm}$ is a nonradial function in $HC(\Omega^N_{\varphi\psi})_0$, and $B^N_{\varphi\psi} \in \tilde{O}^N_{HD}$.

The proof of Theorem 2.11 is herewith complete.

NOTES TO §2. For Riemann surfaces, the relations $O_{HB} \subset O_{HD} = O_{HC}$ were established in Sario [4] and Virtanen [1], and the corresponding relations for Riemannian manifolds in Sario-Schiffer-Glasner [1]. The latter, on which the presentation in 2.1 is based, made essential use of a generalization of the fruitful concept of span introduced by Schiffer [1]. The relations were also deduced by Nakai by means of a remarkable generalization of Royden's algebra (for a systematic account, see, e.g., Sario-Nakai [1, p. 154 ff.]).

The strictness of $O_G < O_{HD}$ for Riemann surfaces was first established by Ahlfors in his address before the 1949 Annual Meeting of the American Mathematical Society in New York. An improved version was published by Ahlfors and Royden [1]. The strictness of $O_G < O_{HB} < O_{HD}$ is one of the striking achievements in the theory of harmonic functions and is due to Tôki [1]. The somewhat simpler proof in 2.3 is from Sario [5].

For Riemannian manifolds, the strictness of $O^N_{HB} < O^N_{HD}$ was obtained, several years after the other strict inclusions had been established, in Kwon [1] and Hada-Sario-Wang [1]. The greatly simplified proof given in 2.4-2.9 is based on estimates of f_n in Hada-Sario-Wang [3] and subsequent discussions with Hada.

Theorem 2.11 and the new proof of the strictness of $O^N_{HB} < O^N_{HD}$ based on it were given in Nakai-Sario [14].

§3. THE CLASS $O^N_{HL^p}$

In §§1 and 2, we have shown that

$$O^N_\omega = O^N_G < O^N_{HP} < O^N_{HB} < O^N_{HD} = O^N_{HC}$$

for $N \geq 2$. All these classes had their origin in the theory of Riemann surfaces. We now take up a new class which has no meaning for abstract Riemann surfaces, $O^N_{HL^p}$, $1 \leq p < \infty$. This is the class of Riemannian N-manifolds, $N \geq 2$, on which every harmonic function with a finite L^p norm is constant. We may omit the case $p = \infty$, as it leads us back to $HB = HL^\infty$.

If a Riemannian manifold carries HC functions, one might expect, in view of the above inclusion relations, that it then carry HL^p functions as well, at least for some $p \in [1,\infty)$. However, it turns out that $O^N_{HL^p} \cap \tilde{O}^N_{HC} \neq \emptyset$ for every $p \in [1,\infty)$, $N \geq 2$, in fact, even $\bigcap_p O^N_{HL^p} \cap \tilde{O}^N_{HC} \neq \emptyset$. This is the main result of the present section.

At the opposite end of the scheme, since a parabolic manifold carries no HX functions for $X = P, B, D, C$, one is tempted to conjecture that it carry no HL^p functions either, at least for some p. But again, $\bigcap_p \tilde{O}^N_{HL^p} \cap O^N_G \neq \emptyset$.

More superficial results, which serve to complete the picture, are $\bigcap_p O^N_{HL^p} \cap O^N_G \neq \emptyset$ and $\bigcap_p \tilde{O}^N_{HL^p} \cap \tilde{O}^N_{HC} \neq \emptyset$. Thus there are no inclusion relations between $O^N_{HL^p}$ and any O^N_{HX} or their complements.

3.1. Neither HL^p functions nor HX. For a comprehensive notation, we now let X take also the meaning G, with O^N_{HG} standing for O^N_G. First we exclude both the HL^p and HX functions.

THEOREM. For $1 \leq p < \infty$, and $N \geq 2$,

$$\bigcap_p O^N_{HL^p} \cap O^N_{HX} \neq \emptyset,$$

with $X = G, P, B, D, C$.

Proof. In Cartesian coordinates $x, y^1, \ldots, y^{N-1}$, consider the N-cylinder

$$R = \{|x| < \infty, \quad |y^i| \leq \pi, \quad i = 1,\ldots,N-1\},$$

with the pair of opposite faces $y^i = \pi$ and $y^i = -\pi$ identified, for every i, by a parallel translation perpendicular to the x-axis. Endow R with the Euclidean metric $ds^2 = dx^2 + \Sigma_1^{N-1} dy^{i2}$. For the Laplace-Beltrami operator

$\Delta = d\delta + \delta d$ and a function h_0 of x only, the equation $\Delta h_0 = -h_0'' = 0$ gives the harmonic functions $h_0 = ax + b$. The harmonic measure ω_c of $\{x = c\}$ on $\{0 \le x \le c\}$ with $c > 0$ is x/c, and $\omega_c \to 0$ as $c \to \infty$. The analogue is true for $c < 0$, and therefore $R \in O^N_{HG} \subset O^N_{HX}$.

For a trial solution of the general equation $\Delta h = 0$ take

$$h = f(x) \prod_{i=1}^{N-1} g_i(y^i).$$

Then

$$\Delta h = -\left(f'' \prod_{i=1}^{N-1} g_i + f \sum_{i=1}^{N-1} g_i'' \prod_{k \neq i} g_k\right)$$
$$= -h\left(f''f^{-1} + \sum_{i=1}^{N-1} g_i''g_i^{-1}\right) = 0.$$

Each term in ( ) depends on one variable only and is therefore constant. The eigenvalues $n_i \ge 0$ give the equations $g_i'' = -n_i^2 g_i$ and the eigenfunctions

$$g_{i1} = \cos n_i y^i, \qquad g_{i2} = \sin n_i y^i.$$

We shall use the notation

$$n = (n_1,\ldots,n_{N-1}), \qquad 0 = (0,\ldots,0), \qquad \eta^2 = \sum_{i=1}^{N-1} n_i^2, \qquad \eta \ge 0.$$

Given a function j from n to $\{1,2\}$, set

$$G_{nj} = \prod_{i=1}^{N-1} g_{ij}(n_i).$$

Then $f_n G_{nj}$ is harmonic if f_n satisfies $f_n'' = \eta^2 f_n$, that is, $f_n = e^{\pm\eta x}$ if $\eta \neq 0$. It is readily seen that an arbitrary harmonic function h on R has an expansion

$$h = h_0 + \sum_{n,j}{}' (a_{nj}e^{\eta x} + b_{nj}e^{-\eta x})G_{nj}$$

on $\{|x| = c\}$, hence on $\{|x| \le c\}$ and a fortiori on R; here Σ' extends over all $n \neq 0$ and all j.

Suppose $h \in HL^p$. If some $a_{nj} \neq 0$, take a continuous function $\rho_0(x) \geq 0$ on $(-\infty,\infty)$ with $\operatorname{supp} \rho_0 \subset (0,1)$ and $\int_0^1 \rho_0 dx = 1$. For a number $t > 0$ set $\rho_t(x) = \rho_0(x - t)$ and $\varphi_t = \rho_t G_{nj}$. Then

$$(h,\varphi_t) = c \int_t^{t+1} (a_{nj}e^{\eta x} + b_{nj}e^{-\eta x})\rho_t dx.$$

Here and later c is a constant, not always the same. For a sufficiently large t,

$$|(h,\varphi_t)| > ce^{\eta t} \int_t^{t+1} \rho_t dx = ce^{\eta t},$$

whereas, for $1 < p < \infty$, $p^{-1} + q^{-1} = 1$,

$$\|\varphi_t\|_q = c\left(\int_t^{t+1} \rho_t^q dx\right)^{1/q} = \text{const} < \infty,$$

and for $p = 1$, $\|\varphi_t\|_q = \text{const} < \infty$. The Hölder inequality $|(h,\varphi_t)| \leq \|h\|_p \cdot \|\varphi_t\|_q$ is violated by $|(h,\varphi_t)|/\|\varphi_t\|_q \to \infty$ for $\eta t > 0$, that is, for all $\eta > 0$. We conclude that $a_{nj} = 0$ for all $n \neq 0$. An analogous argument using $t < 0$ and $t \to -\infty$ gives $b_{nj} = 0$ for all $n \neq 0$. Therefore, $h = h_0$. For $1 \leq p < \infty$, we have $\|h_0\|_p = \infty$ unless $a = b = 0$. Thus every $h \in HL^p$ reduces to zero, and $R \in O^N_{HL^p}$.

<u>3.2.</u> HX <u>functions but no</u> HL^p. We now exclude the HL^p functions only.

<u>THEOREM</u>. <u>For</u> $1 \leq p < \infty$, <u>and</u> $N \geq 2$,

$$\bigcap_p O^N_{HL^p} \cap \tilde{O}^N_{HX} \neq \emptyset,$$

<u>with</u> $X = G, P, B, D, C$.

<u>Proof</u>. Consider the N-cylinder

$$R = \{|x| < 1, \quad |y^i| \leq \pi, \quad i = 1,\ldots,N - 1\}$$

with the metric

$$ds^2 = \lambda^2 dx^2 + \lambda^{2/(N-1)} \sum_{i=1}^{N-1} dy^{i2},$$

with $\lambda = \lambda(x) \in C^{\infty}(-1,1)$. The equation

$$\Delta h_0(x) = -\lambda^{-2}(\lambda^2\lambda^{-2}h_0')' = 0$$

gives again $h_0 = ax + b$. Since

$$D(x) = c\int_{-1}^{1} \lambda^{-2}\lambda^2 dx < \infty,$$

we have $x \in HC$, hence $R \in \tilde{O}^N_{HC} \subset \tilde{O}^N_{HX}$ for all X.

Let G_{nj} be as in 3.1. Then $h = f_n(x)G_{nj}$ is harmonic if

$$\Delta h = -\lambda^{-2}\left(f_n''G_{nj} - \lambda^{2(N-2)/(N-1)} \sum_{i=1}^{N-1} n_i^2 f_n G_{nj}\right) = 0,$$

that is,

$$f_n'' = \eta^2\lambda^{2(N-2)/(N-1)} f_n.$$

First we consider the case $N > 2$ and choose

$$\lambda = (1 - x^2)^{-(N-1)/(N-2)}.$$

Then

$$(1 - x^2)^2 f_n'' = \eta^2 f_n.$$

To solve this differential equation, set

$$f_n(x) = (1 - x^2)^{1/2} t_n(z), \quad \frac{dz}{dx} = (1 - x^2)^{-1}.$$

A simple computation yields

$$f_n' = (1 - x^2)^{-1/2}(-xt_n + t_n'),$$

$$f_n'' = (1 - x^2)^{-3/2}(t_n'' - t_n),$$

and our equation is transformed into

$$t_n''(z) = (1 + \eta^2)t_n(z).$$

It is satisfied by $t_n = e^{\pm(1+\eta^2)^{1/2}z}$, and in view of

$$z = \frac{1}{2}\log\frac{1+x}{1-x},$$

we obtain the solutions

$$\begin{cases} f_{n1} = (1+x)^{2^{-1}(1+(1+\eta^2)^{1/2})}(1-x)^{2^{-1}(1-(1+\eta^2)^{1/2})}, \\ f_{n2} = (1+x)^{2^{-1}(1-(1+\eta^2)^{1/2})}(1-x)^{2^{-1}(1+(1+\eta^2)^{1/2})}. \end{cases}$$

For $k = 1,2$, the function $f_{nk}G_{nj}$ is harmonic, and an arbitrary harmonic function h can be written

$$h = \sum_n\sum_j\sum_k a_{njk}f_{nk}G_{nj},$$

where the summation with respect to n now includes $n = 0 = (0,\dots,0)$.

Suppose $h \in HL^p$. If some $a_{nj1} \neq 0$, take numbers $0 < \alpha < \beta < 1$ and a continuous function $\rho_0 \geq 0$ on $(-\infty,\infty)$ with $\operatorname{supp}\rho_0 \subset (\alpha,\beta)$. For a number $t \in (0,1)$, set $\rho_t(x) = \rho_0((1-x)/t)$, $\varphi_t = \rho_t G_{nj}$. Now $\operatorname{supp}\rho_t \subset (1-\beta t, 1-\alpha t)$, and as $t \to 0$,

$$(h,\varphi_t) \sim c\int_{1-\beta t}^{1-\alpha t}(1-x)^{2^{-1}(1-(1+\eta^2)^{1/2})}(1-x^2)^{-2(N-1)/(N-2)}\rho_t dx,$$

which, by virtue of $\int_{1-\beta t}^{1-\alpha t}\rho_t dx = O(t)$, gives

$$(h,\varphi_t) \sim ct^{2^{-1}(1-(1+\eta^2)^{1/2})-2(N-1)/(N-2)+1}.$$

On the other hand, if $p > 1$, then

$$\|\varphi_t\|_q \sim c\left(\int_{1-\beta t}^{1-\alpha t}\rho_t^q(1-x^2)^{-2(N-1)/(N-2)}dx\right)^{1/q} = O\left(t^{q^{-1}(1-2(N-1)/(N-2))}\right).$$

In view of $1 - q^{-1} = p^{-1}$, we have again a violation of the Hölder inequality $|(h,\varphi_t)| \leq \|h\|_p\cdot\|\varphi_t\|_q$ for

$$\frac{1}{p}\left(-\frac{2(N-1)}{N-2}+1\right) < \frac{1}{2}((1+\eta^2)^{1/2}-1),$$

that is, for every $\eta \geq 0$. If $p = 1$, then $\|\varphi_t\|_q = \text{const} < \infty$ and we have a contradiction for

$$\frac{1}{2}(1-(1+\eta^2)^{1/2}) - \frac{2(N-1)}{N-2} + 1 < 0,$$

hence again for every $\eta \geq 0$. We infer that all $a_{nj1} = 0$.

If some a_{nj2} is $\neq 0$, we choose $\rho_t(x) = \rho_0((x+1)/t)$ and obtain the same estimates as above for $|(h,\varphi_t)|$ and $\|\varphi_t\|_q$ and $t \to 0$. Thus all $a_{nj2} = 0$, hence $h \equiv 0$, and $R \in O^N_{HL^p}$ for $N > 2$.

For $N = 2$, choose

$$\lambda = (1 - x^2)^{-1}.$$

In the conformal metric $ds = \lambda|dx|$, the harmonic functions are the same as in the Euclidean metric, and we obtain as in 3.1

$$h = h_0 + \sum_{n,j}{}' (a_{nj}e^{nx} + b_{nj}e^{-nx})G_{nj},$$

where now n stands for n_1. Suppose $h \in HL^p$. If $a_{nj}e^n + b_{nj}e^{-n} \neq 0$ for some $n > 0$, then for $\rho_t(x) = \rho_0((1-x)/t)$, $\varphi_t = \rho_t G_{nj}$, $t \to 0$, we have

$$(h,\varphi_t) = c\int_{1-\beta t}^{1-\alpha t} (a_{nj}e^{nx} + b_{nj}e^{-nx})\rho_t(1-x^2)^{-2}dx$$

$$\sim c_1 t^{-2}\int_{1-\beta t}^{1-\alpha t} \rho_t dx = c_2 t^{-1},$$

whereas, for $1 < p < \infty$,

$$\|\varphi\|_q = c\left(\int_{1-\beta t}^{1-\alpha t} \rho_t^q (1-x^2)^{-2}dx\right)^{1/q} = O(t^{-1/q}).$$

In view of $-1 < -q^{-1}$, we again have a violation of the Hölder inequality and conclude that $a_{nj}e^n + b_{nj}e^{-n} = 0$ for all $n > 0$. An analogous reasoning by means of $\rho_t(x) = \rho_0((x+1)/t)$, $t \to 0$, gives $a_{nj}e^{-n} + b_{nj}e^{n} = 0$. As a consequence, $a_{nj} = b_{nj} = 0$ for all $n > 0$, that is, $h = h_0$. Since

$$\|h_0\|_p = c\left(\int_{-1}^{1} |ax + b|^p (1 - x^2)^{-2} dx\right)^{1/p} = \infty$$

unless $a = b = 0$, we obtain $h \equiv 0$, hence $R \in O^2_{HL^p}$ for $1 < p < \infty$.

The conclusion remains valid for $p = 1$, since $\|\varphi_t\|_\infty = \text{const} < \infty$, and the contradiction is for $n > 0$ such that $t^{-1} > O(1)$, hence again $a_{nj} = b_{nj} = 0$ for all $n > 0$. The reasoning for $n = 0$ is the same as before.

<u>3.3. HL^p functions but no HX.</u> Next we exclude the HX functions only.

<u>THEOREM. For</u> $1 \le p < \infty$ <u>and</u> $N \ge 2$,

$$\bigcap_p \tilde{O}^N_{HL^p} \cap O^N_{HX} \ne \emptyset,$$

with $X = G, P, B, D, C$.

<u>Proof.</u> Take the N-cylinder

$$R = \{|x| < \infty, \quad |y^i| \le \pi, \quad i = 1,\ldots,N-1\}$$

with the metric

$$ds^2 = \lambda^2 dx^2 + \lambda^{2/(N-1)} \sum_{i=1}^{N-1} dy^{i2},$$

with $\lambda = \lambda(x) \in C^\infty(-\infty,\infty)$. As in 3.2, $h_0(x) = ax + b$, and we conclude as in 3.1 that $R \in O^N_{HG} \subset O^N_{HX}$.

Choose $\lambda = e^{-x^2}$. Then for $1 \le p < \infty$,

$$\|x\|_p = c\left(\int_{-\infty}^{\infty} |x|^p e^{-2x^2} dx\right)^{1/p} < \infty,$$

and the theorem follows.

It is obvious that $\bigcap_p \tilde{O}^N_{HL^p} \cap \tilde{O}^N_{HX} \ne \emptyset$ for $1 \le p < \infty$, $N \ge 2$, and $X = G, P, B, D, C$. In fact, on the N-cylinder

$$R = \{|x| < 1, \quad |y| \leq \pi, \quad i = 1,\dots,N-1\}$$

with the Euclidean metric, the function $x + 1$ belongs to $HL^p \cap HC$.

We combine our results:

The classes

$$O^N_{HL^p} \cap O^N_{HX}, \quad O^N_{HL^p} \cap \tilde{O}^N_{HX}, \quad \tilde{O}^N_{HL^p} \cap O^N_{HX}, \quad \tilde{O}^N_{HL^p} \cap \tilde{O}^N_{HX}$$

are all nonvoid for $1 \leq p < \infty$, $N \geq 2$, and $X = G, P, B, D, C$.

3.4. A test for HL^p functions. Again we turn from the general existence problem to the question of the dependence of the existence on the metric in the concrete case of the Poincaré N-ball B^N_α. We shall apply our findings in §4.

We shall make use of the following auxiliary result. Recall that every harmonic function h on B^N_α has an expansion in spherical harmonics,

$$h(r,\theta) = \sum_{n=0}^{\infty} f_n(r) S_n(\theta).$$

LEMMA. For $1 \leq p < \infty$ and $N \geq 2$, $B^N_\alpha \in O^N_{HL^p}$ if and only if every nonzero $h \in H(B^N_\alpha)$ satisfies $f_n \notin L^p$ for all $n > 0$.

Proof. For the necessity, suppose $f_n \in L^p$ for some h and some $n > 0$. Then $f_n S_n \in HL^p$ for some nonzero S_n, and $B^N_\alpha \in \tilde{O}^N_{HL^p}$.

For the sufficiency, suppose the condition satisfied and take a nonconstant $h = \Sigma_0^\infty f_n S_n$. Then $f_{n_0} S_{n_0} \neq \text{const}$ for some $n_0 > 0$. Set $h = f_{n_0} S_{n_0} + \Sigma' f_n S_n$, where Σ' excludes $n = n_0$. Since $f_{n_0} \notin L^p$, there exists, for $p^{-1} + q^{-1} = 1$, a function $\varphi \in L^q$ such that $|\int f_{n_0} \varphi dx| = \infty$. By virtue of $\varphi S_{n_0} \in L^q$,

$$\left| \int_{B^N_\alpha} h\varphi S_{n_0} dx \right| = \left| \int_{B^N_\alpha} \left(\sum_0^\infty f_n S_n\right) \varphi S_{n_0} dx \right|$$

$$= \left| \int_{B^N_\alpha} \left[f_{n_0} S_{n_0} \varphi S_{n_0} + \left(\Sigma' f_n S_n\right) \varphi S_{n_0} \right] dx \right|,$$

where the absolute and uniform convergence of the series on compact subsets and the orthogonality of the S_n on ω gives for $B_{r_0} = \{r < r_0 < 1\}$,

$$\int_{B_\alpha^N} (\Sigma' f_n S_n)\varphi S_n dx = \lim_{r_0 \to 1} \int_{B_{r_0}} (\Sigma' f_n S_n)\varphi S_n dx = 0.$$

It follows that

$$\left| \int_{B_\alpha^N} h\varphi S_{n_0} dx \right| \geq c \left| \int_{B_\alpha^N} f_{n_0} \varphi dx \right| = \infty,$$

and therefore, $h \notin L^p$.

3.5. HL^p functions on the Poincaré N-ball. By means of Lemma 3.4, we shall be able to characterize the Poincaré N-balls B_α^N that belong to $O_{HL^p}^N$. We retain the division into the three cases of Lemma 2.7 and continue considering only dimension $N > 2$.

THEOREM. *In the case* $\alpha > 1/(N - 2)$, $B_\alpha^N \in O_{HL^p}^N$ *if and only if*

$$\alpha \geq \frac{p + 1}{(N - 2)p - N} \quad \text{and} \quad p > \frac{N}{N - 2}.$$

In the case $\alpha = 1/(N - 2)$, $B_\alpha^N \in \tilde{O}_{HL^p}^N$ *for all* $p \in [1,\infty)$.

In the case $\alpha < 1/(N - 2)$, $B_\alpha^N \in O_{HL^p}^N$ *if and only if* $\alpha \leq -1/N$, *independently of* $p \in [1,\infty)$.

Proof. In the first case, suppose the conditions in the theorem hold. For $h = \Sigma f_n S_n \in H(B_\alpha^N)$ and every $n > 0$, Lemma 2.7 gives $f_n(r) \sim c(1 - r)^{1-(N-2)\alpha}$. But

$$\|(1 - r)^{1-(N-2)\alpha}\|_p^p \approx c \int_0^1 (1 - r)^{[1-(N-2)\alpha]p}(1 - r)^{N\alpha} dr = \infty.$$

By Lemma 3.4, $B_\alpha^N \in O_{HL^p}^N$. Conversely, if $B_\alpha^N \in O_{HL^p}^N$, then $f_n \notin L^p$ for each $h \in H(B_\alpha^N)$, $n > 0$. Since $f_n \sim c(1 - r)^{1-(N-2)\alpha}$, the above relation gives the conditions of the theorem in the first case.

In the second case, Lemma 2.7 gives for a nonconstant $h \in H(B_\alpha^N)$ and any $n > 0$, $f_n(r) \sim -c \log(1 - r)$. Therefore,

$$\|f_n S_n\|_p^p < c \int_0^1 |\log(1 - r)|^p (1 - r)^{N\alpha} dr < \infty,$$

and $f_n S_n \in HL^p$.

In the third case, observe that, by the maximum principle applied to a nonzero $f_n S_n \in H(B_\alpha^N)$, $n > 0$, $|f_n|$ is a nondecreasing function and $|f_n| > \varepsilon > 0$ for $r > r_0$, say. Therefore,

$$\|f_n\|_p^p \geq c \int_{r>r_0} (1 - r)^{N\alpha} dr = \infty$$

for $\alpha \leq -1/N$, and we have $B_\alpha^N \in O^N_{HL^p}$. For $\alpha \in (-1/N, 1/(N - 2))$, Lemma 2.8 gives the existence of a nonconstant $h \in HB(B_\alpha^N)$. We have

$$\|h\|_p^p \leq c \int_0^1 (1 - r)^{N\alpha} dr < \infty,$$

hence $B_\alpha^N \in \tilde{O}^N_{HL^p}$.

NOTES TO §3. The class $O^N_{HL^p}$ of Riemannian N-manifolds was introduced in Sario [6], and Theorems 3.1 - 3.3 established in Sario-Wang [13]. All results on HL^p functions on the Poincaré N-ball B_α^N, in particular Lemma 3.4 and Theorem 3.5 are due to Chung [1]. He has also shown that, in contrast with the general case, there do exist relations between O^N_{HX} and $O^N_{HL^p}$ in the class $\{B_\alpha^N\}$ for varying α.

§4. Completeness and harmonic degeneracy

We know that the plane does not carry nonconstant HX functions for any $X = G, P, B, D, C, L^p$. Intuitively, we could think of this being a consequence of the "smallness" of the ideal boundary of the plane, a single point. There is some analogy with a rod being able to support a soap film, but if the rod shrinks to a needle, the film will no longer be held up by it. It would seem natural to assume that, more generally, completeness of a Riemannian manifold makes its ideal boundary too "small" to support HX functions. We shall show, however, that completeness and harmonic degeneracy are totally unrelated, except perhaps for the open problem on the existence of HL^p functions on complete manifolds. We continue considering only dimension $N > 2$.

4.1. Complete and degenerate or neither. In this section, let C^N be the class of complete Riemannian N-manifolds, characterized by an infinite distance from a point of the manifold to its ideal boundary. The complement of C^N with respect to the totality of Riemannian N-manifolds will be denoted by $\tilde{C}^N$.

THEOREM. $C^N \cap O^N_{HX} \neq \emptyset$ and $\tilde{C}^N \cap \tilde{O}^N_{HX} \neq \emptyset$ for $X = G, P, B, D, C, L^p$ with $1 \leq p < \infty$, and $N > 2$.

Proof. The Euclidean N-cylinder

$$R = \{|x| < \infty, \ |y^i| \leq \pi, \quad i = 1,\ldots,N-1\}$$

discussed in 3.1 is clearly complete, and we know that it belongs to $O^N_{HG} \cap O^N_{HL^p}$, hence to all O^N_{HX}.

Note that no Poincaré N-ball B^N_α belongs to $C^N \cap O^N_{HG}$. In fact, B^N_α is complete if and only if $\alpha \leq -1$, the distance from the center to the boundary being $\int_0^1 (1 - r^2)^\alpha dr$. We recall from 2.10 that for $\{B^N_\alpha\}$, $O^N_{HG} = O^N_{HP} = O^N_{HB}$, $O^N_{HD} = O^N_{HC}$, and obtain for $N > 2$,

$$B^N_\alpha \in \begin{cases} \tilde{C}^N \cap O^N_{HG} & \text{for no } \alpha, \\ C^N \cap O^N_{HD} & \text{for } \alpha \leq -1, \\ C^N \cap O^N_{HL^p} & \text{for } \alpha \leq -1,\ 1 \leq p < \infty. \end{cases}$$

Here and below we only discuss such membership of B^N_α in a class determined by C^N and $O^N_{HL^p}$ which is valid for all $p \in [1,\infty)$.

For the complementary classes, the Euclidean N-ball gives at once $\tilde{C}^N \cap \tilde{O}^N_{HX} \neq \emptyset$

for all X and all p. The scheme for the Poincaré N-ball, $N > 2$, reads

$$B_\alpha^N \in \begin{cases} \tilde{C}^N \cap \tilde{O}_{HG}^N & \text{if} \quad -1 < \alpha < 1/(N-2), \\ \tilde{C}^N \cap \tilde{O}_{HD}^N & \text{if} \quad -1/(N-2) < \alpha < 1/(N-2), \\ \tilde{C}^N \cap \tilde{O}_{HL^p}^N & \text{if} \quad -1/N < \alpha \leq 1/(N-2). \end{cases}$$

4.2. Not complete but degenerate. More interesting than the above cases is the existence of a manifold that is not complete but nevertheless fails to carry HX functions.

THEOREM. $\tilde{C}^N \cap O_{HX}^N \neq \emptyset$ for $X = G, P, B, D, C, L^p$ with $1 \leq p < \infty$, and $N > 2$.

Proof. Here the Poincaré N-ball covers all cases:

$$B_\alpha^N \in \begin{cases} \tilde{C}^N \cap O_{HG}^N & \text{if} \quad \alpha \geq 1/(N-2), \\ \tilde{C}^N \cap O_{HD}^N & \text{if} \quad \alpha \geq 1/(N-2), \text{ or } -1<\alpha\leq -1/(N-2), \\ \tilde{C}^N \cap O_{HL^p}^N & \text{if} \quad -1 < \alpha \leq -1/N. \end{cases}$$

4.3. Complete but nondegenerate. The most intriguing is the existence of manifolds which are complete but carry HX functions all the same.

THEOREM. $C^N \cap \tilde{O}_{HX}^N \neq \emptyset$ for $X = G, P, B, D, C,$ and $N > 2$.

Proof. The Poincaré N-ball only gives the cases $X = G, P, B$:

$$B_\alpha^N \in C^N \cap \tilde{O}_{HG}^N \quad \text{if} \quad \alpha \leq -1.$$

The following example includes also $X = C$ (hence D). Endow the N-cylinder

$$R = \{|x| < \infty, \quad |y^i| \leq 1, \quad i = 1,\dots,N-1\}$$

with the metric

$$ds^2 = dx^2 + e^{2x^2/(N-1)} \sum_{i=1}^{N-1} dy^{i2}.$$

For $h(x) \in H(R)$, $\Delta h = -e^{-x^2}(e^{x^2} h')' = 0$, hence

$$h(x) = c \int_0^x e^{-t^2} dt \in HB.$$

The Dirichlet integral is

$$D(h) = c \int_{-\infty}^{\infty} e^{-2x^2} e^{x^2} dx < \infty.$$

NOTES TO §4. The question on whether or not completeness implies harmonic degeneracy was raised by Chern. The question was answered for $X = G, P, B$, and D, in the negative, in Nakai-Sario [1]. The counterexamples above are new. The manifold R in 4.3 is from Sario [8].

Chung has recently constructed an (unpublished) counterexample showing that $C^N \cap \tilde{O}^N_{HL^p} \neq \emptyset$ for $p = 1$. For $1 < p < \infty$, the problem of nonvoidness of this class is open at this writing.

CHAPTER II

QUASIHARMONIC FUNCTIONS

The deflection of a membrane under a uniform load of unit density satisfies the equation $\Delta q = 1$. We call the solutions of this equation <u>quasiharmonic</u> functions. The present chapter is devoted to existence problems of these important functions on Riemannian manifolds.

The inclusion relations between quasiharmonic null classes show striking differences from those between harmonic null classes: For $N \geq 2$,

$$O_G^N < O_{QP}^N < O_{QB}^N \cap O_{QD}^N \begin{matrix} < O_{QB}^N < \\ \\ < O_{QD}^N < \end{matrix} O_{QB}^N \cup O_{QD}^N = O_{QC}^N,$$

whereas there are no inclusion relations between O_{QB}^N and O_{QD}^N. For L^p classes, we now do have inclusion relations:

$$O_{QL^1}^N < O_{QD}^N, \qquad O_{QL^p}^N < O_{QC}^N$$

for $p \geq 1$. Moreover,

$$O_{QPL^1}^N = O_{QD}^N, \qquad O_{QPBL^1}^N = O_{QC}^N.$$

In §1 we first deduce some useful tests for the null classes and establish the inclusion relations for $X = P, B, D, C$. The next section is devoted to $O_{QL^p}^N$ and its relations with the O_{QX}^N. Using the tests in §1 and directly estimating the quantities in them, we obtain in §3 complete characterizations of the classes O_{QX}^N of Poincaré N-balls B_α^N. A reader not interested in this methodological aspect may obtain a more rapid access to these characterizations by means of the technique described in §6 (cf. below). The same is true, mutatis mutandis, of §§4 and 5. In §4 we show that there exist on B_α^N functions in QX if and only if a specific quasiharmonic function $s(r)$, which exists on every B_α^N, has the property X. Accordingly, we call $s(r)$ the characteristic quasiharmonic function. We show in

§5 that $s(r)$ has the additional property of being bounded from above on every B^N_α. This result gives rise to the class O^N_{QN} of Riemannian N-manifolds for which the class QN of nonpositive quasiharmonic functions is void, and to the question as to whether or not $O^N_{QN} = \emptyset$ for all N.

An integral form of the characteristic s is introduced in §6. This form leads to an alternate proof of the characterizations of the classes O^N_{QX} of Poincaré N-balls, and also gives directly $s \leq 0$. The last section, §7, is devoted to the question of relations between harmonic and quasiharmonic null classes. Here, especially in proving $O^N_{HX} \cap \tilde{O}^N_{QY} \neq \emptyset$, we have some of the most challenging counterexamples in classification theory.

The present chapter gives a substantial sample of one of the goals of this book: to explore and compare several methods. It is only after such a comparison that the virtues of the simplest method, often obtained after the more elaborate one, can be fully understood and appreciated.

§1. QUASIHARMONIC NULL CLASSES

Let P, B, D, C again be the classes of functions which are positive, bounded, Dirichlet finite, or bounded and Dirichlet finite, respectively. Denote by Q the class of quasiharmonic functions, and by QX the subclasses $Q \cap X$ with $X = P, B, D,$ or C. The class of Riemannian N-manifolds R with $QX(R) = \emptyset$ is denoted by O^N_{QX}. We start with criteria for these null classes.

1.1. Tests for quasiharmonic null classes. Given a Riemannian manifold $R \in \tilde{O}^N_G$, let $g(x,y)$ be its harmonic Green's function with pole y, normalized by $\int_\alpha *dg = -1$, where α is the boundary of a "small" goedesic ball B about y, positively oriented with respect to B. We recall that if the integral

$$Gf(x) = \int_R g(x,y)*f(y) = \lim_{\Omega \to R} \int_\Omega g_\Omega(x,y)*f(y)$$

converges for a function $f \in C^\infty(R)$ as R is exhausted by regular subregions Ω containing $\bar{B}$, then

$$\Delta Gf = f$$

(e.g., Hörmander [1] or Narasimhan [1]). In particular, $G1 \in Q$.

We shall say that $G1 \in P$ or $G1 \notin P$ according as the integral

$$G1(x) = \int_R *g(x,y)$$

converges or not.

For $f_1, f_2 \in C^\infty(R)$, set

$$G(f_1,f_2) = \int_{R\times R} g(x,y)*f_1(x)*f_2(y)$$

whenever the integral converges.

We shall show that if the particular quasiharmonic function $G1$ is not in X, then neither is <u>any</u> other quasiharmonic function:

<u>THEOREM</u>. $R \in O^N_{QX}$ <u>if and only if</u> $G1 \notin X$, <u>where</u> $X = P, B, D, C$, <u>and</u> $N \geq 2$. <u>Moreover</u>,

$$D(G1) = G(1,1).$$

<u>Proof</u>. Only the sufficiency and the equality need a proof. Given $q \in Q(R)$, let $h_\Omega \in H(\Omega) \cap C(\overline{\Omega})$ with $h_\Omega = q$ on $\partial\Omega$. Then

$$\int_{\partial\Omega-\partial B} (q - h_\Omega)*dg_\Omega - g_\Omega * d(q - h_\Omega)$$

$$= -\int_{\Omega-B} (q - h_\Omega)*\Delta g_\Omega - g_\Omega*\Delta(q - h_\Omega).$$

As B shrinks to y, we obtain the Riesz decomposition on Ω,

$$q = h_\Omega + G_\Omega 1.$$

Suppose there exists a $q \in QP$. Since $h_\Omega|\partial\Omega = q|\partial\Omega > 0$, we have $h_\Omega > 0$ on $\overline{\Omega}$. In view of $G_\Omega 1$ increasing with Ω, h_Ω decreases, and the limiting function

$h = \lim_\Omega h_\Omega$ exists. So does, therefore, the limiting function $G1 = \lim_\Omega G_\Omega 1 = q - h \in P$.

If $q \in QB$, then $q - \inf_R q \in QP$, and the above proof yields $G1 \in B$.

For the Dirichlet integral over Ω, Stokes' formula gives

$$D_\Omega(G_\Omega 1) = \int_{\partial\Omega} G_\Omega 1 * dG_\Omega 1 + \int_\Omega G_\Omega 1 * \Delta G_\Omega 1 = G_\Omega(1,1)$$

and

$$D_\Omega(G_\Omega 1, h_\Omega) = \int_{\partial\Omega} G_\Omega 1 * dh_\Omega + \int_\Omega G_\Omega 1 * \Delta h_\Omega = 0.$$

Therefore,

$$D_\Omega(q) = D_\Omega(h_\Omega) + G_\Omega(1,1)$$

for any $q \in Q$. If $q \in D$, then

$$D(G1) = \lim_{\Omega \to R} D_\Omega(G_\Omega 1) = G(1,1) = D(q) - D(h)$$

and $G1 \in D$.

If $q \in QC$, then by the above, $G1 \in C$.

<u>1.2.</u> <u>Green's functions but no</u> QP. We shall discuss the inclusion relations in the order given in the introduction to the present chapter.

<u>THEOREM</u>. <u>The strict inclusion</u> $O_G^N < O_{QP}^N$ <u>is valid for every</u> $N \geq 2$.

<u>Proof</u>. Since $R \in \tilde{O}_{QP}^N$ implies $G1 \in P$, g exists and we have $O_G^N \subset O_{QP}^N$. This can also be seen from the fact that a $q \in QP$ is positive superharmonic, hence entails the existence of g. To prove $O_G^N < O_{QP}^N$, it thus suffices to find a hyperbolic N-manifold with $QP = \emptyset$. The simple example of the Euclidean N-space does not qualify for $N = 2$, whereas the N-cylinder

$$R = \{0 < x < 1, \ |y^i| \leq 1, \ i = 1,\ldots,N-1\}$$

with the metric

$$ds^2 = (x^{-2} + (1 - x)^{-2})dx^2 + (x^{-2} + (1 - x)^{-2})^{1/(N-1)} \sum_{i=1}^{N-1} dy^{i2}$$

belongs to $\tilde{O}_G^N \cap O_{QP}^N$ for all $N \geq 2$. In fact,

$$\Delta h(x) = -g^{-1/2}(g^{1/2} g^{xx} h')' = 0$$

gives $h'' = 0$, hence $h = ax + b$. The harmonic measure is nonconstant, and $R \in \tilde{O}_G^N$. The function

$$q(x) = \log x + \log(1 - x)$$

is quasiharmonic. If $G1 \in P$, then by the symmetry of R with respect to the y^i coordinates, $G1$ is a function of x only, and $\Delta G1 = 1$ gives

$$G1(x) = q(x) + h(x)$$

for some $h = ax + b$. In view of $q(x) + h(x) \to -\infty$ as $x \to 0$ and $x \to 1$, we have a violation of $G1 \in P$. Therefore, $R \in O_{QP}^N$.

<u>1.3.</u> QP <u>functions but no</u> QB ∪ QD. The next step is to prove:

<u>THEOREM.</u> <u>The strict inclusion</u> $O_{QP}^N < O_{QB}^N \cap O_{QD}^N$ <u>is valid for every</u> $N \geq 2$.

<u>Proof.</u> The inclusion $O_{QP}^N \subset O_{QB}^N \cap O_{QD}^N$ is again trivial by Theorem 1.1. For the strictness, consider the N-cylinder

$$R = \{0 < x < 1, \quad |y^i| \leq 1, \quad i = 1,\ldots,N - 1\}$$

with the metric

$$ds^2 = x^{-3}dx^2 + x^{3/(N-1)} \sum_{i=1}^{N-1} dy^{i2}.$$

Clearly $q = x^{-1} \in QP$, and the most general $h(x) \in H$ is

$$h(x) = ax^{-2} + b.$$

Since $G1$ exists,

$$G1 = q + h = ax^{-2} + x^{-1} + b$$

for some a, b. This function is unbounded, hence $R \in O^N_{QB}$.

Let D_x stand for the Dirichlet integral over the subregion of R from x to 1 and denote by c a constant, not always the same. Then

$$D_x(q) = c\int_x^1 x^{-4}x^3\,dx = -c\log x,$$

$$D_x(h) = c\int_x^1 x^{-6}x^3\,dx = c(1 - x^{-2}),$$

and $D_x(G1) > |D_x(q) - D_x(h)|$, which is unbounded as $x \to 0$. Therefore, $R \in O^N_{QD}$.

1.4. QD functions but no QB. We proceed to show:

THEOREM. The strict inclusion $O^N_{QB} \cap O^N_{QD} < O^N_{QB}$ is valid for every $N \geq 2$.

Proof. Now we endow the N-cylinder

$$R = \{0 < x < 1, \ |y^i| \leq 1, \ i = 1,\ldots,N-1\}$$

with the metric

$$ds^2 = x^{-2}dx^2 + x^{2/(N-1)}\sum_{i=1}^{N-1} dy^{i2}.$$

We have

$$q = \log x^{-1} \in Q$$

with

$$D(q) = c\int_0^1 x^2x^{-2}dx < \infty.$$

Therefore, $R \in \tilde{O}^N_{QD}$.

The general $h(x) \in H$ is

$$h(x) = ax^{-1} + b.$$

For some a, b,

$$Gl = \log x^{-1} + ax^{-1} + b,$$

which is unbounded. A fortiori, $R \in O^N_{QB}$.

<u>1.5.</u> QB <u>functions but no</u> QD. For $N > 2$, the Poincaré N-ball B^N_α will provide us with the strictness of $O^N_{QB} \cap O^N_{QD} < O^N_{QD}$. In fact, we shall see in 3.9 that $B^N_\alpha \in O^N_{QB}$ if and only if $\alpha \notin (-1, 1/(N-2))$, and $B^N_\alpha \in O^N_{QD}$ if and only if $\alpha \notin (-3/(N+2), 1/(N-2))$. Therefore, $B^N_\alpha \in \tilde{O}^N_{QB} \cap O^N_{QD}$ is characterized by $\alpha \in (-1, -3/(N+2)]$. Here we give two other examples which furnish the proof for all N.

<u>THEOREM</u>. <u>The strict inclusion</u> $O^N_{QB} \cap O^N_{QD} < O^N_{QD}$ <u>is valid for every</u> $N \geq 2$.

<u>Proof</u>. The first example is the N-cylinder

$$R = \{|x| < \infty, \quad |y| \leq 1, \quad |z^i| \leq 1, \quad i = 1, \ldots, N-2\}$$

with the metric

$$ds^2 = (1 + x^4)^{-1} dx^2 + (1 + x^4)(1 + x^6)^2 dy^2 + \sum_{i=1}^{N-2} dz^{i2}.$$

It is readily verified that the function

$$q(x) = -\int_0^x (1 + s^4)^{-1}(1 + s^6)^{-1} \int_0^s (1 + r^6) dr \, ds$$

belongs to Q. As $|x| \to \infty$, $q'(x) \sim cx^{-3}$ and

$$q(x) = O(1).$$

Consequently, $R \in \tilde{O}^N_{QB}$.

Suppose there exists a $q \in QD$. Take a function $\varphi_0 \in C^\infty_0(-\infty, \infty)$, $\varphi_0 \geq 0$, $\operatorname{supp} \varphi_0 \subset (0,1)$. For $t > 0$, set $\varphi_t(x) = \varphi_0(x - t)$. Then $\operatorname{supp} \varphi_t \subset (t, t+1)$,

$$(1, \varphi_t) = c \int_t^{t+1} (1 + x^6) \varphi_t \, dx > ct^6,$$

and

$$D(\varphi_t) = c\int_t^{t+1} \varphi_t'^2(1 + x^4)(1 + x^6)dx = O(t^{10}).$$

Therefore, $(1,\varphi_t)/D(\varphi_t)^{1/2} \to \infty$ as $t \to \infty$. This violates Stokes' formula, which gives $(\Delta q,\varphi_t) = (1,\varphi_t) < D(q)^{1/2}D(\varphi_t)^{1/2}$. We conclude that $R \in O_{QD}^N$.

Our second example for $\tilde{O}_{QB}^N \cap O_{QD}^N \neq \emptyset$ is the N-cylinder

$$R = \{0 < x < 1, \quad |y^i| \leq 1, \quad i = 1,\ldots,N - 1\}$$

with the metric

$$ds^2 = x^{-1}\, dx^2 + x^{-5/(N-1)} \sum_{i=1}^{N-1} dy^{i2}.$$

The general solution of $\Delta q = 1$ is

$$q = ax^3 + \tfrac{1}{2}x + b,$$

which is bounded, hence $R \in \tilde{O}_{QB}^N$. Since $G1 = q$ for some (a,b) and

$$D(G1) = D(q) = c\int_0^1 (3ax^2 + \tfrac{1}{2})^2 xx^{-3}\, dx = \infty,$$

we have $R \in O_{QD}^N$.

<u>1.6.</u> QC <u>functions if</u> QB <u>and</u> QD. It remains to show that the equality

$$O_{QB}^N \cup O_{QD}^N = O_{QC}^N$$

is valid for every $N \geq 2$.

Trivially, $O_{QB}^N \cup O_{QD}^N \subset O_{QC}^N$. If there exist $q_1 \in QB$ and $q_2 \in QD$, then $G1$ belongs to both B and D, hence to C.

<u>1.7.</u> <u>No relations between</u> O_{QB}^N <u>and</u> O_{QD}^N. We note in closing that there are no inclusion relations between O_{QB}^N and O_{QD}^N for any $N \geq 2$. In fact, we know from 1.4 that $O_{QB}^N \cap \tilde{O}_{QD}^N \neq \emptyset$ and from 1.5 that $\tilde{O}_{QB}^N \cap O_{QD}^N \neq \emptyset$. Moreover, $O_{QB}^N \cap O_{QD}^N \neq \emptyset$ by virtue of O_{QP}^N being contained in this intersection. Finally,

$\tilde{O}^N_{QB} \cap \tilde{O}^N_{QD} \neq \emptyset$ is trivial in view of the Euclidean N-ball.

1.8. Summary. We collect our results:

The strict inclusions

$$O^N_G < O^N_{QP} < O^N_{QB} \cap O^N_{QD} \begin{matrix} < O^N_{QB} < \\ \\ < O^N_{QD} < \end{matrix} O^N_{QB} \cup O^N_{QD} = O^N_{QC}$$

are valid for every $N \geq 2$. There are no inclusion relations between O^N_{QB} and O^N_{QD} for any N.

NOTES TO §1. The quasiharmonic classification was introduced in Sario [6], and the tests in Theorem 1.1 established in Nakai-Sario [6]. In the case of 2-manifolds, the relations $O^2_{QP} < O^2_{QB}$; $O^2_{QP} < O^2_{QD}$; $O^2_{QB} \not< O^2_{QD}$; $O^2_{QD} \not< O^2_{QB}$; $O^2_{QB} < O^2_{QC}$; and $O^2_{QD} < O^2_{QC}$ were also proved in Nakai-Sario [6]. The proofs were based on the identity $\Delta q = \lambda^{-2}\Delta_e q$, valid for $N = 2$, between Δ, the Euclidean Δ_e, and the conformal metric $ds = \lambda(z)|dz|$. The relations in 1.8 were established in Sario [7].

§2. THE CLASS $O^N_{QL^p}$

We now take up the class $O^N_{QL^p}$ of Riemannian N-manifolds R for which $QL^p(R) = \emptyset$. First we show that, in interesting contrast with the case of harmonic functions, we now do have some inclusion relations: $O^N_{QL^1} < O^N_{QD}$ and $O^N_{QL^p} < O^N_{QC}$ for $p \geq 1$. The essence of this section consists in showing that there are no other relations between $O^N_{QL^p}$ and any other quasiharmonic null class O^N_{QX}. In particular, there exist Riemannian manifolds which are parabolic but nevertheless carry QL^p functions for every $p \geq 1$; and there exist those which, despite carrying even QD functions, fail to carry any QL^p functions for any $p > 1$.

2.1. Inclusions for QL^p. We start by establishing the inclusion relations:

THEOREM. For $N \geq 2$,

$$O^N_{QL^1} < O^N_{QD}, \qquad O^N_{QL^p} < O^N_{QC},$$

with $1 \leq p < \infty$.

Proof. The inclusion $O^N_{QL^1} \subset O^N_{QD}$ is immediate. In fact, if $R \in \tilde{O}^N_{QD}$, then $G1 \in QD$, and

$$\|G1\|_1 = \int_R *G1 = G(1,1) = D(G1) < \infty.$$

Therefore, $R \in \tilde{O}^N_{QL^1}$.

To prove $O^N_{QL^p} \subset O^N_{QC}$ for $1 < p < \infty$, suppose $R \in \tilde{O}^N_{QC}$. For $R_0 = \{x \in R | G1(x) > 1\}$,

$$\int_{R_0} *1 < \int_{R_0} *G1 < G(1,1) = D(G1) < \infty$$

and, in view of $G1 < M < \infty$,

$$\|G1\|_p^p = \int_R *(G1)^p = \int_{R_0} *(G1)^p + \int_{R-R_0} *(G1)^p \leq M^p \int_{R_0} *1 + G(1,1) < \infty.$$

Therefore, $G1 \in QL^p$, and $R \in \tilde{O}^N_{QL^p}$.

The strictness of both inclusions will be a consequence of $\tilde{O}^N_{QL^p} \cap O^N_{QX} \neq \emptyset$ for $1 \leq p < \infty$ to be proved in 2.3.

2.2. Equalities for QL^1. Can the gaps $O^N_{QD} - O^N_{QL^1}$ and $O^N_{QC} - O^N_{QL^p}$ be closed if we impose further boundedness conditions on QL^p functions? This is indeed so for $p = 1$:

THEOREM. For $N \geq 2$,

$$O^N_{QPL^1} = O^N_{QD}, \qquad O^N_{QPBL^1} = O^N_{QC}.$$

Proof. We already know that $R \in \tilde{O}^N_{QD}$ implies $G1 \in QPL^1$. Conversely, suppose there exists a $q \in QPL^1$. Then by the Riesz decomposition $q = h + G1$ in 1.1,

$$D(G1) = G(1,1) = \|G1\|_1 < \|q\|_1 < \infty,$$

hence $R \in \tilde{O}^N_{QD}$. The proof of $O^N_{QPBL^1} = O^N_{QC}$ is analogous.

2.3. QL^p functions but no QX. Are there any other inclusion relations between $O^N_{QL^p}$ and the other quasiharmonic null classes O^N_{QX}? In 2.3 - 2.6, we shall give a complete answer: the relations in Theorems 2.1 and 2.2 are the only ones between the classes $O^N_{QL^p}$ and O^N_{QX} for $1 \le p < \infty$, and $X = G, P, B, D, C$, where, for economy of notation, we let O^N_{QG} stand for O^N_G.

We start with the following relations which also establish the strictness of the inclusions in Theorem 2.1.

THEOREM. *For* $1 \le p < \infty$ *and* $N \ge 2$,

$$\bigcap_p \tilde{O}^N_{QL^p} \cap O^N_{QX} \ne \emptyset,$$

with $X = G, P, B, D, C$.

Proof. Endow the N-cylinder

$$R = \{|x| < \infty, \quad |y^i| \le 1, \quad i = 1,\ldots,N-1\}$$

with the metric

$$ds^2 = e^{-x^2}dx^2 + e^{-x^2/(N-1)} \sum_{i=1}^{N-1} dy^{i2}.$$

The quasiharmonic equation

$$\Delta q(x) = -e^{x^2}(e^{-x^2}e^{x^2}q')' = 1$$

is satisfied by

$$q(x) = -\int_0^x \int_0^t e^{-s^2}\, ds\, dt.$$

To estimate

$$\|q\|_p^p = c\int_{-\infty}^{\infty} \left|\int_0^x \int_0^t e^{-s^2}\, ds\, dt\right|^p e^{-x^2}\, dx,$$

we set $a = \int_0^\infty e^{-s^2}\, ds$ and obtain

$$\|q\|_p^p < ca^p \int_{-\infty}^{\infty} |x|^p e^{-x^2} dx < \infty.$$

Therefore, $R \in \tilde{O}^N_{QL^p}$ for all p.

The harmonic equation $\Delta h(x) = 0$ gives $h = ax + b$. In particular, the harmonic measure ω on $|x| > c > 0$ of the ideal boundary of R is of this form, and we have $\omega = b$. Hence, $R \in O^N_G \subset O^N_{QX}$ for all X.

2.4. <u>Neither or both QL^p and QX.</u> We shall continue the proof that there are no relations between $O^N_{QL^p}$ and O^N_{QX} other than Theorems 2.1 and 2.2.

<u>THEOREM</u>. <u>For</u> $1 \leq p < \infty$ <u>and</u> $N \geq 2$,

$$\bigcap_p O^N_{QL^p} \cap O^N_{QX} \neq \emptyset, \qquad \bigcap_p \tilde{O}^N_{QL^p} \cap \tilde{O}^N_{QX} \neq \emptyset,$$

<u>with</u> $X = G, P, B, D, C$.

<u>Proof</u>. For the first relation, consider the N-cylinder

$$R = \{|x| < \infty,\ |y^i| \leq \pi,\ i = 1,\ldots,N-1\}$$

with the Euclidean metric. The harmonic functions of x are $h(x) = ax + b$, hence the harmonic measure ω on $|x| > c > 0$ is constant, and $R \in O^N_G \subset O^N_{QX}$ for all X.

To show that $R \in O^N_{QL^p}$, note that a general quasiharmonic function is of the form $q(x,y) = q_0(x) + h(x,y)$, $y = (y^1,\ldots,y^{N-1})$, where $q_0(x) = -\frac{1}{2}x^2$ and $h(x,y) \in H$ has a representation $h_0(x) + \Sigma_n' f_n(x)G_n(y)$, convergent absolutely and uniformly on compact subsets. Here $h_0(x) = ax + b$, and the $G_n(y)$ are products of sines and cosines of multiples of the y^i as in I.3.1, the index n now standing for the double index nj.

Suppose there exists a $q \in QL^p$. Then $(\cdot, q)$ is a bounded linear functional on $L^{p'}$, with $1/p + 1/p' = 1$. The characteristic function φ_i of the subregion $\{x \in [i, i+1]\}$ belongs to $L^{p'}$, with $\|\varphi_i\|_{p'}$ independent of i. But

$$(\varphi_i, q) = \int_R \varphi_i * q = \int_y \int_i^{i+1} (q_0 + h_0 + \textstyle\sum_n' f_n G_n) dx dy,$$

with $dy = dy^1 \dots dy^{N-1}$, and therefore,

$$(\varphi_i, q) = c \int_i^{i+1} (q_0 + h_0) dx = c_1 x^3 + c_2 x^2 + c_3 x \Big|_i^{i+1},$$

which is unbounded as $i \to \pm\infty$. This contradiction shows that $R \in O^N_{QL^p}$.

The relation $\bigcap_p \tilde{O}^N_{QL^p} \cap \tilde{O}^N_{QX} \neq \emptyset$ is trivial by virtue of the Euclidean N-ball.

2.5. QB functions but no QL^p. In view of the inclusion relations $O^N_{QL^1} \subset O^N_{QD}$, and $O^N_{QL^p} \subset O^N_{QC}$, the following result for all $p \geq 1$ can only be valid for $X = G, P, B$. For $p > 1$, the case $X = D$ will be discussed in 2.6.

THEOREM. *For* $1 \leq p < \infty$ *and* $N \geq 2$,

$$\bigcap_p O^N_{QL^p} \cap \tilde{O}^N_{QX} \neq \emptyset$$

with $X = G, P, B$.

Proof. Let R be the N-space with the metric

$$ds^2 = dr^2 + \psi(r)^{2/(N-1)} \sum_{i=1}^{N-1} \gamma_i(\theta) d\theta^{i2},$$

where $\psi \in C^\infty[0,\infty)$,

$$\psi(r) = \begin{cases} r^{N-1} & \text{for } r \leq \frac{1}{2}, \\ e^{e^r} e^r & \text{for } r \geq 1, \end{cases}$$

and the γ_i are trigonometric functions of $\theta = (\theta^1, \dots, \theta^{N-1})$ such that the metric is Euclidean on $\{r \leq \frac{1}{2}\}$. The function

$$q_0(r) = -\int_0^r \psi^{-1}(t) \int_0^t \psi(s) ds dt$$

satisfies the quasiharmonic equation $\Delta q_0 = -\psi^{-1}(\psi q_0')' = 1$. For $r > 1$,

$$q_0(r) = q_0(1) - \int_1^r e^{-e^t}e^{-t}\left(a + \int_1^t e^{e^s}e^s ds\right)dt$$

$$= q_0(1) - \int_1^r (e^{-t} + a_1 e^{-e^t}e^{-t})dt = O(1)$$

as $r \to \infty$. Therefore, $q_0 \in QB$, and $R \in \tilde{O}^N_{QB} < \tilde{O}^N_{QP} < \tilde{O}^N_G$.

We next prove that $R \in O^N_{QL^1}$. Suppose there exists a $q \in QL^1$. Then $|(q,e^{-r})| < \infty$. On the other hand, we may write $q = q_0 + c + k$, $k \in H$, $k(0) = 0$, and obtain

$$(q,e^{-r}) = (q_0 + c,\ e^{-r}) = a + b\int_1^\infty (q_0 + c)e^{-r}e^{e^r}e^r dr.$$

Set

$$c_0 = -q_0(\infty) = \int_0^\infty \psi^{-1}\int_0^t \psi\ dsdt.$$

If $c \neq c_0$, then $\lim_{r\to\infty}(q_0(r) + c) = d \neq 0$, and $|(q,e^{-r})| = \infty$. If $c = c_0$, then for $r > 1$,

$$q_0(r) + c = \int_r^\infty (e^{-t} + a_1 e^{-e^t}e^{-t})dt$$

and

$$(q,e^{-r}) = a + b\int_1^\infty (q_0 + c)e^{e^r}dr$$

$$= a + b\int_1^\infty\left(e^{-r} + a_1\int_r^\infty e^{-e^t}e^{-t}dt\right)e^{e^r}dr$$

$$= a + b\int_1^\infty e^{-r}e^{e^r}dr + a_1 b\int_1^\infty\left(\int_r^\infty e^{-e^t}e^{-t}dt\right)e^{e^r}e^r e^{-r}dr.$$

Since

$$\lim_{r\to\infty}\left(\int_r^\infty e^{-e^t}e^{-t}dt\right)e^{e^r}e^r = \lim_{r\to\infty}\frac{-e^{-e^r}e^{-r}}{-e^{-e^r}(1 + e^{-r})} = 0,$$

we have

$$(q,e^{-r}) = a + b\int_1^\infty e^{-r}e^{e^r}dr + a_1 b\int_1^\infty o(1)e^{-r}dr.$$

The last integral converges, the first diverges, and therefore, $|(q,e^{-r})| = \infty$. This contradiction shows that $R \in O^N_{QL^1}$.

To see that $R \in O^N_{QL^p}$ for $p > 1$, let p' be determined by $1/p + 1/p' = 1$. For a radial function $\varphi \in C^\infty(R)$ with

$$\varphi|\{r \geq 1\} = (e^{-e^r}e^{-r}r^{-2})^{1/p'},$$

we have

$$\|\varphi\|^{p'}_{p'} = a + b\int_1^\infty e^{-e^r}e^{-r}r^{-2}e^{e^r}e^r dr < \infty,$$

hence $\varphi \in L^{p'}$. If there exists a $q \in QL^p$, then $|(q,\varphi)| < \infty$. But $(q,\varphi) = (q_0 + c,\varphi)$, and if $c \neq c_0$, the integrand in $(q_0 + c,\varphi)$ is asymptotically

$$c_1(e^{-e^r}e^{-r}r^{-2})^{1/p'}e^{e^r}e^r = c_1(e^{e^r}e^r)^{1/p}r^{-2/p'},$$

so that $|(q,\varphi)| = \infty$.

In the case $c = c_0$, we observe that for $r > 1$,

$$q_0(r) + c = e^{-r} + a_1\int_r^\infty e^{-e^t}e^{-t}dt,$$

where

$$\int_r^\infty e^{-e^t}e^{-t}dt < e^{-e^r}\int_r^\infty e^{-t}dt = e^{-e^r}e^{-r},$$

and therefore, $q_0(r) + c \sim e^{-r}$ as $r \to \infty$. It follows that the integrand in $(q_0 + c,\varphi)$ is asymptotically

$$e^{-r}(e^{-e^r}e^{-r}r^{-2})^{1/p'}e^{e^r}e^r = (e^{e^r})^{1/p}(e^{-r}r^{-2})^{1/p'},$$

and we have again $|(q,\varphi)| = \infty$. This contradiction shows that $R \in O^N_{QL^p}$.

2.6. QD functions but no QL^p, $p > 1$. Can the relation $O^N_{QL^p} < O^N_{QC}$ for $p > 1$ be improved to $O^N_{QL^p} < O^N_{QD}$? That the answer is in the negative, is the essence of the following theorem, which will complete our claim that Theorems 2.1 and 2.2 are the only relations between $O^N_{QL^p}$ and any O^N_{QX}. Note that, in contrast with 2.3 - 2.5, the counterexample will now depend on p, so that $\bigcap_p$ cannot be used. We do not know whether or not $\bigcap_p O^N_{QL^p} \cap \tilde{O}^N_{QD} \neq \emptyset$.

THEOREM. For $1 < p < \infty$ and $N \geq 2$,

$$O^N_{QL^p} \cap \tilde{O}^N_{QD} \neq \emptyset.$$

Proof. Take the N-cylinder

$$R = \{0 < x < 1,\ |y^i| \leq \pi,\ \ i = 1,\dots,N-1\}$$

with the metric

$$ds^2 = x^{-2\alpha}dx^2 + x^{2\beta/(N-1)} \sum_{i=1}^{N-1} dy^{i2},$$

where $\alpha = \alpha(p)$, $\beta = \beta(p)$ are constants to be specified later.

The function

$$q_0(x) = -(\beta - \alpha + 1)^{-1}(-2\alpha + 2)^{-1}\, x^{-2\alpha+2}$$

satisfies the quasiharmonic equation

$$\Delta q_0 = -x^{\alpha-\beta}(x^{\beta-\alpha}x^{2\alpha}q_0')' = 1$$

provided

$$\beta - \alpha + 1 \neq 0,\quad \alpha \neq 1.$$

The Dirichlet integral is

$$D(q_0) = c\int_0^1 q_0'^2 x^{2\alpha}x^{\beta-\alpha}dx = c_1\int_0^1 x^{\beta-3\alpha+2}dx < \infty$$

if

$$\beta - 3\alpha + 3 > 0.$$

For the L^p norm we have

$$\|q_0\|_p^p = c \int_0^1 x^{(-2\alpha+2)p} x^{\beta-\alpha} dx = \infty$$

if

$$\beta - (2p + 1)\alpha + 2p + 1 \leq 0.$$

An inspection of the last two inequalities shows that $p = 1$ is ruled out. For

$$p > 1, \quad \alpha > \tfrac{3}{2}, \quad \beta \in (3(\alpha - 1), (2p + 1)(\alpha - 1)],$$

all four inequalities are satisfied. In particular, $R \in \tilde{O}^N_{QD}$.

The exponent $\beta - \alpha$ in the volume element is positive, and the constant function 1 belongs to $L^{p'}$, the conjugate space of L^p. Suppose there exists a $q \in QL^p$. Then $|(q,1)| < \infty$.

As in 2.4, we write $q = q_0 + h$, $h \in H$, with

$$h(x,y) = h_0(x) + \sum_n{}' f_n(x)G_n(y).$$

The harmonic equation $\Delta h_0(x) = -x^{\alpha-\beta}(x^{\beta-\alpha}x^{2\alpha}h_0')' = 0$ is satisfied by

$$h_0(x) = ax^{-\alpha-\beta+1} + b.$$

Suppose first $a \neq 0$. Since $-2\alpha + 2 > -\alpha - \beta + 1$,

$$q_0(x) + h_0(x) \sim h_0(x) \quad \text{as} \quad x \to 0.$$

It follows that the integrand in $(q,1) = (q_0 + h_0,1)$ is asymptotically $x^{-\alpha-\beta+1+\beta-\alpha} = x^{-2\alpha+1}$. A fortiori, $|(q,1)| = \infty$, a contradiction.

Now let $a = 0$, $h_0 = b$. Since

$$\varphi(x) = x^{(-2\alpha+2)/p'} \in L^{p'},$$

$|(q,\varphi)| < \infty$. On the other hand,

$$|(q,\varphi)| = |(q_0 + b,\varphi)| = a_1 + b_1 \int_0^1 x^{-2\alpha+2} x^{(-2\alpha+2)/p'} x^{\beta-\alpha} dx = \infty$$

if

$$-2(\alpha - 1)(1 + \frac{1}{p'}) + \beta - (\alpha - 1) \leq 0,$$

that is,

$$\beta \leq (3 + \frac{2}{p'})(\alpha - 1).$$

Since $2p + 1 > 3 + 2/p'$ for $p > 1$, the choice

$$\beta \in (3(\alpha - 1), (3 + \frac{2}{p'})(\alpha - 1)]$$

gives the contradiction $|(q,\varphi)| = \infty$ while preserving the earlier inequalities.

<u>2.7. QL^p functions</u>, $p > 1$, <u>but no</u> QL^1. Since $O^N_{QL^1}$ is contained in O^N_{QD} while $O^N_{QL^p}$, $p > 1$, is not, a natural question arises: Are there any inclusion relations between $O^N_{QL^1}$ and $O^N_{QL^p}$, $p > 1$? We shall show that the answer is in the negative. This is seen from the relations $\bigcap_{p\geq 1} O^N_{QL^p} \neq \emptyset$, $\bigcap_{p\geq 1} \tilde{O}^N_{QL^p} \neq \emptyset$, $O^N_{QL^p} \cap \tilde{O}^N_{QD} \neq \emptyset$ for $p > 1$, and $O^N_{QL^1} \cap \bigcap_{p>1} \tilde{O}^N_{QL^p} \neq \emptyset$. The first three relations were established in 2.4 and 2.6. Here we shall prove the last one.

<u>THEOREM</u>. <u>For</u> $1 < p < \infty$ <u>and</u> $N \geq 2$,

$$O^N_{QL^1} \cap \bigcap_p \tilde{O}^N_{QL^p} \neq \emptyset.$$

<u>Proof</u>. On the N-cylinder

$$R = \{1 < x < \infty, \quad |y^i| \leq \pi, \quad i = 1,\ldots,N - 1\}$$

with the metric

$$ds^2 = x^{-3}dx^2 + x^{3/(N-1)} \sum_{i=1}^{N-1} dy^{i2},$$

the function $q_0(x) = x^{-1}$ satisfies the quasiharmonic equation

$$\Delta q_0 = -(x^3 q_0')' = 1.$$

Clearly, $q_0 \in L^p$ for $p > 1$.

Suppose there exists a $q \in QL^1$. Then $|(q,1)| < \infty$. In the same manner as in 2.4, write $q = q_0 + ax^{-2} + b + \Sigma_n' f_n G_n$. We have the contradiction $|(q_0 + ax^{-2} + b, 1)| = c + |(q_0 + b, 1)| = \infty$, and therefore $R \in O^N_{QL^1}$.

<u>2.8.</u> <u>Summary</u>. We collect our results:

<u>The quasiharmonic null classes satisfy the following strict inclusion relations, valid for all</u> $N \geq 2$ <u>and</u> $p \geq 1$:

$$\begin{array}{ccccccc}
 & & < O^N_{QB} < & & & & \\
O^N_G < O^N_{QP} < O^N_{QB} \cap O^N_{QD} & & & O^N_{QB} \cup O^N_{QD} & = & O^N_{QC} & = O^N_{QPBL^1}. \\
 & & < O^N_{QD} < & & & \vee & \\
 & & \parallel & & & O^N_{QL^p} & \\
 & & O^N_{QPL^1} > O^N_{QL^1} & & & &
\end{array}$$

<u>There are no other inclusion relations.</u>

<u>NOTES TO</u> §2. Theorems 2.1 and 2.3 - 2.6 were proved in Chung-Sario-Wang [2]. Theorems 2.2 and 2.7 are new.

We bring here to the attention of the reader the open problem referred to at the beginning of 2.6: Is $\bigcap_{p>1} O^N_{QL^p} \cap \tilde{O}^N_{QD} \neq \emptyset$?

<u>§3. QUASIHARMONIC FUNCTIONS ON THE POINCARÉ N-BALL</u>

To study the dependence of the null classes on the metric, we again turn to the illuminating case of the Poincaré N-ball B^N_α. We shall deduce the complete characterizations summarized in Theorem 3.9.

We shall first establish these characterizations in the illuminative case $N = 3$. The reasoning is then generalized to an arbitrary dimension.

Our approach in the present section is based on direct estimates of Green's potentials. A reader not interested in this methodological aspect may omit the present section without loss of continuity, a shorter proof of Theorem 3.9 being given in §6, after the characteristic quasiharmonic function on B_α^N has been introduced.

3.1. Parabolicity. On a hyperbolic Riemannian 3-manifold $B = \{r < r_0 \le \infty\}$, $|x| = r$, with the metric $ds = \lambda(r)|dx|$, $\lambda \in C^\infty$, the harmonic Green's function with pole 0 is

$$g(x,0) = \frac{1}{4\pi}\int_r^{r_0} \frac{d\rho}{\rho^2\lambda(\rho)} .$$

In fact, for $f(r) \in C^2$,

$$\Delta f = -\frac{1}{\lambda^3 r^2 \sin\psi}\frac{\partial}{\partial r}(\lambda^3 r^2 \sin\psi\cdot\lambda^{-2}f')$$

$$= -\frac{1}{\lambda^2}[f'' + (\frac{2}{r} + \frac{\lambda'}{\lambda})f'],$$

which vanishes if and only if

$$f(r) = c\int_r^{r_0} \frac{d\rho}{\rho^2\lambda(\rho)},$$

with $c = (4\pi)^{-1}$ determined by the flux $\int_{r=const} {*dg(x,0)} = -1$. We have

$$B_\alpha^3 \in O_G^3 \Leftrightarrow \alpha \ge 1.$$

Indeed, $\int_r^1 \rho^{-2}(1 - \rho^2)^{-\alpha}d\rho = \infty$ for some and hence every r if and only if $\alpha \ge 1$.

3.2. Potentials. For $f \in C(B)$, set

$$G_B f(x) = \int_B g(x,y)*f(y).$$

Given $x \in R$, $|x| = r$, let

$$B_1 = B_1(x) = \{y \mid |y| < r\},$$

$$B_2 = B_2(x) = \{y \mid r < |y| < r_0\},$$

and denote by $V_{B_1}(x)$ the Riemannian volume of $B_1(x)$.

<u>LEMMA</u>. $G_B 1(x) = g(x,0)V_{B_1}(x) + G_{B_2}1(0)$.

<u>Proof</u>. For $\rho \in (0,r_0)$, set $\beta_\rho = \{|y| = \rho\}$. A function $h \in H\{|y| < \rho\} \cap C\{|y| \leq \rho\}$ enjoys the mean value property $h(0) = -\int_{\beta_\rho} h(y) * dg(y,0)$, where $*dg(y,0) = -|\mathrm{grad}_\lambda g(y,0)| dS(y)$, with $dS(y)$ the Riemannian area element of β_ρ. By 3.1,

$$|\mathrm{grad}_\lambda g(y,0)|^{-1} = \lambda(\rho)|\mathrm{grad}\ g(y,0)|^{-1} = 4\pi\rho^2\lambda^2(\rho),$$

and therefore,

$$\begin{aligned} G_{B_1}1(x) &= \int_{\rho=0}^{r} \int_{\beta_\rho} g(x,y)dS(y)\lambda(\rho)d\rho \\ &= 4\pi \int_0^r \left(-\int_{\beta_\rho} g(x,y)*dg(y,0)\right)\rho^2\lambda^3(\rho)d\rho \\ &= 4\pi \int_0^r g(x,0)\rho^2\lambda^3(\rho)d\rho \\ &= g(x,0)\int_0^r S(\rho)\lambda(\rho)d\rho, \end{aligned}$$

where $S(\rho)$ is the Riemannian area of β_ρ. We conclude that

$$G_{B_1}1(x) = g(x,0)V_{B_1}(x).$$

Set $g_0 = g(y,0)$, $g_1 = g(y,x)$, and take level surfaces $\delta_0 = \{g_0 = c_0\}$, $\delta_1 = \{g_1 = c_1\}$, which shrink to 0 and x, respectively, as the constants $c_0, c_1 \to \infty$. We orient β_ρ, δ_0 and δ_1 positively with respect to the regions bounded by them and obtain for $|x| < \rho$ by Stokes' formula,

$$\int_{\beta_\rho - \delta_0 - \delta_1} g_1 * dg_0 - g_0 * dg_1 = 0.$$

Here

$$\int_{-\delta_0} g_1 {*dg_0} - g_0 {*dg_1} = \int_{-\delta_0} g_1 {*dg_0} \to g_1(0)$$

as $c_0 \to \infty$, and

$$\int_{-\delta_1} g_1 {*dg_0} - g_0 {*dg_1} = \int_{-\delta_0} g_0 {*dg_1} \to -g_0(x)$$

as $c_1 \to \infty$. By the symmetry of the Green's function, $g_1(0) - g_0(x) = 0$ and therefore,

$$\int_{\beta_\rho} g(y,x){*dg(y,0)} = \int_{\beta_\rho} g(y,0){*dg(y,x)}$$

$$= g(\rho,0)\int_{\beta_\rho} {*dg(y,x)} = -g(\rho,0),$$

where $g(\rho,0)$ stands for $g(y,0)$, $y \in \beta_\rho$. On multiplying by $|\mathrm{grad}_\lambda g(y,0)|^{-1} = 4\pi\rho^2\lambda^2(\rho) = S(\rho)$, we obtain

$$\int_{\beta_\rho} g(y,x)dS(y) = g(\rho,0)S(\rho).$$

It follows that

$$G_{B_2}1(x) = \int_{\rho=r}^{r_0} \int_{\beta_\rho} g(y,x)dS(y)\lambda(\rho)d\rho$$

$$= \int_r^{r_0} g(\rho,0)S(\rho)\lambda(\rho)d\rho,$$

that is,

$$G_{B_2}1(x) = G_{B_2}1(0).$$

3.3. Bounds for the Green's function. Take $r_0 = 1$ and denote by D_1 and D_2 the subsets $\{r \le \frac{1}{2}\}$ and $\{\frac{1}{2} < r < 1\}$ of B_α^3. Observe that $1 - r \le 1 - r^2 < 2(1 - r)$. If $x \in D_1$, there exist constants $c, c' > 0$, dependent on α only, such that

$$c'(\tfrac{1}{r} - 2) < \int_r^{1/2} \frac{1}{\rho^2(1 - \rho^2)^\alpha}\, d\rho < c(\tfrac{1}{r} - 2).$$

We shall again use the same symbols for various constants, not always the same. Suppose g exists, that is, $\alpha < 1$. Then if $x \in D_2$,

$$c'(1 - r)^{1-\alpha} < \int_r^1 \frac{1}{\rho^2(1 - \rho^2)^\alpha}\, d\rho < c(1 - r)^{1-\alpha}$$

for some $c, c' > 0$ dependent on α. Set $d = \int_{1/2}^1 \rho^{-2}(1 - \rho^2)^{-\alpha} d\rho$. We shall make use of the estimates of $g(x,0)$:

<u>If</u> $x \in D_1$,

$$c'(\tfrac{1}{r} - 2) + d < g(x,0) < c(\tfrac{1}{r} - 2) + d.$$

<u>If</u> $x \in D_2$,

$$c'(1 - r)^{1-\alpha} < g(x,0) < c(1 - r)^{1-\alpha}.$$

<u>3.4.</u> <u>Bounds for the potential</u> $G_{B_1}1$. If $x \in D_1$, the Riemannian volume $V_{B_1}(x) = c \int_0^r \rho^2(1 - \rho^2)^{3\alpha} d\rho$ of B_1 has the bounds

$$c'r^3 < V_{B_1}(x) < cr^3.$$

If $x \in D_2$, suppose first $\alpha \neq -1/3$. Then

$$c'(1 - r)^{1+3\alpha} + d' < V_{B_1}(x) < c(1 - r)^{1+3\alpha} + d.$$

If $\alpha = -1/3$,

$$c' \log(1 - r) + d' < V_{B_1}(x) < c \log(1 - r) + d.$$

Since $G_{B_1}1(x) = g(x,0)V_{B_1}(x)$, we have:

<u>If</u> $x \in D_1$,

$$c'r^2 + d'r^3 < G_{B_1}1(x) < cr^2 + dr^3.$$

<u>If</u> $x \in D_2$,

$$c'(1-r)^{2+2\alpha} + d'(1-r)^{1-\alpha} < G_{B_1}1(x) < c(1-r)^{2+2\alpha} + d(1-r)^{1-\alpha}$$

<u>for</u> $\alpha \neq -1/3$, <u>and</u>

$$c'(1-r)^{4/3}\log(1-r) + d'(1-r)^{4/3}$$

$$< G_{B_1}1(x) < c(1-r)^{4/3}\log(1-r) + d(1-r)^{4/3}$$

<u>for</u> $\alpha = -1/3$.

<u>3.5. Bounds for the potential</u> $G_{B_2}1$. If $x \in D_2$,

$$c'\int_r^1 (1-\rho)^{1-\alpha}\rho^2(1-\rho)^{3\alpha}d\rho < \int_{B_2} *g(y,0) < c\int_r^1 (1-\rho)^{1-\alpha}\rho^2(1-\rho)^{3\alpha}d\rho$$

and

$$c'\int_r^1 (1-\rho)^{1+2\alpha}d\rho < G_{B_2}1(x) < c\int_r^1 (1-\rho)^{1+2\alpha}d\rho.$$

For $\alpha \leq -1$, the left-hand side is ∞, and therefore $G_{B_2}1 \equiv \infty$. For $\alpha \in (-1,1)$,

$$c'(1-r)^{2+2\alpha} < G_{B_2}1(x) < c(1-r)^{2+2\alpha}.$$

If $x \in D_1$, then for $\alpha \in (-1,1)$,

$$\int_r^{1/2} [c'(\rho^{-1}-2) + d']\rho^2(1-\rho)^{3\alpha}d\rho + e'$$

$$< G_{B_2}1(x) < \int_r^{1/2} [c(\rho^{-1}-2) + d]\rho^2(1-\rho)^{3\alpha}d\rho + e.$$

<u>If</u> $\alpha \leq -1$, $G_{B_2}1 \equiv \infty$ <u>for all</u> x. <u>If</u> $\alpha \in (-1,1)$,

$$c' < G_{B_2}1(x) < c$$

<u>for</u> $x \in D_1$, <u>and</u>

$$c'(1-r)^{2+2\alpha} < G_{B_2}1(x) < c(1-r)^{2+2\alpha}$$

for $x \in D_2$.

3.6. Bounds for the potential $G_B 1$. In view of $G_B = G_{B_1} + G_{B_2}$, we obtain:

LEMMA. *If* $\alpha \leq -1$, $G_B \equiv \infty$. *If* $\alpha \in (-1,1)$,

$$c' r^2 + d' r^3 + e' < G_B 1(x) < c r^2 + d r^3 + e$$

for $x \in D_1$,

$$c'(1-r)^{2+2\alpha} + d'(1-r)^{1-\alpha} + e' < G_B 1(x) < c(1-r)^{2+2\alpha} + d(1-r)^{1-\alpha} + e$$

for $x \in D_2$, $\alpha \neq -1/3$, *and*

$$c'(1-r)^{4/3}\log(1-r) + d'(1-r)^{4/3} < G_B 1(x) < c(1-r)^{4/3}\log(1-r) + d(1-r)^{4/3}$$

for $x \in D_2$, $\alpha = -1/3$.

3.7. Null classes of the Poincaré 3-balls. We can now characterize the following classes:

(a) $B_\alpha^3 \in \tilde{O}_G^3 \Leftrightarrow \alpha < -1$,

(b) $B_\alpha^3 \in \tilde{O}_{QP}^3 \Leftrightarrow \alpha \in (-1,1)$,

(c) $B_\alpha^3 \in \tilde{O}_{QB}^3 \Leftrightarrow \alpha \in (-1,1)$,

(d) $B_\alpha^3 \in \tilde{O}_{QD}^3 \Leftrightarrow \alpha \in (-3/5,1)$,

(e) $B_\alpha^3 \in \tilde{O}_{QC}^3 \Leftrightarrow \alpha \in (-3/5,1)$.

Relation (a) was proved in 3.1. The estimates for $G_B 1$ in 3.6 together with Theorem 1.1 give (b) and (c). Since by Theorem 1.1, (e) is a consequence of (c) and (d), we only have to prove (d).

We know that $\alpha < 1$ is necessary. By Theorem 1.1, $B_\alpha^3 \in O_{QD}^3$ is equivalent to $G_B(1,1) = \infty$, where

$$G_B(1,1) = \int_{B\times B} g(x,y)*1(x)*1(y) = \sum_{i,j=1}^{2} \int_{D_i\times D_j} g(x,y)*1(x)*1(y).$$

By the energy principle (e.g., Sario-Nakai [1, p. 324 ff.]), $G_B(1,1) < \infty$ if and only if

$$G_{D_1}(1,1) = \int_{D_1\times D_1} g(x,y)*1(x)*1(y) < \infty$$

and

$$G_{D_2}(1,1) = \int_{D_2\times D_2} g(x,y)*1(x)*1(y) < \infty.$$

In view of 3.4, we have for $\alpha \in (-1,1)$

$$c' \int_0^{1/2} (\rho^2 + d'\rho^3)\rho^2(1-\rho)^{3\alpha} d\rho$$

$$< G_{D_1}(1,1) < c \int_0^{1/2} (\rho^2 + d\rho^3)\rho^2(1-\rho)^{3\alpha} d\rho,$$

$$c' \int_{1/2}^{1} [(1-\rho)^{2+5\alpha} + d'(1-\rho)^{1+2\alpha}] d\rho$$

$$< G_{D_2}(1,1) < c \int_{1/2}^{1} [(1-\rho)^{2+5\alpha} + d(1-\rho)^{1+2\alpha}] d\rho$$

if $\alpha \neq -1/3$, and

$$c' \int_{1/2}^{1} [(1-\rho)^{1/3} \log(1-\rho) + d'(1-\rho)^{1/3}] d\rho$$

$$< G_{D_2}(1,1) < c \int_{1/2}^{1} [(1-\rho)^{1/3} \log(1-\rho) + d(1-\rho)^{1/3}] d\rho$$

if $\alpha = -1/3$. Clearly, the bounds in the first and third estimates above are finite for all α. We conclude that a necessary and sufficient condition for $G_B(1,1) < \infty$ is $\alpha > -3/5$. Statement (d) follows.

<u>3.8.</u> <u>Arbitrary dimension</u>. The above results extend immediately to all

dimensions $N \geq 3$. The harmonic equation for $f(r)$ takes the form

$$\Delta f = -\frac{1}{\lambda^2}\left[f'' + \left(\frac{N-1}{r} + \frac{(N-2)\lambda'}{\lambda}\right)f'\right],$$

so that $f(r)$ is harmonic if and only if

$$f(r) = c\int_r^{r_0} \frac{d\rho}{\rho^{N-1}\lambda^{N-2}(\rho)}.$$

On the region $\{x|r < r_0\}$,

$$g(x,0) = \frac{1}{A_N}\int_r^{r_0} \frac{d\rho}{\rho^{N-1}\lambda^{N-2}(\rho)},$$

where A_N is the Euclidean area of the unit sphere in N-space.

$B_\alpha^N \in O_G^N$ if and only if $\alpha \geq 1/(N-2)$.

Lemma 3.2 remains valid without change, the only modification of the proof being the replacement of $4\pi\rho^2\lambda^2$ by $A_N\rho^{N-1}\lambda^{N-1}$.

Inequalities 3.3 take the following form: If $x \in D_1$,

$$\frac{c'}{r^{N-2}} + d' < g(x,0) < \frac{c}{r^{N-2}} + d.$$

If $x \in D_2$,

$$c'(1-r)^{1-(N-2)\alpha} < g(x,0) < c(1-r)^{1-(N-2)\alpha}.$$

The bounds for the volume of $B_1(x)$ are now

$$c'r^N < V_{B_1}(x) < cr^N$$

for $x \in D_1$,

$$c'(1-r)^{1+N\alpha} + d' < V_{B_1}(x) < c(1-r)^{1+N\alpha} + d$$

for $x \in D_2$, $\alpha \neq -1/N$, and

$$c' \log(1-r) + d' < V_{B_1}(x) < c \log(1-r) + d$$

for $\alpha = -1/N$. Consequently, inequalities 3.4 now read: If $x \in D_1$,

$$c'r^2 + d'r^N < G_{B_1}1(x) < cr^2 + dr^N.$$

If $x \in D_2$,

$$c'(1 - r)^{2+2\alpha} + d'(1 - r)^{1-(N-2)\alpha}$$
$$< G_{B_1}1(x) < c(1 - r)^{2+2\alpha} + d(1 - r)^{1-(N-2)\alpha}$$

for $\alpha \neq -1/N$, and

$$c'(1 - r)^{2-2/N} \log(1 - r) + d'(1 - r)^{2-2/N}$$
$$< G_{B_1}1(x) < c(1 - r)^{2-2/N} \log(1 - r) + d(1 - r)^{2-2/N}$$

for $\alpha = -1/N$.

Inequalities 3.5 are the same as before, with $(-1,1)$ replaced by $(-1, 1/(N - 2))$, and the new form of Lemma 3.6 is obtained by adding up the bounds in the new forms of inequalities 3.4 and 3.5.

3.9. Null classes of the Poincaré N-balls. We are ready to state:

THEOREM. The following characterizations are valid for every N, including $N = 2$:

(a) $B_\alpha^N \in \tilde{O}_G^N \Leftrightarrow \alpha < 1/(N - 2)$,

(b) $B_\alpha^N \in \tilde{O}_{QP}^N \Leftrightarrow \alpha \in (-1, 1/(N - 2))$,

(c) $B_\alpha^N \in \tilde{O}_{QB}^N \Leftrightarrow \alpha \in (-1, 1/(N - 2))$,

(d) $B_\alpha^N \in \tilde{O}_{QD}^N \Leftrightarrow \alpha \in (-3/(N + 2), 1/(N - 2))$,

(e) $B_\alpha^N \in \tilde{O}_{QC}^N \Leftrightarrow \alpha \in (-3/(N + 2), 1/(N - 2))$.

Proof. First suppose $N > 2$. Again, only (d) needs proving. Suppose $B_\alpha^N \in \tilde{O}_{QP}^N$. By the above estimates,

$$c'\int_0^{1/2}(\rho^2 + d'\rho^N)\rho^{N-1}(1-\rho)^{N\alpha}d\rho$$

$$< G_{D_1}(1,1) < c\int_0^{1/2}(\rho^2 + d\rho^N)\rho^{N-1}(1-\rho)^{N\alpha}d\rho,$$

$$c'\int_{1/2}^{1}[(1-\rho)^{2+(N+2)\alpha} + d'(1-\rho)^{1+2\alpha}]d\rho$$

$$< G_{D_2}(1,1) < c\int_{1/2}^{1}[(1-\rho)^{2+(N+2)\alpha} + d(1-\rho)^{1+2\alpha}]d\rho$$

for $\alpha \neq -1/N$, and

$$c'\int_{1/2}^{1}[(1-\rho)^{1-2/N}\log(1-\rho) + d'(1-\rho)^{1-2/N}]d\rho$$

$$< G_{D_2}(1,1) < c\int_{1/2}^{1}[(1-\rho)^{1-2/N}\log(1-\rho) + d(1-\rho)^{1-2/N}]d\rho$$

for $\alpha = -1/N$. Thus the bounds for $G_{D_1}(1,1)$ and $G_{D_2}(1,1)$ are finite if and only if $\alpha > -3/(N+2)$. By $O^N_{QP} \subset O^N_{QD}$, we have (d).

By obvious modifications of the proofs, we see that for $N = 2$ the bound in (a) is eliminated, and so are the upper bounds in (b) - (e). This is to be expected in view of the simplifying circumstance that harmonicity no longer depends on λ and is thus reduced to the Euclidean case. Our results are thus valid for every N.

We observe in passing that Theorem 3.9 provides us with new counterexamples showing the strictness of some of the inclusion relations in 1.8.

NOTES TO §3. The Poincaré N-ball was introduced, and Theorem 3.9 established with the method used above, in Sario-Wang [4].

§4. CHARACTERISTIC QUASIHARMONIC FUNCTION

The characterizations of quasiharmonic null classes in Theorem 1.1 depend on properties of the function Gl, which may or may not exist. We shall now introduce, on the Poincaré N-ball B^N_α, a fixed radial quasiharmonic function $s(r)$ which exists on B^N_α for every N, α. We shall show that necessary and sufficient for $B_\alpha^N \in \tilde{O}_{QX}^N$ with $X = P, B, D,$ or C is that $s(r) \in X$. Accordingly, we shall call $s(r)$ the <u>characteristic quasiharmonic function</u> of the Poincaré N-ball.

<u>4.1.</u> <u>Existence</u>. Here we introduce the function $s(r)$ by means of power series. Again, a reader not interested in methodology may omit §§ 4-5 and proceed directly to the integral representation of $s(r)$ in §6.

<u>THEOREM</u>. <u>For every $N \geq 2$ and every α, there exists on the Poincaré N-ball B_α^N the characteristic quasiharmonic function $s(r)$ defined by</u>

$$s(r) = -\sum_{i=0}^{\infty} b_i r^{2i+2},$$

<u>where</u> $b_0 = 1/(2N)$ <u>and, for</u> $i > 0$,

$$b_i = \frac{1}{2N}\prod_{j=1}^{i} p_j + \sum_{j=1}^{i-1} q_j \prod_{k=j+1}^{i} p_k + q_i,$$

<u>with</u>

$$p_i = \frac{2i[2i + (N-2)(2\alpha+1)]}{(2i+2)(2i+N)}, \qquad q_i = \frac{\Pi_{j=1}^{i}(j - 2\alpha - 2)j^{-1}}{(2i+2)(2i+N)}.$$

<u>Proof.</u> The quasiharmonic equation

$$\begin{aligned}\Delta s(r) &= -g^{-1/2}\frac{\partial}{\partial r}(g^{1/2}g^{rr}s') \\ &= -\lambda^{-2}\left[s'' + \left(\frac{N-1}{r} + \frac{(N-2)\lambda'}{\lambda}\right)s'\right] = 1\end{aligned}$$

takes the form

$$\begin{aligned}r^2(1-r^2)s'' + r[(N-1)(1-r^2) \\ -2(N-2)\alpha r^2]s' + r^2(1-r^2)^{2\alpha+1} = 0.\end{aligned}$$

On substituting $s(r) = -\Sigma_0^\infty b_i r^{2i+2}$, we obtain

$$- \sum_{i=0}^{\infty} (1 - r^2)(2i + 2)(2i + 1)b_i r^{2i+2} - \sum_{i=0}^{\infty} (N - 1)(2i + 2)b_i r^{2i+2}$$

$$+ \sum_{i=0}^{\infty} [(N - 1) + 2(N - 2)\alpha](2i + 2)b_i r^{2i+4}$$

$$+ r^2 + r^2 \sum_{i=1}^{\infty} \left(\prod_{j=1}^{i} \frac{j - 2\alpha - 2}{j} \right) r^{2i} = 0.$$

We change by unity the summation index in the coefficient of r^{2i+4}:

$$\sum_{i=0}^{\infty} (2i + 2)(2i + N)b_i r^{2i+2} = \sum_{i=1}^{\infty} 2i[2i + (N - 2)(2\alpha + 1)]b_{i-1} r^{2i+2}$$

$$+ r^2 + \sum_{i=1}^{\infty} \left(\prod_{j=1}^{i} \frac{j - 2\alpha - 2}{j} \right) r^{2i+2} = 0.$$

Equating the coefficient of r^2 to 0 yields $b_0 = 1/(2N)$. The coefficient of r^{2+2i} for $i > 0$ gives in the notation of the Theorem,

$$b_i = p_i b_{i-1} + q_i,$$

which by induction provides us with the desired result.

4.2. Characteristic property. We shall show:

THEOREM. The Poincaré N-ball B_α^N belongs to $\tilde{O}_{QX}^N$ for $X = P, B, D,$ or C if and only if the characteristic quasiharmonic function $s(r)$ belongs to X.

If we utilize the properties of Gl, the proof is immediate. In fact, only the necessity needs verification. Suppose B_α^N belongs to some $\tilde{O}_{QX}^N$. Then by Theorem 1.1, the function $Gl(x) = \int_{B_\alpha^N} *g(x,y)$ exists and belongs to X. In view of the radial nature of B_α^N, the function Gl is radial, that is, $Gl(x_1) = Gl(x_2)$ if $|x_1| = |x_2|$. By virtue of $\Delta Gl(r) = \Delta s(r) = 1$, the function $h = Gl - s$ is radial harmonic, and the maximum principle gives $s = Gl + \text{const}$. Since Gl

belongs to X, so does s.

From a methodological and "computational" view point, we are, however, also interested in estimating $s(r)$ by means of power series, without utilizing G1. We shall do this for the case $X = B$. This will also prepare us for the proof, in §5, of the boundedness of s from above.

In view of Theorem 3.9, we only have to show that if $\alpha \in (-1, 1/(N-2))$, then $s \in B$. The proof will be given in 4.3 - 4.6.

4.3. Estimating $\Pi\, p_j$. To estimate b_i given in Theorem 4.1, we start with $\Pi\, p_j$. Let i_0 be any integer such that

$$i_0 \geq 1 - \alpha(N-2) - \frac{N}{2}.$$

Further conditions on i_0 will be imposed in the course of our reasoning.

LEMMA. For $\alpha < 1/(N-2)$ and $i > i_0$,

$$\prod_{j=i_0+1}^{i} p_j < \frac{i_0+1}{i+1}\left(\frac{2i_0+N+2}{2i+N+2}\right)^{1-\alpha(N-2)}.$$

Proof. In p_i, consider the factor

$$\delta_i = \frac{2i + (N-2)(2\alpha+1)}{2i+N} = 1 - \frac{2[1-\alpha(N-2)]}{2i+N}.$$

For $\alpha < 1/(N-2)$ and $i > i_0$, we have $0 < \delta_i < 1$ and

$$\log \delta_i < -\frac{2[1-\alpha(N-2)]}{2i+N} < 0.$$

Therefore,

$$\log \prod_{j=i_0+1}^{i} \delta_j < -2[1-\alpha(N-2)] \int_{i_0+1}^{i+1} \frac{dx}{2x+N}$$

and

$$\prod_{j=i_0+1}^{i} \delta_j < \left(\frac{2i_0+N+2}{2i+N+2}\right)^{1-\alpha(N-2)}.$$

In view of

$$p_i = \frac{i}{i + 1} \delta_i,$$

the Lemma follows.

4.4. Estimating q_i. To proceed with the estimation of b_i, we now utilize also the condition $\alpha > -1$ and impose on i_0 the additional requirement

$$i_0 \geq 2(\alpha + 1).$$

In the sequel c will stand for a positive constant, not always the same.

LEMMA. For $\alpha \in (-1, 1/(N - 2))$ and $i > i_0$,

$$|q_i| < \frac{c}{(2i + 2)(2i + N)} \left(\frac{i_0 + 1}{i + 1}\right)^{2(\alpha+1)} .$$

Proof. For $j > i_0$,

$$0 < 1 - \frac{2(\alpha + 1)}{j} < 1,$$

and therefore,

$$\log \prod_{j=i_0+1}^{i} \left(1 - \frac{2(\alpha + 1)}{j}\right) < -2(\alpha + 1) \sum_{i_0+1}^{i} \frac{1}{j} < -2(\alpha + 1) \int_{i_0+1}^{i+1} \frac{dx}{x}.$$

This gives

$$\prod_{j=i_0+1}^{i} \frac{j - 2\alpha - 2}{j} < \left(\frac{i_0 + 1}{i + 1}\right)^{2(\alpha+1)},$$

hence the Lemma.

4.5. Estimating b_i. We now come to the main step in estimating b_i. It will be necessary to consider separately the cases $\alpha \in (-1/N, 1/(N - 2))$, $\alpha = -1/2$, and $\alpha \in (-1,0) - \{-1/2\}$.

LEMMA. For $\alpha \in (-1/N, 1/(N - 2))$, and $i > i_0$,

$$|b_i| < c\left(\frac{1}{i}\right)^{2-\alpha(N-2)} + d\left(\frac{1}{i}\right)^{3/2-(1/2)\alpha(N-2)} + e\left(\frac{1}{i}\right)^{2(\alpha+2)},$$

where c, d, e are positive constants.

Proof. By the definition of b_i in Theorem 4.1,

$$b_i = b_{i_0} \prod_{j=i_0+1}^{i} p_j + \sum_{j=i_0+1}^{i-1} q_j \prod_{k=j+1}^{i} p_k + q_i,$$

and by Lemma 4.3,

$$|b_{i_0} \prod_{j=i_0+1}^{i} p_j| < \frac{c}{i+1}\left(\frac{1}{2i+N+2}\right)^{1-\alpha(N-2)} < c\left(\frac{1}{i}\right)^{2-\alpha(N-2)}.$$

In view of Lemmas 4.3 and 4.4,

$$|q_j \prod_{k=j+1}^{i} p_k| < \frac{c}{(2j+2)(2j+N)}\left(\frac{1}{j+1}\right)^{2(\alpha+1)} \cdot \frac{j+1}{i+1}\left(\frac{2j+N+2}{2i+N+2}\right)^{1-\alpha(N-2)}.$$

For $\alpha \in (-1/N, 1/(N-2))$,

$$1 - \alpha(N-2) < 2(1+\alpha).$$

We, therefore, may and do require of i_0 further that for $j > i_0$,

$$\frac{(2j+N+2)^{1-\alpha(N-2)}}{(j+1)^{2(1+\alpha)}} < 2^{1-\alpha(N-2)}.$$

Then

$$|\sum_{j=i_0+1}^{i-1} q_j \prod_{k=j+1}^{i} p_k| < \frac{c}{2i+2}\left(\frac{1}{2i+N+2}\right)^{1-\alpha(N-2)} \sum_{j=i_0+1}^{i-1} \frac{1}{2j+N},$$

where

$$\sum_{j=i_0+1}^{i-1} \frac{1}{2j+N} < \int_{i_0}^{i-1} \frac{dx}{2x+N} = \frac{1}{2}\log\frac{2i+N-2}{2i_0+N}.$$

Accordingly, we impose on i_0 the additional condition that for $i > i_0$,

$$\left(\frac{1}{i}\right)^{(1/2)[1-\alpha(N-2)]} \log \frac{2i + N + 2}{2i_0 + N} < 1.$$

Then

$$\left| \sum_{j=i_0+1}^{i-1} q_j \prod_{k=j+1}^{i} p_k \right| < c\left(\frac{1}{i}\right)^{3/2-(1/2)\alpha(N-2)}.$$

It is immediate by Lemma 4.4 that

$$|q_i| < c\left(\frac{1}{i}\right)^{2(\alpha+2)}.$$

We combine the above estimates and obtain the Lemma.

4.6. Boundedness of $s(r)$. We come to the concluding step in proving Theorem 4.2 for $X = B$ by means of direct estimates of $s(r)$:

LEMMA. *For $\alpha \in (-1, 1/(N-2))$, the characteristic quasiharmonic function $s(r)$ of the Poincaré N-ball is bounded.*

Proof. For $\alpha \in (-1/N, 1/(N-2))$, all three exponents in Lemma 4.5 are > 1, and therefore,

$$|s(r)| < \left| \sum_{i=0}^{\infty} b_i r^{2i+2} \right| < \sum_{i=0}^{\infty} |b_i| < \infty.$$

The case $\alpha = -\frac{1}{2}$ is simple, as all $q_i = 0$, and by the definition of p_i in Theorem 4.1,

$$|b_i| = |b_{i_0}| \prod_{j=i_0+1}^{i} p_j < |b_{i_0}| \frac{(2i_0+2)^2}{(2i+2)(2i+N)} < c\left(\frac{1}{i}\right)^2,$$

whence $\Sigma_0^\infty |b_i| < \infty$.

It remains to consider the case $\alpha \in (-1,0) - \{-\frac{1}{2}\}$. We obtain at once

$$p_k < \frac{2k(2k + N - 2)}{(2k + 2)(2k + N)} ,$$

$$\prod_{k=j+1}^{i} p_k < \frac{(2j + 2)(2j + N)}{(2i + 2)(2i + N)} ,$$

and by Lemma 4.4,

$$|q_j| < \frac{c}{(2j + 2)(2j + N)} \cdot \left(\frac{1}{j + 1}\right)^{2(\alpha+1)}$$

for $j > i_0$. Therefore,

$$\sum_{j=i_0+1}^{i-1} \left| q_j \prod_{k=j+1}^{i} p_k \right| < \frac{c}{(2i + 2)(2i + N)} \int_{i_0}^{i-1} \frac{dx}{(x + 1)^{2(\alpha+1)}} ,$$

where the integral has the value

$$\frac{1}{-2\alpha - 1}[i^{-2\alpha-1} - (i_0 + 1)^{-2\alpha-1}]$$

since $\alpha \neq -\frac{1}{2}$. As a consequence,

$$\sum_{j=i_0+1}^{i-1} \left| q_j \prod_{k=j+1}^{i} p_k \right| < c\left(\frac{1}{i}\right)^{3+2\alpha} + d\left(\frac{1}{i}\right)^{2} .$$

Similarly,

$$\left| b_{i_0} \prod_{j=i_0+1}^{i} p_j \right| < c\left(\frac{1}{i}\right)^{2}$$

and

$$|q_i| < c\left(\frac{1}{i}\right)^{2\alpha+4} .$$

Since all exponents in these estimates are > 1, $s(r)$ is bounded and we have the Lemma.

The proof of Theorem 4.2 is herewith complete.

NOTES TO §4. The characteristic quasiharmonic function $s(r)$ was introduced in Sario-Wang [6]. In §§5-6, it will provide us with a convenient tool in further problems of quasiharmonic classification.

§5. NEGATIVE CHARACTERISTIC

The characteristic quasiharmonic function $s(r)$ discussed in §4 has another important property: normalized by an additive constant, it is negative. It exists on the Poincaré N-ball B^N_α for every N, α.

5.1. Negative quasiharmonic functions. We introduce the class of negative quasiharmonic functions,

$$QN = \{q \in Q,\ q \leq 0\},$$

where we include the equality sign since a superharmonic function may take on its supremum. The corresponding null class of Riemannian N-manifolds is

$$O^N_{QN} = \{R \mid QN(R) = \emptyset\}.$$

There will be no confusion in using the same natural symbol N in its two entirely different meanings.

THEOREM. *The Poincaré N-ball carries QN functions,*

$$B^N_\alpha \in \tilde{O}^N_{QN},$$

for all N, α.

Once again, we are intrigued by the methodological problem of giving the proof by means of power series. A very short proof in terms of the integral form of the characteristic quasiharmonic function will be given in 6.1.

Here we shall discuss dimension $N = 3$ only. As in 3.8, the modification for $N > 3$ offers no difficulty.

For convenient use in the course of the proof, we start by summarizing some steps from the beginning of §4 in the present case $N = 3$.

The function

$$s(r) = -\sum_{i=0}^{\infty} b_i r^{2i+2}$$

with $\Delta s = 1$ on B_α^3 has $b_0 = 1/6$, and the b_i, $i > 0$, are determined by the recursive formula

$$b_i = p_i b_{i-1} + q_i,$$

where

$$p_i = \frac{2i(2i + 1 + 2\alpha)}{(2i + 2)(2i + 3)},$$

$$q_i = \frac{\Pi_{j=1}^{i}(j - 2\alpha - 2)j^{-1}}{(2i + 2)(2i + 3)} .$$

If in the expansion

$$b_i = b_0 \prod_{j=1}^{i} p_j + \sum_{j=1}^{i-1} q_j \prod_{k=j+1}^{i} p_k + q_i,$$

we use the notation $b_i = \Sigma_{j=0}^{i} \beta_{ij}$ with

$$\begin{cases} \beta_{i0} = b_0 \prod_{j=1}^{i} p_j, \quad \beta_{ii} = q_i, \\ \beta_{ij} = q_j \prod_{k=j+1}^{i} p_k \quad \text{for } 1 \leq j \leq i - 1, \end{cases}$$

then for a fixed i_0 and all $i > i_0$,

$$b_i = b_{i_0} \prod_{j=i_0+1}^{i} p_j + \sum_{j=i_0+1}^{i} \beta_{ij}.$$

5.2. Dependence on α. The signs of p_i and q_i will be instrumental. For a given α, we set

$$i_p = \max\{i \mid i < -\alpha - \tfrac{1}{2}\}, \quad i_q = \max\{i \mid i < 2\alpha + 2\}.$$

The following immediate observations are compiled here for easy reference:

If $\alpha > -3/2$, then all $p_i > 0$. If $\alpha = -3/2$, then $p_1 = 0$ and $p_i > 0$ for $i > 1$. If $\alpha < -3/2$, then $p_i < 0$ for $i \leq i_p$, and $p_i \geq 0$ for $i > i_p$, with equality at most for $i = i_p + 1$.

If $\alpha < -1/2$, then all $q_i > 0$. If $\alpha = -1/2$, then all $q_i = 0$. If $\alpha > -1/2$ and $i \leq i_q$, then $q_i > 0$ for i even and $q_i < 0$ for i odd. If $\alpha > -1/2$ and $i > i_q$, then $q_i \geq 0$ for i_q even, and $q_i \leq 0$ for i_q odd.

These rules motivate the division of our discussion in the sequel into the cases $\alpha < -3/2$; $-3/2 \leq \alpha \leq -1/2$; and $\alpha \geq 1$. If $\alpha \in (-1,1)$, then by Theorem 3.9, there exist functions $q \in QB$, and

$$q - \sup_{B_\alpha^3} q \in QN.$$

Thus it will suffice to consider the above three cases.

We shall first show, in 5.3 and 5.4, that $b_i > 0$ for all sufficiently large i, and then in 5.5 that the series $s = -\Sigma_0^\infty b_i r^{2i+2}$ converges, hence $s - \sup s \in QN$.

<u>5.3.</u> <u>Case</u> $\alpha < -3/2$. By the expansion of b_i at the end of 5.1, we have for $i > i_p$,

$$b_i = b_{i_p} \prod_{j=i_p+1}^{i} p_j + \sum_{j=i_p+1}^{i} \beta_{ij},$$

where $b_{i_p} = \Sigma_{j=0}^{i_p} \beta_{i_p j}$.

<u>LEMMA</u>. <u>For</u> $\alpha < -3/2$ <u>and</u> $i \geq i_p$,

$$b_i > 0 \quad \underline{\text{and}} \quad \sum_{i=0}^{\infty} b_i = \infty.$$

<u>Proof</u>. We first prove that $b_{i_p} > 0$. Set

$$\delta_{i_p j} = \frac{\beta_{i_p j}}{\beta_{i_p, j-1}}, \quad 2 \leq j \leq i_p,$$

with $\beta_{i_p j} = q_j \Pi_{k=j+1}^{i_p} p_k$. We have

$$\delta_{i_p j} = \frac{q_j}{q_{j-1} p_j} < 0$$

and

$$|\delta_{i_p j}| = 1 + \frac{4j^2 - 2(\alpha + 1)(j + 1)}{-j(2j + 1 + 2\alpha)} > 1 + \frac{4j^2 + j + 1}{-j(2j + 1 + 2\alpha)} .$$

Therefore,

$$|\delta_{i_p j}| > 1 \quad \text{for} \quad 2 \leq j \leq i_p.$$

Suppose first i_p even. Then

$$b_{i_p} = \beta_{i_p 0} + \sum_{j=1}^{i_p/2} \left(\beta_{i_p, 2j-1} + \beta_{i_p, 2j}\right).$$

Since $\beta_{i_p i_p} = q_{i_p} > 0$, we see by the properties of $\delta_{i_p j}$ that each sum in parentheses is > 0. The same is true of $\beta_{i_p 0} = b_0 \Pi_1^{i_p} p_j$, as each $p_j < 0$, and we conclude that $b_{i_p} > 0$.

If i_p is odd, we first observe that

$$\delta_{i_p 1} = \frac{\beta_{i_p 1}}{\beta_{i_p 0}} = \frac{q_1}{b_0 p_1} = 3 \cdot \frac{-1 - 2\alpha}{3 + 2\alpha} < 0$$

for $\alpha < -3/2$, and

$$|\delta_{i_p 1}| = 3\left(1 + \frac{2}{-3 - 2\alpha}\right) > 3.$$

Since $\beta_{i_p 0} < 0$ and $\beta_{i_p 1} > 0$, we have $\beta_{i_p 0} + \beta_{i_p 1} > 0$. From $|\delta_{i_p j}| > 1$ for $2 \leq j \leq i_p$, we obtain $\beta_{i_p 2j} + \beta_{i_p, 2j+1} > 0$ for $1 \leq j \leq \frac{1}{2}(i_p - 1)$. Therefore,

$$b_{i_p} = \sum_{j=0}^{(i_p-1)/2} \left(\beta_{i_p 2j} + \beta_{i_p, 2j+1}\right) > 0.$$

Inequality $b_i > 0$ for $i \geq i_p$ is a direct consequence of the expression for b_i at the beginning of 5.3.

To prove $\Sigma_0^\infty b_i = \infty$, let $s = s_1 + s_2$ with

$$s_1 = -\sum_{i=0}^{i_p-1} b_i r^{2i+2}, \qquad s_2 = -\sum_{i=i_p}^{\infty} b_i r^{2i+2}.$$

Here $s_1 \in B$ and $|s_2| < \Sigma_{i_p}^\infty b_i$. If this sum converges, we have $s_2 \in B$, hence $s \in QB$, a contradiction since $\alpha \notin (-1,1)$. This proves the Lemma.

Note that the condition on α in this Lemma cannot be suppressed, as e.g., $\alpha = 0$ gives $b_i = 0$ for $i \geq 1$.

5.4. Other cases. The remaining two cases of α are covered in the following

LEMMA. If $-3/2 \leq \alpha \leq -1/2$,

$$b_i > 0 \quad \text{for all} \quad i.$$

If $\alpha \geq 1$, there exists an $i_0 \geq i_q$ such that

$$b_i > 0 \quad \text{for all} \quad i \geq i_0$$

and

$$\sum_{i=0}^{\infty} b_i = \infty.$$

Proof. For $\alpha = -3/2$, $p_1 = 0$, $p_i > 0$ if $i > 1$. For $-3/2 < \alpha \leq -1/2$, all $p_i > 0$. For $\alpha = -1/2$, all $q_i = 0$. For $-3/2 \leq \alpha < -1/2$, all $q_i > 0$. For $-3/2 \leq \alpha \leq -1/2$, we therefore have $\beta_{i0} \geq 0$, $\beta_{ij} \geq 0$, $j > 1$, and the first part of the Lemma follows.

If $\alpha \geq 1$, suppose that there exists an i_0 such that $b_i \leq 0$ for $i > i_0$. Since $\alpha \notin (-1,1)$, s is unbounded and

$$-s + \sum_{i=0}^{i_0} |b_i| \in QP,$$

a contradiction. We conclude that there exist infinitely many $b_i > 0$. In particular, there is some $i_0 \geq i_q$ such that $b_{i_0} > 0$.

For $i > i_0$,

$$b_i = b_{i_0} \prod_{j=i_0+1}^{i} p_j + \sum_{j=i_0+1}^{i} \beta_{ij}.$$

Each $p_j > 0$, hence the first term on the right is > 0. If i_q is even, then $q_j \geq 0$ for $j > i_q$, and $\beta_{ij} \geq 0$ for $i,j > i_q$. Therefore, $b_i > 0$ for $i > i_0$. If i_q is odd, then $q_j \leq 0$ for $j > i_q$. Suppose $b_{i_1} \leq 0$ for some $i_1 > i_q$. Then

$$b_{i_1+1} = p_{i_1+1} b_{i_1} + q_{i_1+1} \leq 0$$

and by induction, we infer that $b_i \leq 0$ for $i \geq i_1$, in violation of the fact that there are infinitely many $b_i > 0$. Consequently, $b_i > 0$ for $i > i_q$.

The proof that $\Sigma_0^\infty b_i = \infty$ is the same as for Lemma 5.3.

Note that Lemma 5.4 cannot be sharpened to $b_i > 0$ for all $i \geq 0$, since, e.g., $b_1 = -\alpha/15$.

We have established that, in all cases, $b_i > 0$ for all but a finite number of i. It remains to show that the series $\Sigma_0^\infty b_i r^{2i+2}$ converges.

<u>5.5.</u> <u>Convergence</u>. We claim:

<u>LEMMA</u>. <u>For</u> $\alpha \notin (-1,1)$, $\Sigma_{i=0}^\infty b_i r^{2i+2} < \infty$.

<u>Proof</u>. Since for $\alpha \in (-1,1)$, $B_\alpha^3 \in \tilde{O}_{QB}^3 \subset \tilde{O}_{QN}^3$, only the above cases $\alpha \leq -1$ and $\alpha \geq 1$ are of interest.

First suppose $\alpha \leq -1$. The ratio of successive terms being $b_{i+1}r^2/b_i$, it suffices to show that $b_{i+1}/b_i \to 1$ as $i \to \infty$. We have

$$\frac{b_{i+1}}{b_i} = p_{i+1} + \frac{q_{i+1}}{b_i},$$

where $p_{i+1} \to 1$. We shall show that $q_{i+1}/b_i \to 0$, that is, for any positive integer n, fixed henceforth, there exists an i_n such that $b_i/q_{i+1} > n$ for $i \geq i_n$. For $i > i_p$,

$$\frac{b_i}{q_{i+1}} = \frac{b_{i_p}}{q_{i+1}} \prod_{j=i_p+1}^{i} p_j + \sum_{j=i_p+1}^{i-1} \frac{q_j}{q_{i+1}} \prod_{k=j+1}^{i} p_k + \frac{q_i}{q_{i+1}} ,$$

where $b_{i_p} > 0$. Note that the case $-3/2 \leq \alpha \leq -1$ is included, for then $b_{i_p} = b_0 = 1/6$. Since $p_j \geq 0$ for $j > i_p$, with equality at most for $j = i_p + 1$, and since $q_j > 0$ for all j, we obtain for $\alpha \leq -1$ and $i \geq i'_n = i_p + n + 1$,

$$\frac{b_i}{q_{i+1}} \geq f(i) = \sum_{j=i-n}^{i-1} \frac{q_j}{q_{i+1}} \prod_{k=j+1}^{i} p_k + \frac{q_i}{q_{i+1}} .$$

It suffices to show that the function $f(i)$ introduced herewith dominates n for all sufficiently large i.

Since $f(i)$ and hence $f'(i)$ are rational in i, there exists an i''_n such that $f'(i)$ is of constant sign and $f(i)$ is monotone for $i \geq i''_n$. Observe that

$$\frac{q_i}{q_{i+1}} = \frac{i + 1}{i - 1 - 2\alpha} \cdot \frac{(2i + 4)(2i + 5)}{(2i + 2)(2i + 3)} \to 1$$

as $i \to \infty$, and so does each q_j/q_{i+1} for $i - n \leq j \leq i - 1$. Moreover, each $p_k \to 1$, and therefore, $f(i) \to n + 1$, with the convergence monotone for $i \geq i''_n$. We conclude that there exists an $i_n \geq \max(i'_n, i''_n)$ such that

$$f(i) > n \quad \text{for} \quad i \geq i_n .$$

This completes the proof of the Lemma for $\alpha \leq -1$.

Now suppose $\alpha \geq 1$. If i_q is even, then $q_i \geq 0$ for $i > i_q$. Since each $p_i > 0$, the argument in the case $\alpha \leq -1$ continues to hold, with i_p replaced by i_0 of Lemma 5.4. If i_q is odd, then $q_i \leq 0$ for $i > i_q$. Again, each $p_i > 0$, and since by Lemma 5.4, $b_i > 0$ for $i \geq i_0$, we have

$$0 < \frac{b_{i+1}}{b_i} \leq p_{i+1} \to 1.$$

This gives the Lemma, and the proof of Theorem 5.1 is complete.

5.6. The class O^N_{QN}. The Poincaré N-ball has already rendered us the great service of providing a sample of every null class, and its complement, that we have discussed thus far. Here it gives us the following general result.

THEOREM. The relations

$$\tilde{O}^N_{QN} \cap O^N_{HX} \neq \emptyset, \qquad \tilde{O}^N_{QN} \cap \tilde{O}^N_{HX} \neq \emptyset$$

hold for $N \geq 2$ and $X = G, P, B, D, C, L^p, p \geq 1,$ and the relations

$$\tilde{O}^N_{QN} \cap O^N_{QX} \neq \emptyset, \qquad \tilde{O}^N_{QN} \cap \tilde{O}^N_{QX} \neq \emptyset$$

for $N \geq 2$ and $X = P, B, D, C.$

For the class $O^N_{QL^p}$, we refer to 6.4.

For $N > 2$, the Theorem is a direct consequence of Theorems I.2.10, I.3.5, I.3.9 and II.5.1. For $N = 2$, the Euclidean plane gives the first and third relations, the Euclidean disk the second and fourth relations.

In the first part of the Theorem, we have anticipated the topic of §7, relations between harmonic and quasiharmonic null classes, as we shall not return to the class O^N_{QN}, the theory of QN functions being totally undeveloped.

NOTES TO §5. Theorems 5.1 and 5.6 were proved in Sario-Wang [11]. It was recently shown in Nakai-Sario [21] that there exist (both parabolic and hyperbolic) manifolds in O^N_{QN}. This is of interest in view of Theorem 5.1 and the fact that even the Euclidean plane, which fails to carry any other functions considered in classification theory, possesses the function $-4^{-1}r^2$ in QN. In general, developing a theory of QN functions appears to be a challenging task.

§6. INTEGRAL FORM OF THE CHARACTERISTIC

The characteristic quasiharmonic function $s(r)$ was defined and its properties explored in §§4 and 5 by means of power series. We now give $s(r)$ the form of an integral and produce new proofs of the characterizations of the classes O^N_{QX} of the Poincaré N-balls B^N_α. The integral form of the characteristic also serves to give a simple necessary and sufficient condition for B^N_α to belong to the class

$O^N_{QL^p}$ of Riemannian N-manifolds which do not carry quasiharmonic functions with a finite L^p norm. All proofs are valid at once for every $N \geq 2$.

6.1. Integral form. The characteristic quasiharmonic function $s(r)$ of the Poincaré N-ball B^N_α, $N \geq 2$, has the integral representation

$$s(r) = -\int_0^r (1 - t^2)^{-(N-2)\alpha} t^{-N+1} \int_0^t (1 - \tau^2)^{N\alpha} \tau^{N-1} d\tau dt.$$

In fact, the expression q, say, on the right satisfies the quasiharmonic equation

$$\Delta q(r) = -(1 - r^2)^{-N\alpha} r^{-N+1}[(1 - r^2)^{(N-2)\alpha} r^{N-1} q'(r)]' = 1.$$

By the maximum principle for harmonic functions, q differs from the characteristic quasiharmonic function s of §4 by a constant. In view of $s(0) = q(0) = 0$, we have $s = q$.

The first immediate consequence of this observation is that s is not only bounded from above, as proved in §5, but actually nonpositive:

THEOREM. The characteristic quasiharmonic function of the Poincaré N-ball satisfies

$$s \leq 0,$$

for all N and all α.

In fact, the integrands in the above expression of $s(r)$ are nonnegative.

6.2. Characterization of O^N_{QP} and O^N_{QB}. We shall first give an alternate proof of parts (b) and (c) of Theorem 3.9, established by means of estimating the Green's function:

$$B^N_\alpha \in \begin{cases} \tilde{O}^N_{QP} \Leftrightarrow \alpha \in (-1, 1/(N - 2)), & N \geq 2, \\ \tilde{O}^N_{QB} \Leftrightarrow \alpha \in (-1, 1/(N - 2)), & N \geq 2. \end{cases}$$

First assume $\alpha \in (-1, 1/(N-2))$. Since $\tilde{O}^N_{QB} \subset \tilde{O}^N_{QP}$, it suffices to show that $B^N_\alpha \in \tilde{O}^N_{QB}$. For $\alpha \in (-1/N, 1/(N-2))$, $\int_0^t (1-\tau^2)^{N\alpha} \tau^{N-1} d\tau$ is bounded as $t \to 1$, and $\sim ct^N$ as $t \to 0$. Therefore,

$$|s(r)| = O\left(\int_0^r (1-t)^{-(N-2)\alpha} dt\right) \in B$$

as $r \to 1$. For $\alpha = -1/N$, $\int_0^t (1-\tau^2)^{N\alpha} \tau^{N-1} d\tau = O(\log(1-t))$ as $t \to 1$, and

$$|s(r)| = O\left(\int_0^r (1-t)^{-(N-2)\alpha} \log(1-t) dt\right) \in B.$$

For $-1 < \alpha < -1/N$, $\int_0^t (1-\tau^2)^{N\alpha} \tau^{N-1} d\tau \sim c(1-t)^{N\alpha+1}$ with $c \neq 0$, and $|s(r)| \sim c \int_0^r (1-t)^{2\alpha+1} dt \in B$. We conclude that $B^N_\alpha \in \tilde{O}^N_{QB}$ for $\alpha \in (-1, 1/(N-2))$.

If $\alpha \notin (-1, 1/(N-2))$, let c stand for a positive constant. For $\alpha \geq 1/(N-2)$, a case not occurring if $N = 2$, we let c absorb bounded factors and obtain

$$s(r) \sim -c \int_0^r (1-t)^{-(N-2)\alpha} dt \to -\infty$$

as $r \to 1$. For $\alpha \leq -1$,

$$s(r) \sim -c \int_0^r (1-t)^{-(N-2)\alpha} (1-t)^{N\alpha+1} dt$$

$$\sim -c \int_0^r (1-t)^{2\alpha+1} dt \to -\infty.$$

Thus $s(r) \to -\infty$ for $\alpha \notin (-1, 1/(N-2))$.

Every quasiharmonic function $q(r,\theta)$ on B^N_α can be represented as $s(r) + h(r,\theta)$ with $h \in H$. By the minimum principle for harmonic functions, there exists, for each $0 \leq r < 1$, a point $(r, \theta(r))$ such that $h(r, \theta(r)) \leq h(0)$, where 0 stands for the origin. As $r \to 1$,

$$q(r, \theta(r)) \leq s(r) + h(0) \to -\infty,$$

hence $q \notin P$, and we have $B^N_\alpha \in O^N_{QP} \subset O^N_{QB}$.

6.3. Characterization of O^N_{QD} and O^N_{QC}. We proceed to give alternate proofs of parts (d) and (e) of Theorem 3.9:

$$B^N_\alpha \in \begin{cases} \tilde{O}^N_{QD} \Leftrightarrow \alpha \in (-3/(N+2),\ 1/(N-2)), & N \geq 2, \\ \tilde{O}^N_{QC} \Leftrightarrow \alpha \in (-3/(N+2),\ 1/(N-2)), & N \geq 2. \end{cases}$$

The Dirichlet integral is

$$D(s) = \int_{B^N_\alpha} *|\operatorname{grad} s|^2 = \int_{B^N_\alpha} *g^{rr}s'^2$$

$$= c\int_0^1 (1-r^2)^{-2\alpha}[(1-r^2)^{-(N-2)\alpha}\, r^{-N+1}\int_0^r (1-t^2)^{N\alpha}\, t^{N-1}dt]^2(1-r^2)^{N\alpha}r^{N-1}dr$$

$$= c\int_0^1 (1-r^2)^{-(N-2)\alpha}\, r^{-N+1}\left[\int_0^r (1-t^2)^{N\alpha}\, t^{N-1}dt\right]^2 dr.$$

For $\alpha \in (-1/N,\ 1/(N-2))$, $[\quad]^2$ is bounded as $r \to 1$, and $\sim cr^{2N}$ as $r \to 0$. Therefore,

$$D(s) \approx c\int_0^1 (1-r)^{-(N-2)\alpha}dr < \infty.$$

For $\alpha = -1/N$, $[\quad]^2 \sim c[\log(1-r)]^2$ as $r \to 1$, and $(1-r^2)^{-(N-2)\alpha}r^{-N+1} \sim (1-r)^{-(N-2)\alpha}$, so that again $D(s) < \infty$. For $\alpha \in (-3/(N+2),\ -1/N)$, $[\quad]^2 \sim c(1-r)^{2N\alpha+2}$ and

$$D(s) \approx c\int_0^1 (1-r)^{N\alpha+2\alpha+2}dr < \infty.$$

We have shown that $D(s) < \infty$ for $\alpha \in (-3/(N+2),\ 1/(N-2))$.

For $\alpha \leq -3/(N+2)$,

$$D(s) \approx c\int_0^1 (1-r)^{N\alpha+2\alpha+2}dr = \infty.$$

For $\alpha \geq 1/(N-2)$, a case that does not occur if $N = 2$, we absorb the bounded factor in c and obtain

$$D(s) \approx c \int_0^1 (1 - r)^{-(N-2)\alpha} dr = \infty.$$

Thus $D(s) = \infty$ for $\alpha \notin (-3/(N + 2), 1/(N - 2))$. Every quasiharmonic function on B_α^N has the form $q(r,\theta) = s(r) + h(r,\theta)$ with $h(r,\theta) = \Sigma_n f_n(r)S_n(\theta) \in H$, the summation including $n = (0,\ldots,0)$. By the Dirichlet orthogonality of spherical harmonics to radial functions, we conclude that $D(q) = \infty$ as well. This completes the proof of the characterization of O_{QD}^N.

The characterization of O_{QC}^N is immediate in view of the relation $\tilde{O}_{QC}^N = \tilde{O}_{QB}^N \cap \tilde{O}_{QD}^N$ in 1.6.

6.4. <u>Class $O_{QL^p}^N$ and the characteristic function.</u> The above reasoning provides us with a new proof of not only Theorem 3.9 but also of Theorem 4.2, which characterizes the class $\tilde{O}_{QX}^N$ of Poincaré N-balls in terms of the property $s \in X$ of the characteristic quasiharmonic function, with $X = P, B, D, C$. We shall now show that the class $O_{QL^p}^N$ of Poincaré N-balls which do not carry QL^p functions, $p \geq 1$, also has a simple characterization in terms of s:

<u>THEOREM.</u> <u>The Poincaré N-ball belongs to</u> $\tilde{O}_{QL^p}^N$, $1 \leq p < \infty$, <u>if and only if</u> $s(r) + c \in L^p$ <u>for some constant</u> c.

<u>Proof.</u> Only the necessity needs a proof.

Suppose $s(r) + c \notin L^p$ for any constant c. Every quasiharmonic function has the form $q(r,\theta) = s(r) + c + h(r,\theta)$, where $h(r,\theta)$, with $h(0) = 0$, is a harmonic function with an expansion $\Sigma_n' f_n S_n$. For $1/p + 1/p' = 1$, there exists a function $\varphi(r) \in L^{p'}$ such that $|\int \varphi(r) * (s(r) + c)| = \infty$. Since $\int \varphi * f_n S_n = 0$ for $n \neq (0,\ldots,0)$,

$$\left| \int \varphi(r) * q(r,\theta) \right| = \left| \int \varphi(r) * (s(r) + c) \right| = \infty,$$

and we have the desired conclusion $q(r,\theta) \notin L^p$.

<u>NOTES TO §6.</u> Integral forms of quasiharmonic functions were introduced in

Sario [6]. The one for the Poincaré N-ball in 6.1 is due to Hada (unpublished) and Chung [1]. The latter has also characterized the Poincaré N-balls in $O^N_{QL^p}$ in terms of N and α, and deduced relations between $O^N_{QL^p}$ and $O^N_{QL^t}$.

§7. HARMONIC AND QUASIHARMONIC DEGENERACY OF RIEMANNIAN MANIFOLDS

The harmonic and quasiharmonic classifications of Riemannian manifolds have been largely brought to completion. In summary, we have the following diagrams:

$$\begin{array}{ccccccccc} & & & & O^N_{HL^p} & & & & \\ O^N_G & < & O^N_{HP} & < & O^N_{HB} & < & O^N_{HD} & = & O^N_{HC} \end{array}$$

and

$$\begin{array}{ccccccccc} & & & & & < O^N_{QB} < & & \\ O^N_G & < & O^N_{QP} & < & O^N_{QB} \cap O^N_{QD} & & O^N_{QC} \\ & & & & & < O^N_{QD} < & & \\ & & & & & \vee \quad\quad \vee & & \\ & & & & & O^N_{QL^1} \quad O^N_{QL^p} & & \end{array},$$

where $p > 1$. The absence of a relation symbol means that there is no inclusion.

A natural question arises: Is there any relation between O^N_{HX} and O^N_{QY}? In this section, we shall show that, except for a case in which the problem is open, the answer is in the negative for any two such classes. Explicitly,

$$\tilde{O}^N_{HX} \cap \tilde{O}^N_{QY} \neq \emptyset, \qquad O^N_{HX} \cap O^N_{QY} \neq \emptyset,$$

$$\tilde{O}^N_{HX} \cap O^N_{QY} \neq \emptyset, \qquad O^N_{HX} \cap \tilde{O}^N_{QY} \neq \emptyset,$$

for $X, Y = P, B, D, C, L^p$ with $1 \leq p < \infty$, and $N \geq 2$, the exception being that we do not know whether or not $O^2_{HX} \cap \tilde{O}^2_{QY} \neq \emptyset$ for $X = P, B, D, C$ and $Y = P, B, D, C, L^p$. We shall establish the four relations in the above order of increasing challenge. The fourth relation is one of the most intriguing in all classification theory: there exist Riemannian N-manifolds, of every dimension

$N > 2$, which admit even QC functions but fail to carry any HX functions.

The first two of the above relations will be proved in 7.1, the third relation, for $X, Y = P, B, D, C$ in 7.2, for $X = L^p$ and $Y = P, B, D, C$ in 7.3, for $X = P, B, D, C$ and $Y = L^p$ in 7.4, and for $X = L^p$ and $Y = L^t$ in 7.5.

The proof of the fourth relation, $O_{HX}^N \cap \tilde{O}_{QY}^N \neq \emptyset$, is divided into 7.6 - 7.11. After some preliminaries in 7.6, we discuss the cases $X = L^p$, $Y = P, B, D, C, L^t$, and $N \geq 2$ in 7.7. The restriction to dimension $N > 2$ starts with 7.8, where the Riemannian N-manifold R is constructed which will serve as an example for $X = P, B, D, C$ and $Y = P, B, D, C, L^p$. In 7.9, we discuss the rate of growth of harmonic functions on R. The estimates so obtained are used to prove in 7.10 that $R \in O_{HX}^N$ for $X = P, B, D, C$. The concluding step of the proof, the relation $R \in \tilde{O}_{QY}^N$ for $Y = P, B, D, C, L^p$, is taken in 7.11.

<u>7.1. HX and QY functions, or neither</u>. We start with the simple cases:

<u>THEOREM</u>. <u>For</u> $X, Y = P, B, D, C, L^p$, $1 \leq p < \infty$, <u>and</u> $N \geq 2$,

$$\tilde{O}_{HX}^N \cap \tilde{O}_{QY}^N \neq \emptyset, \qquad O_{HX}^N \cap O_{QY}^N \neq \emptyset.$$

<u>Proof</u>. The first relation is trivial in view of the Euclidean N-ball B_0^N. As to the second relation, we know from I.3.1 that the Euclidean N-cylinder $R = \{|x| < \infty, |y^i| \leq \pi, i = 1,\ldots,N - 1\}$ belongs to O_{HX}^N for $X = P, B, D, C, L^p$, and from II.2.4 that it belongs to O_{QY}^N, $Y = P, B, D, C, L^p$. Here we give an independent proof of $O_{HX}^N \cap O_{QY}^N \neq \emptyset$ for all X, Y by showing that the Euclidean N-space E^N belongs to this class.

By means of the Poisson integral and Harnack's inequality, we see at once that every $h \in HP$ in the Euclidean N-space E^N reduces to a constant (cf. I.1.11). Therefore, $E^N \in O_{HX}^N$ for $X = P, B, D, C$.

To show that $E^N \in O_{HL^p}^N$ we first consider the case $p = 1$. Suppose there exists a nonconstant $h \in HL^1$. It has a unique representation $h = \Sigma_0^\infty r^n S_n$, with the $S_n = S_n(\theta)$ spherical harmonics, and $(r,\theta) = (r,\theta^1,\ldots,\theta^{N-1})$ polar coordinates. For some $n_0 > 0$, $S_{n_0} \not\equiv 0$. Take a function $\rho(r) \in C[0,\infty)$ with $\rho(r) = 1/r$ for

$r \geq 1$ and set $\varphi = \rho S_{n_0}$. Since $\varphi \in B$, we have $\|h\varphi\|_1 < \infty$. On the other hand,

$$\|h\varphi\|_1 \geq |(h\rho, S_{n_0})| = c\int_0^\infty r^{n_0}\rho r^{N-1}\,dr$$

$$= a + c\int_1^\infty r^{n_0+N-2}\,dr = \infty.$$

The contradiction shows that $E^N \in O^N_{HL^1}$.

Now let $p > 1$ and take p' with $1/p + 1/p' = 1$. Suppose there exists a nonconstant $h \in HL^p$. In the expansion $h = \Sigma_0^\infty r^n S_n$, $S_{n_0} \not\equiv 0$ for some $n_0 > 0$. Let $\rho(r) \in C[0,\infty)$ with

$$\rho(r) = r^{-(N+1)/p'} \quad \text{for} \quad r \geq 1,$$

and set $\varphi = \rho S_{n_0}$. Since

$$\|\rho\|_{p'}^{p'} = a + c\int_1^\infty r^{-(N+1)}r^{N-1}\,dr < \infty,$$

$\|\varphi\|_{p'} < \infty$ and $|(h,\varphi)| < \infty$. We again have a contradiction:

$$|(h,\varphi)| = \left|a + c\int_1^\infty r^{n_0}r^{-(N+1)/p'}r^{N-1}\,dr\right|$$

$$= \left|a + c\int_1^\infty r^{n_0+N/p-1/p'-1}\,dr\right| = \infty.$$

A fortiori, $E^N \in O^N_{HL^p}$ for $p > 1$ as well.

To prove $E^N \in O^N_{QX}$ for $X = P, B, D, C$, it suffices to establish $E^N \in O^N_{QP}$. Since

$$q_0 = -(2N)^{-1}r^2 \in Q,$$

every $q \in Q$ can be written $q = q_0 + h$ with some $h \in H$. We are to show that $q \notin P$. Set

$$q = q_0 + h(0) + k, \qquad k \in H, \qquad k(0) = 0,$$

where $h(0)$, $k(0)$ are the values at the origin. By the mean value theorem there exists, for every r_n, a $\theta_n = (\theta_n^1,\ldots,\theta_n^{N-1})$ such that $k(r_n,\theta_n) = 0$. If $\{r_n\}_0^\infty$ is an increasing sequence with $r_n \to \infty$, then

$$q(r_n,\theta_n) = -\frac{1}{2N} r_n^2 + h(0) \to -\infty,$$

and therefore $q \notin P$.

It remains to show that $E^N \in O^N_{QL^p}$. Again we start with $p = 1$. For $q \in Q$, $q = q_0 + h(0) + k$, we have by $k(0) = 0$, $\int_{E^N} *k = 0$, and therefore,

$$\|q\|_1 \geq \left| \int_{E^N} *(q_0 + h(0)) \right|.$$

The integrand (with respect to $dr d\theta^1 \ldots d\theta^{N-1}$) is $\sim cr^2 r^{N-1}$, hence $\|q\|_1 = \infty$, and $E^N \in O^N_{QL^1}$.

In the case $p > 1$, the choice $\varphi(r) \in C[0,\infty)$, $\varphi(r) = r^{-(N+1)/p'}$ for $r \geq 1$ gives $\|\varphi\|_{p'} < \infty$. If there exists a $q \in QL^p$, $q = q_0 + h(0) + k$, then by $(k,\varphi) = 0$, the integrand in $(q,\varphi) = (q_0 + h(0),\varphi)$ is asymptotically

$$cr^2 r^{-(N+1)/p'} r^{N-1} = cr^{N/p-1/p'+1}.$$

The exponent dominates N/p, hence $|(q,\varphi)| = \infty$, in violation of $\|\varphi\|_{p'} < \infty$.

The proof of the Theorem is herewith complete.

<u>7.2.</u> HX <u>functions but no</u> QY. We now take up the third relation stated in the middle of the introduction to §7:

<u>THEOREM</u>. <u>For</u> $X, Y = P, B, D, C, L^p$, $1 \leq p < \infty$, <u>and</u> $N \geq 2$,

$$\tilde{O}^N_{HX} \cap O^N_{QY} \neq \emptyset.$$

The proof will be given in 7.2 - 7.5. In the present 7.2, we consider the cases $X, Y = P, B, D, C$. It suffices to show that

$$\tilde{O}^N_{HD} \cap O^N_{QP} \neq \emptyset.$$

Consider the N-cylinder

$$R = \{|x| < \infty, \ |y^i| \leq 1, \ i = 1,\ldots,N-1\}$$

with a metric in which each g^{ij} depends on x only, and set

$$\rho(x) = g(x)^{1/2}, \qquad \sigma(x) = g^{11}(x).$$

Then $h_0(x)$ is harmonic if and only if

$$\Delta h_0 = -\rho^{-1}(\rho\sigma h_0')' = 0,$$

that is,

$$h_0(x) = c\int_a^x \rho^{-1}\sigma^{-1}\,dt.$$

The Dirichlet integral over the subspace from $-x$ to x is

$$D_x(h_0) = c\int_{-x}^{x} h_0'^2\sigma\rho dt = c\int_{-x}^{x}\rho^{-1}\sigma^{-1}\,dt.$$

We know from Theorem 1.1 that a manifold R belongs to $\tilde{O}^N_{QP}$ if and only if the potential $G1(\xi) = \int_R *g(\xi,\eta)$ of the constant function 1 exists at some, and hence every, $\xi \in R$. In the present case this potential depends on the x-coordinate $x(\xi)$ of ξ only, and we can use the notation $G1(x)$. Suppose $G1(x)$ exists. Since $\Delta G1(x) \equiv 1$,

$$G1(x) = q_0(x) + h_0(x),$$

where

$$q_0(x) = -\int_0^x \rho^{-1}\,\sigma^{-1}\int_0^t \rho ds dt$$

is quasiharmonic by $\Delta q_0 = -\rho^{-1}(\rho\sigma q_0')' = 1$. We seek functions $\rho(x)$ which satisfy

$$\int_{-x}^{x}\rho^{-1}\sigma^{-1}dt \in B$$

and

$$\int_0^x \rho^{-1}\sigma^{-1}\int_0^t \rho ds dt \to \infty \quad \text{as} \quad x \to \infty.$$

The choice $\rho = 1$, $\sigma = 1 + x^2$ gives

$$\int_0^x \rho^{-1}\sigma^{-1}\,dt = \int_0^x (1 + t^2)^{-1}\,dt = \arctan x \in HD,$$

hence $R \in \tilde{O}^N_{HD}$. An arbitrary $h_0(x) \in H$ is

$$h_0(x) = a \arctan x + b \in B,$$

and

$$q_0(x) = -\int_0^x (1 + t^2)^{-1} t\,dt = -\frac{1}{2}\log(1 + x^2),$$

which tends to $-\infty$ as $|x| \to \infty$. Therefore $G1 = q_0 + h_0 \to -\infty$, $G1 \notin P$, and consequently $R \in O^N_{QP}$.

<u>7.3.</u> HL^p <u>functions but no</u> QY. We continue the proof of Theorem 7.2 by showing next:

<u>For</u> $1 \le p < \infty$, $Y = P, B, D, C$, <u>and</u> $N \ge 2$,

$$\tilde{O}^N_{HL^p} \cap O^N_{QY} \neq \emptyset.$$

Let R be the N-space, with coordinates $(r,\theta) = (r,\theta^1,\ldots,\theta^{N-1})$, endowed with the metric

$$ds^2 = \varphi(r)dr^2 + \psi(r)^{1/(N-1)} \sum_{i=1}^{N-1} \gamma_i(\theta)d\theta^{i2},$$

where φ, ψ are positive functions in $C^\infty[0,\infty)$ with

$$\varphi(r) = \begin{cases} 1 & \text{for } r < \frac{1}{2}, \\ e^{-r^2} & \text{for } r > 1, \end{cases} \qquad \psi(r) = \begin{cases} r^2 & \text{for } r < \frac{1}{2}, \\ e^{-r^2} & \text{for } r > 1, \end{cases}$$

and the γ_i are trigonometric functions of θ such that the metric is Euclidean on $\{r < \frac{1}{2}\}$. To show that $R \in \tilde{O}^N_{HL^p}$, let $h = f(r)S_n(\theta)$, $n > 0$, be a harmonic function on R, with S_n a spherical harmonic. Set $\Gamma(\theta) = \prod_{i=1}^{N-1} \gamma_i(\theta)$. We have

$$0 = \Delta h = \Delta f \cdot S_n + f\Delta S_n$$

$$= -(\varphi\psi\Gamma)^{-1/2}\frac{\partial}{\partial r}[(\varphi\psi\Gamma)^{1/2}\varphi^{-1}f']S_n$$

$$+ f\left\{-(\varphi\psi\Gamma)^{-1/2}\sum_{i=1}^{N-1}\frac{\partial}{\partial\theta^i}\left[(\varphi\psi\Gamma)^{1/2}\psi^{-1/(N-1)}\gamma_i^{-1}\frac{\partial S_n}{\partial\theta^i}\right]\right\}$$

$$= -(\varphi\psi)^{-1/2}(\varphi^{-1/2}\psi^{1/2}f')'S_n$$

$$+ f\left\{-r^{1-N}\Gamma^{-1/2}\sum_{i=1}^{N-1}\frac{\partial}{\partial\theta^i}\left[r^{N-1}\Gamma^{1/2}r^{-2}\gamma_i^{-1}\frac{\partial S_n}{\partial\theta^i}\right]\right\}\psi^{-1/(N-1)}r^2.$$

Here $\{\ \}$ is the Laplacian of S_n in the Euclidean metric, $\Delta_e S_n = n(n + N - 2)r^{-2}S_n = \eta^2 r^{-2}S_n$ (cf. I.2.5), and we obtain

$$(\varphi^{-1/2}\psi^{1/2}f')' - \eta^2(\varphi\psi)^{1/2}\psi^{-1/(N-1)}f = 0.$$

The restriction to $\{r > 1\}$ of any solution of this differential equation is some solution of

$$f'' - \eta^2 e^{-(N-2)r^2/(N-1)}f = 0$$

on $\{r > 1\}$. Since for $N > 2$,

$$\int^\infty re^{-(N-2)r^2/(N-1)}\,dr < \infty,$$

a theorem by Haupt [1] and Hille [1] (see also Cesari [1]) gives for every solution f,

$$f(r) \sim ar + b$$

with some a, b. A fortiori,

$$\|h\|_p^p = c\int_0^\infty |f|^p(\varphi\psi)^{1/2}\,dr = c_1 + c\int_1^\infty |f|^p e^{-r^2}\,dr < \infty$$

for $N > 2$, with c a constant, not always the same. In the case $N = 2$ the differential equation reduces to $f'' - \eta^2 f = 0$, and the solution is $f = ae^{\eta r} + be^{-\eta r}$. Therefore, $\|h\|_p^p < \infty$, and we have $R \in \tilde{O}^N_{HL^p}$ for every N.

We proceed to show that $R \in O^N_{QY}$. The function

$$q_0(r) = -\int_0^r \varphi(t)^{1/2}\psi(t)^{-1/2}\int_0^t \varphi(s)^{1/2}\psi(s)^{1/2}ds\,dt + c$$

satisfies the quasiharmonic equation

$$\Delta q_0 = -(\varphi\psi)^{-1/2}(\varphi^{-1/2}\psi^{1/2}q_0')' = 1.$$

Since $\varphi^{1/2}\psi^{-1/2} = 1$ for $r > 1$, $-q_0$ increases at the rate of r as $r \to \infty$.

An arbitrary $q(r,\theta) \in Q$ can be written

$$q(r,\theta) = q_0(r) + h(r,\theta),$$

where $h \in H$ with $h = 0$ at $r = 0$, the additive constant in q_0 suitably chosen. Pick a sequence $\{(r_n,\theta_n)\}_1^\infty$ such that $r_n \to \infty$ and $h(r_n,\theta_n) = 0$. Then $q(r_n,\theta_n) = q_0(r_n) \to -\infty$ and $q \notin QP$, that is, $R \in O^N_{QP} \subset O^N_{QY}$ for $Y = P, B, D, C$.

<u>7.4.</u> HX <u>functions but no</u> QL^p. In the proof of Theorem 7.2, we now come to the following step:

<u>For</u> $X = P, B, D, C$, $1 \le p < \infty$, <u>and</u> $N \ge 2$,

$$\tilde{O}^N_{HX} \cap O^N_{QL^p} \ne \emptyset.$$

Take the N-cylinder

$$R = \{x > 1, \quad |y^i| \le \pi, \quad i = 1,\ldots,N-1\}$$

with the metric

$$ds^2 = dx^2 + x^{2\alpha/(N-1)}\sum_{i=1}^{N-1} dy^{i2},$$

with α a constant > 1.

To show that $R \in \tilde{O}^N_{HX}$, it suffices to consider $X = C$. The function $x^{-\alpha+1}$ is in HB, by virtue of

$$\Delta x^{-\alpha+1} = -x^{-\alpha}(x^{\alpha}(x^{-\alpha+1})')' = 0.$$

Moreover,

$$D(x^{-\alpha+1}) = c\int_1^{\infty}(x^{-\alpha})^2 x^{\alpha}dx < \infty.$$

Therefore, $R \in \tilde{O}^N_{HC}$.

To prove $R \in O^N_{QL^p}$, we first take $p = 1$. Since

$$q_0(x) = -\tfrac{1}{2}(\alpha + 1)^{-1}x^2 \in Q,$$

every $q \in Q$ can be written

$$q = q_0 + ax^{-\alpha+1} + b + \Sigma' f_n(x)G_n(y),$$

where $f_nG_n \in H$ and G_n ranges over all products of sines and cosines of multiples of the y^i as in 2.4, with $n \neq (0,\ldots,0)$. In view of $\int_R f_n{*}G_n = 0$,

$$\|q\|_1 \geq \left|\int_R {*}(q_0 + ax^{-\alpha+1} + b)\right|.$$

The integrand on the right with respect to $dxdy^1 \ldots dy^{N-1}$ is $\sim cx^{2+\alpha}$, hence $\|q\|_1 = \infty$ and $R \in O^N_{QL^1}$.

Now let $p > 1$, and $1/p + 1/p' = 1$. The function $\varphi(x) = x^{-(\alpha+p')/p'}$ belongs to $L^{p'}$ by virtue of

$$\int_R {*}|\varphi|^{p'} = c\int_1^{\infty} x^{-(\alpha+p')}x^{\alpha}dx < \infty.$$

If there exists a $q \in L^p$, then $|(q,\varphi)| < \infty$. On the other hand,

$$|(q,\varphi)| = \left|\int_R (q_0 + ax^{-\alpha+1} + b){*}x^{-(\alpha+p')/p'}\right| ,$$

where the integrand is asymptotically $x^{\alpha/p+1}$. A fortiori, $|(q,\varphi)| = \infty$, and the contradiction gives $R \in O^N_{QL^p}$.

<u>7.5.</u> HL^p <u>functions but no</u> QL^t. We come to the concluding step in the proof

of Theorem 7.2:

<u>For</u> $1 \leq p < \infty$, $1 \leq t < \infty$, <u>and</u> $N \geq 2$,

$$\tilde{O}^N_{HL^p} \cap O^N_{QL^t} \neq \emptyset.$$

First fix t and consider the manifold

$$R = \{x > 1, \quad |y^i| \leq \pi, \quad i = 1,\ldots,N-1\}$$

with the metric

$$ds^2 = e^{-x/t}dx^2 + e^{(2x+x/t)/(N-1)} \sum_{i=1}^{N-1} dy^{i2}.$$

The function $h_0(x) = e^{-x-x/t}$ belongs to HL^p, and $R \in \tilde{O}^N_{HL^p}$.

To see that $R \in O^N_{QL^t}$, note that $q_0(x) = te^{-x/t} \in Q$. Every $q \in Q$ has the form

$$q = q_0 + ah_0 + b + \Sigma' f_n G_n.$$

For $t = 1$,

$$\|q\|_1 \geq c \left| \int_1^\infty (q_0 + ah_0 + b)e^x \, dx \right|.$$

If $b \neq 0$, then the integrand is $\sim be^x$, and $\|q\|_1 = \infty$. If $b = 0$, the dominating term in the integrand is $q_0 e^x$, so that again $\|q\|_1 = \infty$, and $R \in O^N_{QL^1}$.

In the case $t > 1$, take t' such that $1/t + 1/t' = 1$. The function $\varphi(x) = e^{-x/t'}x^{-1}$ belongs to $L^{t'}$. If there exists a $q \in QL^t$, then $|(q,\varphi)| < \infty$. But

$$|(q,\varphi)| = |(q_0 + ah_0 + b + \Sigma' f_n G_n, \varphi)| = |(q_0 + ah_0 + b, \varphi)|.$$

If $b \neq 0$, the integrand with respect to $dxdy^1 \ldots dy^{N-1}$ is $\sim be^{-x/t'}x^{-1}e^x = be^{x/t}x^{-1}$, and we have the contradiction $|(q,\varphi)| = \infty$. If $b = 0$, the integrand is asymptotically

$$ce^{-x/t-x/t'}x^{-1}e^x = cx^{-1},$$

and again we have divergence. Therefore, $R \in O^N_{QL^t}$.

The proof of Theorem 7.2 is herewith complete.

7.6. QY functions but no HX. We turn to our final, and most challenging, step in proving that, possibly with some exceptions in the case $N = 2$, there are no inclusion relations between any O^N_{HX} and any O^N_{QY}: the fourth relation in the middle of the introduction to §7.

THEOREM. For $N > 2$,

$$O^N_{HX} \cap \tilde{O}^N_{QY} \neq \emptyset,$$

with $X, Y = P, B, D, C, L^p$, $1 \leq p < \infty$. For $N = 2$, the same is true for $X = L^p$, $Y = P, B, D, C, L^t$, $1 \leq p < \infty$, $1 \leq t < \infty$.

The proof will be presented in 7.6 - 7.11, in the order given in the introduction to §7.

In view of the inclusion relations between the harmonic null classes and those between the quasiharmonic null classes, summarized in the introduction to §7, the problem of establishing the nonvoidness of the class $O^N_{HX} \cap \tilde{O}^N_{QY}$ consists in finding a manifold which belongs to the "narrow" gap between O^N_G and O^N_{HP}, yet carries QY functions.

For $X = P, B, D, C$, $Y = P, B, D, C, L^p$, and $1 \leq p < \infty$, the question remains open if $N = 2$.

Before starting a systematic proof of the Theorem, we observe in passing that the Poincaré N-balls provide us with the following partial solutions:

$$B^N_\alpha \in O^N_{HD} \cap \tilde{O}^N_{QC}, \quad N \geq 5,$$

for $-3/(N + 2) < \alpha \leq -1/(N - 2)$, and

$$B^N_\alpha \in O^N_{HD} \cap \tilde{O}^N_{QB}, \quad N \geq 4,$$

for $-1 < \alpha \leq -1/(N - 2)$. This follows from Theorems I.2.10 and II.3.9.

7.7. QY functions but no HL^p. We start the proof of Theorem 7.6 with the case $X = L^p$:

For $1 \leq p < \infty$, $Y = P, B, D, C, L^t$, $1 \leq t < \infty$, and $N \geq 2$,

$$O^N_{HL^p} \cap \tilde{O}^N_{QY} \neq \emptyset.$$

Since the O^N_{QY} for $Y = P, B, D, L^t$ are contained in O^N_{QC}, it suffices to find a manifold in $O^N_{HL^p} \cap \tilde{O}^N_{QC}$.

Let R be the N-space with the metric

$$ds^2 = \varphi(r)dr^2 + \psi(r)^{1/(N-1)} \sum_{i=1}^{N-1} \gamma_i d\theta^{i2},$$

$\varphi, \psi \in C^\infty[0,\infty)$, where

$$\varphi(r) = \begin{cases} 1 & \text{for } r < \frac{1}{2}, \\ e^{-r} & \text{for } r > 1, \end{cases} \qquad \psi(r) = \begin{cases} r^2 & \text{for } r < \frac{1}{2}, \\ e^r & \text{for } r > 1, \end{cases}$$

and the γ_i are trigonometric functions of θ such that the metric is Euclidean on $\{r < \frac{1}{2}\}$. To see that $R \in O^N_{HL^p}$, let $h(r,\theta)$ be an arbitrary harmonic function on R. It has an expansion in spherical harmonics, $h(r,\theta) = \Sigma_0^\infty f_n(r)S_n(\theta)$. Let $n_0 \geq 0$ be such that $f_{n_0} \not\equiv 0$. By the maximum principle applied to the harmonic function $f_{n_0}S_{n_0}$, $|f_{n_0}| \geq c_0 > 0$ on $(1,\infty)$.

For $p = 1$, let $g(r) \in C^\infty[0,\infty)$, $0 < g < 1$, with $g(r) = 1/(2r)$ for $r > 1$. Then

$$\|h\|_1 \geq c\left| \int_R {*hgS_{n_0}} \right|$$

$$\geq c_1 + c_2\left| \int_1^\infty f_{n_0} g dr \right| \geq c_1 + c_0 c_2 \left| \int_1^\infty g dr \right| = \infty,$$

that is, $h \notin L^1$.

For $p > 1$, take p' with $1/p + 1/p' = 1$. The function $gS_{n_0} \in L^{p'}$ gives a linear functional $(\cdot, gS_{n_0})$ on L^p. But

$$|(h, gS_{n_0})| = \left| c\int_0^\infty f_{n_0} g dr \right| = \infty,$$

and therefore, $h \notin L^p$. A fortiori, $R \in O^N_{HL^p}$ for all p.

It remains to show that $R \in \tilde{O}^N_{QC}$. Clearly, the function

$$q_0(r) = -\int_0^r \varphi(t)^{1/2}\psi(t)^{-1/2}\int_0^t \varphi(s)^{1/2}\psi(s)^{1/2}dsdt$$

is in Q. It grows at the rate of $-\int_0^r e^{-t}(c + t)dt$ and is therefore bounded. Moreover,

$$D(q_0) = c + c_1\int_1^\infty q_0'^2\, e^r dr$$

$$= c + c_1\int_1^\infty e^{-2r}\left(c_2 + \int_1^r ds\right)^2 e^r dr < \infty,$$

hence $q_0 \in QC$, and $R \in \tilde{O}^N_{QC}$.

7.8. The manifold. We come to the most essential step in the proof of Theorem 7.6:

For $X = P, B, D, C$, $Y = P, B, D, C, L^p$, $1 \leq p < \infty$, *and* $N > 2$,

$$O^N_{HX} \cap \tilde{O}^N_{QY} \neq \emptyset.$$

Take the plane $R^2 = \{0 \leq r < \infty,\ 0 \leq \theta^1 \leq 2\pi\}$ and the $(N - 2)$-torus $T^{N-2} = \{0 \leq \theta^i \leq 2\pi,\ i = 2,\dots,N - 1\}$, and consider the N-manifold

$$R = R^2 \times T^{N-2} = \{0 \leq r < \infty,\ 0 \leq \theta^i \leq 2\pi,\ i = 1,\dots,N - 1\},$$

$N > 2$, with the metric

$$ds^2 = \varphi(r)^2dr^2 + \sum_{i=1}^{N-1} \psi_i(r)^2d\theta^{i2},$$

where $\varphi, \psi_1, \dots, \psi_{N-1}$ are positive C^∞ functions satisfying

$$\varphi(r) = \begin{cases} 1 & \text{for } r < \tfrac{1}{2}, \\ r^{-2} & \text{for } r > 1; \end{cases}$$

$$\psi_1(r) = r \quad \text{for } r < \tfrac{1}{2},$$

$$\psi_i(r) = 1 \quad \text{for } r < \tfrac{1}{2}, \quad i \neq 1.$$

Note that on the subspace R^2, the metric for $r < \frac{1}{2}$ is the Euclidean $ds^2 = dr^2 + r^2(d\theta^1)^2$.

The following further requirements on the ψ_i, $i = 1,\ldots,N-1$, are in terms of an auxiliary function ψ on $\{1 < r < \infty\}$ and a certain partition $\{I_{ij} | i \neq j;\ i,j = 1,\ldots,N-1\}$ of $\{1 < r < \infty\}$, both to be presently defined:

$$\psi_i(r) = \begin{cases} \psi(r) & \text{for } r \in I_{ij}, \\ 1/\psi(r) & \text{for } r \in I_{ji}, \\ 1 & \text{for } r \notin I_{ij} \cup I_{ji}. \end{cases}$$

For the definition of I_{ij}, divide the interval $I^n = (n, n+1]$, $n \geq 1$, into $(N-1)(N-2)$ equal semiopen subintervals, open on the left, closed on the right. Since we can choose a one-to-one correspondence between the numbers $1,2,\ldots,(N-1)(N-2)$ and the ordered pairs $\{(i,j) | i \neq j;\ i,j = 1,\ldots,N-1\}$, we can index these subintervals in the form I^n_{ij}. Let $I_{ij} = \bigcup_{n=1}^{\infty} I^n_{ij}$.

We define $\psi(r) \in C^\infty(1,\infty)$ on each interval I^n_{ij} by dividing I^n_{ij} into five equal semiopen subintervals I_1, I_2, I_3, I_4, and I_5, in this order, each open on the left, closed on the right, and setting

$$\psi(r) = \begin{cases} 1 & \text{for } r \in I_1, \\ r^2 & \text{for } r \in I_3, \\ 1 & \text{for } r \in I_5, \\ \geq 1 & \text{for } r \in I_2 \cup I_4. \end{cases}$$

Every $r > 1$ is in exactly one I^n_{ij}. Thus $\prod_{i=1}^{N-1} \psi_i(r) = \psi(r)/\psi(r) = 1$, and, in the volume element, $g^{1/2} = \varphi \prod \psi_i = r^{-2}$ for $r > 1$.

Our Riemannian manifold R is thus well defined.

7.9. Rate of growth of harmonic functions. We shall show that R excludes nonconstant HX functions while it carries QY functions.

As in 2.4, every harmonic function $h(r,\theta) = h(r,\theta^1,\ldots,\theta^{N-1})$ can be expanded into a series $\Sigma_n f_n(r)G_n(\theta)$, convergent absolutely and uniformly on compact sets, with

$$G_n(\theta) = \pm \prod_{i=1}^{N-1} \begin{matrix}\sin\\ \cos\end{matrix}\, n_i\theta^i,$$

where we now choose the sign of G_n such that $f_n(r) \geq 0$. We shall see that f_n is of constant sign for $r > 0$, so that this convention is legitimate. In the sequel we write simply fG for $f_n G_n$.

LEMMA. If $h = fG$ is a nonconstant harmonic function, then $f(0) = 0$, $|f(r)|$ is strictly increasing, and, for some constant $c > 0$ and all sufficiently large r,

$$|f(r)| > cr.$$

Proof. By the maximum principle for harmonic functions, G is not constant. The fact that $f(0) = 0$ follows from $h(0) = c \int_{r=r_0} fGd\theta = 0$ for a nonconstant G. If $|f|$ were not strictly increasing, then fG would violate the maximum principle. With the above convention on the sign of G_n, we thus have $f(r) > 0$ for $r > 0$. We shall estimate the rate of growth of f as $r > 1$ increases.

Since $\operatorname{grad} f \cdot \operatorname{grad} G = 0$, we have

$$\Delta(fG) = (\Delta f)G + f\Delta G = -r^2(r^2 f')'G + \left(\sum_{i=1}^{N-1} n_i^2 \psi_i^{-2}\right) fG = 0.$$

A fortiori,

$$(r^2 f')' = r^{-2}\left(\sum_{i=1}^{N-1} n_i^2 \psi_i^{-2}\right) f.$$

The right-hand side being positive,

$$(r^2f')' > cr^{-2}\sum_{i=1}^{N-1} n_i^2\psi_i^{-2},$$

with $c > 0$. As G is not constant, there is an i_0 such that $n_{i_0} \neq 0$, and we have

$$(r^2f')' > cr^{-2}\psi_{i_0}^{-2},$$

hence

$$r^2f' - f'(1) > c\int_1^r r^{-2}\psi_{i_0}^{-2}\,dr.$$

Since f is strictly increasing, $f'(1) \geq 0$, and

$$f'(r) > cr^{-2}\int_1^r r^{-2}\psi_{i_0}^{-2}\,dr.$$

Recall that $\psi_{i_0}^{-2}(r) = r^4$ for $r \in I^n_{ji_03}$ with $j = 1,2,\ldots,N-1$, $j \neq i_0$, where the index 3 indicates the middle subinterval I_3 of $I^n_{ji_0}$. It follows that

$$\int_1^r r^{-2}\psi_{i_0}^{-2}\,dr > \sum_{n=1}^{[r-1]}\ \sum_{j=1;j\neq i_0}^{N-1}\int_{I^n_{ji_03}} t^2dt,$$

where $[r - 1]$ is the largest integer $\leq r - 1$. We obtain, for some $0 \leq d < 1$,

$$\int_{I^{[r-1]}_{ji_03}} t^2dt \geq \frac{1}{3}\,t^3\Big|_{[r-1]+d}^{[r-1]+d+(5(N-1)(N-2))^{-1}}$$

$$= \frac{(5(N-1)(N-2)([r-1]+d)+1)^3 - (5(N-1)(N-2)([r-1]+d))^3}{3(5(N-1)(N-2))^3}$$

$$> \frac{(5(N-1)(N-2)([r-1]+d))^2}{(5(N-1)(N-2))^3} > a(r-2)^2.$$

For some $b > 0$ and $r \geq 3$, say, this is $> br^2$, so that $f'(r) > cr^{-2}r^2 = c$. Therefore,

$$f(r) \geq cr + f(1) > cr.$$

The proof of the Lemma is herewith complete.

7.10. Exclusion of HX functions. We claim that our manifold R has the property

$$R \in O^N_{HX}, \quad X = P, B, D, C.$$

It suffices to show that $HP = \{c \in P\}$. Suppose there exists a nonconstant $h \in HP$ on R. In the expansion $h(r,\theta) = c + \Sigma_{n\neq 0} f_n G_n$, let the term $f_1 G_1$, say, be nonconstant. We divide Σ into its positive and negative parts, $\Sigma = \Sigma^+ - \Sigma^-$. Since Σ is bounded from below, Σ^- is bounded, and so is $\int_\theta G_1 \Sigma^- d\theta$, a function of r. As a consequence, $r^{-1/2} \int_\theta G_1 \Sigma^- d\theta \to 0$ as $r \to \infty$.

Since $\int_\theta \Sigma f_n G_n d\theta = \Sigma \int_\theta f_n G_n d\theta = 0$, we have $\int_\theta \Sigma^+ d\theta = \int_\theta \Sigma^- d\theta$. Hence the function

$$\left| \int_\theta G_1 \Sigma^+ d\theta \right| \leq \int_\theta |G_1 \Sigma^+| d\theta \leq c \int_\theta \Sigma^+ d\theta$$

$$= c \int_\theta \Sigma^- d\theta$$

is bounded in r, and $r^{-1/2} \int_\theta G_1 \Sigma^+ d\theta \to 0$ as $r \to \infty$.

We have a contradiction: for all sufficiently large r,

$$\left| r^{-1/2} \left(\int_\theta G_1 \Sigma^+ d\theta - \int_\theta G_1 \Sigma^- d\theta \right) \right| = \left| r^{-1/2} \int_\theta G_1 \Sigma \, d\theta \right|$$

$$= |c r^{-1/2} f_1(r)| > |c r^{1/2}| \to \infty.$$

7.11. Construction of QY functions. It remains to show that

$$R \in \tilde{O}^N_{QX}, \quad X = P, B, D, C, L^p.$$

It suffices to find a $q \in QC$. The function

$$q(r) = -\int_0^r g(t)^{-1/2}\varphi(t)^2 \int_0^t g(s)^{1/2} ds dt$$

has this property. In fact, it satisfies the quasiharmonic equation

$$\Delta q = -g^{-1/2}(g^{1/2}\varphi^{-2}q')' = 1.$$

For $s > 1$, $g(s)^{1/2} = s^{-2}$, so that $\int_0^t g(s)^{1/2}ds$ is bounded in t. For $t > 1$, $g(t)^{-1/2}\varphi(t)^2 = t^{-2}$, and therefore, $q \in QB$. Moreover,

$$D(q) = \int_R q'^2 g^{rr} g^{1/2} dr d\theta$$

$$= \int_{\{r\leq 1\}} + \int_{\{r>1\}} \leq c + c_1 \int_{\{r>1\}} r^{-4} r^4 r^{-2} dr < \infty.$$

Relations 7.8 have thus been demonstrated, the proof of Theorem 7.6 is complete, and all four relations stated in the middle of the introduction to §7 have been established.

NOTES TO §7. Theorem 7.1, relations 7.2 for $X, Y \neq L^p$, and relations 7.4 and 7.5 were proven in Chung-Sario [1], relations 7.3 and 7.7 in Chung-Sario [2], and relations 7.8 - 7.11 in Chung [2].

The problem that remains open on the relations between harmonic and quasiharmonic null classes is to determine whether $O^2_{HX} \cap \tilde{O}^2_{QY} \neq \emptyset$ for $X = P, B, D, C$, $Y = P, B, D, C, L^p$, $1 \leq p < \infty$. It would suffice to find a 2-manifold in $O^2_{HP} \cap \tilde{O}^2_{QC}$.

A more general problem is: Can a proof be given to show $O^N_{HX} \cap \tilde{O}^N_{QY} \neq \emptyset$ simultaneously for all X, Y and all $N \geq 2$?

CHAPTER III

BOUNDED BIHARMONIC FUNCTIONS

Harmonic and quasiharmonic functions, discussed in Chapters I and II, are special cases of biharmonic functions. The rest of this book is devoted to general biharmonic functions with various boundedness properties. We start with bounded biharmonic functions.

The basic class in all classification theory is the class O_G^N of parabolic N-manifolds. The first question with any new class is: How is it related to O_G^N? Here we have a striking contrast with the case of harmonic and quasiharmonic functions: there are no inclusion relations between O_G^N and the class $O^N_{H^2B}$ of Riemannian N-manifolds which do not carry bounded nonharmonic biharmonic functions. In particular, there do exist manifolds, of any dimension, which are parabolic but nevertheless carry H^2B functions. These topics are the content of §1.

Another interesting contrast with harmonic functions is that an isolated point is not a removable singularity for bounded biharmonic functions. Typically, $\cos 2\theta$ and $\sin 2\theta$ are in H^2B on the punctured plane. An interesting problem, taken up in §2, is to find the generators of H^2B on the punctured N-space E_0^N. Here we encounter a fascinating phenomenon: there are no H^2B functions on E_0^N if $N > 3$.

In §3 we ask: Can H^2B functions be brought in even for $N > 3$ by replacing the Euclidean metric of the punctured N-space by $ds = r^\alpha |dx|$, α a constant? Interestingly enough, although there are infinitely many α for which $H^2B \neq \emptyset$ if $N = 2$ and $N = 3$, and in fact $\dim H^2B \to \infty$ as $|\alpha| \to \infty$, there nevertheless are no H^2B functions for any α if $N > 3$.

The Poincaré N-balls in $O^N_{H^2B}$ are characterized in §4, and the question of completeness as related to $O^N_{H^2B}$ is discussed in §5.

An illuminating generalization of the results in §§2-3 to polyharmonic functions is carried out in §6. This is the only context in the present book in which we discuss polyharmonic functions.

§1. PARABOLICITY AND BOUNDED BIHARMONIC FUNCTIONS

We shall show that the existence of bounded nonharmonic biharmonic functions is in no way related to the existence of harmonic Green's functions.

1.1. Parabolic with H^2B functions. A function u is, by definition, biharmonic, if $\Delta^2 u = \Delta(\Delta u) = 0$. We denote the class of nonharmonic biharmonic functions by

$$H^2 = \{u \mid \Delta u \in H - \{0\}\},$$

and the class of Riemannian N-manifolds that do not carry bounded H^2 functions by $O^N_{H^2B}$.

THEOREM. For $N \geq 2$,

$$O^N_G \cap \tilde{O}^N_{H^2B} \neq \emptyset,\ \tilde{O}^N_G \cap O^N_{H^2B} \neq \emptyset,\ O^N_G \cap O^N_{H^2B} \neq \emptyset,\ \tilde{O}^N_G \cap \tilde{O}^N_{H^2B} \neq \emptyset.$$

The proof will be given in 1.1-1.6. We start with

$$O^N_G \cap \tilde{O}^N_{H^2B} \neq \emptyset.$$

Let E^2_0 be the punctured plane $\{0 < r < \infty,\ 0 \leq \theta^1 \leq 2\pi\}$, and T^{N-2} the (N-2)-torus $\{0 \leq \theta^i \leq 2\pi,\ i = 2,\dots,N-1\}$. Consider the product space

$$R = E^2_0 \times T^{N-2} = \{0 < r < \infty,\ 0 \leq \theta^i \leq 2\pi,\ i = 1,\dots,N-1\}$$

with the Euclidean metric

$$ds^2 = dr^2 + r^2(d\theta^1)^2 + \sum_{i=2}^{N-1} d\theta^{i2}.$$

The harmonic measure $\omega(r)$ satisfies

$$\Delta\omega(r) = -\frac{1}{r}(r\,\omega'(r))' = 0,$$

hence $\omega(r) = a \log r + b$. This is bounded only if $\omega = b$, and therefore $R \in O^N_G$.

The function $u = \cos 2\theta^1$ satisfies

$$\Delta u = -r^{-1}\frac{\partial}{\partial\theta^1}(rr^{-2}\frac{\partial}{\partial\theta^1}(\cos 2\theta^1)) = 4r^{-2}\cos 2\theta^1$$

and

$$\Delta^2 u = -4r^{-1}\{\frac{\partial}{\partial r}[r\frac{\partial}{\partial r}(r^{-2}\cos 2\theta^1)] + \frac{\partial}{\partial\theta^1}[r^{-1}\frac{\partial}{\partial\theta^1}(r^{-2}\cos 2\theta^1)]\}$$
$$= -4r^{-1}(4r^{-3}\cos 2\theta^1 - 4r^{-3}\cos 2\theta^1) = 0,$$

so that $u \in H^2B$. Thus $R \in O_G^N \cap \tilde{O}^N_{H^2B}$.

1.2. Hyperbolic 2-manifolds without H^2B functions. To prove the second relation in Theorem 1.1, we first establish it in the case $N = 2$:

$$\tilde{O}_G^2 \cap O^2_{H^2B} \neq \emptyset.$$

The reason for treating the case $N = 2$ separately will become apparent when we proceed to the case $N > 2$.

Take the Poincaré disk

$$B_\alpha^2 = \{r < 1 \mid ds = (1 - r^2)^\alpha |dx|, \ \alpha \ \text{constant}\}.$$

Endowing the disk $\{r < 1\}$ with a conformal metric does not change harmonicity (cf. I.1.11), so that $-(2\pi)^{-1}\log r$ continues being a Green's function, and $R \in \tilde{O}_G^2$ for every α.

Suppose there exists a $u \in H^2B$. The harmonic function $h = \Delta u$ has an expansion

$$h = \sum_{n=0}^{\infty} r^n(a_n \cos n\theta + b_n \sin n\theta)$$

on B_α^2. Take a function $\rho(r) \in C_0^\infty[0,1)$, $\rho \geq 0$, $\operatorname{supp} \rho \subset (\beta,\gamma)$, where β, γ are constants with $0 < \beta < \gamma < 1$. For $0 < t < 1$, the function $\rho_t(r) = \rho((1-r)/t)$ has $\operatorname{supp} \rho_t \subset (1 - \gamma t, 1 - \beta t)$. If some $a_n \neq 0$, let $\varphi_t = \rho_t \cos n\theta$. Then

$$\Delta\varphi_t = \Delta\rho_t \cdot \cos n\theta + \rho_t \Delta \cos n\theta,$$

which on $\operatorname{supp} \varphi_t$ is

$$-(1-r^2)^{-2\alpha}(\rho_t'' + r^{-1}\rho_t' - n^2 r^{-2}\rho_t)\cos n\theta .$$

For $s = (1-r)/t$, we obtain as $t \to 0$,

$$(1,|\Delta\varphi_t|) \ge \left|c\int_{1-\gamma t}^{1-\beta t} (1-r^2)^{-2\alpha}(\rho_t'(r) + r^{-1}\rho_t'(r) - n^2 r^{-2}\rho_t(r))(1-r^2)^{2\alpha} r\,dr\right|$$

$$\sim \left|\int_\beta^\gamma (c_1 t^{-2}\rho'(s) + c_2 t^{-1}\rho'(s) + c_3\rho(s))t\,ds\right| .$$

This is asymptotically ct^{-1}, whereas

$$|(h,\varphi_t)| \sim c\int_{1-\gamma t}^{1-\beta t} \rho_t\, t^{2\alpha}\,dr \sim ct^{2\alpha+1} .$$

Thus we have a violation of the inequality

$$|(h,\varphi_t)| = |(\Delta u,\varphi_t)| = |(u,\Delta\varphi_t)| \le c|(1,|\Delta\varphi_t|)|$$

if $2\alpha+1 < -1$, that is, $\alpha < -1$. For such α, this contradiction gives $a_n = 0$. An analogous reasoning with $\varphi_t = \rho_t \sin n\theta$ shows that $b_n = 0$ as well. A fortiori, $h = 0$, and $B_\alpha^2 \in \tilde{O}_G^2 \cap O^2_{H^2B}$ for $\alpha < -1$.

The simplicity of the above example lies in the fact that, for $N = 2$, a harmonic function has the same expansion as in the Euclidean case. For $N > 2$, characterizing the Poincaré N-balls in $O^N_{H^2B}$ is an intricate problem which we shall not solve until §4. Here we shall use for $N > 2$ a different space; it does not qualify as an example for $N = 2$.

1.3. <u>Hyperbolic space E_α^N for</u> $N > 2$. We proceed to the second relation of Theorem 1.1 for $N > 2$,

$$\tilde{O}_G^N \cap O^N_{H^2B} \ne \emptyset .$$

Consider the punctured N-space with Riemannian metric,

$$E_\alpha^N = \{x = (r,\theta^1,\ldots,\theta^{N-1}) \mid 0 < r < \infty,\ ds = r^\alpha|dx|\},$$

$$ds^2 = r^{2\alpha}\,dr^2 + r^{2\alpha+2}\sum_{i=1}^{N-1}\gamma_i(\theta)d\theta^{i2},$$

where α is a constant and the γ_i are trigonometric functions of $\theta = (\theta^1, \ldots, \theta^{N-1})$ as in I.2.4. For later reference we include the case $N = 2$:

<u>LEMMA</u>. <u>For</u> $N = 2$, $E_\alpha^N \in O_G^N$ <u>for every</u> α. <u>For</u> $N > 2$, $E_\alpha^N \in \tilde{O}_G^N$ <u>if and only if</u> $\alpha \neq -1$.

<u>Proof</u>. For $N = 2$, a conformal metric has no effect on harmonic functions, and E_α^2 is parabolic if and only if the Euclidean E_0^2 is. On E_0^2, $h(r) \in H$ if and only if $h(r) = a \log r + b$, so that the harmonic measure $\omega(r)$ is constant for each boundary component of E_0^2, hence $E_0^2 \in O_G^2$.

For $N > 2$, set $\gamma = (\Pi_1^{N-1} \gamma_i)^{1/2}$. The harmonic measure $\omega(r)$ satisfies

$$\Delta\omega = -r^{-N+1-N\alpha} \gamma^{-1} \frac{\partial}{\partial r} (r^{N-1+(N-2)\alpha} \gamma \omega') = 0,$$

so that

$$\omega(r) = a \int_1^r r^{-N+1-(N-2)\alpha} dr + b.$$

The integral is unbounded both for $r \to 0$ and $r \to \infty$, hence $a = 0$, $\omega = \text{const}$, and $E_\alpha^N \in O_G^N$ if and only if $\alpha = -1$. We have proved the Lemma.

The fact that $E_\alpha^2 \in O_G^2$ for all α is the reason for our having discussed H^2B functions separately in the case $N = 2$.

It will be convenient to choose $\alpha = -3/4$. Then the harmonic equation $\Delta h(r) = -r^{-N/4+1}(r^{(N+2)/4}h')' = 0$ has a solution

$$\sigma(r) = \begin{cases} \log r, & N = 2, \\ r^{-(N-2)/4}, & N > 2, \end{cases}$$

and the general solution is $a\sigma + b$.

This time, we choose to consider expansions in the fundamental spherical harmonics $S_{nm}(\theta)$, $n = 1, 2, \ldots$, and $m = 1, \ldots, m_n$ (cf. I.2.4). These will be explicitly used in 3.6. Note that we do not include $n = 0$ in our notation S_{nm}, as we treat constants among radial functions. We wish to find an $f(r)$ such that $f(r)S_{nm}$ is harmonic.

<u>For</u> $N \geq 2$ <u>and any</u> α,

$$f(r)S_{nm} \in H(E_\alpha^N)$$

<u>if and only if</u>

$$f(r) = ar^{p_n} + br^{q_n},$$

<u>where</u> a,b <u>are arbitrary constants and</u>

$$p_n = \tfrac{1}{2}\{-(N-2)(\alpha+1) + [(N-2)^2(\alpha+1)^2 + 4n(n+N-2)]^{1/2}\}$$

$$q_n = \tfrac{1}{2}\{-(N-2)(\alpha+1) - [(N-2)^2(\alpha+1)^2 + 4n(n+N-2)]^{1/2}\}.$$

For the proof, we recall from I.2.5 that the Laplace-Beltrami operator Δ on E_α^N gives

$$\Delta S_{nm} = n(n + N - 2)r^{-2\alpha-2} S_{nm},$$

and our equation takes the form

$$\Delta(f(r)S_{nm}) = \Delta f(r) \cdot S_{nm} + f(r)\Delta S_{nm},$$

$$= - r^{-2\alpha}\{f''(r) + [N - 1 + (N - 2)\alpha]r^{-1}f'(r)$$

$$- n(n + N - 2)r^{-2} f(r)\}S_{nm} = 0.$$

For a trial solution of $\{ \} = 0$, we choose $f(r) = r^p$, p constant. A substitution gives

$$p(p-1)r^{p-2} + [N - 1 + (N-2)\alpha]pr^{p-2} - n(n+N - 2)r^{p-2} = 0,$$

which has the solutions $p = p_n, q_n$ given in the statement. Since r^{p_n}, r^{q_n} are linearly independent, the general solution is $ar^{p_n} + br^{q_n}$, as asserted.

In our special case $\alpha = - 3/4$, we have

$$p_n, q_n = \tfrac{1}{2}\{- \frac{N-2}{4} \pm [\frac{(N-2)^2}{16} + 4n(n + N - 2)]^{1/2}\}.$$

For a fixed r, any harmonic function $h(r,\theta)$ on $E^N_{-3/4}$ restricted to the (Euclidean) r-sphere has an eigenfunction expansion

$$h(r,\theta) = f_0(r) + \sum_{n=1}^{\infty} \sum_{m=1}^{m_n} f_{nm}(r) S_{nm}(\theta).$$

Given $0 < r_1 < r_2 < \infty$, choose constants a_{nm}, b_{nm}, a, b, such that for $i = 1,2$,

$$a_{nm} r_i^{p_n} + b_{nm} r_i^{q_n} = f_{nm}(r_i),$$

$$a\sigma(r_i) + b = f_0(r_i).$$

Then h has the expansion

$$h = \sum_{n=1}^{\infty} \sum_{m=1}^{m_n} (a_{nm} r^{p_n} + b_{nm} r^{q_n}) S_{nm} + a\sigma(r) + b$$

on $r = r_1$ and $r = r_2$, hence by the harmonicity, on $r_1 \leq r \leq r_2$. The uniqueness is verified by choosing $0 < r_1' < r_1 < r_2 < r_2' < \infty$, which gives on $\{r_1' \leq r \leq r_2'\}$ an expansion that on $r_1 \leq r \leq r_2$ must coincide with the one above.

1.4. Biharmonic expansions on $E^N_{-3/4}$. We continue the proof of the second relation of Theorem 1.1 for $N > 2$, and for later reference we include here the case $N = 2$. A straightforward computation shows that the quasiharmonic equation $\Delta q(r) = 1$ has a solution $s(r) = -8N^{-1}r^{1/2}$, the general solution being $s + a\sigma + b$. Similarly, the biharmonic equation $\Delta u(r) = \sigma(r)$ has a solution

$$\tau(r) = \begin{cases} s(r)(\log r - 4), & N = 2, \\ -2 \log r, & N = 4, \\ \frac{8}{N-4} r^{-(N-4)/4}, & N \neq 2,4, \end{cases}$$

and the general solution is $\tau + a\sigma + b$. Set $P_n = N/8 + p_n$, $Q_n = N/8 + q_n$. Since $\Delta(r^{p_n} S_{nm}) = \Delta(r^{q_n} S_{nm}) = 0$, the equation $\Delta u = r^{p_n} S_{nm}$ is satisfied by

$$u_{nm} = -\frac{1}{P_n} r^{p_n + 1/2} S_{nm},$$

and the equation $\Delta v = r^{q_n} S_{nm}$ by

$$v_{nm} = -\frac{1}{Q_n} r^{q_n+1/2} S_{nm}.$$

Given a biharmonic function u on $E^N_{-3/4}$, let

$$\Delta u = \sum_{n=1}^{\infty} \sum_{m=1}^{m_n} (a_{nm} r^{p_n} + b_{nm} r^{q_n}) S_{nm} + a\sigma(r) + b,$$

and set (cf. 4.10)

$$u_0 = \sum_{n=1}^{\infty} \sum_{m=1}^{m_n} (a_{nm} u_{nm} + b_{nm} v_{nm}) + a\tau(r) + bs(r).$$

Then $u = u_0 + k$, where k is a harmonic function

$$k = \sum_{n=1}^{\infty} \sum_{m=1}^{m_n} (c_{nm} r^{p_n} + d_{nm} r^{q_n}) S_{nm} + c\sigma(r) + d.$$

1.5. Exclusion of H^2B functions on $E^N_{-3/4}$. We are ready to complete the proof of the second relation of Theorem 1.1 for $N > 2$. In the present 1.5 we again include the case $N = 2$. We shall show that

$$E^N_{-3/4} \in O^N_{H^2B} \quad \text{for} \quad N \geq 2.$$

Let $u \in H^2B(E^N_{-3/4})$. Take a function $\rho \in C_0^\infty[0,\infty)$ with $\rho \geq 0$, $\operatorname{supp} \rho \subset (1,2)$, and set $\rho_t(r) = \rho(r+1-t)$ for $t \geq 1$. We make use of $|(u,\varphi)| \leq \sup|u|(1,|\varphi|)$ for all $\varphi \in L^1$, in particular for the class of functions $\varphi_t = \rho_t(r) S_{nm}$. By the orthogonality of $\{S_{nm}\}$,

$$(u,\varphi_t) = \int_t^{t+1} (c_1 a_{nm} r^{p_n+1/2} + c_2 b_{nm} r^{q_n+1/2} + c_3 c_{nm} r^{p_n} + c_4 d_{nm} r^{q_n}) \rho_t(r) r^{N/4-1}\, dr.$$

Note that $\int_t^{t+1} \rho_t(r)dr$ is constant as $t \to \infty$. If some $a_{nm} \neq 0$, then

$$(u,\varphi_t) \sim ct^{p_n+N/4-1/2}, \qquad (1,|\varphi_t|) = O(t^{N/4-1}).$$

A fortiori, we have a contradiction for n such that $p_n + N/4 - \frac{1}{2} > N/4 - 1$, that is, $p_n > -\frac{1}{2}$. Since $p_n > 0$ for all n, we obtain $a_{nm} = 0$ for all n,m.

If some $c_{nm} \neq 0$, we infer by $q_n + \frac{1}{2} < p_n$ for all n,N, that

$$(u,\varphi_t) \sim ct^{p_n+N/4-1}$$

as $t \to \infty$. Every n such that $p_n > 0$ is ruled out, and we conclude that $c_{nm} = 0$ for all n,m.

Now choose $\rho_t(r) = \rho(r/t)$, with ρ as before, and $0 < t \leq 1$. Then $\operatorname{supp} \rho_t \subset (t,2t)$ and $\int_t^{2t} \rho_t(r)dr = ct$. If some $d_{nm} \neq 0$, then

$$(u,\varphi_t) \sim ct^{q_n+N/4}, \qquad (1,|\varphi_t|) \sim O(t^{N/4})$$

as $t \to 0$. The inequality $q_n + N/4 < N/4$ gives a contradiction for n with $q_n < 0$, so that $d_{nm} = 0$ for all n,m. In the same manner we see that $b_{nm} = 0$ if $q_n + \frac{1}{2} < 0$, that is, for all n,m.

Thus the function u reduces to $a\tau(r) + bs(r) + c\sigma(r) + d$. Since u is bounded but τ, s, σ are linearly independent and unbounded, we have $a = b = c = 0$, and u is a constant.

The proof of the second relation of Theorem 1.1 for all $N \geq 2$ is herewith complete.

<u>1.6. Parabolic manifolds without H^2B functions.</u> To establish the third relation of Theorem 1.1, we only have to show:

$$E_{-1}^N \in O_{H^2B}^N \quad \text{for } N \geq 2.$$

The proof arrangement is the same as in 1.3-1.5, and we only point out the changes. We now have $\sigma(r) = \log r$ for every N, $p_n = -q_n = [n(n+N-2)]^{1/2}$, and the expansion of a harmonic function h is as before. As to biharmonic functions, $s(r) = -\frac{1}{2}(\log r)^2$, $\tau(r) = -\frac{1}{6}(\log r)^3$, both for every N, and

$$u_{nm} = -\frac{1}{2p_n} r^{p_n} \log r \cdot S_{nm}, \qquad v_{nm} = \frac{1}{2p_n} r^{-p_n} \log r \cdot S_{nm}.$$

With this notation, there is again no change in the expansion of a biharmonic function u.

If some $a_{nm} \neq 0$, we have for $\varphi_t = \rho_t(r)S_{nm}$, $\rho_t(r) = \rho(r+1-t)$, with ρ as before,

$$(u,\varphi_t) \sim c\int_t^{t+1} r^{p_n} \log r \cdot \rho_t(r) r^{-1} dr \sim ct^{p_n-1} \log t,$$

$$(1,|\varphi_t|) = O(t^{-1})$$

as $t \to \infty$. Therefore, $a_{nm} = 0$ for $p_n - 1 > -1$, that is, for all n,m. That $c_{nm} = 0$ for all n,m is concluded in the same manner.

Now choose $\rho_t(r) = \rho(r/t)$, $t \to 0$. If some $d_{nm} \neq 0$, then

$$(u,\varphi_t) \sim ct^{-p_n}, \qquad (1,|\varphi_t|) = O(1).$$

Thus all u with $-p_n < 0$ are ruled out, and we have $d_{nm} = 0$ for all n,m. Similarly, all $b_{nm} = 0$.

The function u again reduces to the radial terms of its expansion, and as before we infer that u is constant. We have proved the third relation of Theorem 1.1.

In view of the Euclidean N-ball B_0^N, the relation $\tilde{O}_G^N \cap \tilde{O}_{H^2B}^N \neq \emptyset$ is trivial. The proof of Theorem 1.1 is herewith complete.

NOTES TO §1. For $N = 2$, Theorem 1.1 was proved in Nakai-Sario [10] by means of complex analysis techniques. For arbitrary Riemannian N-manifolds, the Theorem was established in Sario-Wang [3]. The counterexamples in 1.1 and 1.2 are new.

§2. GENERATORS OF BOUNDED BIHARMONIC FUNCTIONS

We have seen in 1.1 that $\cos 2\theta$ is a bounded biharmonic function on the punctured plane E_0^2. An interesting problem is to find all such functions on E_0^2 or, more generally, on E_0^N.

We shall show that the vector space of bounded biharmonic functions on E_0^2 is generated by 1, $\cos 2\theta$, and $\sin 2\theta$, whereas on E_0^3, the generators are the spherical harmonics of degree 0 and 1, that is, the functions 1, $\sin\theta\cos\psi$, $\sin\theta\sin\psi$, and $\cos\theta$, with θ the angle between the radius vector and the z-axis. One might expect that an analogous result holds on E_0^N. Surprisingly, there are no nonconstant bounded biharmonic functions on any E_0^N with $N > 3$.

In the present section we do not use the symbol H^2B, as we include the harmonic functions.

2.1. Generators on the punctured plane. First we show:

THEOREM. The vector space of bounded biharmonic functions on the punctured Euclidean plane E_0^2 is generated by 1, $\sin 2\theta$, and $\cos 2\theta$.

Proof. Let u be a bounded biharmonic function on E_0^2. In 2.2, we shall systematically discuss representations of biharmonic functions by means of two harmonic functions, and take here the liberty of anticipating Lemma 2.2 in the present case $N = 2$. The function u has a representation

$$u = h + r^2 k + r \log r \cdot (c \cos \theta + d \sin \theta)$$

with $h, k \in H$, and c, d constants. Since every $h \in H$ has an expansion

$$h = \sum_{n=-\infty}^{\infty} r^n (a_n \cos n\theta + b_n \sin n\theta) + a \log r,$$

we obtain

$$u = \sum_{n=-\infty}^{\infty} [r^n (a_n \cos n\theta + b_n \sin n\theta) + r^{n+2}(c_n \cos n\theta + d_n \sin n\theta)]$$

$$+ (a + br^2) \log r + r \log r \cdot (c \cos n\theta + d \sin \theta).$$

We are to show that all coefficients are zero except perhaps a_0, c_{-2}, and d_{-2}.

Suppose $c_n \neq 0$ for some $n \geq 0$. Take a function $\rho(r) \in C_0^\infty(0,\infty)$, $\rho \geq 0$, $\operatorname{supp} \rho \subset (0,1)$, and set $\rho_t(r) = \rho(r - t)$ with $t \in (0,\infty)$. For $\varphi_t = \rho_t \cos n\theta$,

$$(u, \varphi_t) = c \int_t^{t+1} [r^n(a_n + c_n r^2) + r^{-n}(a_{-n} + c_{-n} r^2) + c' r \log r] \rho_t(r) r \, dr,$$

with $c' = 0$ if $n \neq 1, -1$. As $t \to \infty$,

$$(u, \varphi_t) \sim ct^{n+3}$$

and

$$(1, |\varphi_t|) = c \int_t^{t+1} \rho_t(r) r \, dr = O(t).$$

Thus we have a violation of $|(u, \varphi_t)| \leq \sup|u| \cdot (1, |\varphi_t|)$ for $n + 3 > 1$, and we conclude that $c_n = 0$ for all $n \geq 0$. On replacing $\rho_t(r) \cos n\theta$ by $\rho_t(r) \sin n\theta$, we similarly obtain $d_n = 0$ for all $n \geq 0$.

Suppose $a_n \neq 0$ for some $n > 1$; in view of the term $c'r \log r$, we do not consider here $n=1$. For $\varphi_t = \rho_t \cos n\theta$, $(u,\varphi_t) \sim ct^{n+1}$ and $(1,|\varphi_t|) = O(t)$ as $t \to \infty$, and we infer that $a_n = 0$ for $n > 1$. Similarly, we obtain $b_n = 0$ for $n > 1$ on replacing $\rho_t(r)\cos n\theta$ by $\rho_t(r)\sin n\theta$. Thus the representation of u reduces to

$$u = \sum_{n=-\infty}^{1} r^n(a_n \cos n\theta + b_n \sin n\theta) + \sum_{n=-\infty}^{-1} r^{n+2}(c_n \cos n\theta + d_n \sin n\theta)$$
$$+ (a + br^2)\log r + r \log r \cdot (c \cos\theta + d \sin\theta).$$

Suppose $a_n \neq 0$ for some $n < 0$. Choose constants $0 < \beta < \gamma < \infty$ and a nonnegative C_0^∞ function $\rho(r)$ with $\operatorname{supp} \rho \subset (\beta,\gamma)$. For the testing functions take $\varphi_t = \rho_t \cos n\theta$, $\rho_t(r) = \rho(r/t)$. Since $\int_{\beta t}^{\gamma t} \rho_t(r)dr = ct$, we have

$$(u,\varphi_t) = \begin{cases} c\int_{\beta t}^{\gamma t} r^n(a_n + c_n r^2)\rho_t(r)r\,dr \sim ct^{n+2} & \text{for } n<-1, \\ c\int_{\beta t}^{\gamma t} (a_1 r + a_{-1}r^{-1} + c_{-1}r + cr\log r)\rho_t(r)r\,dr \sim ct & \text{for } n=-1, \end{cases}$$

and

$$(1,|\varphi_t|) = c\int_{\beta t}^{\gamma t} \rho_t(r)r\,dr = O(t^2),$$

as $t \to 0$. We conclude that $a_n = 0$ for all $n < 0$. Similarly, we obtain $b_n = 0$ for $n < 0$ on replacing $\rho_t(r)\cos n\theta$ by $\rho_t(r)\sin n\theta$.

Finally suppose $c_n \neq 0$ for $n < -2$; because of the term a_0, we do not consider here $n = -2$. As $t \to 0$ we obtain $(u,\varphi_t) \sim ct^{n+4}$ and $(1,|\varphi_t|) = O(t^2)$. Hence $c_n = 0$ for n such that $n+4<2$, that is, all $n < -2$. Similarly, $d_n = 0$ for $n < -2$. Thus the representation of the biharmonic function u reduces further to

$$u = a_0 + (c_{-2} \cos 2\theta - d_{-2} \sin 2\theta) + r[(a_1 + c_{-1})\cos\theta + (b_1 - d_{-1})\sin\theta]$$
$$+ (a + br^2)\log r + r\log r \cdot (c \cos\theta + d \sin\theta);$$

Since u is bounded, we obtain $a_1 + c_{-1} = b_1 - d_{-1} = a = b = c = d = 0$ on letting $r \to \infty$. Therefore,

$$u = a_0 + c_{-2} \cos 2\theta - d_{-2} \sin 2\theta .$$

The proof of the Theorem is herewith complete.

In view of the linear independence of 1, $\cos 2\theta$, and $\sin 2\theta$, the vector space of bounded biharmonic functions on E_0^2 has dimension 3.

2.2. Biharmonic expansions on the punctured N-space. We proceed to find the generators of the space of bounded biharmonic functions on the punctured N-space E_0^N. We know from classical potential theory (e.g., Brelot [1]) that every harmonic function h on E_0^N, $N \geq 2$, has a series expansion in spherical harmonics,

$$h = as + \sum_{n=0}^{\infty} r^n \sum_{m=1}^{m_n} a_{nm} S_{nm} + \sum_{n=1}^{\infty} r^{-n-N+2} \sum_{m=1}^{m_n} b_{nm} S_{nm},$$

absolutely and uniformly convergent on compact sets. Here $m_n = 2$ for all $n > 0$ if $N = 2$, and $m_n = 2n+1$ if $N = 3$ (e.g., Müller [1, p. 4]). The fundamental singularity s at the origin is $-\log r$ for $N = 2$, and r^{-N+2} for $N > 2$.

Clearly $r^2 h \in H^2$ whenever $h \in H$, since

$$\Delta(r^2 h) = -2Nh - 4r \frac{\partial h}{\partial r}$$

is harmonic. The converse, that every $u \in H^2$ is of the form $r^2 h$ for some h H up to an additive harmonic function, is also true for the Euclidean 3-space E^3 (e.g., Bergman-Schiffer [1, p. 241]). In contrast, $h + r^2 k$ must be modified on E_0^N if $N = 2$ or 4:

LEMMA. *Every biharmonic function* u *on the punctured N-space* E_0^N, $N \geq 2$, *can be written as*

$$u = h + r^2 k + r^2 k_1 ,$$

with $h, k \in H$ *and*

$$k_1 = \begin{cases} r^{-1} \log r \cdot (a \cos\theta + b \sin\theta), & N = 2, \\ c\, r^{-2} \log r, & N = 4, \\ 0, & N \neq 2,4. \end{cases}$$

Proof. By the above expansion of harmonic functions, we can write $\Delta u \in H(E_0^N)$ as

$$\Delta u = \sum_{n=0}^{\infty} \sum_{m=1}^{m_n} (a_{nm} r^n + b_{nm} r^{-n-N+2}) S_{nm} + d \log r,$$

with the constant $d = 0$ if $N \neq 2$ and $b_{01} = 0$ if $N = 2$. We shall discuss separately the cases $N \neq 2,4$; $N = 2$; and $N = 4$.

Case $N \neq 2,4$. The convergence of the expansion of Δu is readily seen to imply that of

$$k = \sum_{n=0}^{\infty} \sum_{m=1}^{m_n} \left(\frac{-1}{4n+2N} a_{nm} r^n + \frac{1}{4n+2N-8} b_{nm} r^{-n-N+2} \right) S_{nm}.$$

Since each term is harmonic, the same is true of k. An easy computation yields

$$\Delta(r^2 k) = -2Nk - 4r \frac{\partial k}{\partial r} = \Delta u.$$

Thus $u - r^2 k$ is a harmonic function h,

$$u = h + r^2 k.$$

Case $N = 2$. We now choose

$$k = \sum_{n=0}^{\infty} \sum_{m=1}^{m_n} \frac{-1}{4n+4} a_{nm} r^n S_{nm} + \sum_{n=2}^{\infty} \sum_{m=1}^{m_n} \frac{1}{4n-4} b_{nm} r^{-n} S_{nm} + \frac{1}{4} d(1 - \log r)$$

and

$$k_1 = -\frac{1}{4} r^{-1} \log r \cdot (b_{11} \cos\theta + b_{12} \sin\theta).$$

Again $k \in H(E_0^2)$ and therefore,

$$\Delta(r^2 k) = -4k - 4r \frac{\partial k}{\partial r} = \sum_{n=0}^{\infty} \sum_{m=1}^{m_n} a_{nm} r^n S_{nm} + \sum_{n=2}^{\infty} \sum_{m=1}^{m_n} b_{nm} r^{-n} S_{nm} + d \log r.$$

By virtue of

$$\Delta(r^2 k_1) = r^{-1}(b_{11} \cos \theta + b_{12} \sin \theta),$$

we have

$$\Delta(u - r^2 k - r^2 k_1) = 0,$$

that is,

$$u = h + r^2 k + r^2 k_1$$

for some harmonic function h.

Case $N = 4$. Take

$$k = \sum_{n=0}^{\infty} \sum_{m=1}^{m_n} \frac{-1}{4n+8} a_{nm} r^n S_{nm} + \sum_{n=1}^{\infty} \sum_{m=1}^{m_n} \frac{1}{4n} b_{nm} r^{-n-2} S_{nm}$$

and

$$k_1 = -\frac{1}{4} b_{01} r^{-2} \log r.$$

By the harmonicity of k, we again obtain

$$\Delta(r^2 k + r^2 k_1) = \Delta u,$$

and some harmonic function h gives

$$u = h + r^2 k + r^2 k_1.$$

The proof of the Lemma is herewith complete.

By virtue of this Lemma, every biharmonic function on E_0^N, $N \geq 2$, can be expanded into a series in spherical harmonics.

2.3. Generators on the punctured 3-space. Let (r, θ, ψ) be the polar coordinates of the 3-space, with θ the angle between the radius vector and the z-axis.

THEOREM. *The vector space of bounded biharmonic functions on the punctured Euclidean 3-space E_0^3 is generated by* 1, $\sin\theta \cos\psi$, $\sin\theta \sin\psi$, *and* $\cos\theta$.

Proof. Suppose there exists a bounded biharmonic function u on E_0^3. It has an expansion

$$u = \sum_{n=0}^{\infty} \sum_{m=1}^{2n+1} (a_{nm} r^n + b_{nm} r^{-n-1} + c_{nm} r^{n+2} + d_{nm} r^{-n+1}) S_{nm}.$$

If $c_{nm} \neq 0$ for some (n,m), choose the testing functions $\varphi_t(r,\theta,\psi) = \rho_t(r) S_{nm}(\theta,\psi)$ with $\rho_t(r) = \rho(r-t)$ and $\rho(r) \in C_0^\infty(0,\infty)$, $\rho \geq 0$, supp $\rho \subset (0,1)$. As $t \to \infty$,

$$(u,\varphi_t) = c\int_t^{t+1} (a_{nm}r^n + b_{nm}r^{-n-1} + c_{nm}r^{n+2} + d_{nm}r^{-n+1})\rho_t(r)r^2 dr \sim ct^{n+4}$$

and $(1,|\varphi_t|) = O(t^2)$. This violates $|(u,\varphi_t)| < c(1,|\varphi_t|)$ for $n+4 > 2$, and we have $c_{nm} = 0$ for all $n \geq 0$. Suppose $a_{nm} \neq 0$ for some (n,m) with $n > 0$. Then $(u,\varphi_t) \sim ct^{n+2}$ and therefore, $a_{nm} = 0$ for all $n > 0$.

If $b_{nm} \neq 0$ for some $n \geq 0$ and some m, take $\varphi_t(r,\theta,\psi) = \rho_t(r) S_{nm}(\theta,\psi)$, $\rho_t(r) = \rho(r/t)$, $\rho \in C_0^\infty(0,\infty)$, $\rho \geq 0$, supp $\rho \subset (\beta,\gamma)$, $0 < \beta < \gamma < \infty$. As $t \to 0$,

$$(u,\varphi_t) \sim ct^{-n+2}$$

and $(1,|\varphi_t|) = O(t^3)$. Therefore, $b_{nm} = 0$ for all n with $-n+2 < 3$, that is, $n \geq 0$. If $d_{nm} \neq 0$ for some (n,m) with $n > 1$, we have a contradiction for $-n+4 < 3$, and obtain $d_{nm} = 0$ for all $n > 1$. Hence the representation reduces to

$$u = a_0 + d_0 r + \sum_{m=1}^{3} d_{1m} S_{1m}.$$

Since u is bounded, $d_0 = 0$. The homogeneous harmonic polynomial $rS_1(\theta,\psi) = r\Sigma_1^3 d_{1m} S_{1m}$ of degree 1 is generated by $x = r \sin\theta \cos\psi$, $y = r\sin\theta \sin\psi$, and $z = r\cos\theta$. The Theorem follows.

2.4. Nonexistence for $N > 3$. In all other cases, that is, $N > 3$, we obtain the following somewhat unexpected result:

THEOREM. *There exist no nonconstant bounded biharmonic functions on the punctured Euclidean N-space E_0^N for $N > 3$.*

Proof. Suppose there exists a bounded biharmonic function u on E_0^N. It has an expansion

$$u = \sum_{n=0}^{\infty} \sum_{m=1}^{m_n} [r^n(a_{nm} + b_{nm}r^2) + r^{-n-N+2}(c_{nm} + d_{nm}r^2)]S_{nm} + e \log r,$$

with $c_{om} = 0$ for $N = 4$, and $e = 0$ for $N > 4$. If $b_{nm} \neq 0$ for some (n,m), set $\varphi_t(r,\theta) = \rho_t(r)S_{nm}(\theta)$ with $\rho_t(r) = \rho(r-t)$ as before. As $t \to \infty$,

$$(u,\varphi_t) \sim ct^{n+N+1} \quad \text{and} \quad (1,|\varphi_t|) = O(t^{N-1}).$$

We have a contradiction for $n + N + 1 > N - 1$, and obtain $b_{nm} = 0$ for all $n \geq 0$. Similarly, $a_{nm} = 0$ for all $n > 0$, and u reduces to

$$u = a_0 + \sum_{n=0}^{\infty} \sum_{m=1}^{m_n} r^{-n-N+2}(c_{nm} + d_{nm}r^2)S_{nm} + e \log r.$$

If $c_{nm} \neq 0$ for some (n,m), set $\varphi_t(r,\theta) = \rho_t(r)S_{nm}$ with $\rho_t(r) = \rho(r/t)$ as before. As $t \to 0$,

$$(u,\varphi_t) \sim ct^{-n+2} \quad \text{and} \quad (1,|\varphi_t|) = O(t^N).$$

Therefore, the contradiction is for $n > 2 - N$. Since $N > 3$, $c_{nm} = 0$ for all $n \geq 0$.

Finally, if $d_{nm} \neq 0$ for some (n,m), $(u,\varphi_t) \sim ct^{-n+4}$ as $t \to 0$, whence the contradiction for $n > 4 - N$. If $N > 4$, then $d_{nm} = 0$ for all $n \geq 0$. If $N = 4$, then $d_{nm} = 0$ for all $n > 0$. Thus $u = c + e \log r$ for all $N > 3$. By the boundedness of u, $e = 0$, and u is constant.

The proof of the Theorem is herewith complete.

NOTES TO §2. The problem of generators of bounded biharmonic functions was introduced and Theorems 2.1, 2.3, and 2.4 proved in Sario-Wang [2].

§3. INDEPENDENCE ON THE METRIC

We shall show that even the metric $ds = r^{\alpha}|dx|$ fails to bring in any nonconstant bounded biharmonic functions to the punctured N-space with $N > 3$, regardless of what α is chosen.

3.1. Radial harmonic and biharmonic functions. We return to the N-manifold considered in 1.3,

$$E_\alpha^N = \{x = (r,\theta^1,\ldots,\theta^{N-1}) \mid 0 < r < \infty,\ ds = r^\alpha |dx|,\quad \alpha \ \text{const}\}.$$

We claim:

THEOREM. *If $N > 3$, there exist no nonconstant bounded biharmonic functions on E_α^N for any α.*

The proof will be given in 3.1-3.5. First we consider radial harmonic functions.

The harmonic equation $\Delta h(r) = 0$ *has a solution*

$$\sigma(r) = \begin{cases} \log r & \text{for } N = 2,\ \text{any } \alpha;\ \text{and any } N,\ \alpha = -1, \\ r^{-(N-2)(\alpha+1)} & \text{for } N > 2,\ \alpha \neq -1. \end{cases}$$

The general solution is $a\sigma(r) + b$.

In fact, the metric is

$$ds^2 = r^{2\alpha} dr^2 + r^{2\alpha+2} \sum_{i=1}^{N-1} \gamma_i(\theta) d\theta^{i2},$$

where $\theta = (\theta^1,\ldots,\theta^{N-1})$, and $\gamma_1,\ldots,\gamma_{N-1}$ are trigonometric functions of θ. Set $\gamma = (\Pi_1^{N-1} \gamma_i)^{1/2}$. For $h(r) \in C^2$,

$$\Delta h(r) = -\frac{1}{r^{N-1+N\alpha}\gamma} \frac{\partial}{\partial r} (r^{N-1+(N-2)\alpha} \gamma h'(r))$$

$$= -r^{-2\alpha}\{h''(r) + [N-1+(N-2)\alpha] r^{-1} h'(r)\}.$$

Thus $h(r) \in H(E_\alpha^N)$ if and only if

$$h(r) = c \int_a^r \frac{dr}{r^{N-1+(N-2)\alpha}} .$$

Next we consider radial quasiharmonic functions.

The quasiharmonic equation $\Delta q(r) = 1$ *has a solution*

$$s(r) = \begin{cases} -\dfrac{1}{2N(\alpha+1)^2} r^{2\alpha+2} & \underline{\text{for}}\ \ N \geq 2,\ \alpha \neq -1. \\ -\tfrac{1}{2}(\log r)^2 & \underline{\text{for}}\ \ N \geq 2,\ \alpha = -1. \end{cases}$$

<u>The general solution is</u> $s(r) + a\sigma(r) + b$.

For $\alpha \neq -1$,

$$\Delta r^{2\alpha+2} = -r^{-2\alpha}\{(2\alpha+2)(2\alpha+1)r^{2\alpha} + [N-1+(N-2)\alpha](2\alpha+2)r^{2\alpha}\}$$
$$= -2N(\alpha+1)^2.$$

For $\alpha = -1$,

$$\Delta(\log r)^2 = -r^2[2r^{-2}(1-\log r) + 2\log r \cdot r^{-2}] = -2.$$

We proceed to radial biharmonic functions. Straightforward computations of Δ as above yield:

<u>The biharmonic equation</u> $\Delta u(r) = \sigma(r)$ <u>has a solution</u>

$$\tau(r) = \begin{cases} s(r)(\log r - \dfrac{1}{\alpha+1}) & \underline{\text{for}}\ \ N = 2,\ \alpha \neq -1, \\ -\dfrac{1}{2\alpha+2}\log r & \underline{\text{for}}\ \ N = 4,\ \alpha \neq -1, \\ \dfrac{1}{2(N-4)(\alpha+1)^2} r^{-(N-4)(\alpha+1)} & \underline{\text{for}}\ \ N \neq 2,4,\ \alpha \neq -1, \\ -\dfrac{1}{6}(\log r)^3 & \underline{\text{for any}}\ \ N,\ \alpha = -1. \end{cases}$$

<u>The general solution is</u> $\tau(r) + a\sigma(r) + b$.

<u>3.2</u>. <u>Nonradial harmonic functions</u>. In preparation for finding biharmonic functions for which Δ gives $r^{p_n}S_{nm}$ or $r^{q_n}S_{nm}$, with p_n, q_n as in 1.3, set

$$P_n = \tfrac{1}{2}N(\alpha+1) + p_n, \quad Q_n = \tfrac{1}{2}N(\alpha+1) + q_n$$

for any N, α, n. Recall the convention $n > 0$ in 1.3. First we consider those n for which P_n or Q_n vanishes.

$P_n = 0$ <u>if and only if</u>

$$\alpha = \begin{cases} -1 - n & \text{for } N = 2, \text{ any } n, \\ -1 - [\frac{4}{3} n(n+1)]^{1/2} & \text{for } N = 3, \text{ any } n, \end{cases}$$

<u>and</u> $P_n \neq 0$ <u>for</u> $N > 3$, <u>any</u> α, n.

$Q_n = 0$ <u>if and only if</u>

$$\alpha = \begin{cases} -1 + n & \text{for } N = 2, \text{ any } n, \\ -1 + [\frac{4}{3} n(n+1)]^{1/2} & \text{for } N = 3, \text{ any } n, \end{cases}$$

<u>and</u> $Q_n \neq 0$ <u>for</u> $N > 3$, <u>any</u> α, n.

Since

$$P_n = \alpha + 1 + \tfrac{1}{2}[(N-2)^2(\alpha+1)^2 + 4n(n+N-2)]^{1/2},$$

$P_n = 0$ implies

$$N(4 - N)(\alpha + 1)^2 = 4n(n + N - 2).$$

If $N > 3$, there are no roots since our $n > 0$. If $N \leq 3$,

$$\alpha = -1 - \left(\frac{4n(n+N-2)}{N(4-N)}\right)^{1/2} ,$$

and the statement follows for P_n. For Q_n, the proof is the same, with the signs of the square roots reversed.

<u>3.3</u>. <u>Nonradial biharmonic functions</u>. Define μ, ν by

$$P_\mu = 0, \qquad Q_\nu = 0.$$

<u>The equation</u> $\Delta u = r^{p_\mu} S_{\mu m}$ <u>has a solution</u>

$$u_{\mu m} = -\frac{1}{2(\alpha+1)} r^{(2-N/2)(\alpha+1)} \log r \cdot S_{\mu m},$$

<u>and the equation</u> $\Delta v = r^{q_\nu} S_{\nu m}$ <u>has a solution</u>

$$v_{\nu m} = -\frac{1}{2(\alpha+1)} r^{(2-N/2)(\alpha+1)} \log r \cdot S_{\nu m}.$$

By $P_\mu = Q_\nu = 0$,

$$p_\mu = q_\nu = -\tfrac{1}{2} N(\alpha + 1).$$

Therefore,

$$\Delta(r^{(2-N/2)(\alpha+1)} \log r \cdot S_{\mu m}) = \Delta(r^{2\alpha+2} \log r) r^{p_\mu} S_{\mu m}$$

$$-2 \operatorname{grad}(r^{2\alpha+2} \log r) \cdot \operatorname{grad} r^{p_\mu} S_{\mu m}.$$

The first term on the right is

$$[\Delta r^{2\alpha+2} \cdot \log r - 2 \operatorname{grad} r^{2\alpha+2} \cdot \operatorname{grad} \log r + r^{2\alpha+2} \Delta \log r] r^{p_\mu} S_{\mu m}$$

$$= [-2N(\alpha+1)^2 \log r - 2(2\alpha+2) - (N-2)(\alpha+1)] r^{p_\mu} S_{\mu m}$$

$$= -(\alpha+1)[2N(\alpha+1)\log r + N + 2] r^{p_\mu} S_{\mu m}.$$

The second term on the right in the previous equation is

$$-2[\operatorname{grad}(r^{2\alpha+2} \log r) \cdot \operatorname{grad} r^{p_\mu}] S_{\mu m} = -2[(2\alpha+2)\log r + 1] p_\mu r^{p_\mu} S_{\mu m}$$

$$= -(\alpha+1)[-2N(\alpha+1)\log r - N] r^{p_\mu} S_{\mu m}.$$

In summary,

$$\Delta(r^{2\alpha+2+p_\mu} \log r \cdot S_{\mu m}) = -2(\alpha+1) r^{p_\mu} S_{\mu m}.$$

The proof for ν is the same.

We proceed to $n \neq \mu, \nu$. Here as well as above, we shall only need particular solutions at this stage.

<u>For</u> $N \geq 2$, $\alpha \neq -1$, $n \neq \mu$, <u>the equation</u> $\Delta u = r^{p_n} S_{nm}$ <u>has a solution</u>

$$u_{nm} = -\frac{1}{4(\alpha+1)P_n} r^{p_n+2\alpha+2} S_{nm}.$$

<u>For</u> $N \geq 2$, $\alpha = -1$, <u>any</u> n, <u>a solution is</u>

$$u_{nm} = -\frac{1}{2p_n} r^{p_n} \log r \cdot S_{nm}.$$

<u>For</u> $N \geq 2, \alpha \neq -1, n \neq \nu,$ <u>the equation</u> $\Delta v = r^{q_n} S_{nm}$ <u>has a solution</u>

$$v_{nm} = -\frac{1}{4(\alpha+1)Q_n} r^{q_n+2\alpha+2} S_{nm}.$$

<u>For</u> $N \geq 2, \alpha = -1,$ <u>any</u> $n,$ <u>a solution is</u>

$$v_{nm} = \frac{1}{2p_n} r^{-p_n} \log r \cdot S_{nm}.$$

For the proof, suppose first $\alpha \neq -1$. Since $\Delta(r^{p_n} S_{nm}) = 0$, we have

$$\begin{aligned}\Delta(r^{p_n+2\alpha+2} S_{nm}) &= (\Delta r^{2\alpha+2} \cdot r^{p_n} - 2 \operatorname{grad} r^{2\alpha+2} \cdot \operatorname{grad} r^{p_n}) S_{nm} \\ &= [-2N(\alpha+1)^2 r^{p_n} - 4(\alpha+1)p_n r^{p_n}] S_{nm} \\ &= -4(\alpha+1)[\tfrac{1}{2} N(\alpha+1) + p_n] r^{p_n} S_{nm}.\end{aligned}$$

For $\alpha = -1$, since $\Delta \log r = \Delta(r^{p_n} S_{nm}) = 0$,

$$\Delta(r^{p_n} \log r \cdot S_{nm}) = -2(\operatorname{grad} \log r \cdot \operatorname{grad} r^{p_n}) S_{nm} = -2p_n r^{p_n} S_{nm}.$$

The proof for v_n is the same when we observe that

$$q_n = -p_n = -[n(n+N-2)]^{1/2} \quad \text{if} \quad \alpha = -1.$$

<u>3.4</u>. <u>Harmonic and biharmonic expansions</u>. In the same fashion as in 1.3, we have:

For any α, and $N \geq 2$, every $h \in H(E_\alpha^N)$ has an expansion

$$h = \sum_{n=1}^{\infty} \sum_{m=1}^{m_n} (a_{nm} r^{p_n} + b_{nm} r^{q_n}) S_{nm} + a\sigma(r) + b,$$

where σ is the radial harmonic function of 3.1. The series converges absolutely and uniformly on compact subsets of E_α^N.

We compile the basic nonradial and radial biharmonic functions in 3.3 into the following expansion of an arbitrary biharmonic function on E_α^N, with p_n, q_n as defined in 1.3. This expansion will be the basis of the crucial proof in 3.5.

<u>LEMMA</u>. <u>Let</u> u <u>be a biharmonic function on</u> E_α^N, $N \geq 2$, <u>with</u>

$$\Delta u = \sum_{n=1}^{\infty} \sum_{m=1}^{m_n} (a_{nm} r^{p_n} + b_{nm} r^{q_n}) S_{nm} + a\sigma(r) + b.$$

<u>If</u> $\alpha \neq -1$, <u>then</u>

$$u = -\frac{1}{4(\alpha+1)} \left(\sum_{n \neq \mu} \sum_{m=1}^{m_n} \frac{1}{P_n} a_{nm} r^{p_n+2\alpha+2} S_{nm} + \sum_{n \neq \nu} \sum_{m=1}^{m_n} \frac{1}{Q_n} b_{nm} r^{q_n+2\alpha+2} S_{nm} \right)$$

$$- \frac{1}{2(\alpha+1)} r^{(2-N/2)(\alpha+1)} \log r \cdot \left(\sum_{m=1}^{m_\mu} a_{\mu m} S_{\mu m} + \sum_{m=1}^{m_\nu} b_{\nu m} S_{\nu m} \right)$$

$$+ \sum_{n=1}^{\infty} \sum_{m=1}^{m_n} (c_{nm} r^{p_n} + d_{nm} r^{q_n}) S_{nm} + a\tau(r) + bs(r) + c\sigma(r) + d.$$

<u>If</u> $\alpha = -1$, <u>then</u>

$$u = -\frac{1}{2p_n} \sum_{n=1}^{\infty} \sum_{m=1}^{m_n} (a_{nm} r^{p_n} - b_{nm} r^{-p_n}) \log r \cdot S_{nm}$$

$$+ \sum_{n=1}^{\infty} \sum_{m=1}^{m_n} (c_{nm} r^{p_n} + d_{nm} r^{-p_n}) S_{nm} + a\tau(r) + bs(r) + c\sigma(r) + d.$$

In both cases, if u_0 is the sum of those terms which do not involve constants c_{nm}, d_{nm}, c, or d, then $\Delta u_0 = \Delta u$, and c_{nm}, d_{nm}, c, d are determined by the expansion of the harmonic function $k = u - u_0$.

The absolute and uniform convergence on compact sets of all expansions in the Lemma is entailed by that of the expansion of Δu.

<u>3.5</u>. <u>Nonexistence of</u> H^2B <u>functions for</u> $N > 3$. We are ready to complete the proof of Theorem 3.1.

Let $u \in H^2B(E_\alpha^N)$, $N > 3$. It has an expansion given in Lemma 3.4. Since u is bounded, so is $(u, S_{nm})_\theta$ for every (n,m). We know from 3.2 that there are no μ, ν with $P_\mu = 0$ or $Q_\nu = 0$. Therefore,

$$(u, S_{nm})_\theta = \begin{cases} Aa_{nm} r^{p_n+2\alpha+2} + Bb_{nm} r^{q_n+2\alpha+2} + Cc_{nm} r^{p_n} + Dd_{nm} r^{q_n} & \text{for } \alpha \neq -1, \\ (Aa_{nm} r^{p_n} + Bb_{nm} r^{-p_n}) \log r + Cc_{nm} r^{p_n} + Dd_{nm} r^{-p_n} & \text{for } \alpha = -1, \end{cases}$$

with $A,B,C,D \neq 0$, $p_n > 0$, $q_n < 0$, and $p_n + 2\alpha + 2 \neq 0$, $q_n + 2\alpha + 2 \neq 0$ for $N > 3$. We conclude that $a_{nm} = b_{nm} = c_{nm} = d_{nm} = 0$ for every (n,m). The expansion of u thus reduces to $u = a\tau(r) + bs(r) + c\sigma(r) + d$, where $\tau(r)$, $s(r)$ and $\sigma(r)$ are the radial biharmonic, quasiharmonic and harmonic functions given in 3.1. For any N,α, these functions are linearly independent and unbounded on $(0,\infty)$. Therefore, the coefficients a,b,c must all vanish, and $u = d$.

The proof of Theorem 3.1 is herewith complete.

The significance of the Theorem is brought forth by the solution of the existence problem on an E_α^N for the lower dimensions. We proceed to it.

__3.6__. H^2B __functions on__ E_α^2 __and__ E_α^3. We shall show that for $N = 2,3$, there exist infinitely many α such that E_α^N carries bounded nonharmonic biharmonic functions. We know from 1.3 and 1.6 that $E_{-1}^N \in O_G^N \cap O_{H^2B}^N$ for $N \geq 2$. We may, therefore, dispense with the case $\alpha = -1$.

__THEOREM__. __For__ $n = 1,2,\ldots,$

$$E_\alpha^2 \in \tilde{O}_{H^2B}^2 \quad \underline{\text{if}} \quad \alpha = -1 \mp \tfrac{1}{2} n,$$

$$E_\alpha^3 \in \tilde{O}_{H^2B}^3 \quad \underline{\text{if}} \quad \alpha = -1 \mp [\tfrac{1}{2} n(n+1)]^{1/2}.$$

__Explicitly, for these__ N, α, __and__ n,

$$S_{nm} \in H^2B(E_\alpha^N) \quad \underline{\text{for all}} \quad m = 1,\ldots,m_n$$

__and__

$$\Delta S_{nm} = cr^p S_{nm}.$$

__Here__ c __is a constant and__ $p = p_n, q_n$ __according as the__ $-$ __or__ $+$ __sign is used in the above expressions of__ α.

__Proof__. By 3.2, $P_n \neq 0$, $Q_n \neq 0$ for the above α. In view of 3.3, we have for $N \geq 2$, $\alpha \neq -1$, any n,m,

$$\Delta(r^{p+2\alpha+2} S_{nm}) = cr^p S_{nm}.$$

Here we require $p + 2\alpha + 2 = 0$, that is,

$$(3 - \tfrac{N}{2})(\alpha + 1) = \mp \tfrac{1}{2}[(N - 2)^2(\alpha + 1)^2 + 4n(n + N - 2)]^{1/2},$$

$$\alpha = -1 \mp \left(\frac{n(n + N - 2)}{8 - 2N}\right)^{1/2},$$

with the upper signs corresponding to $p = p_n$, the lower signs to $p = q_n$.

As a special case of this Theorem, the functions $\sin 2\theta$, $\cos 2\theta$ belong to H^2B not only on the Euclidean E_0^2, and the functions $\sin\theta \cos\psi$, $\sin\theta \sin\psi$, $\cos\theta$ to H^2B not only on the Euclidean E_0^3, as was shown in 2.1 and 2.3, but also on the non-Euclidean E_{-2}^2 and E_{-2}^3, respectively.

We expect that these functions continue being the generators of the vector space of bounded biharmonic functions on E_α^2 and E_α^3, but we have not carried out the study.

Theorem 3.6 throws new light on the result in 2.4 that for $N > 3$ and the specific $\alpha = 0$, we have $E_0^N \in O_{H^2B}^N$. In fact, it now appears (cf. proof of relation 1.5) that even for $N = 2,3$ only some exceptional values of α permit H^2B functions. Theorem 3.1 goes deeper, as α is allowed to vary freely.

Recall that $\dim\{S_{nm}\} = m_n$ for each n is determined by the power series

$$(1 + x)(1 - x)^{-N+1} = \sum_{n=0}^{\infty} m_n x^n.$$

In view of this, Theorem 3.6 entails:

<u>COROLLARY</u>. <u>For</u> $N = 2,3$, <u>and</u> α <u>of Theorem</u> 3.6,

$$\underline{\dim}\ H^2B(E_\alpha^N) \to \infty \quad \underline{\text{as}} \quad |\alpha| \to \infty.$$

This is in striking contrast with the nonexistence of H^2B functions on E_α^N for all $N > 3$ and all α.

<u>NOTES TO §3</u>. Theorems 3.1 and 3.6 were established in Sario-Wang [10]. An open problem of some interest is to show that the functions 1, $\cos 2\theta$, $\sin 2\theta$ for $N = 2$, and the functions 1, $\sin\theta \cos\psi$, $\sin\theta \sin\psi$, $\cos\theta$ for $N = 3$ continue being the generators of the vector space of bounded biharmonic functions on E_α^2 and E_α^3, respectively, for α of Theorem 3.6.

§4. BOUNDED BIHARMONIC FUNCTIONS ON THE POINCARÉ N-BALL

The significance of the Poincaré N-ball $B_\alpha^N = \{r < 1 | ds = (1 - r^2)^\alpha |dx|\}$ in classification theory has already manifested itself, both as a source of counterexamples and as a concrete setup for exploring the dependence on the metric of the existence of functions with various properties. We gave a complete characterization of the existence on B_α^N of harmonic Green's functions and of HP, HB, HD, and HC functions in I.2.10, of HL^p functions in I.3.5, and of QP, QB, QD, and QC functions in II.3.9.

In the present section, we shall characterize the existence of H^2B functions on B_α^N. The interest of the problem lies in the fact that the class H^2B is not a Hilbert space.

4.1. Characterizations. The result of our study will be:

THEOREM. A necessary and sufficient condition for

$$B_\alpha^N \in \tilde{O}^N_{H^2B} \quad \underline{\text{is that}} \quad \begin{cases} \alpha > -1 & \underline{\text{for}} \;\; N = 2,3,4, \\ \alpha \in (-1, 3/(N-4)), & \underline{\text{for}} \;\; N > 4. \end{cases}$$

The proof, to be given in 4.1-4.12, will be divided into the following eight cases, which require a variety of different methods:

Case I: $\alpha < -1$.

Case II: $\alpha > 3/(N-4)$.

Case III: $\alpha \in (-1, 1/(N-2))$.

Case IV: $\alpha \in (1/(N-2), 3/(N-4))$, and α is not an integral multiple of $1/(N-2)$.

Case V: $\alpha \in (1/(N-2), 3/(N-4))$, and α is an integral multiple of $1/(N-2)$.

Case VI: $\alpha = 1/(N-2)$.

Case VII: $\alpha = 3/(N-4)$.

Case VIII: $\alpha = -1$.

The solutions in Cases I and II will be based on the use of testing functions and on the self-adjointness of the Laplace-Beltrami operator $\Delta = \delta d$.

Case III is a consequence of what is already known about the existence of bounded quasiharmonic functions on B_α^N.

In Cases IV and V we expand the solutions of a differential equation at the boundary in order to determine their boundedness. In Case V the roots of the indicial equation differ by an integer, and the convergence proof requires more delicate estimates than in Case IV.

Case VI is solved by using the reasoning developed in Cases IV and V.

The most intriguing cases are VII and especially VIII. The absence of Hilbert space methods necessitates the construction of all biharmonic functions and an estimation of their orders of growth.

Throughout our reasoning, we shall make use of the properties of harmonic functions on B_α^N given in Lemmas I.2.5 and I.2.7.

<u>4.2</u>. <u>Case I</u>: $\alpha < -1$. For these smallest values of α, we have:

<u>LEMMA</u>. $B_\alpha^N \in O_{H^2B}^N$ <u>for</u> $\alpha < -1$, $N \geq 2$.

<u>Proof</u>. By 1.2, we may take $N > 2$. Suppose there exists a $u \in H^2B(B_\alpha^N)$, with $\Delta u = h$. For $0 < \beta < \gamma < 1$, choose a function $s \in C_0^\infty[0,1)$, $s \geq 0$, with $\operatorname{supp} s \subset (\beta,\gamma)$, and set $s_t(r) = s((1-r)/t)$, $t > 0$. We now write $h = \Sigma_0^\infty f_n S_n$, where $S_n \not\equiv 0$ for some $n \geq 0$. Set $\varphi_t = s_t S_n$ and $\lambda = (1 - r^2)^\alpha$. Since $\lambda^N \sim c(1 - r)^{N\alpha}$ as $r \to 1$,

$$|(h,\varphi_t)| > c\left|\int_{1-\gamma t}^{1-\beta t} f_n s_t \lambda^N dr\right| > c\int_{1-\gamma t}^{1-\beta t} s_t(1 - r)^{N\alpha}\, dr$$

$$\sim ct^{N\alpha}\int_{1-\gamma t}^{1-\beta t} s_t dr \sim ct^{N\alpha+1}.$$

Here as usual, c is a positive constant, not always the same.

On the other hand,

$$\Delta\varphi_t = -\lambda^{-2}\left[s_t'' + \left(\frac{N-1}{r} + \frac{(N-2)\lambda'}{\lambda}\right)s_t' - n(n+N-2)r^{-2}s_t\right]S_n.$$

It follows that

$$(1,|\Delta\varphi_t|) < t^{(N-2)\alpha}\left(c_1\int_{1-\gamma t}^{1-\beta t}|s_t''|dr + c_2 t^{-1}\int_{1-\gamma t}^{1-\beta t}|s_t'|dr + c_3\int_{1-\gamma t}^{1-\beta t}s_t dr\right)$$

$$= t^{(N-2)\alpha}(O(t^{-1}) + t^{-1}O(1) + O(t))$$

$$= O(t^{(N-2)\alpha-1}).$$

For $\alpha < -1$, $t^{N\alpha+1}$ grows more rapidly than $t^{(N-2)\alpha-1}$ as $t \to 0$. This contradicts $|(h,\varphi_t)| = |(u,\Delta\varphi_t)| < c(1,|\Delta\varphi_t|)$.

4.3. Case II: $\alpha > 3/(N-4)$. For these largest values of α we prove:

LEMMA. $B_\alpha^N \in O_{H^2B}^N$ for $\alpha > 3/(N-4)$, $N > 4$.

Proof. For $\alpha > 1/(N-2)$, $n > 0$, we know from I.2.7 that $f_n \sim c(1-r)^{1-(N-2)\alpha}$ as $r \to 1$, and

$$|(h,\varphi_t)| > c\int_{1-\gamma t}^{1-\beta t} s_t(1-r)^{2\alpha+1}dr \sim ct^{2\alpha+2}.$$

We have a contradiction for $2\alpha+2 < (N-2)\alpha - 1$, that is, $\alpha > 3/(N-4)$, and infer that $h = c$. If $c \neq 0$, then $c^{-1}u$ belongs to the class QB of bounded quasiharmonic functions. But by Theorem II.3.9, $B_\alpha^N \in \tilde{O}_{QB}^N$ if and only if $\alpha \in (-1,1/(N-2))$. Therefore, $c = 0$, and $u \in H$, in violation of $u \in H^2B$.

4.4. Case III: $\alpha \in (-1,1/(N-2))$. This case is trivial:

LEMMA. $B_\alpha^N \in \tilde{O}_{H^2B}^N$ for $\alpha > -1$, $N = 2$; $\alpha \in (-1,1/(N-2))$, $N > 2$.

In fact, by Theorem II.3.9, $B_\alpha^N \in \tilde{O}_{QB}^N \subset \tilde{O}_{H^2B}^N$.

4.5. Case IV: $\alpha \in (1/(N-2),3/(N-4))$, $\alpha \neq m/(N-2)$. Here we have a more interesting case:

LEMMA. If α is not an integral multiple of $1/(N-2)$, then $B_\alpha^N \in \tilde{O}_{H^2B}^N$ for $\alpha > 1/(N-2)$, $N = 3, 4$; $\alpha \in (1/(N-2),3/(N-4))$, $N > 4$.

Proof. We seek functions $g_n(r)$ such that $\Delta(g_n(r)S_n(\theta)) = f_n(r)S_n(\theta)$. On writing the left-hand side explicitly, we see that g_n satisfies

$$r^2 g_n''(r) + r\,\frac{N-1-(N-1+2(N-2)\alpha)r^2}{1-r^2}\,g_n'(r)$$

$$-\,n(n+N-2)g_n(r) = -r^2(1-r^2)^{2\alpha}f_n(r).$$

Since the right-hand side is of the form $r^{n+2}\sigma(r)$, where $\sigma(r)$ is a power series with radius of convergence 1, we are guaranteed a solution $\tilde{g}_n(r) = r^{n+2}\tilde{\sigma}(r)$, with $\tilde{\sigma}(r)$ a power series whose radius of convergence is also 1. However, a priori there is no assurance that $\tilde{g}_n(r)$ is bounded.

In search of a bounded solution, we set $\rho = 1-r$, suppress the subindex n in our notation and obtain

$$L(g) = \rho^2 g''(\rho) + \rho a(\rho)g'(\rho) + b(\rho)g(\rho) = -\rho^2\lambda(\rho)^2 f(\rho),$$

where $a(\rho)$ and $b(\rho)$ are the same as in I.2.7, and we denote $f(r) = f(1-\rho)$ simply by $f(\rho)$, and $g(r) = g(1-\rho)$ by $g(\rho)$. The roots of the indicial equation $q(p) = p(p-1) + (N-2)\alpha p = 0$ are again $p_0 = 0$ and $p_1 = 1-(N-2)\alpha$. Since α is not an integral multiple of $1/(N-2)$, the roots do not differ by an integer. Therefore,

$$f = A\sum_{i=0}^{\infty} c_i\rho^i + B\sum_{i=0}^{\infty}\gamma_i\rho^{1-(N-2)\alpha+i},$$

and the right-hand side of our differential equation takes the form

$$-\rho^{2\alpha+2}(2-\rho)^{2\alpha}f(\rho) = A\sum_{i=0}^{\infty}\tilde{c}_i\rho^{2\alpha+2+i} + B\sum_{i=0}^{\infty}\tilde{\gamma}_i\rho^{3-(N-4)\alpha+i}.$$

If

$$L(g_{n1}) = \sum_{i=0}^{\infty}\tilde{c}_i\rho^{2\alpha+2+i},\quad L(g_{n2}) = \sum_{i=0}^{\infty}\tilde{\gamma}_i\rho^{3-(N-4)\alpha+i},$$

then $g_n = Ag_{n1} + Bg_{n2}$ will be a solution. Let $a(\rho) = \sum_0^\infty \alpha_i\rho^i$, $b(\rho) = \sum_0^\infty \beta_i\rho^i$. The function

$$g_{n1} = \sum_{i=0}^{\infty}\alpha_i\rho^{p_2+i},\quad p_2 = 2\alpha+2,$$

gives

$$L(g_{n1}) = q(p_2)d_0\rho^{p_2} + \sum_{i=1}^{\infty}\left\{q(p_2+i)d_i + \sum_{j=0}^{i-1}[(p_2+j)\alpha_{i-j} + \beta_{i-j}]d_j\right\}\rho^{p_2+i}$$
$$= \sum_{i=0}^{\infty}\tilde{c}_i\rho^{p_2+i}.$$

Therefore,

$$d_0 = \frac{\tilde{c}_0}{q(p_2)}, \qquad d_i = \frac{\tilde{c}_i - \sum_{j=0}^{i-1}[(p_2+j)\alpha_{i-j} + \beta_{i-j}]d_j}{q(p_2+i)},$$

$i = 1,2,\ldots$. That the denominator is never zero is clear since the roots of $q = 0$ are nonpositive whereas $p_2 + i = 2\alpha + 2 + i > 0$ for all $i \geq 0$.

In the same fashion we find a solution $g_{n2} = \sum_0^{\infty}\delta_i\rho^{p_3+i}$, $p_3 = 3-(N-4)\alpha$. In the cases $N = 3,4$, the condition $\alpha < 3/(N-4)$ is not needed to assure that the δ_i's have a nonvanishing denominator $q(3-(N-4)\alpha+i)$ and that g_{n2} is bounded near $\rho = 0$. Thus for these dimensions we have obtained g_n for all $\alpha > 1/(N-2)$, $\alpha \neq m/(N-2)$.

4.6. Case IV (continued). To show that g_n is well defined, we must establish the convergence of $\sum_0^{\infty} d_i\rho^i$ and $\sum_0^{\infty}\delta_i\rho^i$.

Again we shall only consider $\sum_0^{\infty} d_i\rho^i$ since the convergence of $\sum_0^{\infty}\delta_i\rho^i$ follows in the same manner. Choose a fixed $0 < \rho_0 < 1$. By virtue of the analyticity of $a(\rho)$, $b(\rho)$, and $\sum_0^{\infty}\tilde{c}_i\rho^i$ for $0 \leq \rho < 1$, there exists an $M > 0$ such that

$$|\alpha_i| \leq M\rho_0^{-i}, \qquad |\beta_i| \leq M\rho_0^{-i}, \qquad |\tilde{c}_i| \leq M\rho_0^{-i},$$

$i = 0,1,\ldots$. Set $D_0 = |d_0|$, and

$$D_i = \frac{M[\rho_0^{-i} + \sum_{j=0}^{i-1}(p_2+j+1)\rho_0^{j-i}D_j]}{q(p_2+i)}.$$

Since

$$|d_i| \leq \frac{M[\rho_0^{-i} + \sum_{j=0}^{i-1}(p_2+j+1)\rho_0^{j-i}|d_j|]}{q(p_2+i)},$$

We have by a trivial induction, $|d_i| \leq D_i$ for all i. We shall show by the ratio test that $\sum_0^{\infty} D_i\rho^i$ converges for $\rho < \rho_0$. Clearly,

$$q(p_2+i+1)D_{i+1} = M\left[\rho_0^{-i-1} + \sum_{j=0}^{i} (p_2+j+1)\rho_0^{j-i-1} D_j\right]$$

$$= \rho_0^{-1}M\left[\rho_0^{-i} + \sum_{j=0}^{i-1} (p_2+j+1)\rho_0^{j-i}D_j\right] + \rho_0^{-1}M(p_2+i+1)D_i$$

$$= \rho_0^{-1}[q(p_2+i) + M(p_2+i+1)]D_i .$$

Hence,

$$\frac{D_{i+1}\rho^{i+1}}{D_i\rho^i} = \frac{(i+p_2-p_0)(i+p_2-p_1) + M(i+p_2+1)}{(i+p_2+1-p_0)(i+p_2+1-p_1)} \cdot \frac{\rho}{\rho_0},$$

which approaches ρ/ρ_0 as $i \to \infty$.

We would like to say that $g_nS_n \in H^2B(B_\alpha^N)$, but g_nS_n may fail to be biharmonic at the center of B_α^N. However, $\tilde{g}_nS_n$ is biharmonic at $r = 0$. Since g_n and $\tilde{g}_n$ are particular solutions of the same linear differential equation, they differ by a solution of the homogeneous equation. Therefore, in the notation of I.2.7 and 4.5, $\tilde{g}_n = g_n + Cf_{n1} + Df_{n2}$ for appropriate constants C and D. The function f_nS_n with $f_n = Af_{n1} + Bf_{n2}$ is harmonic at $r = 0$, and a fortiori $\hat{g}_nS_n$ with $\hat{g}_n = \tilde{g}_n - Df_n/B$ is biharmonic at $r = 0$. Also, $\hat{g}_n = g_n + (C - AD/B)f_{n1}$ is bounded since both g_n and f_{n1} are bounded. Thus $\hat{g}_nS_n \in H^2B$.

To simplify the notation, we shall henceforth assume that g_n has been normalized so that g_nS_n is biharmonic on all of B_α^N. Furthermore, we observe for later use that $r^{-n}g_n$ is real analytic at $r = 0$.

4.7. Case V: $\alpha \in (1/(N-2), 3/(N-4))$, $\alpha = m/(N-2)$. This is the case in which the roots of the indicial equation differ by an integer.

LEMMA. *If α is an integral multiple of $1/(N-2)$, then $B_\alpha^N \in \tilde{O}_{H^2B}^N$ for* $\alpha > 1/(N-2)$, $N = 3,4$; $\alpha \in (1/(N-2), 3/(N-4))$, $N > 4$.

Proof. In the notation of Lemma 4.5,

$$L(g) = \rho^2 g'' + \rho a(\rho)g'(\rho) + b(\rho)g(\rho) = -\rho^2\lambda(\rho)^2 f(\rho),$$

where this time, by virtue of the proof of I.2.7, $f(\rho)$ is of the form

$$f(\rho) = A\sum_{i=0}^{\infty} c_i\rho^i + B\left(\sum_{i=0}^{\infty} \gamma_i\rho^{1-(N-2)\alpha+i} + c\log\rho\cdot\sum_{i=0}^{\infty} c_i\rho^i\right).$$

Hence,

$$-\rho^2\lambda(\rho)^2 f(\rho) = A\sum_{i=0}^{\infty} \tilde{c}_i\rho^{2\alpha+2+i}$$

$$+ B\left(\sum_{i=0}^{\infty} \tilde{\gamma}_i\rho^{3-(N-4)\alpha+i} + c\log\rho\cdot\sum_{i=0}^{\infty} \tilde{c}_i\rho^{2\alpha+2+i}\right).$$

By the proof of Lemma 4.5, there exist g_{n1}, g_{n2} such that

$$L(g_{n1}) = \sum_{i=0}^{\infty} \tilde{c}_i\rho^{2\alpha+2+i}, \qquad L(g_{n2}) = \sum_{i=0}^{\infty} \tilde{\gamma}_i\rho^{3-(N-4)\alpha+i}.$$

Therefore, if we can find a g_{n3} such that

$$L(g_{n3}) = \log\rho\cdot\sum_{i=0}^{\infty} \tilde{c}_i\rho^{2\alpha+2+i},$$

then

$$g_n = Ag_{n1} + B(g_{n2} + cg_{n3})$$

will be a solution. We shall show that such a g_{n3}, of the form

$$g_{n3} = \log\rho\cdot\sum_{i=0}^{\infty} d_i\rho^{2\alpha+2+i} + \sum_{i=0}^{\infty} \delta_i\rho^{2\alpha+2+i},$$

exists. On substituting this into our equation we obtain

$$\begin{cases} d_0 = \dfrac{\tilde{c}_0}{q(2\alpha+2)}, \qquad \delta_0 = -\dfrac{4\alpha+3+\alpha_0}{q(2\alpha+2)}\,d_0, \\[2ex] d_i = \dfrac{\tilde{c}_i - \sum_{j=0}^{i-1}[(2\alpha+2+j)\alpha_{i-j}+\beta_{i-j}]d_j}{q(2\alpha+2+i)}, \\[2ex] \delta_i = -\dfrac{(4\alpha+3+2i+\alpha_0)d_i + \sum_{j=0}^{i-1}\alpha_{i-j}d_j + \sum_{j=0}^{i-1}[(2\alpha+2+j)\alpha_{i-j}+\beta_{i-j}]\delta_j}{q(2\alpha+2+i)}, \end{cases}$$

since

$$L\left(\log \rho \cdot \sum_{i=0}^{\infty} d_i \rho^{2\alpha+2+i}\right)$$

$$= \log \rho \cdot \sum_{i=0}^{\infty} \left[q(2\alpha+2+i)d_i + \sum_{j=0}^{i-1} ((2\alpha+2+j)\alpha_{i-j} + \beta_{i-j})d_j \right] \rho^{2\alpha+2+i}$$

$$+ \sum_{i=0}^{\infty} \left[(4\alpha+3+2i+\alpha_0)d_i + \sum_{j=0}^{i-1} \alpha_{i-j} d_j \right] \rho^{2\alpha+2+i}$$

and

$$L\left(\sum_{i=0}^{\infty} \delta_i \rho^{2\alpha+2+i}\right)$$

$$= \sum_{i=0}^{\infty} \left[q(2\alpha+2+i)\delta_i + \sum_{j=0}^{i-1} ((2\alpha+2+j)\alpha_{i-j} + \beta_{i-j})\delta_j \right] \rho^{2\alpha+2+i} .$$

<u>4.8</u>. <u>Case V (continued)</u>. Again as in the proof of Lemma 4.5, $\Sigma_0^{\infty} d_i \rho^i$ converges, and it suffices to prove the convergence of $\Sigma_0^{\infty} \delta_i \rho^i$ for $\rho < \rho_0$. Let $M > 0$ be such that $|\alpha_i| \le M\rho_0^{-i}$, $|\beta_i| \le M\rho_0^{-i}$, and $|d_i| \le M\rho_0^{-i}$. Define D_i by $D_0 = |\delta_0|$ and

$$q(2\alpha+2+i)D_i = M \left[(4\alpha+3+\alpha_0+(M+2)i)\rho_0^{-i} + \sum_{j=0}^{i-1} (2\alpha+3+j)\rho_0^{j-i} D_j \right] .$$

We obtain in the same manner as in 4.6 that $|\delta_i| \le D_i$. Moreover,

$$q(2\alpha+3+i)D_{i+1} = \rho_0^{-1} M \left[(4\alpha+3+\alpha_0+(M+2)i)\rho_0^{-i} + \sum_{j=0}^{i-1} (2\alpha+3+j)\rho_0^{j-i} D_j \right]$$

$$+ \rho_0^{-1} M[(M+2)\rho_0^{-i} + (2\alpha+3+i)D_i]$$

$$= \rho_0^{-1} [q(2\alpha+2+i)D_i + M(2\alpha+3+i)D_i + M(M+2)\rho_0^{-i}].$$

Therefore, for $i = 0,1,2,\ldots,$

$$D_{i+1} = \rho_0^{-1}(A_{i+1}D_i + B_{i+1}\rho_0^{-i}),$$

where

$$A_{i+1} = \frac{q(2\alpha+2+i) + M(2\alpha+3+i)}{q(2\alpha+3+i)}, \quad B_{i+1} = \frac{M(M+2)}{q(2\alpha+3+i)} .$$

From this, we see that

$$D_{i+1} = M_{i+1}\rho_0^{-(i+1)},$$

where

$$M_{i+1} = D_0A_1A_2 \cdots A_{i+1} + B_1A_2 \cdots A_{i+1} + \cdots + B_iA_{i+1} + B_{i+1}.$$

Hence,

$$\frac{D_{i+1}\rho^{i+1}}{D_i\rho^i} = \frac{M_{i+1}}{M_i} \cdot \frac{\rho}{\rho_0} = \left(A_{i+1} + \frac{B_{i+1}}{M_i}\right)\frac{\rho}{\rho_0}.$$

But

$$A_{i+1} = \frac{q(2\alpha+2+i) + M(2\alpha+3+i)}{q(2\alpha+3+i)},$$

which converges to 1 as $i \to \infty$. It remains to show that $B_{i+1}/M_i \to 0$ as $i \to \infty$. We have

$$A_{i+1} = \frac{q(2\alpha+2+i) + M(2\alpha+3+i)}{q(2\alpha+3+i)} > \frac{q(2\alpha+2+i)}{q(2\alpha+3+i)},$$

so that

$$B_jA_{j+1}A_{j+2} \cdots A_1 \geq \frac{M(M+2)}{q(2\alpha+2+i)}, \qquad 1 \leq j \leq i.$$

Also,

$$D_0 = K\frac{M(M+2)}{q(2\alpha+2)}$$

for some constant $K > 0$. Consequently,

$$\frac{B_{i+1}}{M_i} = \frac{B_{i+1}}{D_0A_1 \cdots A_i + B_1A_2 \cdots A_i + \cdots + B_{i-1}A_i + B_i}$$

$$< \frac{1}{i+K}\frac{q(2\alpha+2+i)}{q(2\alpha+3+i)},$$

which approaches 0 as $i \to \infty$.

4.9. Case VI: $\alpha = 1/(N-2)$. We can now make use of the reasoning in Cases IV and V to prove:

LEMMA. $B^N_{1/(N-2)} \in \widetilde{O}^N_{H^2B}$, $N > 2$.

Proof. For $\alpha = 1/(N-2)$, the indicial equation has the repeated root 0. Therefore, $f(\rho)$ has the form

$$f(\rho) = A\sum_{i=0}^{\infty} c_i\rho^i + B\left(\sum_{i=0}^{\infty}\gamma_i\rho^{1+i} + \log\rho \cdot \sum_{i=0}^{\infty} c_i\rho^i\right),$$

and

$$-\rho^2\lambda(\rho)^2 f(\rho) = A \sum_{i=0}^{\infty} \tilde{c}_i \rho^{2\alpha+2+i} + B\left(\sum_{i=0}^{\infty} \tilde{\gamma}_i \rho^{2\alpha+3+i} + \log \rho \cdot \sum_{i=0}^{\infty} \tilde{c}_i \rho^{2\alpha+2+i} \right).$$

The existence of a $g(\rho)$ satisfying $L(\rho) = -\rho^2\lambda(\rho)^2 f(\rho)$ follows by taking $g = Ag_{n1} + B(g_{n2}+g_{n3})$ with

$$L(g_{n1}) = \sum_{i=0}^{\infty} \tilde{c}_i \rho^{2\alpha+2+i}, \quad L(g_{n2}) = \sum_{i=0}^{\infty} \tilde{\gamma}_i \rho^{2\alpha+3+i}, \quad L(g_{n3}) = \log \rho \cdot \sum_{i=0}^{\infty} \tilde{c}_i \rho^{2\alpha+2+i}$$

as can be done by virtue of the proofs of Lemmas 4.5 and 4.7.

4.10. Preparation for Cases VII and VIII. The remaining two cases are the most delicate ones. For preparation, we insert here a somewhat more detailed discussion of the convergence of the expansion of a biharmonic function on B_α^N than was done in 1.4 and 3.4. Here it will be applied to the cases $\alpha = 3/(N-4)$ and $\alpha = -1$. We again use the fundamental spherical harmonics $S_{nm}(\theta)$, $m = 1,\ldots,m_n$.

LEMMA. *For all α, every biharmonic function $u(r,\theta)$ on B_α^N has an expansion*

$$u(r,\theta) = \sum_{n=0}^{\infty} \sum_{m=1}^{m_n} (a_{nm} f_n(r) + b_{nm} g_n(r)) S_{nm}(\theta),$$

where $g_n(r)$ satisfies

$$\Delta(g_n(r)S_n(\theta)) = f_n(r)S_n(\theta)$$

and $g_n(r) \neq 0$ for $0 < r < 1$ unless $g_m(r) \equiv 0$.

Proof. That for all α there exists at least one $g_n(r)$ satisfying the hypothesis is clear. For let $\tilde{g}(r)$ be as at the beginning of the proof of Lemma 4.5. (Note that $\tilde{g}(r)$ is well defined not only for the α's considered in Lemma 4.5 but for all α). The function $r^{-n}\tilde{g}(r)$ is bounded near $r = 0$, and $\lim_{r\to1} \tilde{g}_n(r) \leq \infty$ exists. Consequently, since $r^{-n}f_n(r)$ is bounded away from 0, there exists a constant c such that $r^{-n}\tilde{g}_n(r) + c\, r^{-n}f_n(r) \neq 0$, $0 < r < 1$. A fortiori, $g_n(r) = \tilde{g}_n(r) + c\, f_n(r) \neq 0$, $0 < r < 1$.

For $u \in H^2(B_\alpha^N)$, the harmonic function Δu has an expansion

$$\Delta u(r,\theta) = \sum_{n=0}^{\infty} \sum_{m=1}^{m_n} b_{nm} f_n(r) S_{nm}(\theta).$$

For a fixed $0 < r_0 < 1$, since $g_n(r_0) \neq 0$, there exist constants c_{nm} such that

$$\sum_{n=0}^{\infty} \sum_{m=1}^{m_n} c_{nm} g_n(r) S_{nm}(\theta)$$

converges absolutely and uniformly to u on the sphere $S(r_0)$ of radius r_0. Let $B(r_0)$ be the ball bounded by $S(r_0)$, and denote by $g(x,y)$ the Green's function on $B(r_0)$. The Riesz decomposition of $g_n S_{nm}$ reads

$$g_n S_{nm} = h_{nm} + G(f_n S_{nm}),$$

where h_{nm} is the harmonic part, and the potential part

$$G(f_n S_{nm})(x) = \int_{B(r_0)} g(x,y) * f_n(y) S_{nm}(y)$$

vanishes identically on $S(r_0)$. By taking inner products with S_{nm} over $S(r_0)$ on both sides of the decomposition, we obtain for some constant d_{nm},

$$h_{nm} = d_{nm} f_n S_{nm}$$

on $S(r_0)$ and hence on $B(r_0)$. Substituting the decomposition into the expansion of u on $S(r_0)$, we see that $\Sigma_n \Sigma_m c_{nm} h_{nm}$ is harmonic, and converges absolutely and uniformly to u on $S(r_0)$. It follows that

$$\sum_n \sum_m (c_{nm} h_{nm} + b_{nm} G(f_n S_{nm}))$$

$$= \sum_n \sum_m [(c_{nm} - b_{nm}) h_{nm} + b_{nm}(h_{nm} + G(f_n S_{nm}))]$$

$$= \sum_n \sum_m (a_{nm} f_n + b_{nm} g_n) S_{nm},$$

where $a_{nm} = (c_{nm} - b_{nm}) d_{nm}$. Since $\Sigma_n \Sigma_m b_{nm} f_n S_{nm}$ converges absolutely and uniformly on $B(r_0)$, so does $\Sigma_n \Sigma_m b_{nm} G(f_n S_{nm})$. As a consequence, the expansion we have deduced is absolutely and uniformly convergent on $B(r_0)$ and converges to u on $S(r_0)$. On applying Δ to the expansion, we obtain Δu and conclude that the expansion is indeed that of u on $B(r_0)$. That the a_{nm}'s and b_{nm}'s are independent of r_0 follows easily by the uniqueness of the coefficients of an expansion in the S_{nm}'s.

4.11. Case VII: $\alpha = 3/(N-4)$. We are ready to tackle the simpler one of the two remaining cases:

LEMMA. $B^N_{3/(N-4)} \in O^N_{H^2B}$, $N > 4$.

Proof. Again we seek a function $g(\rho)$ such that

$$L(g(\rho)) = \rho^2 g''(\rho) + \rho a(\rho) g'(\rho) + b(\rho) g(\rho) = -\rho^2 \lambda(\rho)^2 f(\rho).$$

The roots 0 and $1-(N-2)\alpha$ of the indicial equation are now distinct and non-positive; the function $f(\rho)$ has the form

$$f(\rho) = A \sum_{i=0}^{\infty} c_i \rho^i + B\Big(\sum_{i=0}^{\infty} \gamma_i \rho^{1-(N-2)\alpha+i} + c \log \rho \cdot \sum_{i=0}^{\infty} c_i \rho^i\Big), \qquad B \neq 0,$$

and

$$-\rho^2 \lambda(\rho)^2 f(\rho) = A \sum_{i=0}^{\infty} \tilde{c}_i \rho^{2\alpha+2+i} + B\Big(\sum_{i=0}^{\infty} \tilde{\gamma}_i \rho^i + c \log \rho \cdot \sum_{i=0}^{\infty} \tilde{c}_i \rho^{2\alpha+2+i}\Big).$$

Let g_{n1} and g_{n3} be as in the proof of Lemma 4.7. We must assure the existence of a g_{n2} such that $L(g_{n2}) = \Sigma_0^\infty \tilde{\gamma}_i \rho^i$. We try

$$g_{n2}(\rho) = \sum_{i=0}^{\infty} d_i \rho^i + d \log \rho \cdot \sum_{i=0}^{\infty} c_i \rho^i, \qquad d_0 = 1.$$

On substituting, we obtain $d = \tilde{\gamma}_0/[(\alpha_0 - 1)c_0]$,

$$d_i = \{\tilde{\gamma}_i - \sum_{j=0}^{i-1} (j\alpha_{i-j} + \beta_{i-j}) d_j - d[(2i+\alpha_0-1)c_i + \sum_{j=0}^{i-1} \alpha_{i-j} c_j]\}/q(i),$$

$i = 1,2,\dots$. That $\Sigma_0^\infty d_i \rho^i$ converges is seen in the same manner as before. It follows that

$$g_n(\rho) = A g_{n1} + B(g_{n2} + c g_{n3})$$

satisfies $\Delta(g_n S_n) = f_n S_n$. Moreover, $g_n(\rho) \sim c \log \rho$ as $\rho \to 0$. Using the remark at the end of 4.6, and arguing as in the remark at the beginning of the proof of Lemma 4.10, we can assume $g_n \neq 0$, $0 < r < 1$. Hence by Lemma 4.10, if $u \in H^2$, then it has an expansion

$$u(r,\theta) = \sum_{n=0}^{\infty} \sum_{m=1}^{m_n} (a_{nm} f_n + b_{nm} g_n) S_{nm},$$

where $b_{nm} \neq 0$ for some $n \geq 0$. Now suppose u is bounded. Then $\int_\theta u\,S_{nm}\,d\theta$ is bounded as a function of r. But

$$\int_\theta u\,S_{nm}\,d\theta = (S_{nm},S_{nm})(a_{nm}f_n(r) + b_{nm}g_n(r))$$

is not bounded since $f_n(\rho) \sim c\,\rho^{1-(N-2)\alpha}$ and $g_n(\rho) \sim c\log\rho$ are not bounded as $\rho \to 0$. Thus we have a contradiction.

4.12. Case VIII: $\alpha = -1$. Solving for g_n was simplest in Case IV, for the hypothesis assured that there was no difficulty with the indicial roots. In Cases V and VI the indicial roots differed by an integer or were repeated; this complicated the form of f_n. However, the difficulty encountered in Case VII was more critical in that the indicial root 0 prevented us from solving for d_0, and thereby required the addition of the term $d(\log\rho)\Sigma_0^\infty c_i\rho^i$ in the expression for g_{n2}. In the remaining case to be now discussed, the indicial roots cause the greatest complication and necessitate a quite involved expression for g_n.

LEMMA. $B_{-1}^N \in O_{H^2B}^N$, $N \geq 2$.

Proof. Once more we look for a g satisfying

$$L(g(\rho)) = -\rho^2\lambda(\rho)^2 f(\rho).$$

Since $\alpha = -1$,

$$f(\rho) = A\sum_{i=0}^\infty c_i\rho^{N-1+i} + B\left(\sum_{i=0}^\infty \gamma_i\rho^i + c\log\rho\cdot\sum_{i=0}^\infty c_i\rho^{N-1+i}\right), \qquad B \neq 0,$$

and

$$-\rho^2\lambda(\rho)^2 f(\rho) = A\sum_{i=0}^\infty \tilde{c}_i\rho^{N-1+i} + B\left(\sum_{i=0}^\infty \tilde{\gamma}_i\rho^i + c\log\rho\cdot\sum_{i=0}^\infty \tilde{c}_i\rho^{N-1+i}\right).$$

That there exists a g_{n1} with $L(g_{n1}) = \Sigma_0^\infty \tilde{c}_i\rho^{N-1+i}$ follows from the proof of Lemma 4.11 and the fact that the indicial roots are 0 and $N-1$. Thus, the present task is to find a g_{n2} such that

$$L(g_{n2}) = \sum_{i=0}^\infty \tilde{\gamma}_i\rho^i + c\log\rho\cdot\sum_{i=0}^\infty \tilde{c}_i\rho^{N-1+i}.$$

We shall express g_{n2} as the sum of three functions,

$$g_{n2} = \Phi_1 + \Phi_2 + \Phi_3 .$$

Let

$$\Phi_1 = e_1 \log \rho \cdot \left(\sum_{i=0}^{\infty} \gamma_i \rho^i + c \log \rho \cdot \sum_{i=0}^{\infty} c_i \rho^{N-1+i} \right) + \sum_{i=0}^{N-2} d_i \rho^i ,$$

where

$$e_1 = \frac{\tilde{\gamma}_0}{(\alpha_0 - 1)\gamma_0} , \qquad d_0 = 1,$$

$$d_i = \{\tilde{\gamma}_i - \sum_{j=0}^{i-1} (j\alpha_{i-j} + \beta_{i-j}) d_j - e_1[(2i + \alpha_0 - 1)\gamma_i + \sum_{j=0}^{i-1} \alpha_{i-j}\gamma_j]\}/q(i),$$

$i = 1,\ldots,N-2$. In view of

$$L\left(\log \rho \cdot \left(\sum_{i=0}^{\infty} \gamma_i \rho^i + c \log \rho \cdot \sum_{i=0}^{\infty} c_i \rho^{N-1+i} \right)\right)$$

$$= c \log \rho \cdot \sum_{i=0}^{\infty} \left[(2N - 3 + 2i) c_i + \sum_{j=0}^{i} \alpha_{i-j} c_j \right] \rho^{N-1+i} + 2c \sum_{i=0}^{\infty} c_i \rho^{N-1+i}$$

$$+ (\alpha_0 - 1)\gamma_0 + \sum_{i=1}^{\infty} \left[(2i + \alpha_0 - 1)\gamma_i + \sum_{j=0}^{i-1} \alpha_{i-j}\gamma_j \right] \rho^i ,$$

and

$$L\left(\sum_{i=0}^{N-2} d_i \rho^i \right) = \sum_{i=1}^{N-2} \left[q(i) d_i + \sum_{j=0}^{i-1} (j\alpha_{i-j} + \beta_{i-j}) d_j \right] \rho^i + \sum_{i=N-1}^{\infty} \sum_{j=0}^{N-2} (j\alpha_{i-j} + \beta_{i-j}) d_j \rho^i ,$$

it follows that

$$L(\Phi_1) = \sum_{i=0}^{N-2} \tilde{\gamma}_i \rho^i + \sum_{i=0}^{\infty} s_i \rho^{N-1+i} + c \log \rho \cdot \sum_{i=0}^{\infty} \sigma_i \rho^{N-1+i} ,$$

with the s_i's and σ_i's constants. Next choose

$$\Phi_2 = e_2 (\log \rho)^2 \sum_{i=0}^{\infty} c_i \rho^{N-1+i} + c \log \rho \cdot \sum_{i=0}^{\infty} \delta_i \rho^{N-1+i}$$

with

$$e_2 = \frac{\tilde{c}_0 - c\sigma_0}{2(2N - 3 + \alpha_0) c_0} , \qquad \delta_0 = 1,$$

$$\delta_i = \frac{\tilde{c}_i - \sigma_i - \Sigma_{j=0}^{i-1}((N-1+j)\alpha_{i-j} + \beta_{i-j})\delta_j - 2e_2[(2N-3+2i+\alpha_0)c_i + \Sigma_{j=0}^{i-1} \alpha d_{i-j} c_j]}{q(N - 1 + i)} .$$

Since

$$L\left((\log\rho)^2\sum_{i=0}^{\infty}c_i\rho^{N-1+i}\right)$$
$$=2\sum_{i=0}^{\infty}c_i\rho^{N-1+i}+2\log\rho\cdot\sum_{i=0}^{\infty}\left[(2N-3+2i+\alpha_0)c_i+\sum_{j=0}^{i-1}\alpha_{i-j}c_j\right]\rho^{N-1+i},$$

and

$$L\left(\log\rho\cdot\sum_{i=0}^{\infty}\delta_i\rho^{N-1+i}\right)=\sum_{i=0}^{\infty}\left[(2N-3+2i+\alpha_0)\delta_i+\sum_{j=0}^{i-1}\alpha_{i-j}\delta_j\right]\rho^{N-1+i}$$
$$+\log\rho\cdot\sum_{i=0}^{\infty}\left[q(N-1+i)\delta_i+\sum_{j=0}^{i-1}((N-1+j)\alpha_{i-j}+\beta_{i-j})\delta_j\right]\rho^{N-1+i},$$

we have

$$L(\Phi_1+\Phi_2)=\sum_{i=0}^{N-2}\tilde{\gamma}_i\rho^i+\sum_{i=0}^{\infty}\tilde{s}_i\rho^{N-1+i}+c\log\rho\cdot\sum_{i=0}^{\infty}\tilde{c}_i\rho^{N-1+i}.$$

Finally, set

$$\Phi_3=e_3\log\rho\cdot\sum_{i=0}^{\infty}c_i\rho^{N-1+i}+\sum_{i=0}^{\infty}\tilde{d}_i\rho^{N-1+i},$$

where

$$e_3=\frac{\tilde{\gamma}_{N-1}-\tilde{s}_0}{(2N-3+\alpha_0)c_0},\qquad \tilde{d}_0=1,$$

$$\tilde{d}_i=\frac{\tilde{\gamma}_{N-1+i}-\tilde{s}_i-\Sigma_{j=0}^{i-1}((N-1+j)\alpha_{i-j}+\beta_{i-j})\tilde{d}_j-e_3[(2N-3+2i+\alpha_0)c_i+\Sigma_{j=0}^{i-1}\alpha_{i-j}c_j]}{q(N-1+i)},$$

$i=1,2,\ldots$. We infer from

$$L\left(\log\rho\cdot\sum_{i=0}^{\infty}c_i\rho^{N-1+i}\right)=\sum_{i=0}^{\infty}\left[(2N-3+2i+\alpha_0)c_i+\sum_{j=0}^{i-1}\alpha_{i-j}c_j\right]\rho^{N-1+i}$$

and

$$L\left(\sum_{i=0}^{\infty}\tilde{d}_i\rho^{N-1+i}\right)=\sum_{i=0}^{\infty}\left[q(N-1+i)\tilde{d}_i+\sum_{j=0}^{i-1}((N-1+j)\alpha_{i-j}+\beta_{i-j})\tilde{d}_j\right]\rho^{N-1+i}$$

that g_{n2} satisfies our equation. Clearly, Φ_1 is well defined. The convergence of Φ_2 and Φ_3 is seen by arguing as before.

As in Lemma 4.11, $g_n\sim c\log\rho$, and Lemma 4.12 follows.

The proof of Theorem 4.1 is herewith complete.

Note that this Theorem, together with Lemma I.2.6, also provides us with another proof of $\tilde{O}_G^N\cap O_{H^2B}^N\neq\emptyset$ established in 1.2-1.5.

NOTES TO §4. The problem of characterizing the Poincaré N-balls in $O^N_{H^2B}$ resisted solution long after the characterization of those in O^N_{HX} and O^N_{QX} for $X = P,B,D$, and even in $O^N_{H^2D}$ had succeeded. The problem was finally solved, and Theorem 4.1 proved, in Hada-Sario-Wang [3]. An interesting problem would be to find a less computational proof.

§5. COMPLETENESS AND BOUNDED BIHARMONIC FUNCTIONS

In I.4.1-I.4.3, we showed that there are no inclusion relations between the class C^N of complete Riemannian N-manifolds and any of the harmonic null classes O^N_{HX}. In the present section, we shall prove that the same is true of C^N and $O^N_{H^2B}$. In particular, there exist complete manifolds that do carry H^2B functions, and non-complete ones that do not.

5.1. Complete but with H^2B functions. We assert:

THEOREM. For $N \geq 2$,

$$C^N \cap \tilde{O}^N_{H^2B} \neq \emptyset.$$

Proof. Take the N-cylinder

$$R = \{|x| < \infty,\ |y^i| \leq 1,\ i = 1,\ldots,N-1\}$$

with the metric

$$ds^2 = \varphi(x)^{-2}dx^2 + \varphi(x)^{4/(N-1)} \sum_{i=1}^{N-1} dy^{i2},$$

where

$$\varphi(x) = (2 + x^2)^{1/2} \log(2 + x^2).$$

The verification of completeness is immediate:

$$\int_0^\infty \varphi(x)^{-1}dx = \int_0^\infty (2+x^2)^{-1/2}(\log(2+x^2))^{-1}dx$$

$$> \tfrac{1}{2}\int_0^\infty (2+x)^{-1}(\log(2+x))^{-1}dx = \tfrac{1}{2}\log\log(2+x)\Big|_0^\infty = \infty,$$

and similarly, $\int_{-\infty}^{0} \varphi(x)^{-1}dx = \infty$. The function

$$u(x) = \int_0^x \varphi(t)^{-3} \int_0^t \varphi(s) \int_0^s \varphi(r)^{-3}dr\,ds\,dt$$

is nonharmonic biharmonic. In fact, $\Delta u = -\varphi^{-1}(\varphi\varphi^2 u')' = -\int_0^x \varphi(r)^{-3}dr$ and $\Delta^2 u = -\varphi^{-1}(\varphi^3(-\varphi^{-3}))' = 0$. To see that u is bounded, it suffices to show that it is so for $x > 0$. The integral

$$\int_0^s \varphi(r)^{-3}dr = \int_0^s (2+r^2)^{-3/2} (\log(2+r^2))^{-3}dr$$

is bounded and, for $t > 0$,

$$\int_0^t \varphi(s) \int_0^s \varphi(r)^{-3}dr\,ds < c\int_0^t (2+s^2)^{1/2} \log(2+s^2)ds$$

$$< 2c \int_0^t (2+s)\log(2+s)ds$$

$$= c[(2+t)^2\log(2+t) - \tfrac{1}{2}(2+t)^2 + \text{const}].$$

We let [ ] stand for the expression in brackets and obtain

$$u(x) < c\int_0^x (2+t^2)^{-3/2}(\log(2+t^2))^{-3}[\quad]dt.$$

The dominating term in the integrand is majorized by

$$\tfrac{1}{2}\, t^{-3}(\log t)^{-3} \cdot (2+t)^2\log(2+t) \sim ct^{-1}(\log t)^{-2},$$

as $t \to \infty$. The integral from 1 to $x > 1$ is, therefore, bounded, so is u for all x, and we have the Theorem.

The following simple example, valid for $N > 2$, is perhaps also of interest. The N-cylinder

$$R = \{|x| < \infty,\ |y| \le \pi,\ |z^i| \le 1,\ i = 1,\dots,N-2\}$$

with the metric

$$ds^2 = dx^2 + e^{-x}\,dy^2 + e^{(2e^x - x)/(N-2)} \sum_{i=1}^{N-2} dz^{i2}$$

is clearly complete. The function $u = \cos y$ belongs to H^2B. In fact,

$$\Delta u = -e^{-e^x+x}(e^{e^x-x}e^x)(-\cos y) = e^x \cos y$$

and

$$\Delta^2 u = -e^{-e^x+x}[(e^{e^x-x}e^x)'\cos y + e^{e^x-x}e^x e^x(-\cos y)] = 0.$$

5.2. Complete and without H^2B functions. The next case is equally simple:

THEOREM. For $N \geq 2$,

$$C^N \cap O^N_{H^2B} \neq \emptyset.$$

Proof. The Euclidean N-space $E^N \in C^N$. Every biharmonic function u has an expansion

$$u = \sum_{n=0}^{\infty} \sum_{m=1}^{m_n} (a_{nm} r^{n+2} + b_{nm} r^n) S_{nm}.$$

If $u \in H^2B$, then $\int_\theta u\, S_{nm} d\theta$ is bounded as a function of r. But

$$\int_\theta u\, S_{nm} d\theta = c(a_{nm} r^{n+2} + b_{nm} r^n)$$

is not bounded in r, a contradiction.

5.3. Remaining cases. Both cases of noncomplete manifolds are immediate:

THEOREM For $N \geq 2$,

$$\tilde{C}^N \cap O^N_{H^2B} \neq \emptyset, \quad \tilde{C}^N \cap \tilde{O}^N_{H^2B} \neq \emptyset.$$

In fact, the punctured N-space $E^N_\alpha = \{0 < r < \infty\}$ with the metric $ds = r^\alpha |dx|$ is in $\tilde{C}^N$ for $\alpha \neq -1$, and we know from 1.5 that $E^N_{-3/4} \in O^N_{H^2B}$. The second relation of the Theorem is trivial in view of the Euclidean N-ball.

NOTES TO §5. Theorems 5.1, 5.2, and 5.3 were established in Sario [8]. It is of interest that the problem of relation to completeness can so readily be settled in the cases O^N_{HX}, O^N_{QX}, and $O^N_{H^2B}$, whereas for $O^N_{H^2D}$, it has thus far offered unsurmountable difficulties. In particular, we do not know whether or not there exist complete N-manifolds which carry H^2D functions. In the present monograph, we do not pursue the completeness problem beyond the present section.

§6. BOUNDED POLYHARMONIC FUNCTIONS

From biharmonic functions we turn in the present section to the more general class of polyharmonic functions of order k, defined by $\Delta^k u = 0$, k an integer ≥ 2. We denote by H^k the subclass of nondegenerate polyharmonic functions of order k,

$$H^k = \{u \mid \Delta^k u = 0,\ \Delta^{k-1} u \neq 0\}.$$

For a given class X of functions, we again set $H^kX = H^k \cap X$, and denote by $O^N_{H^kX}$ and $\tilde{O}^N_{H^kX}$ the classes of Riemannian N-manifolds R with $H^kX(R) = \emptyset$ and $H^kX(R) \neq \emptyset$, respectively.

It was shown in §3 that although E^N_α with $N = 2,3$ carries H^2B functions for infinitely many values of α, it tolerates no H^2B functions for any α if $N > 3$. In the present section, we ask: What can be said about the class H^kB of bounded nondegenerate polyharmonic functions of degree k? The answer turns out to be rewarding and puts the biharmonic case in proper perspective: there exist no H^kB functions on E^N_α for any α if $N \geq 2k$.

For $N < 2k$, there are infinitely many α for which these functions do exist, and for these α the generators of the space H^kB are surface spherical harmonics. In particular, this is true of H^2B functions on the punctured Euclidean 2- and 3-spaces, as was shown in 3.6.

If $H^kB \neq \emptyset$ on a given E^N_α, is the same true of H^hB for any $h > k$? We shall show that, while this is so for every N if the metric of E^N_α is Euclidean, there are values of (N,α) for which it does not hold.

<u>6.1</u>. <u>Main theorem</u>. We start by stating the main result.

<u>THEOREM</u>. $E^N_\alpha \in O^N_{H^kB}$ <u>for all</u> $N \geq 2k$, $k \geq 1$, <u>and all</u> α.

The proof will be given in 6.1-6.3.

First we consider radial functions and show that the equation $\Delta^k u(r) = 0$ has the following general solutions:

If N is odd, or if N is even with $N > 2k$, then for any $\alpha \neq -1$

$$u_k(r) = \sum_{n=0}^{k-1} (a_n r^{(2n-N+2)(\alpha+1)} + b_n r^{2n(\alpha+1)}).$$

If N is even with $N \leq 2k$, then for $\alpha \neq -1$,

$$u_k(r) = \sum_{n=0}^{k-1} (a_n r^{(2n-N+2)(\alpha+1)} + b_n r^{2n(\alpha+1)}) + \sum_{n=0}^{(2k-N)/2} c_n r^{2n(\alpha+1)} \log r .$$

If $\alpha = -1$, then for any N, $u_k(r) = \Sigma_{n=0}^{2k-1} a_n (\log r)^n$.

Since the proofs are similar in all cases, we shall only discuss the case N odd, $\alpha \neq -1$. For $f \in C^2(E_\alpha^N)$,

$$\Delta f(r) = -\frac{1}{r^{N-1+N\alpha}} \frac{d}{dr} (r^{N-1+(N-2)\alpha} f'(r)).$$

The proof will be by induction. In the cases $k = 1,2$, it was given in 3.1. For $k \geq 3$, we have the induction hypothesis

$$-\frac{1}{r^{N-1+N\alpha}} \frac{d}{dr} (r^{N-1+(N-2)\alpha} f'(r)) = \sum_{n=0}^{k-2} (a_n r^{(2n-N+2)(\alpha+1)} + b_n r^{2n(\alpha+1)}).$$

Here and later, a_n, b_n etc. are constants, not always the same. We obtain successively

$$\frac{d}{dr} (r^{N-1+(N-2)\alpha} f'(r)) = \sum_{n=0}^{k-2} (a_n r^{(2n+2)(\alpha+1)-1} + b_n r^{(2n+N)(\alpha+1)-1}),$$

$$r^{N-1+(N-2)\alpha} f'(r) = \sum_{n=0}^{k-2} (a_n r^{(2n+2)(\alpha+1)} + b_n r^{(2n+N)(\alpha+1)}) + c,$$

$$f'(r) = \sum_{n=0}^{k-2} (a_n r^{(2n+2-N)(\alpha+1)+2\alpha+1} + b_n r^{2n(\alpha+1)+2\alpha+1}) + c\, r^{-N-(N-2)\alpha+1},$$

$$f(r) = \sum_{n=0}^{k-1} (a_n r^{(2n+2-N)(\alpha+1)} + b_n r^{2n(\alpha+1)}).$$

<u>6.2</u>. <u>Polyharmonic expansions</u>. Again let $S_{nm} = S_{nm}(\theta)$, $n = 1,2,\ldots$, $m = 1,2,\ldots,m_n$, be the fundamental spherical harmonics. We do not include $n = 0$ in our notation S_{nm}, as we here treat constants as radial harmonic functions. For harmonic functions, we know from 1.3 that $f(r)S_{nm} \in H(E_\alpha^N)$ for any N and α if and only if $f(r) = ar^{p_n} + br^{q_n}$, where a,b are arbitrary constants

and

$$p_n = \tfrac{1}{2}[-(N-2)(\alpha+1) + ((N-2)^2(\alpha+1)^2 + 4n(n+N-2))^{1/2}],$$

$$q_n = \tfrac{1}{2}[-(N-2)(\alpha+1) - ((N-2)^2(\alpha+1)^2 + 4n(n+N-2))^{1/2}].$$

For any $N, \alpha, n > 0, 0 \leq j \leq k-2$, set

$$P_{nj} = (\tfrac{1}{2}N + j)(\alpha+1) + p_n, \quad Q_{nj} = (\tfrac{1}{2}N + j)(\alpha+1) + q_n.$$

Define μ_j, ν_j by $P_{\mu_j j} = 0$, $Q_{\nu_j j} = 0$. We claim that

$$P_{nj} \neq 0 \text{ and } Q_{nj} \neq 0 \text{ for } N \geq 2k, \text{ any } \alpha, n.$$

To see this, we first observe that $P_{nj} = 0$ implies

$$[4(j+1)^2 - (N-2)^2](\alpha+1)^2 = 4n(n+N-2).$$

If $N \geq 2k$,

$$4(j+1)^2 \leq 4k^2 - 8k + 4 \leq (N-2)^2.$$

Since our $n > 0$, there are no roots. The proof for Q_{nj} is identical.

The equation $\Delta u = r^{p_{\mu_j}+(2\alpha+2)j} S_{\mu_j m}$ has a solution

$$u_{\mu_j m} = a r^{p_{\mu_j}+(2\alpha+2)(j+1)} \log r \cdot S_{\mu_j m},$$

and the equation $\Delta v = r^{q_{\mu_j}+(2\alpha+2)j} S_{\nu_j m}$ has a solution

$$v_{\nu_j m} = b r^{q_{\nu_j}+(2\alpha+2)(j+1)} \log r \cdot S_{\nu_j m},$$

with a,b certain constants. We see this by direct computation which is made easier by noting that $r^{p_{\mu_j}} S_{\mu_j m}$ and $r^{q_{\nu_j}} S_{\nu_j}$ are harmonic. In this computation, one observes that multiplying u or v by $r^{2\alpha+2}$ raises the degree of polyharmonicity by unity and $\Delta(r^{2\alpha+2}u) = cu +$ harmonic function.

It is easy to verify that for any N, $\alpha \neq -1$, the equations

$$\Delta u = r^{p_n+(2\alpha+2)j} S_{nm}, \qquad \Delta v = r^{q_n+(2\alpha+2)j} S_{nm}$$

have solutions u_{nm} for $n \neq \mu_j$ and v_{nm} for $n \neq \nu_j$ given by

$$u_{nm} = ar^{p_n+(2\alpha+2)(j+1)} S_{nm}, \qquad v_{nm} = br^{q_n+(2\alpha+2)(j+1)} S_{nm}.$$

In the case $\alpha = -1$, $j \geq 1$, we shall prove that

$$\Delta(r^{p_n}(\log r)^j S_{nm}) = \sum_{i=0}^{j-1} a_i r^{p_n}(\log r)^i S_{nm}$$

for certain constants a_i. In view of $\Delta \log r = \Delta r^{p_n} S_{nm} = 0$,

$$\Delta(r^{p_n}\log r \cdot S_{nm}) = -2(\text{grad } r^{p_n} \cdot \text{grad } \log r) S_{nm} = -2p_n r^{p_n} S_{nm}.$$

A straightforward induction argument completes the proof.

For harmonic functions, we know from 3.4 that given any N, α, every $h \in H(E_\alpha^N)$ has an expansion

$$h = \sum_{n=1}^{\infty} \sum_{m=1}^{m_n} (a_{nm} r^{p_n} + b_{nm} r^{q_n}) S_{nm} + u_1(r).$$

We can now proceed to polyharmonic functions. For any N, any $u \in H^k(E_\alpha^N)$ has an expansion for $\alpha \neq -1$,

$$\begin{aligned} u = & \sum_{j=0}^{k-1} \Big(\sum_{n \neq \mu_j} \sum_{m=1}^{m_n} a_{jnm} r^{p_n+(2\alpha+2)j} S_{nm} + \sum_{n \neq \nu_j} \sum_{m=1}^{m_n} b_{jnm} r^{q_n+(2\alpha+2)j} S_{nm} \Big) \\ & + \sum_{\mu_j} \sum_{i=0}^{J-j} r^{(2\alpha+2)i} \sum_{m=1}^{m_{\mu_j}} c_{\mu_j im} r^{p_{\mu_j}+(2\alpha+2)(j+1)} \log r \cdot S_{\mu_j m} \\ & + \sum_{\nu_j} \sum_{i=0}^{K-j} r^{(2\alpha+2)i} \sum_{m=1}^{m_{\nu_j}} d_{\nu_j im} r^{q_{\nu_j}+(2\alpha+2)(j+1)} \log r \cdot S_{\nu_j m} + u_k(r), \end{aligned}$$

where $J = \max\{j \mid P_{\mu_j j} = 0\}$, $K = \max\{j \mid Q_{\nu_j j} = 0\}$.

If $\alpha = -1$, then

$$u = \sum_{j=0}^{k-1} \sum_{n=1}^{\infty} \sum_{m=1}^{m_n} (a_{jnm} r^{p_n} + b_{jnm} r^{q_n})(\log r)^j S_{nm} + u_k(r).$$

For the proof, let $h = \Delta^{k-1}u$ have the above harmonic expansion. The proper coefficients of u are obtained from preceding formulas. The expansion of h converges for every $r \in (0,\infty)$ and all θ. Therefore,

$$\lim_{n\to\infty} \left|\sum_{m=1}^{m_n} a_{nm}S_{nm}\right|^{1/p_n} = \lim_{n\to\infty} \left|\sum_{m=1}^{m_n} b_{nm}S_{nm}\right|^{-1/q_n} = 0,$$

$$\lim_{n\to\infty} \left|\sum_{m=1}^{m_n} a_{jnm}S_{nm}\right|^{1/p_n} = \lim_{n\to\infty} \left|\sum_{m=1}^{m_n} b_{jnm}S_{nm}\right|^{-1/q_n} = 0,$$

and the expansion of u converges for all (r,θ). We apply the operator Δ^{k-1} term by term and obtain our claim.

6.3. Completion of the proof of the Main Theorem. We continue with the proof of Theorem 6.1 and discuss first the case $\alpha \neq -1$. We recall that, if $j \neq n$ or $k \neq m$, then S_{jk} and S_{nm} are orthogonal,

$$(S_{jk},S_{nm}) = \int_\omega S_{jk}S_{nm}\, d\omega,$$

where ω is the unit sphere about the origin, and $d\omega$ is the Euclidean surface element of ω.

If $u \in H^kB$, then (u,S_{nm}) is bounded for any (n,m). For $\alpha \neq -1$, $N \geq 2k$,

$$(u,S_{nm}) = c\sum_{j=0}^{k-1} (a_{jnm}r^{p_n+(2\alpha+2)j} + b_{jnm}r^{q_n+(2\alpha+2)j}).$$

Because the right-hand side must be bounded for any choice of $r \in (0,\infty)$, either a_{jnm} or $p_n + (2\alpha+2)j$ vanishes, and either b_{jnm} or $q_n + (2\alpha+2)j$ vanishes, for all j. We note that

$$p_n + (2\alpha+2)j = 0, \qquad q_n + (2\alpha+2)j = 0$$

is equivalent to

$$[(4j+2-N)^2 - (N-2)^2](\alpha+1)^2 = 4n(n+N-2).$$

If $N \geq 2k$, $[(4j+2-N)^2 - (N-2)^2](\alpha+1)^2 \leq 0$ and the above system of equations has no solutions by virtue of $n > 0$. Therefore, the coefficients a_{jnm}, b_{jnm}

vanish for all (j,n,m).

We conclude that all terms, except for the constant, in our expansions of u must vanish because, for fixed N,α, they are unbounded. The proof of Theorem 6.1 is completed by using a similar argument for $\alpha = -1$. In this case, the Theorem is true for all N.

6.4. Lower dimensional spaces. We proceed to show that, for certain α, H^kB functions do exist on E_α^N of any dimension $N < 2k$. Examining the proof of Theorem 6.1, we see that it would hold for $N < 2k$ if again the system

$$p_n + (2\alpha + 2)j = 0, \qquad q_n + (2\alpha + 2)j = 0$$

had no solutions; in fact, the terms involving μ_j and ν_j would be eliminated as they are not bounded. Hence, we need only find out when the system has solutions.

THEOREM. *For fixed* N, $\alpha \neq -1$, $N < 2k$, *the generators of* H^kB *are the* S_{nm} *such that the above two equations hold*.

Proof. That the S_{nm} are H^kB functions follows from the equations

$$u_{nm} = ar^{p_n+(2\alpha+2)(j+1)} S_{nm} \quad \text{and} \quad v_{nm} = br^{q_n+(2\alpha+2)(j+1)} S_{nm}$$

in 6.2. By solving the equations $p_n + (2\alpha + 2)j = 0$ and $q_n + (2\alpha + 2)j = 0$ in the form

$$(2j + 1 - \tfrac{1}{2}N)(\alpha + 1) = -\tfrac{1}{2}[(N - 2)^2(\alpha + 1)^2 + 4n(n + N - 2)]^{1/2},$$

we find that the solutions for $j = k - 1$, $n \neq 0$, are

$$\alpha = -1 \pm \left(\frac{n(n + N - 2)}{4k^2 - (4 + 2N)k + 2N} \right)^{1/2} .$$

We have proved the Theorem.

One might suspect that the existence of H^kB functions always implies that of nondegenerate H^hB functions for $h > k$. However, we shall show:

$E_\alpha^N \in \tilde{O}^N_{H^kB}$ <u>implies</u> $E_\alpha^N \in \tilde{O}^N_{H^hB}$ <u>for all</u> $h > k$ <u>and all</u> N <u>if</u> $\alpha = 0$. <u>There exist</u> E_α^N <u>for which this is no longer true</u>.

To see this, suppose $\alpha = 0$. Then

$$2j + 1 - \tfrac{1}{2}N = -\tfrac{1}{2}[(N-2)^2 + 4n(n+N-2)]^{1/2},$$

hence $n = 2j + 2 - N$. If there exists an n satisfying this for $j = k-1$, there also exists an n for all $h \geq k$.

To show that the above is not true for all α, we choose $N = 4$, $n = 1$. For $j = 3$, we then have $\alpha = -1 + 8^{-1/2}$, whereas $j = 4$ should give $6 = n(n+2)$. Since no integer n satisfies this equation, we conclude for the above α that $E_\alpha^N \in O^N_{H^5B} \cap \tilde{O}^N_{H^4B}$.

<u>NOTES TO §6</u>. Theorems 6.1-6.4 were established in Mirsky-Sario-Wang [1].

That there are no inclusion relations between O_G^N and $O^N_{H^kX}$ for $X = B,D,C$ was shown in Mirsky-Sario-Wang [2]. The same was proved to hold between O^N_{HX} and $O^N_{H^kY}$ in Chung-Sario-Wang [1],[3]. A generalization of the biharmonic projection and decomposition to polyharmonic functions was developed in Wang-Sario [1]. An interesting problem would be to develop a general polyharmonic classification theory of Riemannian manifolds. We shall not return to polyharmonic functions in this book.

CHAPTER IV

DIRICHLET FINITE BIHARMONIC FUNCTIONS

Following the same order as for harmonic and quasiharmonic functions, we now take up the class H^2D of Dirichlet finite nonharmonic biharmonic functions. The first question is again: How are O_G^N and $O_{H^2D}^N$ related?

In the present case, the Poincaré N-ball B_α^N will play a crucial role in providing us with counterexamples, and we devote §1 to a characterization of those B_α^N which belong to $O_{H^2D}^N$. We note in passing the elegant case of B_1^N, which belongs to $O_{H^2D}^N$ if and only if $N > 10$. The results on B_α^N together with those on the punctured N-space E_α^N with the metric $ds = r^\alpha |dx|$ will be applied in §2 to show that no inclusion relations exist between O_G^N and $O_{H^2D}^N$ for any $N \geq 2$. In particular, there exist N-manifolds which are parabolic but nevertheless carry H^2D functions.

Minimum solutions of the Poisson equation $\Delta u = h$, that is, solutions which, in a specified sense, vanish at the ideal boundary, are the topic of §3. The problem of their existence and representation as limits is given a complete solution. As a closely related topic, the existence is shown of biharmonic functions which are Dirichlet finite but nevertheless are not harmonizable.

We return to these topics in Chapter VII, where we impose boundedness conditions on the Laplacian as well and discuss the existence of a Riesz representation.

For a more detailed description of the fascinating topics in the present chapter, we refer to the introductions in the sections.

§1. DIRICHLET FINITE BIHARMONIC FUNCTIONS ON THE POINCARÉ N-BALL

H^2D functions on the Poincaré N-ball B_α^N were first discussed in the concrete case $N = 3$. It was shown that $B_\alpha^3 \in \tilde{O}_{H^2D}^3$ if and only if $\alpha > -3/5$, and the question was raised whether the same is true for every N if and only if $\alpha > -3/(N+2)$.

We shall show that this is indeed so if $2 \leq N \leq 6$. However, quite unexpectedly, for $N > 6$ it turns out that $B_\alpha^N \in \tilde{O}^N_{H^2D}$ if and only if $\alpha \in (-3/(N+2), 5/(N-6))$.

We shall first discuss the case $N = 2$, which will require a separate treatment.

<u>1.1.</u> <u>H^2D functions on the Poincaré disk.</u> We recall that on a Riemannian N-manifold R, the mixed Dirichlet integral of functions $f, g \in C^1$ is, by definition,

$$D(f,g) = \int_R df * dg = \int_R g^{ij} \frac{\partial f}{\partial x^i} \frac{\partial f}{\partial x^j} dx$$

and the Dirichlet integral of a function $f \in C^1$ is $D(f) = D(f,f)$. We set

$$H^2D = \{u \in H^2 | D(u) < \infty\}$$

and

$$O^N_{H^2D} = \{R | H^2D(R) = \emptyset\}.$$

For the Poincaré disk

$$B_\alpha^2 = \{x = (r,\theta) | r < 1, \ ds = (1 - r^2)^\alpha |dx|, \ \alpha \text{ const}\},$$

we claim:

<u>LEMMA.</u> $B_\alpha^2 \in \tilde{O}^2_{H^2D} \Longleftrightarrow \alpha > -3/4.$

The proof will be given in 1.1-1.2.

We recall from II.3.9 and II.6.3 that $B_\alpha^2 \in \tilde{O}^2_{QD}$ if (and only if) $\alpha > -3/4$. Therefore,

$$B_\alpha^2 \in \tilde{O}^2_{H^2D} \quad \text{if} \quad \alpha > -3/4.$$

If $\alpha < -3/4$, suppose there exists a $u \in H^2D(B_\alpha^2)$. Since harmonicity on the Euclidean disk $\{r < 1\}$ is not affected by a conformal metric (cf. I.1.11), we have for $\Delta u = h \in H$,

$$h = \sum_{n=0}^{\infty} r^n (a_n \cos n\theta + b_n \sin n\theta).$$

Suppose $a_n \neq 0$ for some $n \geq 0$. For constants $0 < \beta < \gamma < 1$, take a function $\rho(r) \in C_0^\infty[0,1)$, with $\rho \geq 0$, supp $\rho \subset (\beta,\gamma)$, and for $0 < t \leq 1$, set $\rho_t(r) = \rho((1-r)/t)$, $\varphi_t = \rho_t \cos n\theta$. Then as $t \to 0$,

$$|(h,\varphi_t)| = c\int_{1-\gamma t}^{1-\beta t} r^n \rho_t(r)(1-r^2)^{2\alpha} r\,dr$$

$$\sim ct^{2\alpha}\int_{1-\gamma t}^{1-\beta t} \rho_t\,dr = ct^{2\alpha+1}.$$

On the other hand,

$$D(\varphi_t) = \int_{B_\alpha^2}\left[g^{rr}\left(\frac{\partial\varphi_t}{\partial r}\right)^2 + g^{\theta\theta}\left(\frac{\partial\varphi_t}{\partial\theta}\right)^2\right]g^{1/2}drd\theta$$

$$= \int_{B_\alpha^2}\left[\left(\frac{\partial\varphi_t}{\partial r}\right)^2 + r^{-2}\left(\frac{\partial\varphi_t}{\partial\theta}\right)^2\right] rdrd\theta = D_E(\varphi_t),$$

with the subindex E indicating the Euclidean metric. As $t \to 0$,

$$D(\varphi_t) \sim \int_{1-\gamma t}^{1-\beta t} (c_1\rho_t'^2 + c_2\rho_t^2)dr = O(t^{-1}) + O(t),$$

So that $D(\varphi_t)^{1/2} = O(t^{-1/2})$. We have a contradiction with Stokes' formula,

$$|(h,\varphi_t)| = |D(u,\varphi_t)| \leq D(u)^{1/2}D(\varphi_t)^{1/2} = cD(\varphi_t)^{1/2},$$

if $2\alpha + 1 < -\frac{1}{2}$. Therefore, $a_n = 0$ for $n \geq 0$ if $\alpha < -3/4$. On replacing $\rho_t \cos n\theta$ by $\rho_t \sin n\theta$ we see similarly that $b_n = 0$ for $n \geq 0$ if $\alpha < -3/4$. Thus $h = 0$, in violation of $u \notin H$, and we have

$$B_\alpha^2 \in O^2_{H^2D} \quad \text{if} \quad \alpha < -3/4.$$

1.2. Case $\alpha = -3/4$ for $N = 2$. It remains to show that

$$B^2_{-3/4} \in O^2_{H^2D}.$$

We shall again make use of $D = D_E$. Suppose there exists a $u \in H^2D$. We have $\Delta u = h \in H$,

$$h = \sum_{n=0}^{\infty} r^n(a_n \cos n\theta + b_n \sin n\theta).$$

If $a_n^2 + b_n^2 \neq 0$ for some $n \geq 0$, let $t \in (\frac{1}{2},1)$, choose

$$\rho_t(r) = \begin{cases} 2r, & r \in [0,\frac{1}{2}), \\ 1, & r \in [\frac{1}{2},t), \\ 1 - 2(r-t)/(1-t), & r \in [t,\frac{1}{2}(t+1)), \\ 0, & r \in [\frac{1}{2}(t+1),1), \end{cases}$$

and set $\varphi_t = \rho_t(a_n \cos n\theta + b_n \sin n\theta)$. For a number $\tau > 0$, we obtain

$$\int_B h*\tau\varphi_t \geq a(\tau(1-t)^{-1/2} - \tau),$$

where B stands for $B^2_{-3/4}$ and $a > 0$ is a constant independent of t and τ. For the subset $B' = \{\frac{1}{2} < r < 1\}$ of B, there exists a constant $b > 0$ independent of t and τ such that

$$D_{B'}(\tau\varphi_t) \leq b(\tau^2(1-t)^{-1} + \tau^2).$$

Let $\{t_j\}_1^\infty$ be a sequence of numbers in $(\frac{1}{2},1)$ such that $t_{j+1} > \frac{1}{2}(t_j + 1)$ for $j = 1,2,\ldots$. Clearly, $1 - t_j < 2^{-j}$. Take a sequence $\{\tau_j\}_1^\infty$ given by

$$\tau_j = (1 - t_j)^{1/2} j^{-1}.$$

Observe that

$$\sum_{j=1}^\infty \tau_j < \infty, \qquad \sum_{j=1}^\infty \tau_j^2 < \infty.$$

Next consider a sequence $\{\varphi_n\}_1^\infty$ of functions in $C_0D(B)$,

$$\varphi_n(re^{i\theta}) = \sum_{j=1}^n \tau_j \varphi_{t_j}(re^{i\theta}).$$

We obtain

$$\int_B h*\varphi_n \geq a(\sum_{j=1}^n j^{-1} - \sum_{j=1}^n \tau_j).$$

The definition of φ_n gives on B',

$$\left(\frac{\partial\varphi_n}{\partial x^i}\right)^2 = \sum_{j=1}^n \left(\tau_j \frac{\partial\varphi_{t_j}}{\partial x^i}\right)^2,$$

$i = 1,2$, except at points $|z| = t_1,\ldots,t_n$. A fortiori,

$$D_{B'}(\varphi_n) = \sum_{j=1}^{n} D_{B'}(\tau_j \varphi_{t_j}).$$

Thus we obtain, in view of the choice of b,

$$D_{B'}(\varphi_n) \le b(\sum_{j=1}^{n} j^{-2} + \sum_{j=1}^{n} \tau_j^2).$$

On the subset $B'' = \{r < \frac{1}{2}\}$ of B, the function $\psi = \varphi_t | B''$ is independent of t. Set $c = D_{B''}(\psi)$. In view of $\varphi_n | B'' = \sum_1^n \tau_j \psi$, we have

$$D_{B''}(\varphi_n) \le c(\sum_{j=1}^{n} \tau_j)^2.$$

Since the Dirichlet integral in our conformal metric is the same as in the Euclidean metric, we conclude by $D_B(\varphi_n) = D_{B'}(\varphi_n) + D_{B''}(\varphi_n)$ that

$$D_B(\varphi_n) \le b(\sum_{j=1}^{n} j^{-2} + \sum_{j=1}^{n} \tau_j^2) + c(\sum_{j=1}^{n} \tau_j)^2.$$

From these estimates, it follows that

$$\lim_{n\to\infty} \frac{\int_B h * \varphi_n}{D_B(\varphi_n)^{1/2}} = \infty,$$

in violation of Stokes' formula at the end of 1.1 applied to $\{\varphi_n\}$. Therefore, $a_n = b_n = 0$ for $n \ge 0$, and we have $B^2_{-3/4} \in O^2_{H^2D}$.

The proof of Lemma 1.1 is complete.

1.3. H^2D functions on the Poincaré N-ball. For convenient reference, we include Lemma 1.1 in the following main statement:

THEOREM. *The Poincaré N-balls carrying H^2D functions are characterized by*

$$B_\alpha^N \in \tilde{O}^N_{H^2D} \iff \begin{cases} \alpha > -3/(N+2) & \text{for } N \in [2,6], \\ \alpha \in (-3/(N+2), 5/(N-6)) & \text{for } N > 6. \end{cases}$$

As an interesting special case, the unit N-ball with the natural metric $ds = (1-r^2)|dx|$ carries H^2D functions if and only if $N \leq 10$.

The proof of the Theorem will be given in 1.3-1.8, where we may restrict our attention to $N > 2$.

By Theorem II.3.9, we have at once:

$$B_\alpha^N \in \tilde{O}^N_{H^2D} \quad \text{if} \quad \alpha \in (-3/(N+2), 1/(N-2)), \quad N > 2.$$

For $\alpha \notin (-3/(N+2), 1/(N-2))$, it will be necessary to discuss separately five cases:

Case I: $N > 6$, $\alpha > 5/(N-6)$.

Case II: $N > 6$, $\alpha = 5/(N-6)$.

Case III: $N \in [3,6]$, $\alpha \geq 1/(N-2)$; and $N > 6$, $\alpha \in [1/(N-2), 5/(N-6))$.

Case IV: $N > 2$, $\alpha < -3/(N+2)$.

Case V: $N > 2$, $\alpha = -3/(N+2)$.

In the proofs, we shall make use of the orders of growth of harmonic functions established in I.2.7.

<u>1.4</u>. <u>Case I</u>: $\alpha > 5/(N-6)$. We start with those α which are to the right of the interval in Theorem 1.3. First we omit the end point.

<u>LEMMA</u>. <u>If</u> $N > 6$ <u>and</u> $\alpha > 5/(N-6)$, <u>then</u> $B_\alpha^N \in O^N_{H^2D}$.

<u>Proof</u>. Suppose there exists a $u \in H^2D(B_\alpha^N)$. Expand $\Delta u = h$ in fundamental spherical harmonics:

$$h = \sum_{n=0}^{\infty} f_n(r) \sum_{m=1}^{m_n} a_{nm} S_{nm}(\theta).$$

If $a_{nm} \neq 0$ for some $(n,m) \neq (0,1)$, take the testing function $\varphi_t(r,\theta) = \rho_t(r) S_{nm}(\theta)$, where $\rho_t(r) = \rho((1-r)/t)$, $0 < t \leq 1$, and $\rho(r)$ is a fixed function in $C_0^\infty[0,1)$, $\rho \geq 0$, supp $\rho \subset (\beta,\gamma)$, $0 < \beta < \gamma < 1$. By Lemma I.2.7, we obtain as $t \to 0$

$$|(h,\varphi_t)| = c\int_{1-\gamma t}^{1-\beta t} f_n(r)\rho_t(r)(1-r^2)^{N\alpha} r^{N-1}dr$$

$$> c(\beta t)^{2\alpha+1}\int_{1-\gamma t}^{1-\beta t} \rho_t(r)dr$$

$$= ct^{2\alpha+2} \quad .$$

On the other hand,

$$D(\varphi_t) = \int_{B_\alpha^N} * |\text{grad } \varphi_t|^2$$

$$= \int_{1-\gamma t}^{1-\beta t} (1-r^2)^{-2\alpha}(c_1\rho_t'(r)^2 + c_2 r^{-2}\rho_t(r)^2)(1-r^2)^{N\alpha} r^{N-1}dr$$

$$< c(\gamma t)^{(N-2)\alpha}\int_{1-\gamma t}^{1-\beta t} (c_1\rho_t'(r)^2 + c_2\rho_t(r)^2)dr$$

$$= d_1 t^{(N-2)\alpha-1} + d_2 t^{(N-2)\alpha+1} < dt^{(N-2)\alpha-1},$$

where the c's and d's are independent of t. We have a contradiction with the inequality $|(h,\varphi_t)| \le cD(\varphi_t)^{1/2}$ at the end of 1.1 if $2\alpha + 2 - \frac{1}{2}[(N-2)\alpha - 1] < 0$, that is, $\alpha > 5/(N-6)$. Hence, for such α, $a_{nm} = 0$ for all $(n,m) \neq (0,1)$. But $a_{01} \neq 0$ implies $B_\alpha^N \in \tilde{O}_{QD}^N$ which is a violation of $\alpha > 5/(N-6)$.

<u>1.5</u>. <u>Case II</u>: $\alpha = 5/(N-6)$. Next we consider the right end point of the interval in Theorem 1.3.

<u>LEMMA</u>. <u>If</u> $N > 6$, <u>then</u> $B_{5/(N-6)}^N \in O_{H^2D}^N$.

<u>Proof</u>. Choose a decreasing sequence of real numbers $t_j \in (0,1]$ tending to 0 such that

$$1 - \beta t_j < 1 - \gamma t_{j+1}$$

and, as in 1.4, $|(h,\varphi_{t_j})| > ct_j^{2\alpha+2}$ is satisfied for each t_j. Set

$$q_j = \text{sign}(h,\varphi_{t_j}) \cdot j^{-1} t_j^{-2\alpha-2}$$

and choose for the testing functions $\varphi_n = \Sigma_{j=1}^n q_j\varphi_{t_j}$. Then

$$|(h,\varphi_n)| = \left|\sum_{j=1}^{n} q_j(h,\varphi_{t_j})\right| > c \sum_{j=1}^{n} j^{-1}$$

and

$$D(\varphi_n) = \sum_{j=1}^{n} q_j^2 D(\varphi_{t_j}) < c \sum_{j=1}^{n} j^{-2} t_j^{(N-6)\alpha-5}.$$

For $\alpha = 5/(N-6)$, we have $D(\varphi_n) < c \sum_1^n j^{-2}$, which is bounded as $n \to \infty$, whereas $|(h,\varphi_n)| \to \infty$. We conclude that $B^N_{5/(N-6)} \in O^N_{H^2D}$.

1.6. Case III: $\alpha \in [1/(N-2), 5/(N-6))$. We now come to those α in the intervals in Theorem 1.3 for which the existence of H^2D functions was not assured by that of QD functions.

LEMMA. *If* $N \in [3,6]$ *and* $\alpha \geq 1/(N-2)$, *or* $N > 6$ *and* $\alpha \in [1/(N-2), 5/(N-6))$, *then* $B^N_\alpha \in \tilde{O}^N_{H^2D}$.

Proof. First we show that the relation

$$|(h,\varphi)| \leq c\, D(\varphi)^{1/2}$$

for some $h \in H$ and all $\varphi \in C^1_0 D$, proved at the end of 1.1 to be necessary for $B^N_\alpha \in \tilde{O}^N_{H^2D}$, is also sufficient. Let L be the Hilbert space obtained by completing $C^1_0 D$ with respect to $D(\cdot)^{1/2}$. Since the functional (h,φ) defined for $\varphi \in C^1_0 D$ is linear and bounded, it can be extended to a bounded linear functional on L. By the Riesz theorem, there exists a $u \in L$ such that $(h,\varphi) = D(u,\varphi)$ for every $\varphi \in L$, and in particular for every $\varphi \in C^\infty_0$. Since $\Delta G_\Omega h = h$ for every regular subregion Ω, the Green's function g_Ω on Ω, and the potential $G_\Omega h(x) = \int_\Omega g_\Omega(x,y) * h(y)$, we have, in particular,

$$D(u - G_\Omega h, \varphi) = 0$$

for every $\varphi \in C^\infty_0(\Omega)$. By Weyl's lemma, $u - G_\Omega h \in HD(\Omega)$ and therefore, $u \in C^2(\Omega)$, $\Delta u = h$ on Ω, and $u \in D(B^N_\alpha)$.

We shall show that $h(r,\theta) = f_1(r)S_{11}(\theta)$ satisfies the above condition. We expand $\varphi \in C^\infty_0(B^N_\alpha)$ in spherical harmonics,

$$\varphi(r,\theta) = \sum_{n=0}^{\infty} \sum_{m=1}^{m_n} b_{nm}(r)S_{nm}(\theta),$$

and obtain

$$(h,\varphi) = c\int_0^1 b_{11}(r)f_1(r)(1-r^2)^{N\alpha}r^{N-1}dr.$$

On setting $w(r) = f_1(r)(1-r^2)^{N\alpha}r$ and observing that

$$b_{11}(r)\left(\int_r^1 w(s)ds\right)r^{N-2}\Big|_0^1 = 0,$$

we have

$$(h,\varphi) = c\int_0^1 [b_{11}'(r)r + (N-2)b_{11}(r)]\left(\int_r^1 w(s)ds\right)r^{N-3}dr.$$

We insert the factor $(1-r^2)^{(\varepsilon-1)/2}(1-r^2)^{(1-\varepsilon)/2}$ in the integrand and obtain by the Schwarz inequality,

$$(h,\varphi)^2 < c\int_0^1 [b_{11}'(r)r + (N-2)b_{11}(r)]^2(1-r^2)^{1-\varepsilon}\left(\int_r^1 w(s)ds\right)^2 r^{2N-6}dr.$$

Here $[\ ]^2 < 2(N-2)^2(b_{11}'(r)^2r^2 + b_{11}(r)^2)$, so that

$$(h,\varphi)^2 < c\int_0^1 (b_{11}'(r)^2 + r^{-2}b_{11}(r)^2)\,(1-r^2)^{1-\varepsilon}\left(\int_r^1 w(s)ds\right)^2 r^{2N-4}dr.$$

For $\alpha > 1/(N-2)$, Lemma I.2.7 yields

$$\int_r^1 w(s)ds < c(1-r^2)^{2\alpha+2},$$

and we have

$$(h,\varphi)^2 < c\int_0^1 (b_{11}'(r)^2 + r^{-2}b_{11}(r)^2(1-r^2)^{4\alpha+5-\varepsilon}r^{N-1}dr.$$

On the other hand, by the Dirichlet-orthogonality of the S_{nm},

$$D(\varphi) \geq D(b_{11}(r)S_{11}(\theta))$$

$$\geq \int_{B_\alpha^{N*}} \left((1-r^2)^{-2\alpha}\left(\frac{\partial(b_{11}(r)S_{11}(\theta))}{\partial r}\right)^2 + r^{-2}(1-r^2)^{-2\alpha}\left(\frac{\partial(b_{11}(r)S_{11}(\theta))}{\partial\theta^1}\right)^2\right)$$

$$> c\int_0^1 (b_{11}'(r)^2 + r^{-2}b_{11}(r)^2)(1-r^2)^{(N-2)\alpha}r^{N-1}dr.$$

Therefore, $B^N_\alpha \in \tilde{O}^N_{H^2D}$ if $4\alpha + 5 - \varepsilon > (N-2)\alpha$, that is, $5 - \varepsilon > (N-6)\alpha$, as asserted. For $\alpha = 1/(N-2)$, we use $f_1(r) \sim c \log(1-r)$ as in I.2.7, and argue as above.

1.7. Case IV: $\alpha < -3/(N+2)$. Next we consider those α which are to the left of the interval in Theorem 3.1. First we omit the end point. The next two Lemmas, which cover the remaining two cases, parallel Lemmas 1.4 and 1.5. Whereas the estimate $f_n(r) \sim c(1-r^2)^{1-(N-2)\alpha}$ was valid for $\alpha > 1/(N-2)$, the estimate $f_n(r) \sim \text{const} \neq 0$ holds for $\alpha \leq -3/(N+2)$.

LEMMA. If $N > 2$ and $\alpha < -3/(N+2)$, then $B^N_\alpha \in O^N_{H^2D}$.

Proof. Let h, ρ, ρ_t, and φ_t be as in the proof of Lemma 1.4. Since $\lim_{r\to 1} f_n(r) \neq 0$, we have for $t \to 0$,

$$|(h,\varphi_t)| = c\int_{1-\gamma t}^{1-\beta t} f_n(r)\rho_t(r)(1-r^2)^{N\alpha} r^{N-1} dr$$

$$> c(\beta t)^{N\alpha}\int_{1-\gamma t}^{1-\beta t} \rho_t(r)dr = ct^{N\alpha+1} ,$$

whereas

$$D(\varphi_t) < ct^{(N-2)\alpha-1}.$$

Hence the relation $|(h,\varphi_t)| < c\,D(\varphi_t)^{1/2}$ is violated by $\alpha < -3/(N+2)$.

1.8. Case V: $\alpha = -3/(N+2)$. It remains to consider the left end point of the interval in Theorem 1.3.

LEMMA. If $N > 2$, then $B^N_{-3/(N+2)} \in O^N_{H^2D}$.

The proof is identical with that of Lemma 1.5 except that we now set $q_j = \text{sign}(h,\varphi_{t_j})\cdot j^{-1} t_j^{-N\alpha-1}$.

This completes the proof of Theorem 1.3.

1.9. Test for $H^2D \neq \emptyset$. In the proof of the $\tilde{O}^N_{H^2D}$ test at the end of 1.1 and the beginning of 1.6, we made no use of properties of B^N_α, and the reasoning has general validity. We state for later reference:

<u>LEMMA</u>. <u>A Riemannian</u> N-<u>manifold</u>, $N \geq 2$, <u>belongs to</u> $\tilde{O}^N_{H^2D}$ <u>if and only if</u>

$$(h,\varphi)^2 \leq c\,D(\varphi)$$

<u>for some</u> $h \in H$, <u>all</u> $\varphi \in C^1_0D$, <u>and some constant</u> c <u>independent of</u> φ.

<u>NOTES TO §1</u>. Lemma 1.1 was established for $\alpha \neq -3/4$ in Nakai-Sario [9], and for $\alpha = -3/4$ in O'Malla (alias Nakai) [1]. Theorem 1.3 for $N = 3$ was proved in Sario-Wang [5], and for $N \geq 3$ in Hada-Sario-Wang [2]. Lemma 1.9 was introduced in Nakai-Sario [9].

§2. PARABOLICITY AND DIRICHLET FINITE BIHARMONIC FUNCTIONS

The results on the Poincaré N-ball B^N_α obtained in §1 will provide us with some of the counterexamples needed to determine how $O^N_{H^2D}$ is related to O^N_G. Further counterexamples will be given by the punctured N-space $E^N_\alpha = \{0 < r < \infty,$ $ds = r^\alpha|dx|\}$. We showed in III.1.3 that its parabolicity depends heavily on N and α, and in III.3.1 and III.3.6 that the same is true of the existence of H^2B functions on it. We shall prove in 2.1 that, in interesting contrast, there is no dependence on N or α in the case of H^2D functions: $E^N_\alpha \in O^N_{H^2D}$ for all (N,α).

There is one relation that both B^N_α and E^N_α leave open: Is $O^2_G \cap \tilde{O}^2_{H^2D} \neq \emptyset$? In 2.2, we give a simple example, a 2-cylinder with a suitable metric, which settles this question in the affirmative.

All ingredients are now ready to draw, in 2.3, the conclusion: there are no relations between O^N_G and $O^N_{H^2D}$ for any $N \geq 2$.

For $N = 2$, the discovery that there exist manifolds which are parabolic but nevertheless carry H^2D functions has special significance in the development of classification theory. It was the first "intrusion" into the until then "indivisible" class O^N_G. In view of this significance, we also reproduce, in 2.4-2.5, the original proof. Although it was later superseded by the simpler proof in 2.2, the original reasoning offers considerable methodological interest.

We continue with the 2-dimensional case in 2.6 and 2.7, where we discuss dependence of the H^2D degeneracy on a radial metric in the plane. We consider the

2-manifold $C_\alpha = \{r < \infty \,|\, ds = (1+r^2)^\alpha |dz|\}$ and prove that $C_\alpha \in O^2_{H^2D}$ if and only if $\alpha \geq -3/4$.

<u>2.1</u>. <u>No H^2D functions on E^N_α</u>. We consider the punctured N-space $E^N_\alpha = \{0 < r < \infty\}$ with the metric $ds = r^\alpha |dx|$, α constant.

<u>THEOREM</u>. <u>For</u> $N \geq 2$, <u>and every</u> α,

$$E^N_\alpha \in O^N_{H^2D}.$$

<u>Proof</u>. Suppose there exists a $u \in H^2D$. We know from III.3.4 that if $\alpha \neq -1$, then

$$u = r^{2\alpha+2}\Big(\sum_{n\neq\mu}\sum_{m=1}^{m_n} a_{nm} r^{p_n} S_{nm} + \sum_{n\neq\nu}\sum_{m=1}^{m_n} b_{nm} r^{q_n} S_{nm}\Big)$$
$$+ r^{(2-N/2)(\alpha+1)} \log r \cdot \Big(\sum_{m=1}^{m_\mu} a_{\mu m} S_{\mu m} + \sum_{m=1}^{m_\nu} b_{\nu m} S_{\nu m}\Big)$$
$$+ \sum_{n=1}^{\infty}\sum_{m=1}^{m_n} (c_{nm} r^{p_n} + d_{nm} r^{q_n}) S_{nm} + f(r),$$

with $p_n > 0$, $q_n < 0$, μ, ν, a_{nm}, b_{nm}, c_{nm}, d_{nm} constants, $r^{p_n} S_{nm}$, $r^{q_n} S_{nm}$ harmonic functions, and $f(r)$ biharmonic; the summations $\Sigma_{n\neq\mu}$ and $\Sigma_{n\neq\nu}$ do not include $n = 0$. If $\alpha = -1$, then

$$u = \log r \cdot \sum_{n=1}^{\infty}\sum_{m=1}^{m_n} (a_{nm} r^{p_n} + b_{nm} r^{q_n}) S_{nm} + \sum_{n=1}^{\infty}\sum_{m=1}^{m_n} (c_{nm} r^{p_n} + d_{nm} r^{q_n}) S_{nm} + g(r),$$

with $g(r)$ biharmonic. Both series converge uniformly and absolutely on compact subsets. For a fixed n, let u_n be the sum of the terms involving an S_{nm}, and denote by u_0 the radial part of the expansion of u. Then $u = \Sigma_0^\infty u_i$.

We shall show that $D(u) \geq D(u_n)$ for every n. For constants $0 < r_0 < r_1 < \infty$ and the regular subregion $\Omega = \{r_0 < r < r_1\}$ of E^N_α,

$$D_\Omega(u) = D_\Omega(u_n) + D_\Omega(v) + 2D_\Omega(u_n, v),$$

where $v = u - u_n$. Since the above expansions in S_{nm} were deduced in III.3.4 from series of harmonic functions with compact convergence, the convergence of $\Sigma_0^\infty u_i$ and its partial derivatives is also compact (cf. III.4.10) and we have

$$D_\Omega(u_n,v) = \lim_{j\to\infty} D_\Omega(u_n, \sum_{\substack{i=0\\ i\neq n}}^{j} u_i).$$

The Dirichlet orthogonality of spherical harmonics gives $D_\Omega(u_n,v) = 0$, hence $D_\Omega(u) \geq D_\Omega(u_n)$.

On letting $r_0 \to 0$, $r_1 \to \infty$ we obtain an exhaustion $\Omega \to E^N_\alpha$, which gives $D(u) \geq D(u_n)$. A direct computation shows that $D(u_n) = \infty$ for every nonconstant u_n. This contradiction completes the proof of the Theorem.

2.2. H^2D functions on a parabolic 2-cylinder. In addition to the examples provided by B^N_α and E^N_α, we shall need in 2.3 a 2-manifold to show that

$$O^2_G \cap \tilde{O}^2_{H^2D} \neq \emptyset.$$

Consider the 2-cylinder

$$S_\lambda = \{|x| < \infty, |y| \leq 1\}$$

with the metric

$$ds^2 = \lambda(x)^2(dx^2 + dy^2),$$

where λ will be specified later. The conformal metric does not affect harmonicity, and the harmonic measure $\omega(x) = ax + b$ must reduce to $\omega = \text{const}$ for both boundary components $x = -\infty$ and $x = \infty$ of S_λ. Therefore, $S_\lambda \in O^2_G$.

We shall show that, for a suitable λ, the function

$$u = \log\log(e^x + a)$$

belongs to H^2D. Here the constant $a > 1$ is so chosen that

$$a\log(1 + a) = 1.$$

The Euclidean Laplacian

$$\Delta_E u = \frac{e^x[e^x - a\log(e^x + a)]}{(e^x + a)^2[\log(e^x + a)]^2}$$

has the properties

$$\operatorname{sign} \Delta_E u(x) = \operatorname{sign} x, \qquad (\Delta_E u)'(0) > 0.$$

Thus $\Delta_E u/x$ is well defined and positive for all x. Let

$$\lambda^2 = \frac{\Delta_E u}{x}.$$

On the Riemannian manifold S_λ with this choice of λ, the Laplace-Beltrami operator Δ gives

$$\Delta u = x \in H(S_\lambda),$$

hence $u \in H^2(S_\lambda)$. The Dirichlet integral $D(u)$ is independent of λ (cf. I.2.3) and can be taken with respect to the Euclidean metric:

$$D(u) = \int_{-1}^{1}\int_{-\infty}^{\infty} \left(\frac{\partial u}{\partial x}\right)^2 dxdy = 2\int_{-\infty}^{\infty} \left(\frac{e^x}{(e^x+a)\log(e^x+a)}\right)^2 dx$$

$$< 2\int_{-\infty}^{\infty} \frac{(e^x+a)e^x dx}{(e^x+a)^2[\log(e^x+a)]^2} = -\left.\frac{2}{\log(e^x+a)}\right|_{-\infty}^{\infty} < \infty.$$

Thus $S_\lambda \in \tilde{O}^2_{H^2D}$.

<u>2.3</u>. <u>Parabolicity and</u> H^2D <u>degeneracy</u>. We are ready to prove:

<u>THEOREM</u>. <u>For</u> $N \geq 2$, <u>the totality of Riemannian</u> N-<u>manifolds has the following decomposition into four nonempty components</u>:

$$O^N_G \cap O^N_{H^2D} \oplus O^N_G \cap \tilde{O}^N_{H^2D} \oplus \tilde{O}^N_G \cap O^N_{H^2D} \oplus \tilde{O}^N_G \cap \tilde{O}^N_{H^2D}.$$

The unexpectedness of $O^N_G \cap \tilde{O}^N_{H^2D} \neq \emptyset$ will be discussed in 2.5.

<u>Proof</u>. By Lemma III.1.3 and Theorem 2.1,

$$E^N_{-1} \in O^N_G \cap O^N_{H^2D}$$

for $N \geq 2$. By 2.2,

$$S_\lambda \in O^2_G \cap \tilde{O}^2_{H^2D},$$

and by Lemma I.2.6 and Theorem 1.3,

$$B_\alpha^N \in O_G^N \cap \tilde{O}^N_{H^2D}$$

for $N \in [3,6]$, $\alpha \geq 1/(N-2)$, and for $N > 6$, $\alpha \in [1/(N-2), 5/(N-6))$. By Theorem 1.3,

$$B_\alpha^2 \in \tilde{O}_G^2 \cap O^2_{H^2D}$$

for $\alpha \leq -3/4$, and by Lemma III.1.3 and Theorem 2.1,

$$E_\alpha^N \in \tilde{O}_G^N \cap O^N_{H^2D}$$

for $N > 2$, $\alpha \neq -1$. The final component in the decomposition contains again the Euclidean N-ball and, more generally,

$$B_\alpha^N \in \tilde{O}_G^N \cap \tilde{O}^N_{H^2D}$$

for $N \geq 2$, $\alpha \in (-3/(N+2), 1/(N-2))$.

2.4. Another test for $H^2D \neq \emptyset$. We now present the original example showing that $O_G^2 \cap \tilde{O}^2_{H^2D} \neq \emptyset$. We start with a test for the existence of H^2D functions.

For a given $h \in H - \{0\}$, denote by P_hD the class of Dirichlet finite solutions of the Poisson equation $\Delta u = h$,

$$P_hD = \{u \in H^2D \mid \Delta u = h\} .$$

Let H_{PD} be the class of functions $h \in H - \{0\}$ for which the Poisson equation has Dirichlet finite solutions,

$$H_{PD} = \{h \in H - \{0\} \mid P_hD \neq \emptyset\}.$$

For a Riemannian N-manifold R,

$$R \in O^N_{H^2D} \Longleftrightarrow H_{PD} = \emptyset.$$

On a regular subregion Ω of R, let $g_\Omega(x,y)$ be the harmonic Green's function, and set

$$G_\Omega(h,h) = \int_{\Omega\times\Omega} g_\Omega(x,y)*h(x)*h(y).$$

LEMMA. On a Riemannian manifold R, $h \in H_{PD}$ if and only if

$$\sup_{\Omega \subset R} G_\Omega(h,h) < \infty.$$

Proof. Suppose $h \in H_{PD}$ and take a $u \in P_hD$. Let $h_{\Omega u}$ be the harmonic function on Ω with boundary values $u|\partial\Omega$. In the same fashion as in the proof of Theorem II.1.1, we obtain the Riesz decomposition

$$u = h_{\Omega u} + G_\Omega h$$

on Ω, and the equation

$$D_\Omega(u) = D_\Omega(h_{\Omega u}) + G_\Omega(h,h).$$

It follows that $G_\Omega(h,h) < D_\Omega(u) < D(u)$.

Conversely, suppose $\sup_\Omega G_\Omega(h,h) < \infty$. Fix a regular subregion Ω_0 and consider varying regular subregions Ω with $\overline{\Omega}_0 \subset \Omega$. Set $v_\Omega = G_\Omega h$ and denote by w_Ω the harmonic function on Ω_0 with boundary values $v_\Omega|\partial\Omega_0$. Since $\Delta v_\Omega = h$ on Ω_0,

$$v_\Omega = w_\Omega + G_{\Omega_0} h$$

on Ω_0, and

$$D_{\Omega_0}(v_\Omega) \leq D_\Omega(v_\Omega) = G_\Omega(h,h) < \sup_\Omega G_\Omega(h,h) = K.$$

Therefore, $D_{\Omega_0}(w_\Omega) \leq K$. For a fixed $x_0 \in \Omega_0$ and $c_\Omega = v_\Omega(x_0)$,

$$w_\Omega(x_0) - c_\Omega = -G_{\Omega_0} h(x_0) = \text{const}.$$

We claim that $\{w_\Omega - c_\Omega\} \subset HD(\Omega_0)$ contains a sequence which is uniformly convergent on each compact subset of Ω_0. To see this, let $L = \{w \in HD(\Omega_0) \,|\, w(x_0) = 0\}$. Since L is a Hilbert space with respect to $D_{\Omega_0}(\cdot,\cdot)$, it has a reproducing kernel $k(x,y)$,

$$w_\Omega(x) - w_\Omega(x_0) = D_{\Omega_0}(w_\Omega, k(\cdot,x)),$$

and therefore,

$$|w_\Omega(x) - w_\Omega(x_0)|^2 \leq K\, k(x,x).$$

Since $k(x,y)$ is continuous on $\Omega_0 \times \Omega_0$, $\{w_\Omega - w_\Omega(x_0)\}$ contains a sequence uniformly convergent on each compact subset of Ω_0. In view of $w_\Omega - c_\Omega = (w_\Omega - w_\Omega(x_0)) - G_{\Omega_0}h(x_0)$, we have the assertion for $\{w_\Omega - c_\Omega\}$.

Again by $v_\Omega - v_\Omega(x_0) = (w_\Omega - c_\Omega) + G_{\Omega_0}h$, $\{v_\Omega - v_\Omega(x_0)\}$ contains a sequence uniformly convergent on each compact subset of Ω_0. By the diagonal process, we can find an exhaustion $\{\Omega_n\}_1^\infty$ of R such that $\{G_{\Omega_n}h - G_{\Omega_n}h(x_0)\}$ converges uniformly to a function u on each compact subset of R.

By Fatou's Lemma,

$$D(u) \leq \lim_n \inf D_{\Omega_n}(G_{\Omega_n}h) = \lim_n \inf G_{\Omega_n}(h,h) \leq K.$$

For a fixed $\Omega \subset \Omega_n$, $G_{\Omega_n}h - G_{\Omega_n}h(x_0) = h_n + G_\Omega h$, where $h_n \in H(\Omega)$ with boundary values $G_{\Omega_n}h - G_{\Omega_n}h(x_0)$ on $\partial\Omega$. On letting $n \to \infty$ we obtain

$$u = h_{\Omega u} + G_\Omega h$$

on Ω. This shows that $\Delta u = h$ on Ω and a fortiori, $u \in P_hD(R)$, that is, $h \in H_{PD}$. This completes the proof of the Lemma.

From the above proof, we also have:

<u>COROLLARY</u>. $h \in H_{PD}$ <u>if and only if there exists an exhaustion</u> $\{\Omega_n\}$ <u>of</u> R <u>such that</u> $\sup_n G_{\Omega_n}(h,h) < \infty$.

<u>2.5</u>. <u>Original counterexample</u>. If a Riemannian manifold R belongs to O_G^N, then $\lim_{\Omega \to R} g_\Omega = \infty$, and in view of Lemma 2.4, one is tempted to say that H^2D must be void, that is, $O_G^N \subset O_{H^2D}^N$. We shall now reproduce the original counterexample showing for $N = 2$ that this is not the case:

$$O_G^2 \cap \tilde{O}^2_{H^2D} \neq \emptyset.$$

Although the later example given in 2.2 is simpler, the original one is of substantial methodological interest.

Take a bordered Riemann surface S with an analytic border α which is not necessarily compact but nonvoid, and with the property that the double $\hat{S}$ of S about α is parabolic. (For an axiomatic discussion of the concepts in 2.5, see, e.g., Ahlfors-Sario [1].) The simplest example is the half-plane $S = \{x > 0\}$. Let j be the involution of $\hat{S}$, that is, p and $j(p)$ are symmetric about α for $p \in \hat{S}$. Then $\hat{S} = S \cup \alpha \cup j(S)$, and α is fixed pointwise under j.

Take an $s \in H(S \cup \alpha)$ such that $s|\alpha = 0$ and $s|S > 0$. (For the existence of such an s, see, e.g., Parreau [1].) Consider an $h \in C(S)$ such that $h|S = s$ and $h|j(S) = -s \circ j$. By the reflection principle, $h \in H(\hat{S})$.

Take a C^∞ second order differential $P_0(z)dxdy$, $z = x + iy$, on $\hat{S}$ such that $P_0(z)dxdy = P_0(\zeta)d\xi d\eta$ for $j(z) = \zeta = \xi + i\eta$ and $P_0(z) > 0$. For example, $|\text{grad } h(z)|^2$ suitably modified at points where $dh = 0$ qualifies as $P_0(z)$. We can find a C^∞ function $\psi > 0$ on $\hat{S}$ with $\psi \circ j = \psi$ such that $P(z)dxdy = \psi(z)P_0(z)dxdy$ satisfies $\int_{\hat{S}} P(z)dxdy < \infty$. Clearly, $P(z)dxdy$ is a C^∞ second order differential on $\hat{S}$ such that $P(z) > 0$ and $P(z)dxdy = P(\zeta)d\xi d\eta$ for $\zeta = j(z)$.

Let $g(z,\zeta)$ be the Green's function on S. Since $\int_S P(z)dxdy < \infty$, we see that $\varphi_0(z) \equiv \int_S g(z,\zeta)P(\zeta)d\xi d\eta < \infty$ for every $z \in S$. We extend φ_0 continuously to $\hat{S}$ by $\varphi_0(z) = \varphi_0(j(z))$ for each $z \in j(S)$. Then we take a C^∞ function φ on $\hat{S}$ such that $\varphi \geq \varphi_0$ and $\varphi \circ j = \varphi$ on $\hat{S}$. Finally, we consider on $\hat{S}$ a C^∞ Riemannian metric

$$d\sigma(z) = \lambda(z)|dz| = \left(\frac{1}{1+h(z)^2} \cdot \frac{1}{1+\varphi(z)} \ P(z)\right)^{1/2} |dz| .$$

Note that $d\sigma$ is symmetric about α. We let R be the resulting Riemannian manifold $(\hat{S}, d\sigma)$. Harmonicities on $\hat{S}$ and R are identical, and in particular, $R \in O_G^2$. We shall show that $h \in H_{PD}$.

We are to establish the existence of a finite constant K such that

$$\int_{\Omega \times \Omega} g_\Omega(z,\zeta) * h(z) * h(\zeta) \leq K$$

for every regular subregion Ω of R with

$$j(\Omega) = \Omega.$$

Since such Ω's form an exhaustion of R, the existence of K assures, by Corollary 2.4, that $h \in H_{PD}$.

Let $\omega = \Omega \cap S$ and denote by $g_\omega(z,\zeta)$ the Green's function on ω. Observe that $\Omega = \omega \cup \alpha \cup j(\omega)$ and therefore,

$$g_\omega(z,\zeta) = g_\Omega(z,\zeta) - g_\Omega(j(z),\zeta)$$

for $z,\zeta \in \omega$. This relation, though simple, is the crucial point in the proof of the existence of K.

We proceed to evaluate $A = \int_{\Omega\times\Omega} g_\Omega(z,\zeta)*h(z)*h(\zeta)$. Denote by $Q(z,\zeta)$ the integrand, $A = \int_{\Omega\times\Omega} Q(z,\zeta)$, and decompose A into

$$A = \left(\int_{\omega\times\omega} + \int_{j(\omega)\times j(\omega)} + \int_{\omega\times j(\omega)} + \int_{j(\omega)\times\omega}\right) Q(z,\zeta).$$

Since $g_\Omega(j(z),j(\zeta)) = g_\Omega(z,\zeta)$, $h(j(z)) = -h(z)$, $h(j(\zeta)) = -h(\zeta)$, and the volume element $*1$ satisfies $*1(j(z)) = *1(z)$, we see that

$$\int_{j(\omega)\times j(\omega)} Q(z,\zeta) = \int_{\omega\times\omega} Q(z,\zeta), \qquad \int_{j(\omega)\times\omega} Q(z,\zeta) = \int_{\omega\times j(\omega)} Q(z,\zeta).$$

From this and $h(j(z)) = -h(z)$, we obtain

$$A = 2\left(\int_{\omega\times\omega} + \int_{j(\omega)\times\omega}\right) Q(z,\zeta) = 2\int_{\omega\times\omega} (Q(z,\zeta) + Q(j(z),\zeta))$$

$$= 2\int_{\omega\times\omega} (g_\Omega(z,\zeta)*h(z)*h(\zeta) + g_\Omega(j(z),\zeta)*h(j(z))*h(\zeta))$$

$$= 2\int_{\omega\times\omega} (g_\Omega(z,\zeta) - g_\Omega(j(z),\zeta))*h(z)*h(\zeta)$$

$$= 2\int_{\omega\times\omega} g_\omega(z,\zeta)*h(z)*h(\zeta)$$

$$= 2\int_{\omega\times\omega} g_\omega(z,\zeta)\,\frac{h(z)}{1+h(z)^2}\cdot\frac{h(\zeta)}{1+h(\zeta)^2}\cdot\frac{P(z)}{1+\varphi(z)}\cdot\frac{P(\zeta)}{1+\varphi(\zeta)}\,dxdyd\xi d\eta.$$

Since $0 < h/(1+h^2) < 1$ on ω,

$$A < 2\int_{\omega\times\omega} g_\omega(z,\zeta)\,\frac{P(z)}{1+\varphi(z)}\cdot\frac{P(\zeta)}{1+\varphi(\zeta)}\,dxdyd\xi d\eta$$

$$< 2\int_\omega\left(\int_\omega g_\omega(z,\zeta)P(\zeta)d\xi d\eta\right)\frac{P(z)}{1+\varphi(z)}\,dxdy\ .$$

In view of $g_\omega < g$ and $\int_S g(z,\zeta)P(\zeta)d\xi d\eta = \varphi_0(z) \leq \varphi(z)$ for $z \in S$,

$$A < 2\int_\omega\left(\int_S g(z,\zeta)P(\zeta)d\xi d\eta\right)\frac{P(z)}{1+\varphi(z)}\,dxdy$$

$$< 2\int_\omega \frac{\varphi(z)}{1+\varphi(z)}P(z)dxdy$$

$$< 2\int_S P(z)dxdy.$$

If we choose $K = 2\int_S P(z)dxdy = \int_{\hat S} P(z)dxdy$, we have the required bound for $G_\Omega(h,h)$ and conclude that $R \in \tilde{O}^2_{H^2D}$.

2.6. Plane with radial metrics. The discussion in 2.5 only gives what is needed there: the existence of a metric that turns the plane, say, into a 2-manifold in $O^2_G \cap \tilde{O}^2_{H^2D}$. An explicit expression for the metric was not produced. An interesting question now arises: Can the existence of H^2D functions on the plane be characterized if a suitable explicitly expressed family of metrics is given? As a natural analogue of the Poincaré disk, we consider the 2-manifold

$$C_\alpha = \{r < \infty \mid ds = (1+r^2)^\alpha |dz|,\ \ \alpha \text{ constant}\}.$$

We shall show:

THEOREM. $C_\alpha \in O^2_{H^2D}$ if and only if $\alpha \geq -3/4$.

Proof. Denote by Δ_α, dv_α, and grad_α the Laplace-Beltrami operator, the volume element, and the gradient on C_α. Use Δ, dv, grad, and C for the case

$\alpha = 0$. For $\lambda_\alpha(z) = (1 + |z|^2)^\alpha$, we have $\Delta_\alpha = \lambda_\alpha^{-2}\Delta$, $dv_\alpha = \lambda_\alpha^2 dv$, and $\text{grad}_\alpha = \lambda_\alpha^{-1}\text{grad}$. Therefore, $H(C_\alpha) = H(C)$ and $D_{C_\alpha}(\varphi) = D_C(\varphi)$ for a smooth φ. A fortiori, the assertion $C_\alpha \in \tilde{O}^2_{H^2D}$ is equivalent to the Poisson equation

$$\Delta u(z) = \lambda_\alpha(z)^2 h(z)$$

having a Dirichlet finite solution u on C for some harmonic function $h \not\equiv 0$. We denote by $H_\alpha(C)$ the class of such harmonic functions. Clearly, $H_\alpha(C) \cup \{0\}$ forms a vector space.

We have to show that $H_\alpha(C) = \emptyset$ if and only if $\alpha \geq -3/4$.

Expand an h $H(C)$ into its Fourier series

$$h(re^{i\theta}) = \sum_{n=0}^{\infty} r^n(a_n \cos n\theta + b_n \sin n\theta),$$

$b_0 = 0$, and write $m(h) = \sup\{n | a_n^2 + b_n^2 \neq 0\} \leq \infty$. Denote by E_k the class $\{h \in H(C) | m(h) \leq k\}$ for $k = 0,1,2,\ldots$, and set $E_k = \{0\}$ for $k = -1,-2,\ldots$, $E_k' = \{h \in E_k | h \not\equiv 0,\ a_0 = b_0 = 0\}$ for $k = 1,2,\ldots$, and $E_k' = \emptyset$ for $k = 0,-1,-2,\ldots$. First we prove:

<u>LEMMA</u>. <u>If</u> $-4\alpha > k+2 \geq 3$, <u>then</u> $E_k' \subset H_\alpha(C)$.

We only have to show that $r^n \cos n\theta$ and $r^n \sin n\theta$ belong to $H_\alpha(C)$ for every n with $1 \leq n < -4\alpha - 2$. Since the reasoning is the same for both, it suffices to show that $h = r^n \cos n\theta \in H_\alpha(C)$. Let $\varphi \in C_0^\infty(C)$, and expand it into its Fourier series

$$\varphi(re^{i\theta}) = \sum_{n=0}^{\infty} (a_n(r)\cos n\theta + b_n(r)\sin n\theta),$$

where $a_n(r)$ and $b_n(r)$ are all in $C_0^\infty[0,\infty)$. Observe that

$$D_C(\varphi) = \sum_{n=0}^{\infty} \pi\left(\int_0^\infty (a_n'(r)^2 + b_n'(r)^2) r\,dr + n^2 \int_0^\infty (a_n(r)^2 + b_n(r)^2)\frac{dr}{r}\right).$$

On the other hand,

$$(h,\varphi)_\alpha = \int_0^\infty \left(\int_0^{2\pi} \varphi(re^{i\theta})\cos n\theta\, d\theta\right) r^{n+1}(1+r^2)^{2\alpha} dr = \pi \int_0^\infty a_n(r) r^{n+1}(1+r^2)^{2\alpha} dr.$$

By the Schwarz inequality,

$$(h,\varphi)_\alpha^2 \le \pi^2 K_\alpha \int_0^\infty a_n(r)^2\, \frac{dr}{r},$$

where $K_\alpha = \int_0^\infty r^{2n+3}(1+r^2)^{4\alpha}dr$ is finite if and only if $-4\alpha > n+2 \ge 3$. For such n, Lemma 1.9 gives $r^n\cos n\theta \in H_\alpha(C)$. We have proved the Lemma.

2.7. Completion of the proof. We assert:

LEMMA. If $k+3 \ge -4\alpha$, then $H_\alpha(C) \subset E_k'$.

Let the Fourier expansion of $h \in H_\alpha(C)$ be as above, and suppose $a_n^2 + b_n^2 \ne 0$. For $t > 1$, the function

$$\rho_t(r) = \begin{cases} (r - t^{1/2})^2(t - r)^2, & r \in [t^{1/2}, t], \\ 0, & r \in [0,\infty) - [t^{1/2}, t], \end{cases}$$

belongs to $C_0^1[0,\infty)$. Therefore, the function

$$\varphi_t(re^{i\theta}) = \rho_t(r)(a_n\cos n\theta + b_n\sin n\theta)$$

belongs to $C_0^1(C)$. By an easy computation, we find universal positive constants A, B and $t_0 > 1$ such that

$$(h,\tau\varphi_t)_\alpha \ge A\tau t^{6+n+4\alpha}, \qquad D_C(\tau\varphi_t) \le B\tau^2 t^8$$

for every $t > t_0$ and $\tau > 0$. If $6+n+4\alpha > 4$, then $(h,\varphi_t)_\alpha^2/D_C(\varphi_t) \to \infty$, which contradicts the inequality in Lemma 1.9. If $6+n+4\alpha = 4$, then

$$(h,\tau\varphi_t) \ge A\tau t^4, \qquad D_C(\tau\varphi_t) \le B\tau^2 t^8.$$

Let $\{t_\nu\}_0^\infty$ be a sequence of real numbers such that $t_\nu + \nu < t_{\nu+1}^{1/2}$, and consider a sequence $\{\tau_\nu\}_1^\infty$ given by

$$\tau_\nu t_\nu^4 = \nu^{-1},$$

$\nu = 1,2,\ldots$. Then take a sequence $\{\Phi_\mu\}_1^\infty$ of functions in $C_0^1(C)$,

$$\Phi_\mu(re^{i\theta}) = \sum_{\nu=1}^{\mu} \tau_\nu \varphi_{t_\nu}(re^{i\theta}).$$

We infer that

$$(h,\Phi_\mu)_\alpha \geq A \sum_{\nu=1}^{\mu} \nu^{-1}.$$

Furthermore, $(\partial\Phi_\mu/\partial x^i)^2 = \Sigma_{\nu=1}^{\mu}(\tau_\nu \partial\varphi_{t_\nu}/\partial x^i)^2$, $i = 1,2$, and a fortiori, $D_C(\Phi_\mu) = \Sigma_1^\mu D_C(\tau_\nu \varphi_{t_\nu})$. We obtain

$$D_C(\Phi_\mu) \leq B \sum_{\nu=1}^{\mu} \nu^{-2},$$

and it follows that

$$\frac{(h,\Phi_\mu)_\alpha^2}{D_C(\Phi_\mu)} \geq \frac{A^2(\Sigma_1^\mu \nu^{-1})^2}{B\, \Sigma_1^\mu \nu^{-2}} \to \infty$$

as $\mu \to \infty$. Hence n must satisfy $6+n+4\alpha < 4$, that is, $n+2 < -4\alpha \leq k+3$. Then $n \leq k$, and $H_\alpha(C) \subset E_k$. Because of Lemma 2.6,

$$a_0 = h(re^{i\theta}) - \sum_{n=1}^{m(h)} r^n(a_n \cos n\theta + b_n \sin n\theta)$$

must belong to $H_\alpha(C)$ unless $a_0 = 0$. It is easy to find a bounded sequence $\{\varphi_\mu\}_1^\infty \subset C_0^1(C)$ such that φ_μ converges to 1 and $D_C(\varphi_\mu) \to 0$. If $a_0 \in H_\alpha(C)$, then $(a_0,\varphi_\mu)^2 \to (2\pi a_0 \int_0^\infty (1+r^2)^{2\alpha} r\,dr)^2 > 0$; but $D_C(\varphi_\mu) \to 0$ as $\mu \to \infty$, in violation of the inequality in Lemma 1.9. Therefore, $a_0 = 0$ and $h \in E_k'$, that is, $H_\alpha(C) \subset E_k'$, and we have the Lemma.

Suppose that $H_\alpha(C) = \emptyset$. If $-4\alpha > 1+2 = 3$, then by Lemma 2.6, $E_1' \subset H_\alpha(C)$, a contradiction. Therefore, $-4\alpha \leq 3$. Conversely, suppose that $-4\alpha \leq 3$, that is, $0+3 \geq -4\alpha$. By Lemma 2.7, we see that $H_\alpha(C) \subset E_0' = \emptyset$. Thus $H_\alpha(C) = \emptyset$ if and only if $\alpha \geq -3/4$. This completes the proof of Theorem 2.6.

Let u_1 and u_2 be Dirichlet finite solutions of $\Delta u = \lambda_\alpha^2 h$. Then $u_1 - u_2$ is a Dirichlet finite harmonic function on C, that is, $u_1 - u_2 \in HD(C) = R^1$, with R^1 the 1-space. Therefore, the vector space $H^2D(C_\alpha)/R^1$ is isomorphic to $H_\alpha(C) \cup \{0\}$. By Lemmas 2.6 and 2.7, $H_\alpha(C) \cup \{0\} = E_k'(-4\alpha - 2 > k \geq -4\alpha - 3)$. Since $\dim E_k' = 2k$ for $k > 0$ and $= 0$ for $k \leq 0$, we obtain a more precise form of Theorem 2.6:

Let d_α be the dimension of the vector space $H^2D(C_\alpha)/HD(C_\alpha) = H^2D(C_\alpha)/R^1$. If $\alpha \geq -3/4$, then $d_\alpha = 0$. If $\alpha < -3/4$, then $d_\alpha = 2k_\alpha$, with $-4\alpha - 2 > k_\alpha \geq -4\alpha - 3$.

NOTES TO §2. Theorem 2.1 was proved in Sario-Wang [12], and the counterexample in 2.2 given in Sario-Wang [9]. Theorem 2.3 was obtained in Sario-Wang [12]. Lemma 2.4 and its corollary are from Nakai-Sario [9]. The original counterexample 2.5 to show $O_G^2 \cap \tilde{O}^2_{H^2D} \neq \emptyset$ was exhibited in Nakai-Sario [9]. Theorem 2.6 is due to Nakai [7].

We have not explored what the counterpart of Theorem 2.6 is for $N > 2$. Does $\alpha \geq -3/(2N)$ continue to be a necessary and sufficient condition for $C_\alpha^N \in O^N_{H^2D}$?

§3. MINIMUM DIRICHLET FINITE BIHARMONIC FUNCTIONS

The existence of manifolds which are parabolic but nevertheless carry H^2D functions leads to interesting new questions.

Since the solution u of the Poisson equation $\Delta u = h$ on a Riemannian manifold R is determined up to an additive harmonic function, we ask: Can this function be so chosen that u vanishes, in some sense, at the ideal boundary of R? We shall then call u a minimum solution of the Poisson equation. Suppose $h \in H_{PD}$, in the notation of 2.4. Then if $R \in \tilde{O}_G^N$, there exists a unique minimum solution; if $R \in O_G^N$, then every $u \in P_hD$ is a minimum solution. Having observed this in 3.1, we show in 3.2 that for $R \in \tilde{O}_G^N$, the unique minimum solution is $u_h = \lim_{\Omega\to R} G_\Omega h$, whereas for $R \in O_G^N$, there exists for every minimum solution u a set $\{c_\Omega\}$ of constants such that $u = \lim_{\Omega\to R}(G_\Omega h + c_\Omega)$.

In 3.3, we take up the fascinating question of harmonizability. We give an example of a (parabolic) manifold and a biharmonic function on it which, though Dirichlet finite, is not harmonizable.

The problems in this section are closely related to the Riesz representation, which we shall discuss in Chapter VII.

3.1. Existence of minimum solutions. In the present section, CD will stand for the class of continuous Dirichlet finite functions on a Riemannian manifold R, and C_0D for the subclass of CD functions with compact supports. Consider the topology τ on CD given by the simultaneous convergence in the Dirichlet integral and the uniform convergence on each compact subset on R. Denote by P_0 the closure of C_0D with respect to topology τ in CD. The Royden decomposition reads

$$\begin{cases} CD = HD + P_0, & R \in \tilde{O}_G^N, \\ CD = P_0, & R \in O_G^N, \end{cases}$$

where the sum is the vector space direct sum and also the Dirichlet orthogonal sum (e.g., Sario-Nakai [1, p. 162 ff.]).

For an $h \in H_{PD}$, in the notation of 2.4, a function u in P_hD will be called a minimum solution of the Poisson equation $\Delta u = h$ if $u \in P_0$. We assert:

If $R \in \tilde{O}_G^N$, then there exists a unique minimum solution u_h, and

$$P_hD = HD + u_h.$$

If $R \in O_G^N$, then every $u \in P_hD$ is a minimum solution, and for a fixed $u \in P_hD$,

$$P_hD = u + \{\text{const}\}.$$

In fact, if u_1 and u_2 are minimum solutions, then $\Delta(u_1 - u_2) = h - h = 0$ and $u_1 - u_2 \in HD \cap P_0$. Hence $u_1 - u_2$ is either 0 or a constant according as $R \in \tilde{O}_G^N$ or $R \in O_G^N$. The rest is a consequence of Royden's decomposition.

3.2. Minimum solutions as limits. We recall that every u in the class P_h of solutions of the Poisson equation $\Delta u = h$ has the representation $u = h_{\Omega u} + G_\Omega h$ on a regular subregion Ω of R. The term $G_\Omega h$ is the minimum solution on Ω, and one might expect that $G_\Omega h$ converges to the minimum solution on R. In reality, this rarely happens, as we shall see later in a discussion of Riesz representations. But if we restrict our attention to P_hD, this is certainly the case for $R \in \tilde{O}_G^N$. The question is clearly equivalent to the convergence problem of $h_{\Omega u}$.

In general, a real-valued function f on R is said to be harmonizable if $h_{\Omega f}$ is convergent. As a consequence of the Royden decomposition, every $f \in CD$ on $R \in \tilde{O}_G^N$ is harmonizable. In fact, let $f = w + \varphi$ with $w \in HD$ and $\varphi \in P_0$. We only have to show that φ is harmonizable. Let Ω_0 be a regular subregion and $e_\Omega \in H(\Omega - \bar{\Omega}_0) \cap C(\bar{\Omega} - \Omega_0)$ with $e_\Omega | \partial\Omega_0 = 1$ and $e_\Omega | \partial\Omega = 0$ for $\Omega \supset \bar{\Omega}_0$. By Stokes' formula,

$$\inf_{\partial\Omega_0} h_{\Omega|\varphi|} \int_{\partial\Omega_0} |{*de_\Omega}| \le D(h_{\Omega|\varphi|}, e_\Omega).$$

Using the Harnack constant c for $\partial\Omega_0$, we obtain

$$\sup_{\partial\Omega_0} h_{\Omega|\varphi|} \le c \inf_{\partial\Omega_0} h_{\Omega|\varphi|} \le (D_R(|\varphi|)/D_\Omega(e_\Omega))^{1/2} .$$

From this and the relations $D_R(|\varphi|) = D_R(\varphi)$, $\lim_{\Omega\to R} D_\Omega(e_\Omega) > 0$, and $|h_{\Omega\varphi}| \le h_{\Omega|\varphi|}$, it follows that any sequence in $\{h_{\Omega\varphi}\}$ contains a convergent subsequence. Suppose $\{h_{\Omega_n\varphi}\}$ converges to v. By Stokes' formula, $D_\Omega(h_{\Omega_n\varphi} - v) \to 0$. We set $h_{\Omega_n\varphi} = \varphi$ on $R - \Omega_n$, and conclude that $\varphi - h_{\Omega_n\varphi}$ belongs to C_0D and converges to $\varphi - v$ in the τ-topology. Thus $\varphi - v \in P_0$, $v = \varphi - (\varphi - v) \in P_0$, and $v \equiv 0$. Therefore, $\lim_\Omega h_{\Omega\varphi} \equiv 0$, that is, φ is harmonizable.

An immediate consequence of this observation is the first part of the following statement:

<u>THEOREM</u>. <u>If</u> $R \in \tilde{O}_G^N$ and $h \in H_{PD}$, <u>then the unique minimum solution</u> u_h <u>is obtained as</u>

$$u_h = \lim_{\Omega\to R} G_\Omega h,$$

<u>with the convergence uniform on each compact subset of</u> R.

<u>If</u> $R \in O_G^N$ <u>and</u> $h \in H_{PD}$, <u>then for each minimum solution</u> u <u>there exists a set</u> $\{c_\Omega\}$ <u>of real numbers such that</u>

$$u = \lim_{\Omega\to R} (G_\Omega h + c_\Omega)$$

<u>uniformly on each compact subset of</u> R.

<u>Proof</u>. We only need to prove the second part. By the proof of Lemma 2.4, we see that for $x_0 \in R$, $\{G_\Omega h - G_\Omega h(x_0)\}$ or any sequence in it contains a subsequence uniformly convergent on each compact subset of R. Let u_1 and u_2 be two limits. Then since $\Delta(u_1 - u_2) = h - h = 0$, $u_1 - u_2 \in HD = R^1$. But $u_1(x_0) = u_2(x_0) = 0$. Hence $u_1 \equiv u_2$, that is, $\{G_\Omega h - G_\Omega h(x_0)\}$ converges. Let u be a minimum solution and $u_{x_0} = \lim_\Omega(G_\Omega h - G_\Omega h(x_0))$. Then $u = c + u_{x_0}$, with c constant, and $c_\Omega = -G_\Omega h(x_0) + c$

gives the second part of the Theorem.

3.3. A nonharmonizable H^2D function. For $R \in O_G^N$, a function in $CD(R)$ is not necessarily harmonizable. A natural question arises: What can be said about $P_hD(R) \subset CD(R)$? In other words, does $\{G_\Omega h\}$ converge? We shall show:

THEOREM. There exist parabolic Riemannian manifolds and H^2D functions on them which are not harmonizable.

Let $R' \in O_G^N$ and take an $h' \in H_{PD}(R')$. The existence of such (R',h') was shown in 2.2 and 2.5. Choose $x_i' \in R'$ with $u'(x_i') = a_i$, $i = 1,2$, and $a_1 < a_2$, where $u' \in P_{h'}D(R')$. Let $R = R' - \{x_1',x_2'\}$, $h = h'|R$, and $u = u'|R$. Clearly, $h \in H_{PD}$ on $R \in O_G^N$ and $u \in P_hD(R)$.

Let $\{\Omega_n'\}$ be an exhaustion of R'. For $i = 1,2$, take geodesic balls $\{B_{in}\}_{n=1}^\infty$ such that $\overline{B}_{in} \cap \overline{B}_{jm} = \emptyset$ for $i \neq j$, with the x_i' the centers of the B_{in}, and with $\Omega_1' \supset \overline{B}_{in} \supset B_{in} \supset \overline{B}_{i,n+1}$, and $\bigcap_{n=1}^\infty B_{in} = \{x_i'\}$. Set $\Omega_{nmp} = \Omega_n' - \overline{B}_{1m} - \overline{B}_{2p}$. Let $w_{nmp}^{(q)} \in H(\Omega_{nmp}) \cap C(\overline{\Omega}_{nmp})$ for $q = 1,2,3$ such that

$$\begin{cases} w_{nmp}^{(1)}|\partial\Omega_n' = u, & w_{nmp}^{(1)}|\partial B_{1m} = 0, & w_{nmp}^{(1)}|\partial B_{2p} = 0; \\ w_{nmp}^{(2)}|\partial\Omega_n' = 0, & w_{nmp}^{(2)}|\partial B_{1m} = 1, & w_{nmp}^{(2)}|\partial B_{2p} = 0; \\ w_{nmp}^{(3)}|\partial\Omega_n' = 0, & w_{nmp}^{(3)}|\partial B_{1m} = 0, & w_{nmp}^{(3)}|\partial B_{2p} = 1. \end{cases}$$

Set $w_{n\infty\infty}^{(q)} = \lim_{m,p\to\infty} w_{nmp}^{(q)}$, and define $w_{\infty m\infty}^{(q)}$ and $w_{\infty\infty p}^{(q)}$ similarly. These limits exist and

$$w_{n\infty\infty}^{(1)} = h_{\Omega_n' u}, \quad w_{\infty m\infty}^{(1)} = w_{\infty\infty p}^{(1)} = 0;$$
$$w_{n\infty\infty}^{(2)} = 0, \quad w_{\infty m\infty}^{(2)} = 1, \quad w_{\infty\infty p}^{(2)} = 0;$$
$$w_{n\infty\infty}^{(3)} = w_{\infty m\infty}^{(3)} = 0, \quad w_{\infty\infty p}^{(3)} = 1.$$

These relations are trivial except for $w_{\infty m\infty}^{(1)} = w_{\infty\infty p}^{(1)} = 0$. Let $R_0 = R - \overline{B}_{1m}$. Clearly, $R_0 \in \tilde{O}_G^N$. For $\varphi \in C_0^\infty(R')$ such that $\varphi|B_{1,m-1} = 1$ and $\operatorname{supp}\varphi \subset \Omega_1'$, we have $(1-\varphi)u \in CD(R_0)$ and $(1-\varphi)u = u$ on $R - \Omega_1'$. By Royden's decomposition in 3.1, we see that $w^{(1)}_{\infty m\infty}$ exists. Because of $R \in O^N_G$, $w^{(1)}_{\infty m\infty}$ takes on its maximum and minimum on ∂B_{1m}; since these are zero, $w^{(1)}_{\infty m\infty} = 0$. Similarly, $w^{(1)}_{\infty\infty p} = 0$.

Let $x_0 \in \Omega'_1 - \overline{B}_{11} - \overline{B}_{21}$ and $\alpha^{(i)}_m = \sup_{\partial B_{im}} u$, $\beta^{(i)}_m = \inf_{\partial B_{im}} u$. Then $\alpha^{(i)}_m \searrow a_i$ and $\beta^{(i)}_m \nearrow a_i$ as $m \to \infty$. In view of the properties of the limits of the $w^{(q)}_{nmp}$, we can find sequences $\{n_m\}^\infty_{m=1}$ and $\{p_m\}^\infty_{m=1}$ such that $n_1 < n_2 < \ldots$, $p_1 < p_2 < \ldots$, and

$$|w^{(1)}_{n_m m p_m}(x_0)| < \frac{1}{m}, \qquad 1 > w^{(2)}_{n_m m p_m}(x_0) > 1 - \frac{1}{m}, \qquad w^{(3)}_{n_m m p_m}(x_0) < \frac{1}{m}.$$

Let $R_m = \Omega'_{n_m} - \overline{B}_{1m} - \overline{B}_{2p_m}$ and $v_m \in H(R_m) \cap C(\overline{R}_m)$ such that $v_m | \partial R_m = u$. Then $G_{R_m} h = u - v_m$ on R_m and $\{R_m\}$ is an exhaustion of R. Clearly,

$$w^{(1)}_{n_m m p_m} + \beta^{(1)}_m w^{(2)}_{n_m m p_m} + \beta^{(2)}_m w^{(3)}_{n_m m p_m} \le v_m \le w^{(1)}_{n_m m p_m} + \alpha^{(1)}_m w^{(2)}_{n_m m p_m} + \alpha^{(2)}_m w^{(3)}_{n_m m p_m}$$

on R_m. By the choice of the sequences $\{n_m\}$ and $\{p_m\}$, we obtain

$$-\frac{1}{m} + \beta^{(1)}_m\left(1 - \frac{1}{m}\right) \le v_m(x_0) \le \frac{1}{m} + \alpha^{(1)}_m + \alpha^{(2)}_m \frac{1}{m}$$

and therefore,

$$\lim_{m\to\infty} G_{R_m} h(x_0) = u(x_0) - a_1.$$

Similarly, on interchanging the roles of B_{1m} and B_{2p}, we can find an exhaustion $\{R^p\}$ of R such that

$$\lim_{p\to\infty} G_{R^p} h(x_0) = u(x_0) - a_2.$$

In view of $a_1 < a_2$, we conclude that

$$\liminf_{\Omega\to R} G_\Omega h(x_0) < \lim_{\Omega\to R} G_\Omega h(x_0).$$

The proof of Theorem 3.3 is herewith complete.

<u>NOTES TO §3</u>. Theorems 3.2 and 3.3 were proved in Nakai-Sario [9].

In the present section, we made no assumptions on any boundedness properties of h, or nondegeneracy of R other than hyperbolicity. In Chapter VII, we shall discuss the fascinating role played by the Dirichlet finiteness of h and the existence of QP functions on R in the problem of Riesz representation of H^2D functions.

CHAPTER V

BOUNDED DIRICHLET FINITE BIHARMONIC FUNCTIONS

A striking relation in the theory of harmonic functions is that the existence of HD functions implies that of HC = HBD functions, that is, $O^N_{HD} = O^N_{HC}$. Does the corresponding equality hold for biharmonic functions? Here we have one of the most fascinating phenomena that makes the theory of biharmonic functions fundamentally different from that of harmonic functions: The inequality $O^N_{H^2D} < O^N_{H^2C}$ is strict.

In contrast with Chapter IV, we shall here follow the chronological order in the development of the theory and first present, in §1, the original proof of the strict inequality in the case $N = 2$. The proof consists in constructing one of the most delicate counterexamples in all classification theory. Although a shorter proof, valid for every N, was later developed and will be presented in §2, the methodological interest of the original proof is so overwhelming that it warrants inclusion in this monograph.

We also show in §2 that $O^N_{H^2B} < O^N_{H^2C}$. More generally, the classes $O^N_{H^2B} \cap O^N_{H^2D}$, $O^N_{H^2B} \cap \tilde{O}^N_{H^2D}$, $\tilde{O}^N_{H^2B} \cap O^N_{H^2D}$, and $\tilde{O}^N_{H^2B} \cap \tilde{O}^N_{H^2D}$ are all nonvoid, in interesting contrast with the relation $O^N_{HB} < O^N_{HD}$ for harmonic functions.

§1. H^2D FUNCTIONS BUT NO H^2C FOR $N = 2$

Let $\lambda(z)dxdy$ be a strictly positive C^∞ second order differential on a noncompact Riemann surface R. We endow R with the conformal metric $ds = \lambda(z)^{1/2}|dz|$ so as to obtain a Riemannian 2-manifold, for which we retain the same notation R. The biharmonic equation takes the form $\Delta(\lambda^{-1}\Delta u) = 0$, where, as well as in the sequel, Δ stands for the Euclidean Laplacian. This equation is equivalent to the family of Poisson equations $\Delta u(z) = \lambda(z)h(z)$ for harmonic functions h on R.

Therefore, there exist Dirichlet finite solutions of the former equation if and only if there exist those of the latter for some h. For this reason we shall first characterize the existence of Dirichlet finite solutions of the latter, or a slightly more general equation. After this preparation we shall establish our main result: There exists a (parabolic) plane region R and a conformal metric $ds = \lambda^{1/2}|dz|$ on R such that the biharmonic equation $\Delta(\lambda^{-1}\Delta u) = 0$ has H^2D solutions but no H^2C solutions on the Riemannian 2-manifold $R = (R, ds)$.

<u>1.1</u>. <u>Existence of</u> H^2D <u>functions</u>. Let $\mu(z)dxdy$ be a strictly positive Hölder-continuous second order differential on a noncompact Riemann surface R. Denote by $H_\mu(R)$ the class of $h \in H(R)$, $h \not\equiv 0$, such that the Poisson equation

$$\Delta u(z) = \mu(z)h(z)$$

possesses a Dirichlet finite solution on R. If $R' \subset R$, then $H_\mu(R) \subset H_\mu(R')$. We first give complete conditions for an $h \in H(R)$ to belong to $H_\mu(R)$. To this end, let $\{\Omega\}$ be an exhaustion of R by regular subregions. Denote by $g_\Omega(z,\zeta)$ the Green's function on Ω. If $R \in \tilde{O}_G^2$, then the Green's function on R will be denoted by $g_R(z,\zeta)$ or simply $g(z,\zeta)$. We also consider the class $D(R)$ of functions φ such that the Dirichlet integral $D_R(\varphi)$ can be defined and is finite. We then have:

<u>THEOREM</u>. <u>For an</u> $h \in H(R)$, $h \not\equiv 0$, <u>the following three conditions are equivalent by pairs</u>:

(α) $h \in H_\mu(R)$,

(β) $\sup_{\varphi \in C_0 D(R)} \left(\int_R h(z)\varphi(z)\mu(z)dxdy \right)^2 / D_R(\varphi) < \infty$,

(γ) $\sup_\Omega \int_{\Omega\times\Omega} g_\Omega(z,\zeta)h(z)h(\zeta)\mu(z)\mu(\zeta)dxdyd\xi d\eta < \infty$,

<u>for one and hence for every exhaustion</u> $\{\Omega\}$ <u>of</u> R.

<u>Proof</u>. We set

$$(h,\varphi)_\mu = \int_R h(z)\varphi(z)\mu(z)dxdy$$

and denote by $G_\Omega(h,h)_\mu$ the integral in (γ). Clearly, $(\cdot,\cdot)_\mu$ is a semi-inner product and the same is true of $G_\Omega(\cdot,\cdot)_\mu$ because of the energy principle for Green's potentials (e.g., Constantinescu-Cornea [1]).

Suppose first that $h \in H_\mu(R)$ and u is a solution of $\Delta u = \mu h$ with $u \in D(R)$. Since Stokes' formula yields $(h,\varphi)_\mu = (\Delta u,\varphi)_\mu = D_R(u,\varphi)$ for $\varphi \in C_0^\infty(R)$ and for $\varphi \in C_0 D(R)$ by approximation, the Schwarz inequality implies (β).

Conversely, suppose that (β) is valid. Let F be the Hilbert space obtained by the completion of $C_0^\infty(R)$ with respect to the inner product $D_R(\cdot,\cdot)$. By (β), $\varphi \to (h,\varphi)_\mu$ is a bounded linear functional on $C_0^\infty(R)$ considered as a subspace of $F \subset D$, and therefore, it can be extended to F as a bounded linear functional $\varphi \to L(\varphi)$. By the Riesz theorem, there exists a $u \in F$ such that $L(\varphi) = D_R(u,\varphi)$ for every $\varphi \in F$. In particular, $(h,\varphi)_\mu = D_R(u,\varphi)$ for every $\varphi \in C_0^\infty(R)$. Let

$$\bar{u}_\Omega(z) = \int_\Omega g_\Omega(z,\zeta)h(\zeta)\mu(\zeta)d\xi d\eta.$$

Since $\Delta\bar{u}_\Omega = \mu h$, $D_\Omega(u-\bar{u}_\Omega,\varphi) = 0$ for every $\varphi \in C_0^\infty(\Omega)$. By Weyl's lemma, $u-\bar{u}_\Omega \in H(\Omega)$ and a fortiori, u is a solution of $\Delta u = \mu h$ on Ω. Here the arbitrariness of Ω implies that u is a solution of this equation on R, and $u \in D(R)$. Therefore, $h \in H_\mu(R)$, that is, the equivalence of (α) and (β) has been proved.

Once more suppose that $h \in H_\mu(R)$ and u is a solution of $\Delta u = \mu h$ with $u \in D(R)$. Let $\{\Omega\}$ be the exhaustion of R consisting of <u>every</u> regular subregion Ω of R. By a simple application of Stokes' formula, we have

$$u = h_{\Omega u} + \bar{u}_\Omega,$$

where $\bar{u}_\Omega$ has the expression given above, and $h_{\Omega u} \in H(\Omega) \cap C(\bar{\Omega})$ with $h_{\Omega u}|\partial\Omega = u$. Therefore, $D_\Omega(\bar{u}_\Omega) \le D_\Omega(u) \le D_R(u)$. Observe that

$$D_\Omega(\bar{u}_\Omega) = G_\Omega(h,h)_\mu,$$

which is known as the Evans relation; its proof is an easy application of Stokes' formula. Relation (γ) follows.

Conversely, suppose that (γ) is valid for <u>one</u> exhaustion $\{\Omega\}$ of R. In view of $D_\Omega(\bar{u}_\Omega) = G_\Omega(h,h)_\mu$, $\{\bar{u}_\Omega\}$ has a weak limit u in F, which can be seen, as in the proof of the equivalence of (α) and (β), to be a solution of $\Delta u = \mu h$ on R with $u \in D(R)$. Therefore, (α) and (γ) are equivalent.

<u>1.2</u>. <u>Antisymmetric functions</u>. Let S be a Riemann surface, and α a part of the ideal boundary such that (S,α) is a bordered surface with analytic boundary $\alpha \neq \emptyset$, compact or noncompact. Denote by $\hat{S}$ the <u>double</u> of S about α (e.g., Ahlfors-Sario [1, p. 119 ff]). Let j be the involution of $\hat{S}$, that is, a mapping of $\hat{S}$ such that z and $j(z)$ are symmetric about α. An $h \in H(\hat{S})$ is, by definition, antisymmetric if $h(j(z)) = -h(z)$ on $\hat{S}$. Let $\mu(z)dxdy$ be as before and $\zeta = j(z)$. If $\mu(\zeta)d\xi d\eta = \mu(z)dxdy$, then we say that $\mu(z)dxdy$ or μ is symmetric. As a specialization of Theorem 1.1, we have:

<u>THEOREM</u>. <u>Let</u> h <u>be an antisymmetric harmonic function on</u> $\hat{S}$, <u>and</u> $\mu(z)dxdy$ <u>a nonnegative symmetric Hölder-continuous second order differential</u> $\not\equiv 0$ <u>on</u> $\hat{S}$. <u>Suppose</u> $h \geq 0$ <u>on the support of</u> $\mu(z)dxdy$ <u>in</u> S. <u>Then</u> $h \in H_\mu(\hat{S})$ <u>if and only if</u>

$$\int_{S\times S} g_S(z,\zeta)h(z)h(\zeta)\mu(z)\mu(\zeta)dxdyd\xi d\eta < \infty.$$

The significance of this Theorem manifests itself when $\hat{S}$ is parabolic. Observe that S always carries a Green's function since it has an analytic border α. For the proof of the Theorem, let $\{\Omega\}$ be an exhaustion of $\hat{S}$ such that $j(\Omega) = \Omega$, and set $\omega = \Omega \cap S$. The key relation is

$$g_\Omega(z,\zeta) - g_\Omega(j(z),\zeta) = g_\omega(z,\zeta)$$

for z and ζ in ω. From this, the antisymmetry of h, and the symmetry of μ, we deduce

$$G_\Omega(h,h)_\mu = 2G_\omega(h,h)_\mu.$$

By Theorem 1.1, $h \in H_\mu(S)$ is equivalent to the boundedness of $\{G_\Omega(h,h)_\mu\}_\Omega$, and this in turn to the boundedness of $\{G_\omega(h,h)_\mu\}_\Omega$. Since $h(z)h(\zeta) \geq 0$ on the support of $\mu(z)\mu(\zeta)dxdyd\xi d\eta$ in $S\times S$, the boundedness of $\{G_\omega(h,h)_\mu\}_\Omega$ is equivalent to the condition of the Theorem. This completes the proof.

By definition, we excluded the constant function 0 from $H_\mu(R)$, so that $H_\mu(R)$ is not a vector space. However, for R^1, the field of reals, we can at least say that

$$R^1H_\mu(R) + R^1H_\mu(R) - \{0\} \subset H_\mu(R),$$

that is, $H_\mu(R) \cup \{0\}$ is a vector space. The proof is obvious. The following relation will be used repeatedly:

$$\bigcap_{j=1}^{n} H_{\mu_j}(R) \subseteq H_{\Sigma_{j=1}^{n}\mu_j}(R).$$

This is a direct consequence of the definition of $H_\mu(R)$.

1.3. Main Theorem. We are ready to state our main result:

THEOREM. *There exists a (parabolic) plane region R and a conformal metric ds on R such that the Riemannian 2-manifold (R,ds) carries H^2D functions but no H^2C.*

The proof will be given in 1.3-1.9.

Denote by C the finite complex plane $\{|z| < \infty\}$ and by P the right half-plane. The basic surface we are going to consider is the doubly punctured plane

$$C' = C - \{-3,3\},$$

a parabolic region. Set $P' = P - \{3\}$ and observe that

$$C = \hat{P}, \qquad C' = \hat{P}',$$

with respect to ∂P, the imaginary axis. The R in the main theorem will be C'.

1.4. Auxiliary function μ_1. We shall construct on C' a conformal metric $ds = \lambda^{1/2}|dz|$ with $\lambda > 0$, $\lambda \in C^\infty$, such that the biharmonic equation $\Delta(\lambda^{-1}\Delta u) = 0$ has the desired solution. We shall consider six auxiliary functions (differentials)

$\mu_1,\ldots,\mu_6$, the sum of which will be λ. First we consider a positive radial C^∞ function μ_1 on C such that $\mu_1|\{|z|\le 4\} = 1$ and

$$\mu_1(z) = (1 + |z|)^{-(3+\alpha)}, \qquad 0 < \alpha < 1,$$

on $\{|z| \ge 5\}$ in C and also in C'. The elementary functions

$$c(z) = \frac{1}{2}(z + \bar{z}), \qquad s(z) = \frac{1}{2i}(z - \bar{z})$$

will play important roles. First we assert:

<u>LEMMA</u>. <u>The functions</u> c <u>and</u> s <u>belong to</u> $H_{\mu_1}(C) \subset H_{\mu_1}(C')$.

<u>Proof</u>. Since the proof is the same, we will only show that $c \in H_{\mu_1}(C)$. Take an arbitrary $\varphi \in C_0^\infty(C')$ and expand it into its Fourier series

$$\varphi(re^{i\theta}) = \sum_{n=0}^{\infty} (a_n(r)\cos n\theta + b_n(r)\sin n\theta).$$

In the double integral $(c,\varphi)_{\mu_1}$, we first integrate with respect to θ:

$$(c,\varphi)_{\mu_1} = O\left(\int_0^\infty a_1(r)r^2(1 + r)^{-(3+\alpha)}dr\right).$$

By the Schwarz inequality,

$$(c,\varphi)_{\mu_1}^2 \le K\int_0^\infty a_1(r)^2 r^{-1}dr,$$

where

$$K = O\left(\int_0^\infty r^5(1 + r)^{-2(3+\alpha)}dr\right) < \infty.$$

On the other hand, $D_C(\varphi) \ge \pi \int_0^\infty a_1(r)^2 r^{-1}dr$, and a fortiori,

$$(c,\varphi)_{\mu_1}^2 \le K D_C(\varphi).$$

By Theorem 1.1, we conclude that $c \in H_{\mu_1}(C)$.

<u>1.5</u>. <u>Auxiliary function</u> μ_2. Next take a radial nonnegative C^∞ function φ on C with its support in the disk $|z| < \frac{1}{2}$ such that $\varphi \equiv 1$ in a neighborhood of $z = 0$. Fix a decreasing sequence $\{\varepsilon_n\}_1^\infty$ of positive numbers such that $\Sigma_1^\infty \varepsilon_n < \infty$,

and choose an increasing sequence $\{\eta_n\}_1^\infty$ of positive numbers such that $\eta_1 > 6$, $\varepsilon_n(\eta_n - 1) > 1$, and $(\eta_n + 1)^2 < \eta_{n+1} + 1$ for every $n = 1,2,\ldots$. For the function

$$\mu_2(z) = \sum_{n=1}^{\infty} \varepsilon_n \varphi(z - i\eta_n),$$

we claim:

<u>LEMMA</u>. <u>The function</u> c <u>belongs to</u> $H_{\mu_2}(C) \subset H_{\mu_2}(C')$ <u>but the function</u> s <u>does not belong to</u> $H_{\mu_2}(C')$.

<u>Proof</u>. By Theorem 1.2, we only have to show that $G_P(c,c)_{\mu_2} < \infty$. By the energy principle of Green's potentials,

$$(G_P(c,c)_{\mu_2})^{1/2} \leq \sum_{n=1}^{\infty} (G_P(c\varepsilon_n\varphi(\cdot - i\eta_n), c\varepsilon_n\varphi(\cdot - i\eta_n))_1)^{1/2}.$$

The square of the nth term of the sum on the right is given by

$$\varepsilon_n^2 \int_{(|z-i\eta_n|<1)\cap P\times(|\zeta-i\eta_n|<1)\cap P} \log\left|\frac{z+\bar\zeta}{z-\zeta}\right| \cdot x\xi\varphi(z - i\eta_n)\varphi(\zeta - i\eta_n)dxdyd\xi d\eta$$

and by changing the variables from z and ζ to $z - i\eta_n$ and $\zeta - i\eta_n$, we obtain for the above integral the form $\varepsilon_n^2 a^2$ with

$$a = \left(\int_{(|z|<1)\cap P\times(|\zeta|<1)\cap P} \log\left|\frac{z+\bar\zeta}{z-\zeta}\right| \cdot x\xi\varphi(z)\varphi(\zeta)dxdyd\xi d\eta\right)^{1/2} < \infty.$$

Therefore, $(G_P(c,c)_{\mu_2})^{1/2} \leq a \sum_1^\infty \varepsilon_n < \infty$, and $c \in H_{\mu_2}(C)$.

To show that $s \notin H_{\mu_2}(C')$, we denote by Q the upper half-plane and observe that $C' = \hat{Q}$ with respect to $\partial Q - \{-3,3\}$. Since $s > 0$ on Q, we have

$$G_Q(s,s)_{\mu_2} \geq \sum_{n=1}^{\infty} G_Q(s\varepsilon_n\varphi(\cdot - i\eta_n), s\varepsilon_n\varphi(\cdot - i\eta_n))_1.$$

The nth term of the sum on the right is given by

$$\varepsilon_n^2 \int_{(|z-i\eta_n|<\frac{1}{2})\times(|\zeta-i\eta_n|<\frac{1}{2})} \log\left|\frac{z-\bar\zeta}{z-\zeta}\right| \cdot y\eta\varphi(z - i\eta_n)\varphi(\zeta - i\eta_n)dxdyd\xi d\eta$$

$$= \varepsilon_n^2 \int_{(|z|<\frac{1}{2})\times(|\zeta|<\frac{1}{2})} \log\left|\frac{z-\zeta+2i\eta_n}{z-\zeta}\right| \cdot (y+\eta_n)(\eta+\eta_n)\varphi(z)\varphi(\zeta)dxdyd\xi d\eta$$

$$\geq \varepsilon_n^2 \int_{(|z|<\frac{1}{2})\times(|\zeta|<\frac{1}{2})} \log\frac{1}{|z-\zeta|} \cdot (\eta_n-1)(\eta_n-1)\varphi(z)\varphi(\zeta)dxdyd\xi d\eta$$

$$\geq \int_{(|z|<\frac{1}{2})\times(|\zeta|<\frac{1}{2})} \log\frac{1}{|z-\zeta|} \cdot \varphi(z)\varphi(\zeta)dxdyd\xi d\eta \equiv a > 0 .$$

Therefore, $G_Q(s,s)_{\mu_2} \geq \Sigma_1^\infty a = \infty$, and by Theorem 1.2, $s \notin H_{\mu_2}(\hat{Q}) = H_{\mu_2}(C')$.

__1.6__. __Auxiliary functions__ μ_3 __through__ μ_6. Next take a C^∞ radial with respect to the point 3, nonnegative function $\mu_3(z)$ on C' vanishing identically on $|z-3| \geq \frac{1}{2}$ in C' such that

$$\mu_3(z) = (|z-3|\log|z-3|)^{-2}$$

on $0 < |z-3| \leq 1/4$ in C'. Let $\mu_4(z)$ be the function on C' which is symmetric to $\mu_3(z)$ with respect to the imaginary axis.

To define the fifth function $\mu_5(z)$, we take a strictly decreasing sequence $\{r_n\}_0^\infty$ such that $r_0 = 1$ and $\lim_n r_n = 0$. Let $w_n \in H(P - \{|z-3| \leq r_n\}) \cap C(\overline{P})$ such that $w_n|\{|z-3| \leq r_n\} = 1$ and $w_n|\partial P = 0$. Since $\{w_n\}_1^\infty$ converges to 0 uniformly on each compact subset of P', we can choose $\{r_n\}_0^\infty$ such that $D_P(w_n) < 4^{-n}$. We then regularize w_n so as to obtain a C^∞, radial with respect to 3, superharmonic function ω_n such that $\omega_n = w_n$ on $\{|z-3| > r_{n-1}\} \cup \{|z-3| < r_{n+1}\}$ in $\overline{P}$ with

$$D_P(\omega_n) \leq 4^{-n}, \qquad n = 1,2,\ldots .$$

Then define

$$\mu_5(z) = (\sum_{n=1}^{\infty} \Delta\omega_n(z))/c(z) \geq 0 .$$

At each point $z \in P'$, Σ is a finite sum. Outside of $|z-3| \leq 1$ in P', $\mu_5(z) = 0$. Thus we can consider μ_5 to be C^∞ on C'. Let μ_6 be the function on C' which is symmetric to μ_5 with respect to the imaginary axis.

Let ω be an antisymmetric function on C' with respect to ∂P such that

$$\omega(z) = \sum_{n=1}^{\infty} \omega_n(z)$$

on P'. The sequence is convergent because of $D_P(\omega_n) \leq 4^{-n}$ and $D_{C'}(\omega) < \infty$. Actually, $\omega(z)$ is a positive C^∞ superharmonic function on P' such that

$$\lim_{z \in P', z \to 3} \omega(z) = \infty .$$

It is readily seen that Δ and the sum can be interchanged in the calculation of $\Delta\omega$ and therefore, we obtain

$$\Delta\omega(z) = (\mu_5(z) + \mu_6(z))c(z)$$

on C'. In particular, $c \in H_{\mu_5+\mu_6}(C')$.

LEMMA. The function c belongs to $H_{\mu_3+\mu_4+\mu_5+\mu_6}(C')$.

Proof. Since $\bigcap_j H_{\mu_j}(C') \subset H_{\Sigma_j \mu_j}(C')$, we only have to show that $c \in H_{\mu_3+\mu_4}(C')$. However, by Theorem 1.2 it suffices to show that $G_P(c,c)_{\mu_3+\mu_4} < \infty$. But

$$G_P(c,c)_{\mu_3+\mu_4} = G_P(c,c)_{\mu_3} \leq \int_{P\times P} g_P(z,\zeta)\mu_3(z)\mu_3(\zeta)dxdyd\xi d\eta .$$

Thus by Theorem 1.1, we have $G_P(c,c)_{\mu_3+\mu_4} < \infty$ if $a = \sup_{\varphi \in C_0^\infty(P)} (1,\varphi)^2_{\mu_3}/D_P(\varphi) < \infty$. As in the proof of Lemma 1.4, we deduce

$$(1,\varphi)_{\mu_3} = O\left(\int_0^{1/2} a_0(r)(r \log r)^{-2} r dr\right),$$

where $z - 3 = re^{i\theta}$. By the Schwarz inequality,

$$(1,\varphi)^2_{\mu_3} \leq O\left[\left(\int_0^\infty a_0(r)^2 r^{-1} dr\right)\left(\int_0^{1/2} r^{-1}(\log r)^{-4} dr\right)\right] = O\left(\int_0^\infty a_0(r)^2 r^{-1} dr\right) .$$

On the other hand, $D_P(\varphi) \geq 2\pi \int_0^\infty a_0(r)^2 r^{-1} dr$, and a fortiori $a < \infty$.

1.7. Construction of λ. We define a strictly positive C^∞ function λ on C' by

$$\lambda(z) = \sum_{j=1}^{6} \mu_j(z).$$

<u>LEMMA</u>. <u>Let</u> $U = \{|z-3| < 1/4\}$ <u>and</u> $U' = U - \{3\}$. <u>If</u> $h \in H_\lambda(U')$, <u>then</u> $h \in H(U)$.

<u>Proof</u>. Let h_0 be the regular part of h in $V = \{|z-3| < 1/8\}$, that is, $h_0 \in H(V)$. Since $G_V(h_0,h_0)_{\mu_n} \leq k^2 G_V(1,1)_{\mu_n}$ with $k = \sup_V|h_0|$, we conclude that $h_0 \in H_{\mu_n}(V')$ with $V' = V - \{3\}$ if $\sup_{\varphi\in C_0^\infty(V)}(1,\varphi)_{\mu_n}^2/D_V(\varphi) < \infty$. For $n = 2,4$, and 6, this is trivial. For $n = 1$ and 3, it can be shown as in the proof of Lemma 1.6. Observe that $(1,\varphi)_{\mu_5} = (\Delta\omega,\varphi/c)_1 = D_V(\omega,\varphi/c)$. By the Schwarz inequality,

$$(1,\varphi)_{\mu_5}^2 \leq D_V(\omega)\ (O(D_V(\varphi) + (1,\varphi^2)_1)).$$

Clearly, $(1,\varphi^2)_1 \leq O(D_V(\varphi))$ and thus $h_0 \in H_{\mu_n}(V)$ for $n = 1,\ldots,6$. Therefore, $h_0 \in H_\lambda(V)$. For this reason, we may assume that h lacks the regular part in the proof of the above Lemma. Then, in terms of $z - 3 = re^{i\theta}$, h takes the form

$$h(re^{i\theta}) = a \log r + \sum_{n=1}^{\infty} r^{-n}(a_n \cos n\theta + b_n \sin n\theta).$$

We wish to show that $h \equiv 0$.

First suppose that $a_n^2 + b_n^2 \neq 0$ for some n. Let $\rho_{t,t'}(r) = (r - t')(t - r)$ for $t' \leq r \leq t$ and 0 for $r > t$ or $t' > r \geq 0$. Set $\varphi_{t,t'}(re^{i\theta}) = \rho_{t,t'}(r)(a_n \cos n\theta + b_n \sin n\theta)$. By Theorem 1.1, there exists a constant K such that

$$(h,\varphi_{t,t'})_\lambda^2 = O\left(\int_{t'}^{t} r^{-n}\rho_{t,t'}(r)\lambda(r)r dr\right)^2 \leq K D_U(\varphi_{t,t'}).$$

Here the fact that $\lambda(re^{i\theta}) = \lambda(r)$ is important. On letting $t' \to 0$, we obtain

$$\left(\int_0^t r^{-n+2}(t - r)\lambda(r)dr\right)^2 \leq O(D_U(\varphi_t)),$$

where $\varphi_t(re^{i\theta}) = r(t - r)(a_n \cos n\theta + b_n \sin n\theta)$. Since $\mu_3 < \lambda$,

$$\left(\int_0^t r^{-n+2}(t - r)(r \log r)^{-2}dr\right)^2 \leq O(D_U(\varphi_t)).$$

The first integral is $O(t^{-n+2})$ and $D_U(\varphi_t) = O(t^4)$. Hence, $O(t^{-n+2}) \leq O(t^2)$ or $O(t^{-n}) \leq O(1)$, a contradiction since $t^{-n} \to \infty$ as $t \to 0$.

Next assume $a \neq 0$ and take $\varphi(re^{i\theta}) = a \log(\max(r,t))$. Choose the radial (with respect to 3) $\varphi_m \in C_0^\infty(U')$ such that $\sup_U|\varphi_m-\varphi| + D_U(\varphi_m-\varphi)^{1/2} \to 0$. Since

$$(h,\varphi_m)_\lambda^2 = O\left(\int_0^{1/4} \log r \cdot \varphi_m(r) \cdot \lambda(r)rdr\right)^2 \leq K D_U(\varphi_m)$$

and $\mu_3 < \lambda$, we obtain on letting $m \to \infty$

$$O\left(\int_t^{1/4} (\log r)^2 \cdot (r \log r)^{-2}rdr\right)^2 \leq O\left(\int_t^{1/4} r^{-2} \cdot rdr\right)$$

or $O(\log t)^2 \leq O(\log t)$. This means that $O(|\log t|) \leq O(1)$ as $t \to \infty$, a contradiction.

<u>1.8</u>. <u>Characterization of</u> $H_\lambda(C')$. We are ready to establish the following characterization of $H_\lambda(C')$.

<u>LEMMA</u>. <u>The class</u> $H_\lambda(C')$ <u>consists of the functions</u> kc, <u>with</u> k <u>nonzero constants</u>.

<u>Proof</u>. Let $h \in H_\lambda(C')$. Since $h \in H_\lambda(U')$ and $h \in H_\lambda(-U')$, Lemma 1.7 yields $h \in H(C)$. Therefore, h takes the form

$$h(re^{i\theta}) = \sum_{n=0}^{\infty} r^n(a_n\cos n\theta + b_n\sin n\theta).$$

Consider the functions

$$\rho_n(r) = \begin{cases} (r-(\eta_n+1))((\eta_n+1)^2-r), & r \in [\eta_n+1,(\eta_n+1)^2], \\ 0, & r \in [0,\infty) - [\eta_n+1,(\eta_n+1)^2], \end{cases}$$

where $\{\eta_n\}$ is the sequence in 1.5. Assume that $a_m^2 + b_m^2 \neq 0$ for some $m > 1$ and set

$$\varphi_n(re^{i\theta}) = \rho_n(r)(a_m\cos m\theta + b_m\sin m\theta).$$

By Theorem 1.1,

$$(h,\varphi_n)^2_{\mu_1} \le (h,\varphi_n)^2_\lambda \le K\, D_{C'}(\varphi_n)$$

for a constant K and for every $n = 1,2,\ldots$. Observe that $(h,\varphi_n)_{\mu_1} = O(\eta_n^{2(m+1-\alpha)})$ and $D_{C'}(\varphi_n) = O(\eta_n^8)$. Therefore, $O(\eta_n^{2m+2-2\alpha}) \le O(\eta_n^4)$ or $O(\eta_n^{m-1-\alpha}) \le O(1)$. Since $\eta_n \to \infty$ as $n \to \infty$, we must have $m \le 1 + \alpha < 2$, a contradiction. It follows that $h = a_0 + a_1 c + b_1 s$. By Lemmas 1.4, 1.5 and 1.6, $c \in H_\lambda(C')$, and therefore, $a_0 + b_1 s \in H_\lambda(C')$ unless $a_0 + b_1 s \equiv 0$. If $b_1 \neq 0$, we can find a constant $\gamma > 1$ such that $a_0 + b_1 s$ is of constant sign, say > 0, on $Q + \gamma$. Since $H_\lambda(Q+\gamma) = H_{\mu_1+\mu_2}(Q+\gamma)$,

$$G_{Q+\gamma}(a_0 + b_1 s, a_0 + b_1 s)_\lambda \ge G_{Q+\gamma}(a_0 + b_1 s, a_0 + b_1 s)_{\mu_2}.$$

By the proof of Lemma 1.5, the right-hand side is seen to be ∞, a contradiction. Hence $b_1 = 0$. If $a_0 \neq 0$, say $a_0 > 0$, then $G_\Omega(a_0, a_0)_\lambda \to \infty$ as regular regions Ω exhaust C', since $G_\Omega \to \infty$ as $\Omega \to C'$, again a contradiction. Therefore, $a_0 + b_1 s \equiv 0$ and we infer that $h = kc$. The proof of the Lemma is complete.

1.9. Conclusion. We summarize the results in 1.4-1.8:

THEOREM. The Poisson equation $\Delta u = \lambda h$ for $h \in H(C')$ has a Dirichlet finite solution on C' if and only if $h = kc$, with k a nonzero constant. Every Dirichlet finite solution u of $\Delta u = \lambda(kc)$ on C' is unbounded.

Proof. The first statement is nothing but Lemma 1.8. To prove the second statement, we may assume $k = 1$. Let $\tilde{u}(z) = -u(j(z))$, where j is the involution on $\hat{P}'$. Since $\lambda(j(z)) = \lambda(z)$ and $c(j(z)) = -c(z)$, $\Delta\tilde{u} = \lambda c$. Therefore, $\Delta(u-\tilde{u}) = 0$ and $u - \tilde{u} \in D(C')$. By $C \in O^2_G \subset O^2_{HD}$, $u - \tilde{u} = 2d \in R^1$, that is, $u(z) + u(j(z)) = 2d$. In particular, $u(z) = d$ if $z \in \partial P$. Let $\{\Omega\}$ be an exhaustion of C' and $\omega = \Omega \cap P'$. As in the proof of Theorem 1.1,

$$u = h_{\omega u} + \int_\omega g_\omega(\cdot,\zeta) c(\zeta)\lambda(\zeta)\, d\xi\, d\eta$$

on ω. Since $u = d$ on $\partial\omega \cap \partial P$ and $h_{\omega u}$ converges to a Dirichlet finite harmonic function h_0 on P', $h_0 \equiv d$. By the Lebesgue-Fatou theorem,

$$u = d + \int_{P'} g_P(\cdot,\zeta)c(\zeta)\lambda(\zeta)d\xi d\eta$$

$$> d + \int_{P'} g_P(\cdot,\zeta)c(\zeta)\mu_5(\zeta)d\xi d\eta$$

$$= d + \int_{P'} g_P(\cdot,\zeta)\Delta\omega(\zeta)d\xi d\eta$$

$$= d + \omega.$$

We know from 1.6 that for $z \in P'$, $\lim_{z\to 3}\omega(z) = \infty$. Therefore, $\lim_{z\to 3}u(z) = \infty$, and u is unbounded.

This completes the proof of Theorem 1.3.

NOTES TO §1. Theorems 1.1, 1.2, 1.3 and 1.9 were established in Nakai-Sario [7]. A somewhat simpler version of the above proof of Theorem 1.3 was included in Nakai-Sario [10]. Although an entirely different and shorter proof, valid for all N, will be given in §2, we hope that the original proof for $N = 2$, reproduced above, has offered methodological interest.

§2. HIGHER DIMENSIONS

From the result $O^N_{H^2D} < O^N_{H^2C}$ for $N = 2$ in §1, we now proceed to it for an arbitrary N. The method in §1, based on essential use of complex analysis in the plane, does not extend to an arbitrary dimension. We first observe in 2.1 that the relation $O^N_{H^2D} < O^N_{H^2C}$ is given for $N > 4$ by what we already know of the Poincaré N-ball. For $N = 3,4$, and at once for all $N \geq 2$, this relation is deduced by a rather delicate counterexample presented in 2.2-2.7.

The section closes with the more comprehensive statement that there are no inclusion relations between $O^N_{H^2B}$ and $O^N_{H^2D}$.

2.1. Cases $N > 4$ by the Poincaré N-ball. We retain the notation B^N_α for the Poincaré N-ball $\{r < 1 | ds = (1-r^2)^\alpha |dx|\}$.

THEOREM. For $N > 4$, the relation $O^N_{H^2D} < O^N_{H^2C}$ is given by

$$B_\alpha^N \in \tilde{O}^N_{H^2D} \cap O^N_{H^2C} \quad \underline{\text{for}} \quad \begin{cases} N = 5,6 & \underline{\text{if}} \quad \alpha \geq 3/(N-4), \\ N > 6 & \underline{\text{if}} \quad \alpha \in [3/(N-4), 5/(N-6)). \end{cases}$$

Note that the condition on α is only sufficient.

Proof. We know from III.4.1 that

$$B_\alpha^N \in O^N_{H^2B} \quad \text{for} \begin{cases} N \in [2,4] \Longleftrightarrow \alpha \leq -1, \\ N > 4 \Longleftrightarrow \alpha \notin (-1, 3/(N-4)), \end{cases}$$

and from IV.1.3 that

$$B_\alpha^N \in \tilde{O}^N_{H^2D} \quad \text{for} \begin{cases} N \in [2,6] \Longleftrightarrow \alpha > -3/(N+2), \\ N > 6 \Longleftrightarrow \alpha \in (-3/(N+2), 5/(N-6)). \end{cases}$$

The Theorem follows.

The case $N = 2$ having been settled in §1, the cases $N = 3,4$ remain open.

2.2. Arbitrary dimension. We proceed to the general case:

THEOREM. For $N \geq 2$,

$$O^N_{H^2D} < O^N_{H^2C}.$$

The proof will be given in 2.2-2.7.

Consider the N-cylinder

$$R = \{(x, y^1, \ldots, y^{N-1}) \mid |x| < \infty, |y^i| \leq \pi, i = 1, \ldots, N-1\}$$

with the metric

$$ds^2 = \varphi(x)dx^2 + \varphi(x)(dy^1)^2 + \sum_{i=2}^{N-1} dy^{i2},$$

where $\varphi \in C^\infty(-\infty,\infty)$, $\varphi > 0$, $\varphi(x) = \varphi(-x)$, and $\varphi|\{|x| > 1\} = |x|^{-3}$. In the same manner as in I.3.1 and II.2.4, we expand an $h \in H(R)$ into a series $\Sigma_n f_n(x)G_n(y)$, $y = (y^1, \ldots, y^{N-1})$, of harmonic functions f_nG_n, where G_n is a product of functions $G_i = G_{ni}$ of the form $\cos n_iy^i$ or $\sin n_iy^i$, with $n_i \geq 0$, and $n = (n_1, \ldots, n_{N-1})$ does not omit the value $0 = (0, \ldots, 0)$. We drop the subindex n

and study the order of growth of $f(x)$. We shall assume that $f(x) \not\equiv 0$.

By the maximum principle applied to fG with $G \neq$ const, the function f is strictly monotone. For $G =$ const, $\Delta f = 0$ gives

$$f(x) = ax + b.$$

The harmonic measure ω on $|x| > c$ of the ideal boundary of R is constant, and $R \in O_G^N$.

<u>2.3</u>. <u>Special cases of</u> $f(x)G(y)$. If $G = G_1(y^1)$, $n_1 > 0$, then $\Delta(fG_1) = 0$ reads $(f'' - n_1^2 f)G_1 = 0$, hence

$$f(x) = ae^{n_1 x} + be^{-n_1 x}$$

with $a^2 + b^2 \neq 0$. For $i > 1$, we give an asymptotic expansion for f.

<u>LEMMA</u>. <u>If</u> $f(x)\Pi_{i=2}^{N-1} G_i(y^i) \in H$ <u>with</u> $\Pi_{i=2}^{N-1} G_i(y^i) \neq$ const, <u>then</u>

$$f(x) = ax(1 + o(1))$$

<u>with</u> $a \neq 0$, <u>either as</u> $x \to \infty$ <u>or as</u> $x \to -\infty$.

<u>Proof</u>. For $\eta^2 = \Sigma_2^{N-1} n_i^2$, we have

$$\Delta(f \prod_{i=2}^{N-1} G_i) = (-\varphi^{-1} f'' + \eta^2 f) \prod_{i=2}^{N-1} G_i = 0,$$

hence

$$f'' = \eta^2 \varphi f.$$

We now make use of the following theorem of Haupt [1] and Hille [1] (see Cesari [1]).

A necessary and sufficient condition for the differential equation

$$f''(x) = p(x)f(x)$$

on $(0,\infty)$ to have solutions

$$f_1(x) = x(1 + o(1)),$$

$$f_2(x) = 1 + o(1)$$

as $x \to \infty$ is that

$$xp(x) \in L^1(0,\infty).$$

Since $\eta^2\varphi = \eta^2|x|^{-3}$ on $|x| > 1$, the condition is satisfied, and we conclude that

$$f(x) = a_1 x(1 + o(1)) + b_1(1 + o(1))$$

as $x \to \infty$, and

$$f(x) = a_2 x(1 + o(1)) + b_2(1 + o(1))$$

as $x \to -\infty$.

Since $R \in O_G^N \subset O_{HB}^N$, fG is unbounded and the same is true of f. Consequently, $a_1 \neq 0$ or $a_2 \neq 0$.

2.4. General case of $f(x)G(y)$. We now include y^1.

LEMMA. *If* $f(x)G(y) \in H$ *with* $G(y) = \Pi_{i=1}^{N-1} G_i(y^i)$, $n_1 > 0$, *then*

$$f(x) \sim ae^{n_1|x|}, \qquad a \neq 0,$$

either as $x \to \infty$ *or as* $x \to -\infty$.

Proof. The equation $\Delta(fG) = 0$ gives

$$f'' = \varphi(n_1^2\varphi^{-1} + \eta^2)f.$$

If we denote $f(t) = f(x/n_1)$ again by $f(x)$, we obtain

$$f''(x) = (1 + 2c\varphi)f(x),$$

with $c = \frac{1}{2}n_1^{-2}\eta^2$. We now make use of the following theorem of Bellman [1]:

If $p(x) \to 0$ as $x \to \infty$ and $\int_0^\infty p^2 dx < \infty$, then the equation $f'' = (1+p)f$ on $(0,\infty)$ has solutions

$$f_1(x) \sim \exp\left[x + \tfrac{1}{2}\int_{x_0}^x p(x)dx + o(1)\right],$$
$$f_2(x) \sim \exp\left[-\left(x + \tfrac{1}{2}\int_{x_0}^x p(x)dx + o(1)\right)\right].$$

In the present case, $p(x) = 2c|x|^{-3}$ on $\{|x| > 1\}$ satisfies the conditions of Bellman's theorem, and we obtain $f = a_1 f_1 + b_1 f_2$ for $x > 1$ and $f = a_2 f_1 + b_2 f_2$ for $x < -1$, with

$$f_1(x) \sim \exp\left[|x| + c\int_{x_0}^{x} \varphi dx + o(1) \right],$$

$$f_2(x) \sim \exp\left[-\left(|x| + c\int_{x_0}^{x} \varphi dx + o(1) \right) \right].$$

If $a_1 = a_2 = 0$, then $f(x) \to 0$ as $|x| \to \infty$, in violation of the maximum principle. Therefore, either $a_1 \neq 0$ or $a_2 \neq 0$, and in view of the above transformation, we have the Lemma.

2.5. Biharmonic functions of x. We proceed to quasiharmonic and biharmonic functions.

LEMMA. *A solution of* $\Delta q = 1$ *is* $q_0(x) = -\int_0^x \int_0^t \varphi(s) ds dt$. *The general solution of* $\Delta q = c$ *is* $q(x,y) = cq_0(x) + h(x,y)$, *where* h *is harmonic. Every* q *is unbounded.*

A solution of $\Delta^2 u = 0$ *is* $u_0(x) = -\int_0^x \int_{-\infty}^t s\varphi(s) ds dt$, *with* $\Delta u_0 = x$. *It satisfies* $u_0(x) \sim \pm d \log|x|$, $d > 0$, *as* $x \to \pm\infty$, *respectively. The general solution* $c_0 u_0(x) + c_1 q_0(x) + c_2 x + c_3$ *of* $\Delta^2 u(x) = 0$ *is unbounded.*

Proof. A direct computation shows that $\Delta q_0 = 1$ and, therefore, $\Delta q = c$. To see that every q is unbounded, suppose there exists a bounded q. Then the transform $(Tq)(x) = \int_y q(x,y) dy = aq_0(x) + bx + c$ is bounded. Since $q_0 \to -\infty$ as $|x| \to \infty$, whereas bx changes sign with x, we conclude that $a = b = 0$. But $q = c$ is not a quasiharmonic function, and we have the first part of the Lemma.

The idea of the proof of the second part is the same.

2.6. Biharmonic functions $v(x)G(y)$. Before drawing the conclusion on the existence of H^2D functions, we deduce some auxiliary results to prepare for the nonexistence of H^2C functions.

LEMMA. Let $v(x)$ satisfy the equation $\Delta(v(x)G(y)) = f(x)G(y) \in H$ with $fG \neq$ const. Then v is unbounded.

Proof. First we consider the case $G(y) = G_1(y^1)$, $n_1 > 0$. We have

$$(-\varphi^{-1}v'' + \varphi^{-1}n_1^2 v)G_1 = fG_1,$$

hence

$$v'' = n_1^2 v - \varphi f.$$

In 2.3, we observed that $f(x) = ae^{n_1 x} + be^{-n_1 x}$ with $|a| + |b| \neq 0$. We may assume $a \neq 0$; the proofs for the other cases are analogous.

Suppose v is bounded. As $x \to \infty$, $\varphi f \sim ax^{-3}e^{n_1 x}$. Therefore,

$$v'(x) \sim v'(x_0) + \int_{x_0}^{x} (n_1^2 v(s) - as^{-3}e^{n_1 s})ds,$$

where we may choose $x_0 > 1$. It follows that

$$v(x) \sim v(x_0) + v'(x_0)(x - x_0) + \int_{x_0}^{x}\int_{x_0}^{t} (n_1^2 v(s) - as^{-3}e^{n_1 s})dsdt,$$

which is clearly unbounded, a contradiction.

Next we consider the case $G(y) = \Pi_{i=2}^{N-1} G_i(y^i) \neq$ const. Now

$$v'' = \eta^2 \varphi v - \varphi f.$$

By Lemma 2.3,

$$f = ax(1 + o(1)),$$

$a \neq 0$, either as $x \to \infty$ or as $x \to -\infty$. First assume the latter. If v is bounded, we have for $x_2 < x_1 < x < -1$,

$$v'(x) - c = \int_{-\infty}^{x} \eta^2 |s|^{-3} v(s)ds - \int_{-\infty}^{x} |s|^{-3} f(s)ds,$$

$$v(x_1) - v(x_2) = c(x_1 - x_2) + \int_{x_2}^{x_1}\int_{-\infty}^{t} \eta^2 |s|^{-3} v(s)dsdt - \int_{x_2}^{x_1}\int_{-\infty}^{t} |s|^{-3} f(s)dsdt.$$

As $x_2 \to -\infty$, the first integral converges but the second integral, and hence the right-hand side, diverges whether or not $c = 0$. Thus v is unbounded, a contradiction.

If the above estimate for f holds for $x \to \infty$ instead, the reasoning remains valid with obvious modifications.

We come to the general case: $G(y) = \Pi_{i=1}^{N-1} G_i(y^i)$. In view of the above special cases, we may assume $n_1 > 0$ and at least one $n_i > 0$, $i > 1$. We now have

$$v'' = (n_1^2 + \eta^2\varphi)v - \varphi f .$$

By Lemma 2.4, $f \sim ae^{n_1|x|}$ either as $x \to \infty$ or as $x \to -\infty$. We may assume the former. Clearly $|\varphi f| \to \infty$ as $x \to \infty$. If v is bounded, then φf will dominate the right-hand side of the equation. On integrating as in the two special cases, we arrive at the contradiction that v is both bounded and unbounded.

In the remaining case, $G = \text{const}$, Lemma 2.5 shows that v is unbounded.

The Lemma follows.

2.7. Conclusion. We are ready to summarize:

LEMMA. $R \in O^N_{H^2B} \cap \tilde{O}^N_{H^2D}$.

Proof. The function u_0 of Lemma 2.5 belongs to H^2D. In fact,

$$D(u_0) = c\int_{-\infty}^{\infty} u_0'^2\varphi^{-1}\varphi dx = c_1 + c\left(\int_{-\infty}^{-1} + \int_1^{\infty}\right) x^{-2}dx < \infty,$$

hence $R \in \tilde{O}^N_{H^2D}$.

Suppose there exists a $u(x,y) \in H^2B$. Write $u(x,y) = \Sigma_n v_n(x)G_n(y)$ with $G_0(y)$ constant. Some $v_{n_0}G_{n_0}$ is not harmonic, and the transform

$$(Tu)(x) = \int_y uG_{n_0}dy = cv_{n_0}(x)$$

is bounded, in violation of Lemma 2.6.

The proof of Theorem 2.2 is herewith complete.

2.8. No relation between H^2B and H^2D degeneracies. With the strictness of $O^N_{H^2D} < O^N_{H^2C}$ settled, the natural question arises as to whether $O^N_{H^2B} < O^N_{H^2C}$ as well. Both relations are contained in the following comprehensive statement, which contrasts with the relation $O^N_{HB} < O^N_{HD}$ for harmonic functions:

<u>THEOREM</u>. <u>For</u> $N \geq 2$, <u>the classes</u>

$$O^N_{H^2B} \cap O^N_{H^2D}, \quad O^N_{H^2B} \cap \tilde{O}^N_{H^2D}, \quad \tilde{O}^N_{H^2B} \cap O^N_{H^2D}, \quad \tilde{O}^N_{H^2B} \cap \tilde{O}^N_{H^2D}$$

<u>are all nonvoid</u>.

<u>Proof</u>. For the Poincaré N-ball B^N_α, we have from III.4.1 and IV.1.3:

$$B^N_\alpha \in O^N_{H^2B} \quad \text{for} \quad \begin{cases} N \in [2,4] \Longleftrightarrow \alpha \leq -1, \\ N > 4 \Longleftrightarrow \alpha \notin (-1, 3/(N-4)), \end{cases}$$

$$B^N_\alpha \in O^N_{H^2D} \quad \text{for} \quad \begin{cases} N \in [2,6] \Longleftrightarrow \alpha \leq -3/(N+2), \\ N > 6 \Longleftrightarrow \alpha \notin (-3/(N+2), 5/(N-6)). \end{cases}$$

As a consequence,

$$B^N_\alpha \in O^N_{H^2B} \cap O^N_{H^2D} \quad \text{for} \quad \begin{cases} N \in [2,6] \Longleftrightarrow \alpha \leq -1, \\ N > 6 \Longleftrightarrow \alpha \notin (-1, 5/(N-6)), \end{cases}$$

$$B^N_\alpha \in O^N_{H^2B} \cap \tilde{O}^N_{H^2D} \quad \text{for} \quad \begin{cases} N \in [2,4], \quad \text{no } \alpha, \\ N = 5,6 \Longleftrightarrow \alpha \geq 3/(N-4), \\ N > 6 \Longleftrightarrow \alpha \in [3/(N-4), 5/(N-6)), \end{cases}$$

$$B^N_\alpha \in \tilde{O}^N_{H^2B} \cap O^N_{H^2D} \quad \text{for all} \quad N \Longleftrightarrow \alpha \in (-1, -3/(N+2)],$$

$$B^N_\alpha \in \tilde{O}^N_{H^2B} \cap \tilde{O}^N_{H^2D} \quad \text{for} \quad \begin{cases} N \in [2,4] \Longleftrightarrow \alpha > -3/(N+2), \\ N > 4 \Longleftrightarrow \alpha \in (-3/(N+2), 3/(N-4)). \end{cases}$$

Thus the only relation left open by B^N_α is $O^N_{H^2B} \cap \tilde{O}^N_{H^2D} \neq \emptyset$ for $N \in [2,4]$. But by Lemma 2.7, this holds for every $N \geq 2$.

<u>COROLLARY</u>. <u>For</u> $N \geq 2$, $O^N_{H^2B} < O^N_{H^2C}$.

<u>NOTES TO §2</u>. After Theorem 1.3 had given the strictness of $O^N_{H^2D} < O^N_{H^2C}$ for $N = 2$, Theorem 2.1, a simple consequence of results in Hada-Sario-Wang [2], [3], proved it for $N > 4$. Then Chung [3] provided the missing links $N = 3,4$ in Theorem 2.2 by invoking the theorems of Haupt [1], Hille [1], and Bellman [1].

We have not considered the problem, which may have some interest, of characterizing the Poincaré N-balls in $O^N_{H^2C}$.

Whether or not there are relations between the class $O^N_{H^2LP}$, not considered here, and other biharmonic null classes, in particular $O^N_{H^2D}$, is an open question.

We have not discussed the class $O^N_{H^2P}$ of Riemannian N-manifolds which do not carry any positive biharmonic functions. In view of the relations $O^N_{H^2P} = O^N_{H^2N} \subset O^N_{QN}$ and the recent result $O^N_{QN} \neq \emptyset$ in Nakai-Sario [21], this class should offer considerable interest.

CHAPTER VI

HARMONIC, QUASIHARMONIC, AND BIHARMONIC DEGENERACIES

We have discussed harmonic null classes in Chapter I, quasiharmonic null classes in Chapter II, and biharmonic null classes in Chapters III-V. We now turn to interrelations between these three categories of null classes. Clearly, we have here a vast field, and we shall make no attempt at completeness. Several problems, to be listed in the Notes after both sections, will be left open for future research. Classification theory is far from being a closed book.

In §1, we show that there are no inclusion relations between the harmonic null classes O^N_{HX} and the biharmonic null classes $O^N_{H^2Y}$ for $X, Y = B, D$. We then give a number of other X and Y for which this remains true, and present the open problem whether or not this still holds for $X = L^p$ and $Y = C$.

§2 is devoted to relations between corresponding classes of quasiharmonic and biharmonic degeneracies. Here we have the strict inclusions $O^N_{H^2X} < O^N_{QX}$ for $X = B, D, C, L^p$. In 2.4, we present a diagram which gives a bird's-eye view of all these classes.

§1. HARMONIC AND BIHARMONIC DEGENERACIES

We showed in III.§1 and IV.§2 that the harmonic null class O^N_G neither contains nor is contained in either $O^N_{H^2B}$ or $O^N_{H^2D}$. In the present section, we shall see that the same is true of the harmonic null classes O^N_{HB} and O^N_{HD}. In particular, for every N, there exist Riemannian N-manifolds which carry even nonconstant HD functions but fail to carry any H^2B or H^2D functions.

1.1. No relations. We assert:

THEOREM. *For* $N \geq 2$, *and* $X, Y = B, D$, *the classes*

$$O^N_{HX} \cap O^N_{H^2Y},\ O^N_{HX} \cap \tilde{O}^N_{H^2Y},\ \tilde{O}^N_{HX} \cap O^N_{H^2Y},\ \tilde{O}^N_{HX} \cap \tilde{O}^N_{H^2Y}$$

are all nonvoid.

The proof will be given in 1.1-1.3

From Theorems III.1.1 and IV.2.3, we know that $O_G^N \cap O_{H^2Y}^N$ and $O_G^N \cap \tilde{O}_{H^2Y}^N$ are nonvoid. By I.1.4 and I.2.1, this implies that all classes $O_{HX}^N \cap O_{H^2Y}^N$ and $O_{HX}^N \cap \tilde{O}_{H^2Y}^N$ are nonvoid. The relation $\tilde{O}_{HX}^N \cap \tilde{O}_{H^2Y}^N \neq \emptyset$ is again trivial in view of the Euclidean N-ball.

It remains to show that

$$\tilde{O}_{HD}^N \cap O_{H^2Y}^N \neq \emptyset .$$

In the case $N = 2$, we recall from I.1.11 and I.2.3 that a conformal metric does not affect harmonicity or the Dirichlet integral. Therefore, the Poincaré disk $B_\alpha^2 = \{|z| < 1\}$ with the metric $ds = (1 - |z|^2)^\alpha |dz|$ trivially carries HD functions for every α, whereas by III.4.1, it carries no H^2B functions for $\alpha \leq -1$, and by IV.1.1, no H^2D functions for $\alpha \leq -3/4$.

For $N > 2$, we consider the N-cylinder discussed in I.3.2:

$$R = \{(x, y^1, \ldots, y^{N-1}) \,\big|\, |x| < 1,\ |y^i| \leq \pi,\ i = 1, \ldots, N-1\}$$

with the metric

$$ds^2 = \lambda^2 dx^2 + \lambda^{2/(N-1)} \sum_{i=1}^{N-1} dy^{i2} ,$$

$\lambda = \lambda(x) \in C^\infty(-1,1)$. We recall that $R \in \tilde{O}_{HC}^N$ for every λ. In the sequel, we shall first show that $R \in O_{H^2D}^N$ and then that $R \in O_{H^2B}^N$, both for the same suitably chosen λ.

1.2. No H^2D functions. As in I.3.2, take

$$\lambda = (1 - x^2)^{-(N-1)/(N-2)} .$$

To show that $R \in O_{H^2D}^N$, suppose there exists a $u \in H^2D(R)$. Retain the notation of I.§3 and expand $h = \Delta u$ into a series

$$h = \sum_n \sum_j \sum_k a_{njk} f_{nk} G_{nj},$$

where

$$\begin{cases} f_{n1} = (1 + x)^{2^{-1}(1+(1+\eta^2)^{1/2})} (1 - x)^{2^{-1}(1-(1+\eta^2)^{1/2})} , \\ f_{n2} = (1 + x)^{2^{-1}(1-(1+\eta^2)^{1/2})} (1 - x)^{2^{-1}(1+(1+\eta^2)^{1/2})} . \end{cases}$$

For constants $0 < \beta < \gamma < 1$, choose a function $\rho \in C_0^\infty(-\infty,\infty)$, $\rho \geq 0$, $\operatorname{supp} \rho \subset (\beta,\gamma)$. If $a_{nj1} \neq 0$ for some (n,j), set $\rho_t(x) = \rho((1 - x)/t)$, $t > 0$, and $\varphi_t = \rho_t G_{nj}$ for $n \neq 0$, $\varphi_t = \rho_t$ for $n = 0$. Then

$$|(h,\varphi_t)| = c\int_{1-\gamma t}^{1-\beta t} (a_{nj1}f_{n1} + a_{nj2}f_{n2})\rho_t\lambda^2\, dx .$$

As $x \to 1$, $f_{n1} \to \infty$ if $n \neq 0$, and $f_{n1} \to 2$ if $n = 0$, whereas $f_{n2} \to 0$ for each n. Therefore,

$$\lim_{x\to 1} |a_{nj1}f_{n1} + a_{nj2}f_{n2}| > |a_{nj1}| > 0 .$$

For all sufficiently small t,

$$|(h,\varphi_t)| > c\int_{1-\gamma t}^{1-\beta t} \rho_t(1 - x)^{-2(N-1)/(N-2)}\, dx$$

$$> ct^{-2(N-1)/(N-2)+1} = ct^{-N/(N-2)} ,$$

$$D(\varphi_t) = \int_{-\pi}^{\pi} \cdots \int_{-\pi}^{\pi} \int_{-1}^{1} \left(\lambda^{-2}\rho_t'^2 G_{nj}^2 + \lambda^{-2/(N-1)}\rho_t^2 \sum_{i=1}^{N-1} \left(\frac{\partial G_{nj}}{\partial y^i}\right)^2\right)\lambda^2\, dx dy^1 \cdots dy^{N-1}$$

$$< \int_{1-\gamma t}^{1-\beta t} (c_1\rho_t'^2 + c_2(1 - x)^{-2}\rho_t^2)dx = O(t^{-1}) .$$

Since $-2N/(N - 2) < -1$, $(h,\varphi_t)^2/D(\varphi_t) \to \infty$ as $t \to 0$, a contradiction by Lemma IV.1.9. Therefore, $a_{nj1} = 0$ for every (n,j).

If $a_{nj2} \neq 0$ for some (n,j), take ρ as above but $\rho_t(x) = \rho((x + 1)/t)$. Then

$$\lim_{x\to -1} |a_{nj1}f_{n1} + a_{nj2}f_{n2}| > |a_{nj2}| > 0$$

and

$$|(h,\varphi_t)| > ct^{-2(N-1)/(N-2)} \int_{-1+\beta t}^{-1+\gamma t} \rho_t \, dx = ct^{-N/(N-2)} .$$

Since $D(\varphi_t) = O(t^{-1})$ as before, we have again $a_{nj2} = 0$ for all (n,j).

We conclude that $R \in O^N_{H^2D}$.

1.3. No H^2B functions. To see that $R \in O^N_{H^2B}$, suppose there exists a $u \in H^2B(R)$, and expand $h = \Delta u$ as before. If some $a_{nj1} \neq 0$, choose φ_t as in the first case in 1.2, so that again $|(h,\varphi_t)| \sim ct^{-N/(N-2)}$. On the other hand,

$$\Delta\varphi_t = -\lambda^{-2}\left(\rho_t'' G_{nj} + \lambda^{2(N-2)/(N-1)} \sum_{i=1}^{N-1} \rho_t \frac{\partial^2 G_{nj}}{\partial y^{i2}}\right) ,$$

$$(1, |\Delta\varphi_t|) \sim \int_{1-\gamma t}^{1-\beta t} (c_1|\rho_t''| + c_2 t^{-2}\rho_t)dx$$

$$\sim ct^{-1} .$$

Since $-N/(N-2) < -1$, we have a violation of $|(h,\varphi_t)| = |(\Delta u,\varphi_t)| = |(u,\Delta\varphi_t)| < c(1,|\Delta\varphi_t|)$ as $t \to 0$. Therefore, $a_{nj1} = 0$ for all (n,j). If some $a_{nj2} \neq 0$, we again take $\rho_t(x) = \rho((x+1)/t)$ and arrive at a contradiction.

We have shown that $R \in O^N_{H^2B}$, and the proof of Theorem 1.1 is complete.

NOTES TO §1. Theorem 1.1 was proved in Sario-Wang [8]. It remains trivially valid for $X = C$ and can also be proved for $Y = L^p$. The cases $X = L^p$ and $Y = C$ are open and may offer some interest.

§2. CORRESPONDING QUASIHARMONIC AND BIHARMONIC DEGENERACIES

Here we have an elegant, albeit not deep, phenomenon: the inclusion relations $O^N_{H^2X} \subset O^N_{QX}$ are strict for $X = B, D, C, L^p$. We specifically refer the reader to the diagram in 2.4, which clearly shows the positions of the various null classes.

2.1. Strict inclusions. We claim that the corresponding biharmonic and quasiharmonic null classes satisfy:

THEOREM. For $N \geq 2$, and $X = B, D, C, L^p$, $1 \leq p < \infty$,

$$O^N_{H^2X} < O^N_{QX} .$$

The proof will be given in 2.1-2.3.

First we consider the cases $X = B, D, C$. Since $O^N_{H^2X} \subset O^N_{H^2C}$ and, by II.1.8, $O^N_{QP} \subset O^N_{QX}$, it suffices to show that $\tilde{O}^N_{H^2C} \cap O^N_{QP} \neq \emptyset$.

2.2. H^2C functions but no QP. We assert:

LEMMA. For $N \geq 2$,

$$\tilde{O}^N_{H^2C} \cap O^N_{QP} \neq \emptyset .$$

Proof. Consider the N-cylinder

$$R = \{(x, y^1, \ldots, y^{N-1}) \big| x > 1, \ |y^i| \leq \pi, \ i = 1, \ldots, N-1\}$$

with the metric

$$ds^2 = dx^2 + x^{2\alpha/(N-1)} \sum_{i=1}^{N-1} dy^{i2} ,$$

where α is a constant > 5.

To find an H^2C function on R, we first note that the general solution of $\Delta h(x) = -x^{-\alpha}(x^\alpha h')' = 0$ is

$$h(x) = ax^{-\alpha+1} + b ,$$

and the function $u(x) = x^{-\alpha+3}$ is biharmonic. Since $\alpha > 5$, u is bounded. Moreover,

$$D(u) = c \int_1^\infty u'^2 x^\alpha \, dx < \infty ,$$

hence $R \in \tilde{O}^N_{H^2C}$.

To exclude QP functions, we first observe that

$$q_0(x) = -\frac{x^2}{2(\alpha + 1)}$$

is quasiharmonic. Every quasiharmonic $q(x,y)$ can be written $q = q_0 + h$, $h \in H$. In the same manner as in II.2.4, we have the expansion

$$q = q_0 + ax^{-\alpha+1} + b + \Sigma' f_n(x)G_n(y) ,$$

where each $f_nG_n \in H$ and the summation Σ' excludes $n = (0,\ldots,0)$. To see that $q \notin P$, take x_0 so large that $q_0(x_0) + ax_0^{-\alpha+1} + b < 0$. In view of $\int_y \Sigma' f_n(x_0)G_n(y)dy = 0$, there exists a y_0 such that $\Sigma' f_n(x_0)G_n(y_0) = 0$, and we have $q(x_0,y_0) < 0$, $q \notin P$.

2.3. H^2L^p functions but no QL^p. It remains to consider the case $X = L^p$. Again, we take the manifold

$$R = \{(x,y^1,\ldots,y^{N-1})\,|\, x > 1,\ |y^i| \leq \pi,\ i = 1,\ldots,N-1\} ,$$

but now with the metric

$$ds^2 = dx^2 + \lambda(x)^{2/(N-1)} \sum_{i=1}^{N-1} dy^{i2} ,$$

where $\lambda(x) = e^{-x}e^{e^x}$.

To construct an H^2L^p function, first observe that the function $h(x) = e^{-e^x}$ satisfies the harmonic equation $\Delta h = -e^x e^{-e^x}(e^{-x}e^{e^x}h')' = 0$. The equation $\Delta u(x) = e^{-e^x}$ has a solution $u(x) = -\int_x^\infty e^{-e^x}dx$.

We claim that $u \in L^p$, $1 \leq p < \infty$. Since $|u(x)| \geq |u(x)|^p$, it suffices to show that $u \in L^1$. We have

$$\lim_{x\to\infty} (u(x)e^{e^x}) = \lim_{x\to\infty} \frac{u'(x)}{(e^{-e^x})'} = \lim_{x\to\infty} (-e^{-x}) = 0 .$$

Thus $u(x)e^{e^x}$ is bounded and

$$\|u\|_1 = c\int_1^\infty u(x)e^{-x}e^{e^x}\,dx < \infty .$$

To show that $R \in O^N_{QL^p}$, we note that the quasiharmonic equation $\Delta q = 1$ has a solution

$$q_0(x) = \int_x^\infty e^t e^{-e^t}\int_0^t e^{-s}e^{e^s}\,ds dt .$$

Moreover,

$$\lim_{t\to\infty} \frac{\int_0^t e^{-s}e^{e^s}\,ds}{e^{-t}(e^{-t}e^{e^t})} = \lim_{t\to\infty} \frac{e^{-t}e^{e^t}}{(e^{-t} - 2e^{-2t})e^{e^t}} = 1 .$$

Therefore, the integrand in q_0 is asymptotically e^{-t}, and we have $q_0 \sim e^{-x}$. It follows that

$$\|q_0\|_p^p = c\int_1^\infty q_0(x)^p e^{-x}e^{e^x}\,dx = \infty .$$

A general quasiharmonic function of x has the form $q_1(x) = q_0(x) + ae^{-e^x} + b$. We have $q_1(x) \sim e^{-x}$ as $x \to \infty$, hence $q_1(x) \notin L^p$. Therefore, there exists an $L^{p'}$ function $\varphi(x)$ with $p^{-1} + p'^{-1} = 1$, such that

$$(q_1, \varphi) = \int_R q_1 * \varphi = \infty .$$

An arbitrary quasiharmonic function can be written $q(x,y) = q_1(x) + \Sigma_{n\neq 0} f_n(x)G_n(y)$, where we have again used the notation of II.2.4. The above function φ gives

$$(q,\varphi) = \int_R \Big(q_1 + \sum_{n\neq 0} f_n G_n\Big) * \varphi = \int_R q_1 * \varphi = \infty ,$$

and therefore, $q(x,y) \notin L^p$. Thus $R \in O^N_{QL^p}$, and we have Theorem 2.1.

It was recently shown in Chung-Sario-Wang [4] that there are no inclusion relations between $O^N_{QL^p}$ and $O^N_{H^2C}$.

2.4. Summary. We compile the inclusion relations in II.2.8 between the various O^N_{QX} classes, the noninclusions in Theorem V.2.8 between $O^N_{H^2B}$ and $O^N_{H^2D}$, and the relations in Theorem VI.2.1 between $O^N_{H^2X}$ and O^N_{QX} into the following comprehensive diagram, where $p > 1$, and an arrow means strict inclusion:

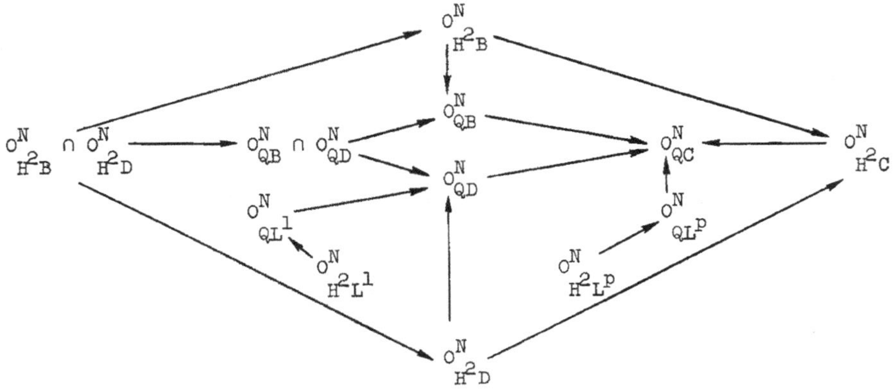

Moreover, $O^N_{QL^1} < O^N_{QPL^1} = O^N_{QD}$ and $O^N_{QB} \cup O^N_{QD} = O^N_{QC} = O^N_{QPBL^1}$.

NOTES TO §2. Theorem 2.1 is new.

Regarding relations between the classes $O^N_{H^2X}$ and the classes O^N_{QY} other than the relations in diagram 2.4, one could successively explore relations (1) between the classes $O^N_{H^2C}$, $O^N_{H^2D}$, $O^N_{H^2B}$, $O^N_{H^2B} \cap O^N_{H^2D}$, and the classes O^N_{QB}, O^N_{QD}, $O^N_{QB} \cap O^N_{QD}$; (2) between the classes $O^N_{H^2X}$ and $O^N_{QL^p}$; and (3) between the class $O^N_{H^2L^p}$ and the above quasiharmonic null classes. Some of the intersections are trivially nonvoid on account of the Euclidean N-ball, Euclidean N-space, or Euclidean N-cylinder; some in view of certain counterexamples in Chapters I-V, in particular those involving the Poincaré N-ball; some by virtue of the strict inclusion relations between the classes O^N_{QY}. For some other classes, new counterexamples have to be devised, e.g., to show that $O^N_{H^2B} \cap O^N_{QD} \neq \emptyset$ and $O^N_{H^2L^p} \cap \tilde{O}^N_{QD} \neq \emptyset$ for $p > 1$. Some of these problems may be quite challenging.

CHAPTER VII

RIESZ REPRESENTATION OF BIHARMONIC FUNCTIONS

In II.§1 and IV.§§2-3, we discussed in passing the Riesz representation of a biharmonic function u as the sum of a harmonic function and the Green potential of Δu. The significance of this representation lies in the fact that it reduces the study of H^2 to that of H, which is more accessible to explicit treatment. For this reason, it is important to distinguish subclasses of H^2 and of Riemannian manifolds R for which the Riesz representation is valid.

A related problem of interest concerns the order of growth of the Laplacian of a biharmonic function. We study this problem in §1 and apply our results to the Riesz representation. That this representation exists for every $u \in H^2D$ with $D(\Delta u) < \infty$ on a Riemannian manifold $R \in \tilde{O}^N_{QP}$ is shown in §2.

In IV.2.4, we introduced the class H_{PD} of harmonic functions h such that the Poisson equation $\Delta u = h$ has solutions with $D(u) < \infty$. We showed in IV.3.2 that, on a hyperbolic R, the unique minimum solution is $u = \lim_{\Omega \to R} G_\Omega h$. A natural question arises: When can the minimum solution be represented as $u = G_R h$? We shall discuss this problem in §3.

The boundary value problem for biharmonic functions or, more generally, (p,q)-biharmonic functions, is the topic of §4. By definition, a function u is (p,q)-biharmonic if it satisfies the equation $(\Delta + q)(\Delta + p)u = 0$. The problem is to find, on a given Riemannian manifold R and for given continuous functions φ and ψ on Wiener's and Royden's p- and q-harmonic boundaries α and β, respectively, a function u with $u|\alpha = \varphi$, $(\Delta + p)u|\beta = \psi$.

§1. METRIC GROWTH OF LAPLACIAN

We study the metric growth of Δu for $u \in H^2D$ with $\Delta u \in C = BD$ on a Riemannian manifold R. The main result to be deduced is that $\int_R g{*}|\Delta u|^2 < \infty$, where g is the harmonic Green's function on R. We apply this result to the existence of a Riesz representation for a biharmonic function.

1.1. The class H^2DC_Δ. Given classes E, F of functions, set

$$H^2EF_\Delta = \{u \in H^2E,\ \Delta u \in F\}\ .$$

We first consider the class H^2DC_Δ.

THEOREM. On a hyperbolic Riemannian manifold R, the metric growth of the Laplacian of $u \in H^2DC_\Delta$ is so slow that

$$\int_R g(x,y)*|\Delta u(y)|^2 < \infty\ .$$

Proof. Fix an $x \in R$ and a geodesic ball $B = B(x,\varepsilon)$. Let $\psi \in C^\infty(R)$ such that $0 \leq \psi \leq 1$, $\psi|(R - \bar{B}) = 1$ and $\psi|B(x,\frac{1}{2}\varepsilon) = 0$. Since $\Delta u \in BD$, the functions

$$\varphi(y) = \psi(y)g(x,y)\Delta u(y), \qquad \varphi_\Omega(y) = \psi(y)g_\Omega(x,y)\Delta u(y)$$

are in the class $BD(R)$, with Ω an arbitrary regular region containing $\bar{B}$, and $g_\Omega(x,y)$ extended to R by $g_\Omega|(R - \Omega) = 0$. Since

$$\lim_{\Omega \to R} D_R(g(x,\cdot) - g_\Omega(x,\cdot)) = 0$$

(cf. Sario-Nakai [1]), we conclude that

$$\lim_{\Omega \to R} D_R(\varphi - \varphi_\Omega) = 0\ .$$

From $d(\varphi_\Omega * du) = d\varphi_\Omega \wedge *du - \varphi_\Omega * \Delta u$ and $\varphi_\Omega|\partial\Omega = 0$, it follows by Stokes' formula that

$$\int_R \psi(y)g_\Omega(x,y)*|\Delta u(y)|^2 = \int_R d\varphi_\Omega \wedge *du\ .$$

By the Schwarz inequality,

$$\int_{R-\bar{B}} g_\Omega(x,y)*|\Delta u(y)|^2 \leq (D_R(\varphi_\Omega)D_R(u))^{1/2}\ .$$

Since $g_\Omega(x,\cdot)$ converges increasingly on R to $g(x,\cdot)$, we conclude that

$$\int_{R-\overline{B}} g(x,y)*|\Delta u(y)|^2 \leq (D_R(\varphi)D_R(u))^{1/2} < \infty .$$

Set $K = \sup_B[(\Delta u)^2 g^{1/2}]$, with g the determinant of the metric tensor. Then on B,

$$g(x,y)*|\Delta u(y)|^2 \leq K\ell_N(y)\, dy^1 \ldots dy^N ,$$

where $\ell_N(y) = c_1|y|^{2-N}$ for $N > 2$, and $\ell_N(y) = c_2 \log(2\varepsilon/|y|)$ for $N = 2$, with the c_i constants. Therefore,

$$\int_{\overline{B}} g(x,y)*|\Delta u(y)|^2 < \infty .$$

We remark that this is valid for every $u \in C^2(\overline{B})$. Thus,

$$\int_R g(x,y)*|\Delta u(y)|^2 = \int_{R-\overline{B}} g(x,y)*|\Delta u(y)|^2 + \int_{\overline{B}} g(x,y)*|\Delta u(y)|^2 < \infty .$$

<u>1.2.</u> <u>The class</u> H^2CD_Δ. From the class H^2DC_Δ we turn to the class H^2CD_Δ consisting of all $u \in H^2C$ with $\Delta u \in D$. This class is technically more difficult to treat than the former. As a counterpart of Theorem 1.1, we shall prove:

<u>THEOREM.</u> <u>On a hyperbolic Riemannian manifold</u> R, <u>the metric growth of the Laplacian of</u> $u \in H^2CD_\Delta$ <u>is so slow that</u>

$$\int_R g(x,y)*|\Delta u(y)|^2 < \infty .$$

The proof will be given in 1.2-1.4.

Fix an $x \in R$ and a geodesic ball $B = B(x,\varepsilon)$. Let Ω be a regular region of R with $\overline{B} \subset \Omega$ and set

$$G_\Omega \Delta u = \int_\Omega g_\Omega(\cdot,y) * \Delta u(y)$$

on Ω. We recall from IV.2.4 the decomposition

$$u = h_\Omega + G_\Omega \Delta u$$

on Ω. Since we are assuming $2 \sup_R |u| = K_1 < \infty$ we have, by the maximum principle, $\sup_\Omega |h_\Omega| \leq \frac{1}{2} K_1$ and consequently, $\sup_\Omega |G_\Omega \Delta u| \leq K_1$. Further immediate consequences of the decomposition are $\Delta G_\Omega \Delta u = \Delta u$ on Ω, $G_\Omega \Delta u | \partial\Omega = 0$, and, by Stokes' formula,

$$D_\Omega(h_\Omega) + D_\Omega(G_\Omega \Delta u) = D_\Omega(u) \leq D_R(u) < \infty .$$

<u>1.3</u>. <u>Auxiliary estimates</u>. Since $g_\Omega(\cdot,\cdot)$ converges increasingly on R to $g(\cdot,\cdot)$, and $g(\cdot,x)$ is bounded on $R - B$,

$$K_2 = \sup_{\Omega \supset \overline{B}} \sup_{y \in \Omega - B} g_\Omega(y,x) < \infty .$$

Stokes' formula applied to $\Delta_y u(y) * (g_\Omega(x,y) d_y (G_\Omega \Delta u)(y))$ on $\Omega - \overline{B}$ reads

$$\int_{\partial\Omega - \partial B} \Delta u * (g_\Omega dG_\Omega \Delta u) = \int_{\Omega - \overline{B}} d(\Delta u * (g_\Omega dG_\Omega \Delta u)) .$$

By $u = h_\Omega + G_\Omega \Delta u$ and the boundedness of $D_\Omega(h_\Omega) + D_\Omega(G_\Omega \Delta u) = D_\Omega(u)$, we see that the $\partial(G_\Omega \Delta u)(y)/\partial y^i$, $i = 1,2,\ldots,N$, are uniformly convergent on $B(x,2\varepsilon)$. Therefore, there exists a constant K_3 such that

$$\left| \int_{\partial B} \Delta u * (g_\Omega dG_\Omega \Delta u) \right| \leq K_3$$

for every Ω. Since

$$\int_{\Omega-\bar{B}} d(\Delta u*(g_\Omega dG_\Omega\Delta u)) = \int_{\Omega-\bar{B}} g_\Omega d\Delta u \wedge *dG_\Omega\Delta u + \int_{\Omega-\bar{B}} \Delta u\cdot dg_\Omega \wedge *dG_\Omega\Delta u$$
$$- \int_{\Omega-\bar{B}} g_\Omega\Delta u*\Delta G_\Omega\Delta u ,$$

we have

$$\int_{\Omega-\bar{B}} g_\Omega*(\Delta u)^2 \le K_3 + \left|\int_{\Omega-\bar{B}} g_\Omega d\Delta u \wedge *dG_\Omega\Delta u\right| + \left|\int_{\Omega-\bar{B}} \Delta u\cdot dg_\Omega \wedge *dG_\Omega\Delta u\right| .$$

<u>1.4.</u> <u>Completion of proof.</u> We shall evaluate the last two terms on the right. By the Schwarz inequality,

$$\left(\int_{\Omega-\bar{B}} g_\Omega d\Delta u \wedge *dG_\Omega\Delta u\right)^2 = \left(\int_{\Omega-\bar{B}} (g_\Omega^{1/2}d\Delta u) \wedge *(g_\Omega^{1/2}dG_\Omega\Delta u)\right)^2$$
$$\le \int_{\Omega-\bar{B}} (g_\Omega^{1/2}d\Delta u) \wedge *(g_\Omega^{1/2}d\Delta u) \cdot \int_{\Omega-\bar{B}} (g_\Omega^{1/2}dG_\Omega\Delta u) \wedge *(g_\Omega^{1/2}dG_\Omega\Delta u)$$
$$\le K_2^2 D_{\Omega-\bar{B}}(\Delta u) D_{\Omega-\bar{B}}(G_\Omega\Delta u) \le K_2^2 D_R(\Delta u) D_R(u) .$$

We set $K_4 = K_2(D_R(\Delta u)D_R(u))^{1/2}$ and obtain

$$\left|\int_{\Omega-\bar{B}} g_\Omega d\Delta u \wedge *dG_\Omega\Delta u\right| \le K_4 .$$

To evaluate the last term in the inequality at the end of 1.3, observe that $\Delta u\cdot dg_\Omega \wedge *dG_\Omega\Delta u = \Delta u\cdot dG_\Omega\Delta u \wedge *dg_\Omega$. Again by Stokes' formula,

$$\int_{\partial\Omega-\partial B} \Delta u\cdot G_\Omega\Delta u*dg_\Omega = \int_{\Omega-\bar{B}} G_\Omega\Delta u\cdot d\Delta u \wedge *dg_\Omega + \int_{\Omega-\bar{B}} \Delta u\cdot dG_\Omega\Delta u \wedge *dg_\Omega .$$

Since $G_\Omega\Delta u$ and the $\partial g_\Omega(z,y)/\partial y^i$, $i = 1,2,\ldots,N$, are uniformly convergent on ∂B, there exists a constant K_5 such that

$$\left|\int_{\partial B} \Delta u\cdot G_\Omega\Delta u*dg_\Omega\right| \le K_5$$

for every Ω. It follows that

$$\left|\int_{\Omega-\bar{B}} \Delta u \cdot dg_{\Omega} \wedge *dG_{\Omega}\Delta u\right| = \left|\int_{\Omega-\bar{B}} \Delta u \cdot dG_{\Omega}\Delta u \wedge *dg_{\Omega}\right|$$

$$\leq K_5 + \left|\int_{\Omega-\bar{B}} G_{\Omega}\Delta u \cdot d\Delta u \wedge *dg_{\Omega}\right| .$$

In the same fashion as at the beginning of 1.4, the Schwarz inequality and our previous estimates yield

$$\left|\int_{\Omega-\bar{B}} G_{\Omega}\Delta u \cdot d\Delta u \wedge *dg_{\Omega}\right|^2 \leq K_1^2 D_{\Omega-\bar{B}}(\Delta u) D_{\Omega-\bar{B}}(g_{\Omega})$$

$$\leq K_1^2 D_R(\Delta u) D_{\Omega-\bar{B}}(g_{\Omega}) .$$

Since $D_{R-\bar{B}}(g(x,\cdot)) = \lim_{\Omega\to R} D_{\Omega-\bar{B}}(g_{\Omega}(x,\cdot))$, there exists a constant K_6 such that $K_5 + K_1(D_R(\Delta u)D_{\Omega-\bar{B}}(g_{\Omega}))^{1/2} \leq K_6$ for every Ω. Consequently,

$$\left|\int_{\Omega-\bar{B}} \Delta u \cdot dg_{\Omega} \wedge *dG_{\Omega}\Delta u\right| \leq K_6$$

and therefore,

$$\int_{\Omega-\bar{B}} g_{\Omega}(x,y)*(\Delta u(y))^2 \leq K_7 ,$$

with $K_7 = K_3 + K_4 + K_6$, for every regular region $\Omega \supset \bar{B}$. On letting $\Omega \to R$, we obtain

$$\int_{R-\bar{B}} g(x,y)*(\Delta u(y))^2 \leq K_7 .$$

We already know that $\int_{\bar{B}} g(x,y)*(\Delta u(y))^2 < \infty$. The proof of Theorem 1.2 is complete.

Consider the measure $d_g x = g(z,x)dx$ on R and the space $L^2(R,d_g x)$. In view of Harnack's inequality, the location of z is immaterial provided it is fixed. Theorems 1.1 and 1.2 can be reformulated as follows:

$$\Delta(H^2DC_\Delta(R)) \cup \Delta(H^2CD_\Delta(R)) \subset L^2(R, d_g x) .$$

1.5. Application to Riesz representation. By II.1.1, $R \in \tilde{O}^N_{QP}$ if and only if $\int_R *g(z,x) < \infty$, that is, $1 \in L^2(R, d_g x)$. We shall show:

THEOREM. *If* $R \in \tilde{O}^N_{QP}$, *then every* $u \in H^2DC_\Delta$ (*resp.* H^2CD_Δ) *has the Riesz representation*

$$u = h + \int_R g(\cdot,y)*\Delta u(y)$$

on R, *with* $h \in HD$ (*resp.* $h \in HC$).

Proof. Let $\langle\ ,\ \rangle$ be the inner product in $L^2(R, d_g y)$. Since 1 and $|\Delta u|$ are in $L^2(R, d_g y)$, the Schwarz inequality yields

$$\int_R g(x,y)*|\Delta u(y)| = \langle 1, |\Delta u|\rangle \le (\langle 1,1\rangle\langle|\Delta u|,|\Delta u|\rangle)^{1/2} < \infty .$$

In view of $u = h_\Omega + G_\Omega \Delta u$ and $D_\Omega(h_\Omega) + D_\Omega(G_\Omega \Delta u) = D_\Omega(u) \le D_R(u) < \infty$, the Theorem follows on letting $\Omega \to R$.

1.6. Dependence on the type of R. One might suspect that the Riesz representation be valid at least for every $H^2BB_\Delta(R)$ without any condition on R. That this is not the case can be seen by the following very simple example.

EXAMPLE. *Let* $R = (R, ds)$ *be the plane region* $1 < |z| < \infty$ *with the metric* $ds^2 = \lambda(z)|dz|^2$, $\lambda(z) = |z|^{-1}$. *Then the function*

$$u(z) = \frac{z + \bar{z}}{2(z\bar{z})^{1/2}}$$

belongs to $H^2BB_\Delta(R)$ (*and actually to* $H^2BC_\Delta(R)$) *but does not admit a Riesz* representation.

Proof. Denote by $\Delta_E = -4\partial^2/\partial z\partial\bar{z}$ the Euclidean Laplacian, with $z = x + iy$, $\partial/\partial z = \frac{1}{2}(\partial/\partial x - i\partial/\partial y)$, $\partial/\partial\bar{z} = \frac{1}{2}(\partial/\partial x + i\partial/\partial y)$. We know from I.1.11 that $H(R) = \{h|\Delta_E h = 0\}$, and from I.2.3 that the Euclidean Dirichlet integral is identical with $D_R(\cdot)$. By a simple computation, we see that $\Delta_E u(z) = \frac{1}{2}(z + \bar{z})/(z\bar{z})^{3/2}$, which in turn gives

$$\Delta u(z) = \frac{1}{2}\left(\frac{1}{z} + \frac{1}{\bar{z}}\right) .$$

Therefore, $\Delta u \in HB(R)$. Moreover,

$$\frac{d\Delta u(z) \wedge *d\Delta u(z)}{*1(z)} = \frac{1}{(z\bar{z})^2} ,$$

and we conclude that $D_R(\Delta u) < \infty$. Since $|u| < 1$, we have $u \in H^2BC_\Delta(R) \subset H^2BB_\Delta(R)$.

To see that u has no Riesz representation, we only have to show that

$$a = \int_{|z|>1} g(w,z)*|\Delta u(z)| = \infty ,$$

where $w \in R$ is fixed with $|w| > 2$. Clearly, there exists a positive number ε such that $g(w,z) \geq \varepsilon$ for $|z| > 2$. Therefore,

$$a \geq \varepsilon\int_{|z|>2} \frac{|z + \bar{z}|}{2z\bar{z}} \frac{1}{|z|} (\tfrac{1}{2}\, i dz \wedge d\bar{z})$$

$$= \varepsilon\int_{|z|>2} \frac{d|z|}{|z|} \int_0^{2\pi} |\sin(\arg z)| d \arg z = \infty .$$

The validity of the Example is herewith established.

Observe that $\bar{R} = \{1 \leq |z| < \infty\}$ can be viewed as a bordered Riemannian manifold with compact border $|z| = 1$ and ideal boundary ∞. The function u is in $H^2(\bar{R})$, and $\Delta u \in H(\bar{R})$. Clearly, Δu is also harmonic at ∞ and $\Delta u(\infty) = 0$ but $\Delta u(z) = -\Delta u(-\bar{z})$. This accounts for the intricate behavior of u

near ∞ and leads to $a = \infty$. Harmonically, the ideal boundary ∞ of R is of a quite simple nature but biharmonically, it is very involved. This shows that biharmonic classification depends heavily on the metric structure of the manifold in addition to its harmonic structure.

On an arbitrary $R \in \tilde{O}^N_{QP}$, every $u \in H^2B_\Delta$ admits a Riesz representation since

$$\int_R g(x,y)*|\Delta u(y)| \le c\int_R g(x,y)*1(y) < \infty ,$$

with $c = \sup_R|\Delta u|$, and the transition from $u = h_\Omega + G_\Omega\Delta u$ to $u = h_R + G_R\Delta u$ is legitimate.

Without the condition $R \in \tilde{O}^N_{QP}$, this conclusion is no longer valid, as is also shown by our example. In fact,

$$\int_{|z|>1} g(w,z)*1(z) \ge \varepsilon\int_{|z|>2} |z|^{-1}(\tfrac{1}{2}\, i dz \wedge d\bar{z}) = 2\pi\varepsilon\int_{|z|>2} d|z| = \infty ,$$

and our R belongs to $\mathcal{O}^2_{QP}$. In this sense, the condition $R \in \tilde{O}^N_{QP}$ is inevitable to assure that every $u \in H^2B_\Delta(R)$ possesses a Riesz representation. The significance of Theorem 1.5 lies in the fact that the same condition is sufficient to admit a Riesz representation of every $u \in H^2CD_\Delta$, with Δu not necessarily bounded.

<u>NOTES TO §1.</u> Theorems 1.1, 1.2, 1.5, and Example 1.6 were given in Nakai-Sario [5].

§2. RIESZ REPRESENTATION

In §1, the main topic was the metric growth of Δu. As an application of our result, we were able to show that for $u \in H^2DC_\Delta$ and $u \in H^2CD_\Delta$, the Riesz representation exists on an $R \in \tilde{O}^N_{QP}$. Can the boundedness requirement on u or Δu be suppressed? The purpose of the present section is to show that the answer is in the affirmative: every $u \in HDD_\Delta$ on an $R \in \tilde{O}^N_{QP}$ admits a Riesz representation.

One might suspect that no condition whatever on R is needed for the validity of the above assertion. That this is not so will be shown by a counterexample.

2.1. Main result. Given a harmonizable function f on R (IV.3.2 and Sario-Nakai [1]), let Ω be a regular region and denote by $\pi_\Omega f$ the harmonic solution of the Dirichlet problem on Ω with boundary values f at $\partial\Omega$. Then the harmonic part of f, $\pi f = \lim_{\Omega\to R} \pi_\Omega f$, exists on R.

Denote by $G = G(R)$ the class of functions f on R such that there exist an $h_R \in H(R)$ and a signed Borel measure σ_f on R with $f = h_R + \int_R g d\sigma_f$ on R,

$$G = \left\{ f \mid f = h_R + \int_R g(\cdot,y)d\sigma_f(y) \right\}.$$

The expression on the right is the Riesz representation of f. Since the potential part $\int_R g(\cdot,y)d\sigma_f(y)$ is the difference of two nonnegative superharmonic functions $\int_R g(\cdot,y)d\sigma_f^{\pm}(y)$ with $\pi \int_R g(\cdot,y)d\sigma_f^{\pm}(y) \equiv 0$, we see that $\pi f = h_R$. By the definition of $g(\cdot,y)$, $d\sigma_f(y) = \Delta f(y)dy$ in the sense of distributions. If $f \in C^2(R)$, then the same is true in the conventional sense. In any case, the representation of f is unique and f is harmonizable.

The main result of this section is the inclusion $H^2DD_\Delta(R) \subset H^2G(R) = H^2(R) \cap G(R)$ for $R \in \tilde{O}^N_{QP}$:

THEOREM. *If $R \in \tilde{O}^N_{QP}$, then every $u \in H^2DD_\Delta$ admits the Riesz representation*

$$u = \pi u + \int_R g(\cdot,y)*\Delta u(y)$$

with πu and Δu in HD.

The proof will be given in 2.2-2.7.

2.2. Frostman-type representation. Let K be a countable union of disjoint noncompact connected oriented C^∞ Riemannian manifolds of the same dimension $N \geq 2$, and suppose that no component of K belongs to O_G^N. Typically, K may be an open subset of R such that each component of K is hyperbolic.

In K we take a regular open set M, that is, one consisting of a finite number of regular subregions of K whose closures are disjoint by pairs. With a $u \in H(\overline{M}) = H(M) \cap C^2(\overline{M})$, we associate a $\hat{u} = \hat{u}_M \in H(K - \overline{M})$ defined by

$$\hat{u}(x) = \inf\{v(x) \mid v \in C(K - M) \cap HP(K - \overline{M}),\ v|\partial M = u\}$$

if $u \geq 0$; for a general u, we set $\hat{u} = \hat{u}^+ - \hat{u}^-$, where $u = u^+ - u^-$ with $u^\pm \in HP(\overline{M})$. Since the boundary values of $\hat{u}$ at ∂M are of class C^1, $\hat{u} \in C^1(K - M)$. If K is connected, then choose a regular region Ω in K with $\Omega \supset \overline{M}$, and let $\hat{u}_\Omega$ be the harmonic function on $\Omega - \overline{M}$ with continuous boundary values u at ∂M and 0 at $\partial\Omega$. It is easy to see that

$$\hat{u}(x) = \lim_{\Omega \to K} \hat{u}_\Omega(x)$$

uniformly on each compact subset of $K - M$. Moreover, if we set $\hat{u}_\Omega = 0$ on $K - \Omega$, then by Stokes' formula,

$$D_{K-\overline{M}}(\hat{u}_\Omega - \hat{u}_{\Omega'}) = D_{\Omega-\overline{M}}(\hat{u}_\Omega) - D_{\Omega'-\overline{M}}(\hat{u}_{\Omega'})$$

for $\Omega \subset \Omega'$, and we conclude that

$$D_{K-\overline{M}}(\hat{u}) = \lim_{\Omega \to K} D_{\Omega-\overline{M}}(\hat{u}_\Omega),$$

with $\hat{u} \in HD(K - \overline{M})$.

By means of $\hat{u}$, we shall establish the following <u>Frostman-type representation</u> of u in potential form:

$$u(x) = \int_{\partial M} g(x,y) {*d}(u(y) - \hat{u}(y))$$

for $x \in \overline{M}$. Here $g(x,y) = g_K(x,y)$ is the Green's function on K, that is, separately on each component K_i of K if $(x,y) \in K_i \times K_i$, and $g(x,y) = 0$ if $(x,y) \in K_i \times K_{i'}$, $i \neq i'$.

To deduce this representation, we may assume that K is connected. First let $x \in M$ and take a geodesic ball $B = B(x,\varepsilon)$ with center x and radius $\varepsilon > 0$ such that $\overline{B} \subset M$. By Stokes' formula,

$$\int_{\partial M-\partial B} (g(x,y)*d_y u(y) - u(y)*d_y g(x,y)) = 0 .$$

On letting $\varepsilon \to 0$, we obtain

$$u(x) = \int_{\partial M} g(x,y)*d_y u(y) - \int_{\partial M} u(y)*d_y g(x,y) .$$

Take a regular region Ω of K with $\overline{M} \subset \Omega$. Stokes' formula applied to the Green's function $g_\Omega(x,y)$, $x \in M$, on Ω, the function $\hat{u}_\Omega$, and the open set $\Omega - \overline{M}$ yields

$$\int_{\partial\Omega-\partial M} (g_\Omega(x,y)*d_y \hat{u}_\Omega(y) - \hat{u}_\Omega(y)*d_y g_\Omega(x,y)) = 0 .$$

Since $\hat{u}_\Omega|\partial M = u$ and $g_\Omega(x,\cdot)|\partial\Omega = \hat{u}_\Omega|\partial\Omega = 0$, we obtain

$$-\int_{\partial M} g_\Omega(x,y)*d_y \hat{u}_\Omega(y) + \int_{\partial M} u(y)*d_y g_\Omega(x,y) = 0 .$$

For every admissible Ω, $\hat{u}_\Omega - \hat{u} = 0$ on ∂M, $\hat{u}_\Omega - \hat{u} \in H(\Omega - \overline{M})$, and $\hat{u}_\Omega - \hat{u}$ converges to 0 uniformly on $\Omega' - M$ as $\Omega \to K$, where Ω' is a fixed regular region in K containing $\overline{M}$. Therefore, $\partial\hat{u}_\Omega/\partial x^i$ converges to $\partial\hat{u}/\partial x^i$, for each i uniformly on the intersection of $\Omega' - \overline{M}$ and a geodesic ball about any point of ∂M. The same is, of course, true of g_Ω. On letting $\Omega \to K$ in the above identity, we thus obtain

$$0 = -\int_{\partial M} g(x,y)*d_y \hat{u}(y) + \int_{\partial M} u(y)*d_y g(x,y)$$

and therefore,

$$u(x) = \int_{\partial M} g(x,y)*d_y u(y) - \int_{\partial M} u(y)*d_y g(x,y)$$

$$= \int_{\partial M} g(x,y)*d(u(y) - \hat{u}(y)) .$$

Thus the Frostman-type representation is valid for $x \in M$.

Since $g(x,y)$ is locally uniformly comparable with the Newtonian kernel ($N \geq 3$) or the logarithmic kernel ($N = 2$) for x and y close to each other (cf. Miranda [1]), it is easy to see that the integral in the above representation of u is a continuous function of x on $\overline{M}$. The same is true of $u(x)$, and the validity of the Frostman-type representation for $x \in M$ implies that for $x \in \overline{M}$.

2.3. Local decomposition. Let K be as in 2.2 and take a regular open set L in K. Suppose $u \in H^2(\overline{L}) = H^2(L) \cap C^4(\overline{L})$. Then we have the local decomposition

$$u = \pi_L u + \int_L g_L(\cdot,y)*\Delta u(y) ,$$

where $\pi_L u \in H(L) \cap C^1(\overline{L})$, $\pi_L u|\partial L = u$, and $g_L(\cdot,y)$ is the Green's function on L.

For the proof of this decomposition, we apply Stokes' formula to functions $u(y) - \pi_L u(y)$ and $g_L(x,y)$ and to the open set $L - \overline{B}$, where $B = B(x,\varepsilon)$, is a geodesic ball about x with radius ε such that $\overline{B} \subset L$:

$$\int_{L-\overline{B}} ((u(y) - \pi_L u(y))*\Delta_y g_L(x,y) - g_L(x,y)*\Delta_y(u(y) - \pi_L u(y)))$$

$$= -\int_{\partial L - \partial B} ((u(y) - \pi_L u(y))*d_y g_L(x,y) - g_L(x,y)*d_y(u(y) - \pi_L u(y))) .$$

On letting $\varepsilon \to 0$, we obtain the desired decomposition.

From this decomposition, it follows by Stokes' formula that

$$D_L(u) = D_L(\pi_L u) + D_L\left(\int_L g_L(\cdot,y)*\Delta u(y)\right)$$

and

$$D_L\left(\int_L g_L(\cdot,y)*\Delta u(y)\right) = \int_{L\times L} g_L(x,y)*\Delta u(x)*\Delta u(y) .$$

In fact, since $\Delta_x \int_L g_L(x,y)*\Delta u(y) = \Delta_x u(x)$ for $x \in L$ and $\int_L g_L(\cdot,y)*\Delta u(y) = 0$ at ∂L, we have

$$\begin{aligned} D_L\left(\int_L g_L(\cdot,y)*\Delta u(y)\right) &= \int_L \left(\int_L g_L(x,y)*\Delta u(y)\right)*\left(\Delta_x \int_L g_L(x,y)*\Delta u(y)\right) \\ &= \int_L \left(\int_L g_L(x,y)*\Delta u(y)\right)*\Delta u(x) \\ &= \int_{L\times L} g_L(x,y)*\Delta u(x)*\Delta u(y) . \end{aligned}$$

2.4. <u>Energy integrals.</u> Let K be as in 2.2 and consider a function $u \in H^2DP_\Delta(K)$. Take a regular open set L in K. Since $D_K(u) < \infty$, u is harmonizable (IV.3.2 and Sario-Nakai [1]), that is,

$$\pi u = \pi_K u = \lim_{L\to K} \pi_L u$$

exists on K. Moreover,

$$\lim_{L\to K} D_K(\pi u - \pi_L u) = 0$$

with $\pi_L u = u$ on K - L. In view of $\Delta u \geq 0$ on K, we deduce from the local decomposition in 2.3 that

$$u = \pi u + \int_K g(\cdot,y)*\Delta u(y)$$

on K, with $g(\cdot,y) = g_K(\cdot,y)$, and

$$D_K(u) = D_K(\pi u) + D_K\left(\int_K g(\cdot,y)*\Delta u(y)\right) .$$

We set $\int_L g_L(\cdot,y)*\Delta u(y) = 0$ on $K - L$ and obtain

$$\lim_{L\to K} D_K\left(\int_K g(\cdot,y)*\Delta u(y) - \int_L g_L(\cdot,y)*\Delta u(y)\right) = 0 .$$

By virtue of $\Delta u \geq 0$ on K, we infer from 2.3 that

$$D_K\left(\int_K g(\cdot,y)*\Delta u(y)\right) = \int_{K\times K} g(x,y)*\Delta u(x)*\Delta u(y) < \infty .$$

The right-hand side is referred to as the <u>energy</u> of Δu on K.

Let M be a regular open subset of K. We shall show that

$$D_K\left(\int_M g(\cdot,y)*\Delta u(y)\right) \leq D_K\left(\int_K g(\cdot,y)*\Delta u(y)\right) .$$

To this end take a sequence $\{M_n\}$ of regular open subsets of K such that $M_n \supset \overline{M}_{n+1} \supset M_{n+1} \supset \overline{M}$ and $\bigcap_1^\infty M_n = M$. Let φ_n be the continuous function defined on K by

$$\varphi_n|\overline{M} = \Delta u, \qquad \varphi_n|K - \overline{M}_n = 0, \qquad \varphi_n \in H(M_n - \overline{M}) .$$

Clearly, $\Delta u \geq \varphi_n \geq \varphi_{n+1}$, and φ_n tends to 0 on $K - M$, to Δu on M.

Since φ_n is locally Hölder continuous, the function $\int_Z g_Z(\cdot,y)*\varphi_n(y)$, where $Z = K$ or a regular open subset of K, is of class C^2 and

$$\Delta_x \int_Z g_Z(x,y)*\varphi_n(y) = \varphi_n(x)$$

for $n = 1,2,\ldots$ (cf. Miranda [1]). By the same reasoning as before, we conclude that

$$D_K\left(\int_K g(\cdot,y)*\varphi_n(y)\right) = \int_{K\times K} g(x,y)*\varphi_n(x)*\varphi_n(y) .$$

Let $x_0 \in K - \overline{M}$, and take a geodesic ball $B = B(x_0,\varepsilon)$ about x_0 with $\overline{B} \subset K - \overline{M}$. The sequence $\{\int_K g(\cdot,y)*\varphi_n(y)|B\}_{n=p}^\infty$ belongs to $H(B)$ for sufficiently large p, and by the Lebesgue-Fatou convergence theorem it converges decreasingly on B to $\int_M g(\cdot,y)*\Delta u(y)$. Therefore,

$$\lim_{n\to\infty} \frac{\partial}{\partial x^i} \int_K g(x,y)*\varphi_n(y) = \frac{\partial}{\partial x^i} \int_M g(x,y)*\Delta u(y)$$

uniformly on B for $i = 1,\ldots,N$.

Next take $x_0 \in M$ and $B = B(x_0,\varepsilon)$ with $\bar{B} \subset M$. Since $\{\int_{M_n-\bar{M}} g(\cdot,y)*\varphi_n(y)\}_1^\infty$ belongs to $H(B)$ and converges decreasingly on B to 0, and

$$\frac{\partial}{\partial x^i} \int_K g(x,y)*\varphi_n(y) - \frac{\partial}{\partial x^i} \int_M g(x,y)*\Delta u(y) = \frac{\partial}{\partial x^i} \int_{M_n-\bar{M}} g(x,y)*\varphi_n(y) ,$$

we conclude further that

$$\lim_{n\to\infty} \frac{\partial}{\partial x^i} \int_K g(x,y)*\varphi_n(y) = \frac{\partial}{\partial x^i} \int_M g(x,y)*\Delta u(y)$$

uniformly on B for $i = 1,\ldots,N$. Therefore,

$$\lim_{n\to\infty} \left|\operatorname{grad} \int_K g(\cdot,y)*\varphi_n(y)\right|^2 = \left|\operatorname{grad} \int_M g(\cdot,y)*\Delta u(y)\right|^2$$

on $K - \partial M$ and, a fortiori, $*1$-almost everywhere on K.

It follows that we can apply Fatou's lemma to deduce the following inequalities:

$$\begin{aligned} D_K\left(\int_M g(\cdot,y)*\Delta u(y)\right) &\leq \liminf_{n\to\infty} D_K\left(\int_K g(\cdot,y)*\varphi_n(y)\right) \\ &= \lim_{n\to\infty} \int_{K\times K} g(x,y)*\varphi_n(x)*\varphi_n(y) \\ &= \int_{M\times M} g(x,y)*\Delta u(x)*\Delta u(y) \\ &\leq \int_{K\times K} g(x,y)*\Delta u(x)*\Delta u(y) \end{aligned}$$

$$= D_K\left(\int_K g(\cdot,y)*\Delta u(y)\right),$$

and we have the desired result.

2.5. Reduction of Theorem 2.1. Let $u \in H^2DD_\Delta(R)$ with $R \in \tilde{O}^N_{QP}$. For a regular region Ω, the local decomposition in 2.3 gives

$$u = \pi_\Omega u + \int_\Omega g_\Omega(\cdot,y)*\Delta u(y)$$

on Ω. Here, $\pi_\Omega u = u$ on $R - \Omega$ and we recall from 2.4 that, since $D_R(u) < \infty$, u is harmonizable:

$$\pi u = \pi_R u = \lim_{\Omega\to R} \pi_\Omega u$$

exists on R and belongs to HD. A fortiori,

$$\lim_{\Omega\to R} \int_\Omega g_\Omega(x,y)*\Delta u(y) = u(x) - \pi u(x)$$

exists for every $x \in R$. Suppose

$$\int_R g(x,y)*|\Delta u(y)| < \infty$$

for one, and, by virtue of Harnack's inequality, for all x in R. Here $g(x,y) = g_R(x,y)$. Extend g_Ω to R by setting $g_\Omega|R - \Omega = 0$. By Lebesgue's convergence theorem, the limit and integration above can be interchanged:

$$u(x) - \pi u(x) = \int_R \lim_{\Omega\to R} g_\Omega(x,y)*\Delta u(y) = \int_R g(x,y)*\Delta u(y) ,$$

and we have Theorem 2.1. Since the converse is trivially true, the Theorem is equivalent to the finiteness property $\int_R g(x,y)*|\Delta u(y)| < \infty$.

If Δu is of constant sign on R, then we may assume that $\Delta u > 0$ on R. Since $\{g_\Omega(\cdot,y)\Delta u(y)\}$ converges increasingly on R to $g(\cdot,y)\Delta u(y)$, the

Lebesgue-Fatou theorem yields

$$\int_R g(x,y)*|\Delta u(y)| = \int_R g(x,y)*\Delta u(y) = u(x) - \pi u(x) < \infty .$$

Therefore, to prove the above finiteness property, we may assume that Δu changes sign on R. Set

$$K = \{x \in R | \Delta u(x) > 0\} .$$

Then $K \neq \emptyset$ and $K < R$. It is clear that K is an open set no component of which belongs to O_G^N. Thus the results in 2.2-2.4 apply to K. In view of

$$\int_R g(x,y)*|\Delta u(y)| = \int_K g(x,y)*\Delta u(y) + \int_{R-K} g(x,y)*(-\Delta u(y)) ,$$

it suffices to prove $\int_K g(x,y)*\Delta u(y) < \infty$ in order to obtain the finiteness property, since the same proof applied to $-u$ gives $\int_{R-K} g(x,y)*(-\Delta u(y)) < \infty$. Again by Harnack's inequality, we only have to establish

$$\int_K g(z,y)*\Delta u(y) < \infty$$

for some point z in $R - \overline{K}$.

Let $t > 0$ and $K_t = \{x \in R | \Delta u(x) > t\}$. Clearly, $\overline{K}_t \subset K$. If $K_{t_0} = \emptyset$ for some $t_0 > 0$, then by $\int_R g(z,y)*1(y) < \infty$,

$$\int_K g(z,y)*\Delta u(y) \leq t_0 \int_K g(z,y)*1(y) \leq t_0 \int_R g(z,y)*1(y) < \infty .$$

Therefore, we may suppose $K_t \neq \emptyset$ for every finite $t > 0$. Fix t_1, t_2 such that

$$0 < t_0 < t_1 < t_2 < \infty .$$

We can find an open set F of R with a smooth relative boundary ∂F such that

$$K_{t_2} \subset F \subset K_{t_1} .$$

By $\int_R g(z,y)*1(y) < \infty$,

$$\int_{K-F} g(z,y)*\Delta u(y) \le \int_{K-K_{t_2}} g(z,y)*\Delta u(y)$$

$$\le t_2 \int_{K-K_{t_2}} g(z,y)*1(y) \le t_2 \int_R g(z,y)*1(y) < \infty .$$

Thus we only have to prove

$$\int_F g(z,y)*\Delta u(y) < \infty$$

in order to obtain $\int_K g(z,y)*\Delta u(y) < \infty$.

Take an exhaustion $\{\Omega_n\}_1^\infty$ of R, that is, a sequence of regular subregions Ω_n of R with $\overline{\Omega}_n \subset \Omega_{n+1}$ and $\bigcup_1^\infty \Omega_n = R$. We assume, moreover, that $\{\Omega_n\}$ is compatible with F, that is, $\Omega_n \cap F$ is a regular open set in R for each $n = 1,2,\ldots$. It is clear that such an exhaustion exists. We set

$$M_n = F \cap \Omega_n .$$

2.6. Royden compactification. At this point we shall make essential use of the Royden compactification R^* of R, that is, the smallest compact Hausdorff space containing R as its open dense subspace such that every continuous Tonelli function with a finite Dirichlet integral on R can be continued to R^* as a $[-\infty,\infty]$-valued continuous function. For a detailed account of R^* we refer the reader to Sario-Nakai [1]. The discussion there is for Riemann surfaces, but the modification for Riemannian manifolds is straightforward.

For a set $X \subset R^*$, we denote by X^* the closure of X in R^*. In particular, $X^* \cap R = \overline{X}$.

Set $v = g(\cdot,z)$, $v_k | \Omega_k = g_{\Omega_k}(\cdot,z)$, and $v_k | R - \Omega_k = 0$. Clearly, $\{v_k\}_1^\infty$ converges increasingly on R to v, and $\lim_k D_R(v - v_k) = 0$. Since $\Delta u \in HD(R)$, we conclude that

$$(R - K_{t_0})^* \cap K_{t_1}^* = \emptyset .$$

By Urysohn's property for R^*, we can find a continuous Dirichlet finite Tonelli function φ on R such that $0 \leq \varphi \leq 1$ on R^*, $\varphi | K_{t_1}^* = 1$, $\varphi | (R - K_{t_0})^* = 0$. Clearly, φv and φv_k are continuous Dirichlet finite Tonelli functions, $\{\varphi v_k\}_1^\infty$ converges increasingly on R to φv, and $\lim_k D_R(\varphi v - \varphi v_k) = 0$. Let π_n be the harmonic projection of continuous Dirichlet finite Tonelli functions on K with respect to M_n:

$$\varphi v_k = \pi_n(\varphi v_k) + q_k, \qquad \varphi v = \pi_n(\varphi v) + q .$$

Here $\pi_n(\varphi v_k) \in HBD(K - \overline{M}_n)$; $\pi_n(\varphi v) \in HBD(K - \overline{M}_n)$; $q_k | M_n \cup \Delta(K) = 0$ with $\Delta(K)$ the Royden harmonic boundary of K; $q | M_n \cup \Delta(K) = 0$; and

$$D_K(\varphi v_k) = D_K(\pi_n(\varphi v_k)) + D_K(q_k), \qquad D_K(\varphi v) = D_K(\pi_n(\varphi v)) + D_K(q) .$$

Similarly,

$$D_K(\varphi v - \varphi v_k) = D_K(\pi_n(\varphi v) - \pi_n(\varphi v_k)) + D_K(q - q_k) .$$

We conclude that $\{\pi_n(\varphi v_k)\}_k$ converges increasingly on K to

$$\lim_{k\to\infty} \pi_n(\varphi v_k) = \pi_n(\varphi v) \geq 0 .$$

On the other hand, it follows from $\hat{u}(x) = \lim_{\Omega\to K} \hat{u}_\Omega(x)$ in 2.2 that

$$\hat{v}_k = \pi_n(\varphi v_k) | K - M_n ,$$

where $\hat{v}_k = \hat{v}_{kM_n}$ is the function on $K - M_n$ associated with $v_k | \overline{M}_n \in H(\overline{M}_n)$ in the manner described in 2.2. Here K may not be connected, but by considering the above formula on each component, the relation $\hat{u}(x) = \lim_{\Omega\to K} \hat{u}_\Omega(x)$ can be applied. By the definition of $\hat{v}$ in 2.2 and by the relation $\pi_n(\varphi v) | \partial M_n = v$, we deduce $\pi_n(\varphi v) | K - M_n \geq \hat{v}$. Since $v \geq v_k$ on $\overline{M}_n$, it is clear that $\hat{v} \geq \hat{v}_k$. From $\lim_k \pi_n(\varphi v_k) = \pi_n(\varphi v) \geq 0$ and $\hat{v}_k = \pi_n(\varphi v_k) | K - M_n$, it now follows that $\hat{v} = \hat{v}_{M_n} = \pi_n(\varphi v) | K - M_n$. In view of the expressions above for $D_K(\varphi v_k)$ and

$D_K(\varphi v)$, we infer that

$$D_{K-M_n}(\hat{v}_{M_n}) \leq c^2 < \infty ,$$

where the constant $c > 0$ is independent of $n = 1,2,\ldots$.

2.7. Completion of the proof of Theorem 2.1. The Frostman-type representation in 2.2 applied to M_n and $v = g(\cdot, z)$ gives

$$g(x,z) = \int_{\partial M_n} g_K(x,y)(*d_y v(y) - *d_y \hat{v}_{M_n}(y))$$

for $x \in M_n$. By Fubini's theorem and the symmetry of g and g_K,

$$\int_{M_n} g(z,x)*\Delta u(x) = \int_{\partial M_n} \left(\int_{M_n} g_K(y,x)*\Delta u(x)\right)(*d_y v(y) - *d_y \hat{v}_{M_n}(y)) .$$

For convenience, we set

$$\bar{g}_n = \int_{M_n} g_K(\cdot,x)*\Delta u(x), \qquad \bar{g} = \int_K g_K(\cdot,x)*\Delta u(x)$$

on K. We recall from 2.4 that

$$u = \pi_K u + \bar{g}, \qquad D_K(u) = D_K(\pi_K u) + D_K(\bar{g}), \qquad D_K(\bar{g}_n) \leq D_K(\bar{g}) .$$

As a consequence,

$$D_{M_n}(\bar{g}_n) \leq D_R(u), \qquad D_{K-M_n}(\bar{g}_n) \leq D_R(u), \qquad D_K(\bar{g}) \leq D_R(u) .$$

Since v belongs to $H(\bar{M}_n)$, Stokes' formula yields

$$\int_{\partial M_n} \left(\int_{M_n} g_K(y,x)*\Delta u(x)\right) *d_y v(y) = \int_{M_n} d\bar{g}_n \wedge *dv .$$

By the Schwarz inequality, we obtain

$$\left| \int_{\partial M_n} \left(\int_{M_n} g_K(y,x)*\Delta u(x)\right) *d_y v(y) \right| \leq D_R(u)^{1/2} D_K(v)^{1/2} .$$

Let L be a regular open subset of K such that $\overline{M}_n \subset L$. Set

$$\overline{g}_{nL}|L = \int_{M_n} g_L(\cdot,x)*\Delta u(x), \qquad \overline{g}_{nL}|K - L = 0 .$$

Since $\lim_{L\to K} g_L(y,\cdot) = g_K(y,\cdot)$ increasingly on M_n for every fixed y in K with $g_L(y,\cdot)|K - L = 0$, the Lebesgue-Fatou theorem yields $\lim_{L\to K} \overline{g}_{nL}(y) = \overline{g}_n(y)$ increasingly for every y in K. Therefore, the convergence is uniform in y on ∂M_n and a fortiori,

$$\int_{\partial M_n}\left(\int_{M_n} g_K(y,x)*\Delta u(x)\right)*d_y\hat{v}_{M_n}(y) = \lim_{L\to K} \int_{\partial M_n} \overline{g}_{nL}(y)*d_y\hat{v}_{M_n}(y) .$$

On the other hand, since $\overline{g}_{nL}|\partial L = 0$, we deduce by Stokes' formula that

$$-\int_{\partial M_n} \overline{g}_{nL}(y)*d_y\hat{v}_{M_n}(y) = \int_{\partial L-\partial M_n} \overline{g}_{nL}(y)*d_y\hat{v}_{M_n}(y) = \int_{L-\overline{M}_n} d\overline{g}_{nL} \wedge *d\hat{v}_{M_n} .$$

By the Schwarz inequality,

$$\left|\int_{\partial M_n} \overline{g}_{nL}(y)*d_y\hat{v}_{M_n}(y)\right| \leq D_{L-\overline{M}_n}(\overline{g}_{nL})^{1/2} D_{L-\overline{M}_n}(\hat{v}_{M_n})^{1/2} .$$

In view of this and the relation at the end of 2.6, we obtain

$$\left|\int_{\partial M_n} \overline{g}_{nL}(y)*d_y\hat{v}_{M_n}(y)\right| \leq cD_L(\overline{g}_{nL})^{1/2} .$$

On applying the relations on D_K in 2.4, we conclude that

$$D_L(\overline{g}_{nL}) \leq D_L\left(\int_L g_L(\cdot,y)*\Delta u(y)\right)$$

$$= \int_{L\times L} g_L(x,y)*\Delta u(x)*\Delta u(y)$$

$$\leq \int_{K\times K} g_K(x,y)*\Delta u(x)*\Delta u(y) = D_K(\overline{g}) .$$

It follows that

$$\left| \int_{\partial M_n} \bar{g}_{nL}(y) * d_y \hat{v}_{M_n}(y) \right| \le c D_L(\bar{g}_{nL})^{1/2}$$

$$\le c D_K(\bar{g})^{1/2} \le c D_R(u)^{1/2}$$

and therefore,

$$\left| \int_{\partial M_n} \left(\int_{M_n} g_K(y,x) * \Delta u(x) \right) * d_y \hat{v}_{M_n}(y) \right|$$

$$= \lim_{L \to K} \left| \int_{\partial M_n} \bar{g}_{nL}(y) * d_y \hat{v}_{M_n}(y) \right| \le c D_R(u)^{1/2} .$$

On combining $M_n = F \cap \Omega_n$ at the end of 2.5 and the above estimates, we obtain

$$\int_{F \cap \Omega_n} g(z,x) * \Delta u(x) \le D_R(u)^{1/2} (c + D_K(v)^{1/2}) = c_0$$

for $n = 1,2,\ldots,$ with c_0 independent of n. Since $\Delta u > 0$ on F, the Lebesgue-Fatou theorem gives the desired inequality

$$\int_F g(z,y) * \Delta u(y) = \lim_{n \to \infty} \int_{F \cap \Omega_n} g(z,y) * \Delta u(y) \le c_0 < \infty .$$

The proof of Theorem 2.1 is herewith complete.

<u>2.8.</u> $H^2 DD_\Delta$ <u>function not in</u> $H^2 G$. The question arises as to whether or not the conclusion $H^2 DD_\Delta \subset H^2 G$ of Theorem 2.1 remains valid without the condition $R \in \tilde{O}^N_{QP}$. We shall show that this is not the case. Explicitly, we shall construct a Riemannian manifold $R \in O^N_{QP}$ and a function $u \in H^2 DD_\Delta$ (in fact, $u \in H^2 CC_\Delta$) such that $u \notin H^2 G$.

Let S be the plane region $\{z \mid e^2 < |z| < \infty\}$. The function

$$\lambda(z) = \frac{1}{|z|} \left(\frac{1}{\log|z|} - \frac{2}{(\log|z|)^3} \right)$$

is clearly C^∞ and strictly positive on S. On the Riemannian manifold R

whose base space is S and metric $\lambda(z)^{1/2}|dz|$, the Laplacian Δ, the volume element $*1$, and the gradient ∇ are related to the corresponding Euclidean Δ_E, $*_E 1$, and ∇_E by

$$\Delta = \lambda^{-1}\Delta_E, \qquad *1 = \lambda *_E 1, \qquad \nabla = \lambda^{-1/2}\nabla_E .$$

In particular, a function is harmonic on S in the Euclidean sense if and only if it is so on R in the Riemannian sense: $H(S) = H(R)$.

Let $g(z,\zeta)$ be the Green's function on R. Since it is also the Green's function on S,

$$g(z,\zeta) = \frac{1}{2\pi}\log\left|\frac{e^4 - \bar{\zeta}z}{e^2(z-\zeta)}\right| .$$

Therefore, there exists a number $a > 0$ and a subregion $R_\rho = \{z \mid \rho < |z| < \infty\}$, $\rho > e^2$, such that $g(z,\zeta) \geq a$ for $(z,\zeta) \in R_\rho \times R_\rho$. To see that

$$R \in O^N_{QP}$$

we only have to prove that $\int_R g(z,\zeta)*1(\zeta) = \infty$. Clearly,

$$\int_R g(z,\zeta)*1(\zeta) \geq \int_{R_\rho} g(z,\zeta)*_E\lambda(\zeta)$$

$$\geq a\int_{R_\rho} *_E\lambda(\zeta)$$

$$\geq 2\pi a\int_\rho^\infty \frac{1}{|z|}\left(\frac{1}{\log|z|} - \frac{2}{(\log|z|)^3}\right)|z|\,d|z| = \infty$$

for $z \in R_\rho$.

Consider the function

$$u(z) = \frac{z+\bar{z}}{2|z|\log|z|}$$

on R. By a simple computation, we obtain

$$\Delta u(z) = \frac{1}{\lambda(z)}\Delta_E u(z) = \frac{1}{2}\left(\frac{1}{z} + \frac{1}{\bar{z}}\right) .$$

This is harmonic on S, and hence on R, as the real part of the analytic function $1/z$. By means of

$$D_R(\cdot) = \int_R *|\nabla \cdot|^2 = \int_S *_E|\nabla_E \cdot|^2 ,$$

it is easy to see that

$$\sup_R |u| + \sup_R |\Delta u| + D_R(u) + D_R(\Delta u) < \infty .$$

Therefore,

$$u \in H^2CC_\Delta \subset H^2DD_\Delta .$$

Finally, we show that u does not admit a Riesz representation on R, that is, $u \notin H^2G$. By 2.5, it suffices to prove that

$$\int_R g(x,y)*|\Delta u(y)| = \infty .$$

Let $z \in S_\rho = \{z \mid \rho < |z| < \infty\}$. We estimate the integral

$$\Lambda = \int_R g(z,\zeta)*|\Delta u(\zeta)| = \int_S g(z,\zeta)*_E|\Delta_E u(\zeta)|$$

$$\geq \int_{S_\rho} g(z,\zeta)*_E|\Delta_E u(\zeta)| \geq a\int_{S_\rho} *_E|\Delta_E u(\zeta)| .$$

Since the quantity $b = a\int_{|z|=1} |\cos \arg z| d \arg z$ is positive,

$$\Lambda \geq b\int_\rho^\infty \{(\log|z|)^{-1} - 2(\log|z|)^{-3}\} \frac{d|z|}{|z|} = \infty .$$

2.9. Dirichlet potentials. We denote by M the class of Dirichlet finite continuous (not necessarily bounded) Tonelli functions f on a Riemannian manifold R. The subclass $M \cap C^\infty$ is dense in M in the norm $\|f\| = \sup_R f + (D_R(f))^{1/2}$. Denote by M_δ the closure of the set of functions in M with compact supports with respect to the topology τ given by the simultaneous convergence in $D_R(\cdot)$ and the uniform convergence on every compact set in R. The Royden harmonic decomposition reads

$$M = HD + M_\delta .$$

The functions f in M_δ are called Dirichlet potentials on R. It is known that, for each $f \in M_\delta$, there exists a positive measure μ on R such that

$$|f| \le \int_R g(\cdot,y)d\mu(y)$$

on R (Constantinescu-Cornea [1]). This suggests a close affiliation between Dirichlet potentials and Green potentials with signed measures. An interesting question is: Under what conditions is a Dirichlet potential a genuine Green potential with a signed measure? That not every Dirichlet potential has this property is of course exemplified by $u - \pi_R u$ on R, with u and R of 2.8. Here not only $u - \pi_R u \in C = BD$ but also $\Delta(u - \pi_R u) \in C$. By means of an obvious modification of the proof of Theorem 2.1, we obtain the following result:

THEOREM. *If $R \in \tilde{O}^N_{QP}$, then every f in M with the property that Δf is in M admits the Riesz representation*

$$f = \pi f + \int_R g(\cdot,y) * \Delta f(y)$$

with πf in HD. In particular, a Dirichlet potential f with Δf in M is the Green potential $\int_R g(\cdot,y)*\Delta f(y)$ of the signed measure $*\Delta f(y)$.

Using the notation $MM_\Delta = \{f \in M | \Delta f \in M\}$, we can give the Theorem the simple form

$$MM_\Delta(R) \subset G(R) \quad \text{for} \quad R \in \tilde{O}^N_{QP}.$$

As shown above, the condition $R \in \tilde{O}^N_{QP}$ cannot be suppressed.

NOTES TO §2. Theorems 2.1 and 2.9 were established in Nakai-Sario [4].

§3. MINIMUM SOLUTIONS AS POTENTIALS

Again we proceed to broader classes of functions and now turn from H^2DD_Δ to H^2D discussed in Chapter IV. Theorem IV.3.2 leads us in a natural manner to the following question on Riesz representation: When can the minimum solution u be represented as $u = G_R h = \int_R g_R * h$, with g_R the Green's function on R? Throughout the present section, we assume that $R \in \tilde{O}_G^N$.

We introduce the class H_{GA} of harmonic functions whose absolute value has a finite Green's potential,

$$H_{GA} = \{h \in H \bigm| G_R|h| < \infty\} .$$

If $G_R|h| < \infty$ at one point, the same is true, by Harnack's inequality, at all points of R. If, in the notation of IV.2.4, $h \in H_{PD}$, then $h \in H_{GA}$ if and only if the minimum solution u has the representation $u = G_R h$. If for a $u \in H^2D$, $\Delta u \in H_{GA}$, then we have the Riesz decomposition

$$u = \pi u + \int_R g_R(\cdot, y) * \Delta u(y) ,$$

and the study of H^2D is greatly simplified. Thus the problem of determining when an $h \in H_{PD}$ belongs to H_{GA} is of great interest.

3.1. Preliminary considerations. The simplest condition in the above problem is: A positive $h \in H_{PD}$ belongs to H_{GA}. In fact, by Theorem IV.3.2, the minimum solution can be written $u = \lim_{\Omega \to R} \int_R g_\Omega(\cdot, y) * h(y)$, where we have set $g_\Omega(\cdot, y) = 0$ for $y \in R - \Omega$. Since $\{g_\Omega(\cdot, y) h(y)\}_\Omega$ increases with Ω, the Lebesgue-Fatou theorem gives $u = G_R h$, and the assertion follows.

Since $h \in HP$ implies that an $h \in H_{PD}$ belongs to H_{GA}, what can be said about $h \in HB$, HD or HC? The answer is: None of the conditions $h \in HB$, $h \in HD$, or $h \in HC$ entails that an $h \in H_{PD}$ belong to H_{GA}. In fact, on the Riemannian manifold $R = \{S, \lambda(z)^{1/2}|dz|\}$ considered in 2.8, both the function

$$h = \frac{1}{2}\left(\frac{1}{z} + \frac{1}{\bar{z}}\right)$$

and the corresponding minimum solution u belong to $C = BD$. The fact that $h \in HC \cap H_{PD}$ was proved in 2.8, and a direct computation shows that, nevertheless, $h \notin H_{GA}$. As to u, the function

$$v(z) = \frac{z + \bar{z}}{2|z|\log|z|}$$

is a solution of $\Delta v = h$ and belongs to H^2C, and by the decomposition in IV.3.1, the minimum solution u also belongs to H^2C.

<u>3.2</u>. <u>Rate of growth</u>. Since $h \in HB$ (or HC) is not sufficient for $h \in H_{PD}$ to be in H_{GA}, we must require something of R in order for h to have this property. An immediate condition is $R \in \tilde{O}^N_{QP}$, which gives $G_R|h| \leq \sup_R |h| G_R 1 < \infty$. The condition $R \in \tilde{O}^N_{QP}$ is thus trivially sufficient for an $h \in HB$ (or HC) to belong to H_{GA}, but whether or not the same condition is also sufficient for an $h \in HD$ to be in H_{GA} is not a trivial question. We shall show that $R \in \tilde{O}^N_{QP}$ does assure this.

To this end, we first consider the rate of growth of an $h \in H_{PD} \cap HD$.

<u>THEOREM</u>. <u>A function</u> $h \in H_{PD} \cap HD$ <u>on</u> $R \in \tilde{O}^N_G$ <u>grows so slowly that for</u> $\varepsilon > 0$,

$$\int_{\{y\in R \,|\, |h(y)|\geq\varepsilon\}} g_R(x,y) * |h(y)| < \infty ,$$

$$\int_{\{y\in R \,|\, |h(y)|\leq\varepsilon\}} g_R(x,y) * |h(y)|^2 < \infty ,$$

<u>for some and hence for all</u> $x \in R$.

<u>Proof</u>. Although the example in 3.1 showed that an $h \in H_{PD} \cap HD$ need not belong to H_{GA}, it is interesting to observe that the above two conditions are very close to $h \in H_{GA}$, that is, $\int_R g_R(x,y) * |h(y)| < \infty$. For an $h \in H_{PD} \cap HD - H_{GA}$ and an arbitrarily small $\varepsilon > 0$, $\int_{|h|\leq\varepsilon} g_R(x,y) * |h(y)| = \infty$, whereas $\int_{|h|\leq\varepsilon} g_R(x,y) * |h(y)|^2 < \infty$.

To prove the Theorem, we fix an $x \in R$ and a geodesic ball B centered at x. Take a $\psi \in C^{\infty}(R)$ such that $0 \leq \psi \leq 1$, $\psi | R - \overline{B} = 1$, and $\psi | B_0 = 0$, where B_0 is a geodesic ball about x with $\overline{B}_0 \subset B$. The truncated function

$$v(x) = \min(\max(h(x), -\varepsilon), \varepsilon)$$

belongs to $C^0BD(R)$. The following relation, though simple, is the crucial point in the proof:

$$vh = \begin{cases} |h|^2 & \text{on } \{|h| \leq \varepsilon\} , \\ \varepsilon |h| & \text{on } \{|h| \geq \varepsilon\} . \end{cases}$$

Let Ω be a regular subregion of R containing $\overline{B}$. We extend $g_{\Omega}(x,y)$ to R by $g_{\Omega}(x,y) = 0$ for $y \in R - \Omega$. The functions

$$\varphi_{\Omega}(y) = \psi(y) g_{\Omega}(x,y) v(y) ,$$

$$\varphi_R(y) = \psi(y) g_R(x,y) v(y)$$

belong to the class $C^0BD(R)$. Since the Dirichlet integral of $g_R(x,\cdot) - g_{\Omega}(x,\cdot)$ tends to zero as Ω exhaust R,

$$\lim_{\Omega \to R} D_R(\varphi_R - \varphi_{\Omega}) = 0 .$$

A minimum solution u satisfies $u \in C^{\infty}D(R)$, $\Delta u = h$. Since $d(\varphi_{\Omega} {*du}) = d\varphi_{\Omega} \wedge {*du} - \varphi_{\Omega} {*h}$ and $C_0^{\infty}D(R)$ is dense in $C_0^0D(R)$, Stokes' formula yields

$$\int_R \psi(y) g_{\Omega}(x,y) v(y) {*h}(y) = \int_R d\varphi_{\Omega} \wedge {*du} .$$

Recall that $vh \geq 0$ and $\psi = 1$ on $R - \overline{B}$, $0 \leq \psi \leq 1$ on $\overline{B}$. By the Schwarz inequality,

$$\int_{R-\overline{B}} g_{\Omega}(x,y) v(y) {*h}(y) \leq (D_R(\varphi_{\Omega}) D_R(u))^{1/2} .$$

Since $g_{\Omega}(x,\cdot)$ converges increasingly on R to $g_R(x,\cdot)$, we infer by the

Lebesgue-Fatou theorem and $D_R(\varphi_R - \varphi_\Omega) \to 0$ that

$$\int_{R-\bar{B}} g_R(x,y)v(y)*h(y) \leq (D_R(\varphi)D_R(u))^{1/2} < \infty .$$

This with the trivial relation $\int_{\bar{B}} g_R(x,y)v(y)*h(y) < \infty$ gives $\int_R g_R(x,y)v(y)*h(y) < \infty$. In view of the above properties of vh, we conclude that

$$\int_{|h|\leq\varepsilon} g_R(x,y)*|h(y)|^2 + \varepsilon\int_{|h|\geq\varepsilon} g_R(x,y)*|h(y)| < \infty ,$$

and the Theorem follows.

3.3. Role of QP functions. As a consequence of Theorem 3.2, we conclude that the condition $R \in \tilde{O}^N_{QP}$ plays the same role for $h \in HD$ as for $h \in HB$:

THEOREM. If $R \in \tilde{O}^N_{QP}$, then every $h \in H_{PD} \cap HD$ belongs to H_{GA}.

Proof. Actually, we can slightly sharpen this: an $h \in HD$ on any R is in H_{GA} if $a_\varepsilon = \int_{|h|\leq\varepsilon} g_R(x,y)*1(y) < \infty$ for some $\varepsilon > 0$. If $R \in \tilde{O}^N_{QP}$, then $a_\varepsilon < \infty$ because of $\int_R g_R(x,y)*1(y) < \infty$.

To prove the Theorem, or its sharper form, we only have to observe that

$$\left(\int_{|h|\leq\varepsilon} g_R(x,y)*|h(y)|\right)^2 \leq \left(\int_{|h|\leq\varepsilon} g_R(x,y)*|h(y)|^2\right)\cdot a_\varepsilon .$$

We shall show later, in 3.5, that $R \in \tilde{O}^N_{QP}$ in the Theorem is not a necessary condition. In this context, it may be interesting to observe the following. Suppose there exists an $h \in H_{PD} \cap HD$. The set $A_\varepsilon = \{y \in R \,\big|\, |h(y)| < \varepsilon\}$ is a neighborhood of the zero set of h and may be considered a rather small region. From the first inequality in Theorem 3.2, we obtain on replacing $|h|$ by ε,

$$\int_{R-A_\varepsilon} g_R(x,y)*1(y) < \infty ,$$

a condition which may be viewed as being close to $R \in \tilde{O}^N_{QP}$. This observation together with the remark after Theorem 3.2 indicate the very delicate situation concerning the membership of an $h \in HD$ in H_{GA}.

3.4. Role of QB functions. In connection with Theorem 3.3, we append here the following statement although it lies somewhat outside the main train of thought in our present discussion:

THEOREM. *If* $R \in \tilde{O}_{QB}^{N}$, *then* $h \in HD$ *belongs to* H_{GA}.

Proof. In comparison with Theorem 3.3, we have here a stronger assumption on R but we are not assuming that $h \in h_{PD}$. We start the proof with the case in which $h \neq 0$ at all points of R. We may suppose $h > 0$ on R. The condition $R \in \tilde{O}_{QB}^{N}$ is equivalent to the existence of a finite constant k such that $G_\Omega 1 \leq G_R 1 \leq k$ on R. As before, we extend g_Ω to R by $g_\Omega = 0$ on $R - \Omega$. Take a geodesic ball B about an arbitrarily fixed $x \in R$. We shall consider only those Ω for which $\overline{B} \subset \Omega$. Recall that $\Delta G_\Omega 1 = 1$ on Ω. We first evaluate

$$a_\Omega \equiv \int_{\Omega-\overline{B}} g_\Omega(x,y)*|h(y)|$$

as $\Omega \nearrow R$ by using the device of expressing a_Ω as

$$a_\Omega = \int_{\Omega-\overline{B}} g_\Omega(x,y)h(y)*\Delta G_\Omega 1(y) .$$

Note that here $h > 0$. By Stokes' formula, $a_\Omega = I_{1\Omega} + I_{2\Omega} + I_{3\Omega}$, where

$$I_{1\Omega} = \int_{\Omega-\overline{B}} G_\Omega 1(y)*\Delta_y(g_\Omega(x,y)h(y)) ,$$

$$I_{2\Omega} = -\int_{\partial B} g_\Omega(x,y)h(y)*dG_\Omega 1(y) ,$$

$$I_{3\Omega} = \int_{\partial B} G_\Omega 1(y)*d(g_\Omega(x,y)h(y)) .$$

We start by estimating $I_{1\Omega}$. By virtue of

$$\Delta_y(g_\Omega(x,y)h(y)) = -2d_y g_\Omega(x,y) \wedge *dh(y),$$

we obtain $I_{1\Omega}^2 \leq 4k^2 D_{\Omega-\bar{B}}(g_\Omega(x,\cdot))D_{\Omega-\bar{B}}(h)$ and consequently,

$$\limsup_{\Omega\to R} I_{1\Omega}^2 \leq 4k^2 D_{R-\bar{B}}(g_R(x,\cdot))D_R(h) < \infty .$$

Since the function $g_\Omega(x,\cdot)$, harmonic on a neighborhood of ∂B, converges uniformly to $g_R(x,\cdot)$ there, it is trivial that $\limsup_{\Omega\to R}(|I_{2\Omega}| + |I_{3\Omega}|) < \infty$ and therefore, $\lim_{\Omega\to R} a_\Omega < \infty$. Clearly, $\int_B g_R(x,y)*h(y) < \infty$ and a fortiori, $\int_R g_R(x,y)*|h(y)| = \int_R g_R(x,y)*h(y) < \infty$, that is, h is in H_{GA}.

We turn to the case in which h is not of constant sign. We may assume that neither $\omega = \{y \in R | h(y) > 0\}$ nor $R - \omega$ is empty. Since the integral of $g_R(x,y)|h(y)|$ over R is the sum of $\int_\omega g_R(x,y)*h(y)$ and $\int_{R-\omega} g_R(x,y)*(-h(y))$, it suffices to show that

$$a \equiv \int_\omega g_R(x,y)*h(y) < \infty ,$$

the proof for $\int_{R-\omega} g_R(x,y)*(-h(y)) < \infty$ being the same. By Harnack's inequality, the validity of $a < \infty$ does not depend on x. We fix $x \in R - \bar{\omega}$. For a regular subregion Ω of R, we set $\Omega' = \Omega \cap \omega$. We shall evaluate

$$a_{\Omega'} = \int_{\Omega'} g_\Omega(x,y)*h(y) .$$

Observe that $g_\Omega(x,\cdot)$ is harmonic on Ω and, therefore, on Ω'. For an arbitrary $\varepsilon > 0$, choose a regular open subset $\Omega'' \subset \Omega'$ such that $g_\Omega(x,y)h(y) < \varepsilon$ for every $y \in \partial\Omega''$. This is certainly possible, because $g_\Omega(x,y)h(y) = 0$ on $\partial\Omega'$. We start by evaluating the integral

$$b_{\Omega''} = \int_{\Omega''} g_\Omega(x,y)*h(y) ,$$

which tends to $a_{\Omega'}$ as Ω'' exhausts Ω'. Again we make use of the relation $\Delta G_{\Omega''}1 = 1$ and obtain

$$b_{\Omega''} = \int_{\Omega''} g_\Omega(x,y)h(y)*\Delta G_{\Omega''}1(y) ,$$

which by Stokes' formula takes the form

$$b_{\Omega''} = \int_{\Omega''} G_{\Omega''}1(y)*\Delta_y(g_\Omega(x,y)h(y))$$

$$- \int_{\partial\Omega''} g_\Omega(x,y)h(y)*dG_{\Omega''}1(y) .$$

The first term on the right tends to $\int_{\Omega'} G_{\Omega'}1(y)*\Delta_y(g_\Omega(x,y)h(y))$ as Ω'' exhausts Ω', since $G_{\Omega''}1$ converges increasingly to $G_{\Omega'}1$. The absolute value of the last term on the right is dominated by

$$\varepsilon \int_{\partial\Omega''} |*dG_{\Omega''}1(y)| = \varepsilon \left| \int_{\partial\Omega''} *dG_{\Omega''}1(y) \right| .$$

By Stokes' formula,

$$\int_{\partial\Omega''} *dG_{\Omega''}1(y) = -\int_{\Omega''} *\Delta G_{\Omega''}1(y) = -\int_{\Omega''} *1 .$$

Therefore,

$$\limsup_{\Omega''\to\Omega'} \left| \int_{\partial\Omega''} g_\Omega(x,y)h(y)*dG_{\Omega''}1(y) \right| \leq \varepsilon \int_{\Omega'} *1 ,$$

that is, the last term in $b_{\Omega''}$ tends to zero as Ω'' exhausts Ω'. It follows that

$$a_{\Omega'} = \int_{\Omega'} G_{\Omega'}1(y)*\Delta_y(g_\Omega(x,y)h(y)) .$$

If $\partial\Omega'$ were sufficiently smooth, then we could derive the above equality simply by applying Stokes' formula to $a_{\Omega'}$. Since

$$a_{\Omega'} = -2\int_{\Omega'} (G_{\Omega'}1)dg_{\Omega}(x,\cdot) \wedge *dh ,$$

the Schwarz inequality yields $a_{\Omega'}^2 \leq 4k^2 D_{\Omega'}(g_{\Omega}(x,\cdot))D_{\Omega'}(h)$ and consequently,

$$\int_{\omega} g_R(x,y)*h(y) = \lim_{\Omega\to R} a_{\Omega'} < \infty .$$

The proof of the Theorem is herewith complete.

3.5. Nonnecessity of $R \in \tilde{O}_{QP}^N$. In view of $\int_{R-A_\varepsilon} g_R(x,y)*1(y) < \infty$, it would seem plausible that the condition $R \in \tilde{O}_{QP}^N$ in Theorem 3.3 is not only sufficient but also necessary. We shall now show, by means of a rather intricate example, that this intuition is misleading. Explicitly:

THEOREM. *There exists an* $R \in O_{QP}^N$ *such that* $H_{PD} \cap HD \neq \emptyset$ *and* $H_{PD} \cap HD \subset H_{GA}$.

Proof. Take a Riemann surface S' such that $\dim HD(S') = 2$ and consequently (cf. Sario-Nakai [1]), $HD(S') = HC(S')$. Since $HD(S')$ contains the real number field R^1 and a nonconstant function h', $h_0 = h' - c$ vanishes at some point of S' if $c \in R^1$ is suitably chosen. We fix a point $z_0 \in S'$ such that

$$h_0(z_0) = 0, \qquad (\text{grad } h_0)(z_0) \neq 0 .$$

Note that $\{1,h_0\}$ forms a basis of $HD(S')$ and

$$\{h \in HD(S')|h(z_0) = 0\} = \{ch_0|c \in R^1\} .$$

For the base space of our manifold R, we choose $S = S' - \{z_0\}$.

To introduce a metric $\lambda(z)^{1/2}|dz|$ on S, we first consider an auxiliary "metric" $\mu_0(z)^{1/2}|dz|$ which may have zeros of μ_0, then modify it to obtain a genuine metric $\mu(z)^{1/2}|dz|$, and finally modify μ to obtain a λ with the desired properties. We shall denote by $g(z,\zeta)$ the Green's function on

$S' \times S'$. Observe that $g(z,\zeta) = g_{S'}(z,\zeta) = g_S(z,\zeta)|S' \times S'$.

As the first step in endowing S with a metric $\lambda(z)^{1/2}|dz|$, we consider the density

$$\mu_0 = -\Delta h_0^2 = 2|\text{grad } h_0|^2 .$$

By virtue of $h_0 \in HBD(S)$, we have $h_0^2 \in C^0BD(S)$, and by the decomposition in IV.3.1, we obtain $h_0^2 = v - \varphi$, where $v \in HBD(S)$ and $\varphi \in P_0(S)$. Since h_0^2 is subharmonic, $\varphi = v - h_0^2$ is a bounded positive potential. In view of the Riesz theorem, it is the Green's potential of the measure $\Delta\varphi(\zeta)d\xi d\eta = \mu_0(\zeta)d\xi d\eta$ with $\zeta = \xi + i\eta$:

$$h_0^2 = v + \int_S g(\cdot,\zeta)\mu_0(\zeta)d\xi d\eta .$$

For the Dirichlet integral we obtain

$$D_S(h_0^2) = D_S(v) + \int_{S\times S} g(z,\zeta)\mu_0(z)\mu_0(\zeta)dxdyd\xi d\eta .$$

It follows that

$$\sup_{z\in S} \int_S g(z,\zeta)\mu_0(\zeta)d\xi d\eta < \infty$$

and also

$$\int_{S\times S} g(z,\zeta)\mu_0(z)\mu_0(\zeta)dxdyd\xi d\eta < \infty .$$

3.6. Construction of the metric. Let $\{z_n\}$, $n = 1,2,\ldots,k \leq \infty$, be the zeros of grad h_0 on S. To obtain a genuine metric from μ_0, we will modify μ_0 at the z_n. To this end, we take parametric disks $U_n' = \{|z - z_n| < 1\}$ about the z_n with $\bar{U}_n' \cap \bar{U}_m' = \emptyset$ for $n \neq m$, and smaller disks $U_n = \{|z - z_n| < r_n < \frac{1}{2}\}$ such that

$$\sup_{z\in S}\int_{U_n} g(z,\zeta)\,d\xi d\eta < 2^{-n}$$

and

$$\int_{U_n\times U_n} g(z,\zeta)\,dxdyd\xi d\eta < 4^{-n} .$$

To show that the first of these conditions can be satisfied, we first observe that the integral on the left is a potential of a mass of density 1 on U_n. By the maximum principle, this potential on S assumes its maximum on $\overline{U}_n$. It, therefore, suffices to consider points $(z,\zeta)\in\overline{U}_n\times\overline{U}_n$. On $U'_n\times U'_n$, we have $g(z,\zeta) = -\frac{1}{2\pi}\log|z-\zeta| + k_n(z,\zeta)$, where k_n is seen by Harnack's inequality to be jointly continuous. Clearly,

$$\left|\int_{U_n} k_n(z,\zeta)\,d\xi d\eta\right| \le \sup_{\overline{U}_n\times\overline{U}_n} |k_n(z,\zeta)|\,\pi r_n^2$$

tends to 0 with r_n. On the other hand, since $|z-\zeta|<1$ on $\overline{U}_n\times\overline{U}_n$, we have $-\log|z-\zeta|>0$. Let $z = re^{i\theta}$ and $\zeta=\rho e^{i\varphi}$. By the Gauss mean value theorem, we obtain $-\int_0^{2\pi}\log|re^{i\theta}-\rho e^{i\varphi}|\,d\varphi = -2\pi\max(\log r, \log\rho)$. Since

$$-\int_{U_n}\log|z-\zeta|\,d\xi d\eta = \int_0^{r_n}\left(-\int_0^{2\pi}\log|re^{i\theta}-\rho e^{i\varphi}|\,d\varphi\right)\rho\, d\rho ,$$

we obtain

$$-\int_{U_n}\log|z-\zeta|\,d\xi d\eta = O(-r_n\log r_n),$$

which tends to zero with r_n. Thus the first condition can be imposed on U_n by taking r_n sufficiently small. Note that if the first condition is satisfied by some r_n, then it is also satisfied by any smaller r_n.

To prove that the second condition is legitimate, we first observe that

$$\left|\int_{U_n\times U_n} k_n(z,\zeta)\,dxdyd\xi d\eta\right| \le \sup_{U_n\times U_n}|k_n(z,\zeta)|\,\pi^2 r_n^4$$

tends to zero with r_n. Furthermore,

$$-\int_{U_n \times U_n} \log|z - \zeta|\, dx dy d\xi d\eta = O(-r_n^2 \log r_n),$$

which tends to zero with r_n. Therefore, the second condition can be imposed on U_n by choosing r_n sufficiently small. Again, making r_n smaller only improves the estimate, and there exists an r_n such that both conditions are satisfied.

Set $U = \cup U_n$. In view of the first condition, we conclude that

$$\sup_{z \in S} \int_U g(z, \zeta) d\xi d\eta < 1 .$$

From the second condition and the triangle inequality for energies, which is a consequence of the energy principle, we obtain

$$\int_{U \times U} g(z, \zeta) dx dy d\xi d\eta < 1 .$$

Take functions $\varphi_n \in C^\infty(S)$ such that $0 \leq \varphi_n \leq 1$, $\varphi_n(z_n) = 1$, and $\varphi_n | S - U_n = 0$. For $\varphi_0 = \Sigma_1^\infty \varphi_n$, it is clear that

$$\sup_{z \in S} \int_S g(z, \zeta)\varphi_0(\zeta) d\xi d\eta < 1$$

and

$$\int_{S \times S} g(z, \zeta)\varphi_0(z)\varphi_0(\zeta) dx dy d\xi d\eta < 1 .$$

The metric $\mu(z)^{1/2}|dz|$ defined by

$$\mu(z) = \mu_0(z) + \varphi_0(z)$$

now satisfies $\mu(z) > 0$. Moreover,

$$\sup_{z \in S} \int_S g(z, \zeta)\mu(\zeta) d\xi d\eta < \infty$$

and similarly,

$$\int_{S\times S} g(z,\zeta)\mu(z)\mu(\zeta)dxdyd\xi d\eta < \infty .$$

To obtain the final metric $\lambda(z)^{1/2}|dz|$, we modify μ near the ideal boundary point z_0 of S. The sole object of this modification is to make the resulting manifold belong to O^N_{QP}. At z_0, we choose a parametric disk $U''_0 = \{|z| < 1\}$ of $S' = S \cup \{z_0\}$ such that $\bar{U}''_0 \cap \bar{U} = \emptyset$, $z(z_0) = 0$, and $\mathrm{Re}(z) = h_0$. Let U'_0 be the punctured disk $0 < |z| < 1/2$ and U_0 the punctured disk $0 < |z| < 1/4$. Take a strictly positive C^∞ density λ on S such that

$$\lambda(z) = |z|^{-2} \quad \text{for} \quad z \in U_0$$

and

$$\lambda|S - U'_0 = \mu .$$

A manifold with the required properties is then $R = (S,\lambda(z)^{1/2}|dz|)$. In fact, $H(R) = H(S)$ and therefore, $g_R(z,\zeta) = g(z,\zeta)$. Consequently, for a fixed $z \in R - \bar{U}_0$, $c = \inf_{\zeta\in U_0} g_R(z,\zeta) > 0$ and a fortiori,

$$(G_R 1)(z) \geq \int_{U_0} g(z,\zeta)\lambda(\zeta)d\xi d\eta \geq c\int_0^{2\pi}\int_0^{1/4} \rho^{-1}d\rho d\varphi = \infty ,$$

that is, $R \in O^N_{QP}$. Had we used μ instead of λ to define R, then R would not belong to O^N_{QP}.

Observe that $D_R(f) = D_S(f) = D_{S'}(f)$ for every f in $D(R) = D(S) = D(S')$ and that a point is removable for HD-functions. Therefore,

$$\dim HD(R) = 2, \qquad HD(R) = HBD(R),$$

and $HD(R)$ is generated by 1 and h_0.

Now we show that $F = H_{PD} \cap HD \neq \emptyset$, or, more precisely,

$$F = \{ch_0 | c \in R^1\} .$$

We start by proving that $h_0 \in F$, that is, $G_R(h_0,h_0) < \infty$. First,

$$\int_{(U_0'-\overline{U}_0)\times(U_0'-\overline{U}_0)} g_R(z,\zeta)|h_0(z)||h_0(\zeta)|\lambda(z)dxdy\lambda(\zeta)d\xi d\eta < \infty .$$

In fact, the integral is dominated by a constant multiple of the integral

$$a = \int_{(U_0'-\overline{U}_0)\times(U_0'-\overline{U}_0)} g_R(z,\zeta)dxdyd\xi d\eta .$$

On $U_0'' \subset R \cup \{z_0\}$, we have $g_R(z,\zeta) = -\frac{1}{2\pi}\log|z-\zeta| + k_0(z,\zeta)$, where k_0 is jointly continuous. As before, we see that the integrals of k_0 and $-\log|z-\zeta|$ over $U_0'' \times U_0''$ are finite and therefore, $a < \infty$.

Next we prove

$$\int_{U_0\times U_0} g_R(z,\zeta)|h_0(z)||h_0(\zeta)|\lambda(z)dxdy\ \lambda(\zeta)d\xi d\eta < \infty .$$

We see that $|h_0(z)| = |x| \leq |z|$ and $\lambda(z) = |z|^{-2}$. If we set $z = re^{i\theta}$ and $\zeta = \rho e^{i\varphi}$, then the above integral is dominated by

$$b = \int_0^{2\pi}\int_0^{1/4}\left(\int_0^{1/4}\left(\int_0^{2\pi} g_R(re^{i\theta},\rho e^{i\varphi})d\varphi\right)d\rho\right)drd\theta .$$

Since $g_R(re^{i\theta},\rho e^{i\varphi}) = -\frac{1}{2\pi}\log|re^{i\theta}-\rho e^{i\varphi}| + k_0(z,\zeta)$ and k_0 is bounded on $U_0 \times U_0$, we have $b = c + \text{const}$, where

$$c = \int_0^{2\pi}\int_0^{1/4}\left(\int_0^{1/4}\left(-\frac{1}{2\pi}\int_0^{2\pi}\log|re^{i\theta} - \rho e^{i\varphi}|\,d\varphi\right)d\rho\right)drd\theta .$$

The finiteness of c follows from the boundedness of the innermost integral, which is seen in the same fashion as in the proof of the relation

$$-\int_{U_n}\log|z-\zeta|d\xi d\eta = O(-r_n\log r_n).$$

The finiteness of the integral we set out to estimate follows.

Finally, the relation

$$\int_{S\times S} g(z,\zeta)\mu(z)\mu(\zeta)dxdyd\xi d\eta < \infty$$

gives

$$\int_{(R-\overline{U}_0')\times(R-\overline{U}_0')} g_R(z,\zeta)|h_0(z)||h_0(\zeta)|\lambda(z)dxdy\lambda(\zeta)d\xi d\eta < \infty.$$

In view of the triangle inequality, a consequence of the energy principle, we conclude that

$$G_R(|h_0|,|h_0|) < \infty .$$

In particular, $G_R(h_0,h_0) < \infty$, that is, $h_0 \in F$, and consequently,

$$F \supset \{ch_0 | c \in R^1\} .$$

Conversely, let $h \in F$. We claim that $h(z_0) = 0$, or more precisely, $\lim_{z\to z_0} h(z) = 0$. If not, we may suppose $h(z_0) > 0$. Then we can find a punctured disk $V_0 \subset U_0$ about z_0 such that $h \geq c > 0$ on V_0 for some $c \in R^1$. Obviously, the membership in H_{PD} is localizable and therefore, $h \in H_{PD}(V_0)$. Since $h > 0$, we see that $h \in H_{GA}(V_0)$:

$$\int_{V_0} g_{V_0}(z,\zeta)h(\zeta)\lambda(\zeta)d\xi d\eta < \infty$$

for every $z \in V_0$. Take a punctured disk $V_1 \subset \overline{V}_1 \subset V_0$ about z_0 and pick a $z \notin \overline{V}_1$. Then there is a $c' \in R^1$ such that $g_{V_0}(z,\zeta) \geq c' > 0$ for $\zeta \in V_1$. We have $\lambda(\zeta) = |\zeta|^{-2}$ on V_1 and therefore,

$$\int_{V_1} g_{V_0}(z,\zeta)h(\zeta)\lambda(\zeta)d\xi d\eta \geq cc' \int_{V_1} |\zeta|^{-2}d\xi d\eta = \infty ,$$

a contradiction. A fortiori, $h(z_0) = 0$, and $h \in \{ch_0 | c \in R^1\}$. We have shown that $F = \{ch_0 | c \in R^1\}$.

It remains to prove that every $h \in F$ is in H_{GA}. It suffices to show that

h_0 has this property. We have

$$\int_R (G_R|h_0|)(z)|h_0(z)|\lambda(z)dxdy < \infty .$$

Since $G_R|h_0|$ is either identically ∞ or finite, and $|h_0(z)| \not\equiv 0$, we conclude that $G_R|h_0| < \infty$.

The proof of Theorem 3.5 is herewith complete.

NOTES TO §3. The class H_{GA} was introduced and Theorems 3.2, 3.3, 3.4 and 3.5 established in Nakai-Sario [9].

§4. BIHARMONIC AND (p,q)-BIHARMONIC PROJECTION AND DECOMPOSITION

Intimately related to the Riesz decomposition are the boundary value problems for biharmonic functions. Typically, such a problem is to find a biharmonic function u with given boundary values of u and Δu. We shall consider that problem not only for biharmonic functions but, more generally, for what we call (p,2)-biharmonic functions, that is, solutions of $\Delta_q \Delta_p u = (\Delta + q)(\Delta + p)u = 0$ for given C^2 functions p, q on a Riemannian manifold R. The compactification of R, and the boundary of R on which the values of u and $\Delta_p u$ are given, will depend on the boundedness properties required of u and $\Delta_p u$. We shall, in fact, use simultaneously different compactifications for u and $\Delta_p u$. Specifically, we ask: Given continuous functions φ and ψ on Wiener's or Royden's p- and q-harmonic boundaries α and β of a Riemannian manifold R, respectively, find a (p,q)-biharmonic function u on R with

$$u|\alpha = \varphi, \qquad \Delta_p u|\beta = \psi .$$

We shall solve this problem by what we call the (p,q)-biharmonic projection and decomposition.

After deducing some preliminary results, we introduce the (p,q)-biharmonic projection and decomposition for various classes of functions and then for suitably restricted Riemannian manifolds. We characterize classes of manifolds with respect to significant subclasses of (p,q)-biharmonic functions by means of the p-harmonic Green's function and the q-elliptic measure on R. The (p,q)-quasiharmonic nondegeneracies of the manifold are the various conditions we impose on R. Finally, we deduce inclusion relations between (p,q)-quasiharmonic null classes of Riemannian manifolds.

For background information on Wiener's and Royden's compactifications and related topics, we refer the reader to, e.g., Sario-Nakai [1].

4.1. Definitions. A nonnegative C^2 function $p(x)$ on a (smooth) noncompact Riemannian manifold R of dimension $N \geq 2$ will be referred to as a density function. A p-harmonic function is a C^2 solution of the equation $\Delta_p u = 0$ with

$$\Delta_p = \Delta + p .$$

We call a C^4 function u (p,q)-biharmonic if it satisfies the equation

$$\Delta_q \Delta_p u = 0 ,$$

and we denote by $H^2_{pq} = H^2_{pq}(R)$ the class of (p,q)-biharmonic functions on R. An important subclass of H^2_{pq} is the class $Q_{pq} = Q_{pq}(R)$ of (p,q)-quasiharmonic functions, that is, C^2 solutions of $\Delta_p u = e_q$, where e_q is the q-elliptic measure on R to be defined below.

Note that for $p \equiv q \equiv 0$, H^2_{pq} and Q_{pq} reduce to the classes $H^2 \cup H$ and Q of biharmonic and quasiharmonic functions.

Let Ω be a regular subregion of R, and $h_{p\Omega 1}$ the continuous function on R which is p-harmonic on Ω and 1 on $R - \Omega$. The limit e_p of the decreasing sequence $\{h_{p\Omega 1}\}$ as $\Omega \to R$ is called the p-elliptic measure of R. Clearly, e_p is nonnegative and p-harmonic on R, with $0 \leq e_p \leq 1$. Explicitly, it is either identically zero or strictly positive. In particular, it is identically 1 if $p \equiv 0$. In the case $p \not\equiv 0$, we call a Riemannian manifold R p-parabolic if $e_p \equiv 0$ and p-hyperbolic if $e_p > 0$.

The harmonic Green's function $g(x,y)$ on R exists on a hyperbolic manifold only. In contrast, the p-harmonic Green's function $g_p(x,y)$ for $p \not\equiv 0$ exists on every Riemannian manifold. Thus on an arbitrary Riemannian manifold R, hyperbolic if $p \equiv 0$, the operator G_p is well defined on the class of continuous functions by

$$G_p f = \int_R g_p(\cdot,y)*f(y) .$$

We are interested in the class

$$F_{pl} = \{f \mid G_p|f| < \infty\} .$$

<u>LEMMA.</u> <u>Let</u> R <u>be an arbitrary Riemannian manifold (hyperbolic if</u> $p \equiv 0$). <u>If</u> $f \in C^\infty \cap F_{pl}$, <u>then</u> $\Delta_p G_p f = f$.

<u>Proof.</u> For every $\varphi \in C_0^\infty$, we have

$$\begin{aligned}\int_R \Delta_p G_p f(x)*\varphi(x) &= \int_R G_p f(x)*\Delta_p \varphi(x) \\ &= \int_R f(x)*G_p\Delta_p\varphi(x) \\ &= \int_R f(x)*\varphi(x) .\end{aligned}$$

Therefore, $\Delta_p G_p f = f$ in the sense of distributions, and the Lemma follows by the hypoellipticity of Δ_p (e.g., Hörmander [1]).

<u>4.2. Potential p-subalgebra.</u> Let M_{pl} be the class of continuous p-harmonizable functions f on R for which there is a continuous p-superharmonic function s_f with $s_f \geq |f|$ on R. Denote by N_{pl} the potential p-subalgebra of M_{pl}, that is, the class of functions in M_{pl} whose p-harmonic part $h_{pf} = \lim_{\Omega\to R} h_{p\Omega f}$ in the Riesz decomposition vanishes identically on R.

<u>LEMMA.</u> <u>Let</u> R <u>be an arbitrary Riemannian manifold (hyperbolic if</u> $p \equiv 0$).

<u>If</u> $f \in C^\infty \cap F_{pl}$, <u>then</u> $G_p f \in N_{pl}$ <u>on</u> R.

<u>Proof</u>. Set $f = f^+ - f^-$ with $f^+ = f \vee 0$ and $f^- = -f \vee 0$. Clearly, $G_p f^+$ and $G_p f^-$ are nonnegative and p-superharmonic on R. In view of $|G_p f| \le G_p f^+ + G_p f^-$, $G_p f \in M_{pl}(R)$.

It remains to show that $h_{pG_pf} = 0$. Let Ω be a regular subregion of R and $g_{p\Omega}(x,y)$ the p-harmonic Green's function on Ω with value zero on $R - \Omega$. For a geodesic ball B_x about $x \in \Omega$ with radius ε, Stokes' formula yields

$$\int_{\partial(\Omega - B_x)} [(G_p f(y) - h_{p\Omega G_p f}(y)) * dg_{p\Omega}(x,y) - g_{p\Omega}(x,y) * d(G_p f(y) - h_{p\Omega G_p f}(y))]$$

$$= \int_{\Omega - B_x} g_{p\Omega}(x,y) * \Delta_p G_p f(y).$$

On letting $\varepsilon \to 0$ and then $\Omega \to R$, we obtain

$$G_p f(x) = h_{pG_pf}(x) + \int_R g_p(x,y) * \Delta_p G_p f(y)$$

and by Lemma 4.1,

$$G_p f = h_{pG_pf} + G_p f .$$

Therefore, $h_{pG_pf} = 0$ and consequently, $G_p f \in N_{pl}(R)$.

<u>4.3</u>. <u>Energy integral</u>. Denote by H_p the class of p-harmonic functions on R, and let $E(u)$ be the energy integral

$$E(u) = \int_R du \wedge *du + \int_R p(x) * u(x)^2 .$$

Consider the real-valued linear operator $G_p(\cdot,\cdot)$ on $C^0 \times C^0$ defined by

$$G_p(f,g) = \int_{R \times R} g_p(x,y) * f(x) * g(y) .$$

LEMMA. *The energy integral is lower semicontinuous:*

$$E(u_0) \leq \underline{\lim}_{n\to\infty} E(u_n)$$

for every sequence $\{u_n\}$ *in* H_p *converging uniformly to* u_0 *on compact subsets of* R.

If $f \in C^\infty$, *then*

$$E(G_p f) = G_p(f,f)$$

whenever the right-hand side is finite.

Proof. For $x_0 \in R$ and a geodesic ball $B \subset R$ about x_0,

$$u_n(x) = -\int_{\partial B} u_n(y) \frac{\partial g_{pB}(x,y)}{\partial n_y} dS_y ,$$

$n = 0,1,2,\ldots,$ with $x \in B$, g_{pB} the p-Green's function on B, $\partial g_{pB}/\partial n$ the inner normal derivative of g_{pB}, and dS the surface element of ∂B. Then

$$\frac{\partial u_n(x)}{\partial x^i} = -\int_{\partial B} u_n(y) \frac{\partial}{\partial x^i} \frac{\partial g_{pB}(x,y)}{\partial n_y} dS_y$$

$$\to -\int_{\partial B} u_0(y) \frac{\partial}{\partial x^i} \frac{\partial g_{pB}(x,y)}{\partial n_y} dS_y = \frac{\partial u_0(x)}{\partial x^i}$$

as $n \to \infty$. Therefore, $\partial u_n(x)/\partial x^i \to \partial u_0(x)/\partial x^i$ uniformly on every geodesic ball as $n \to \infty$. The uniform convergence on compact subsets of R is a consequence of the fact that every compact set can be covered by a finite number of geodesic balls. Clearly,

$$E_\Omega(u_0) = \lim_{n\to\infty} E_\Omega(u_n) \leq \underline{\lim}_{n\to\infty} E(u_n)$$

for every relatively compact set Ω. The first part of the Lemma follows on letting $\Omega \to R$.

For the proof of the second part, let

$$G_{p\Omega}f = \int_R g_{p\Omega}(\cdot,y)*f(y) .$$

We have

$$E(G_{p\Omega}f) = \int_{R\times R} g_{p\Omega}(x,y)*f(x)*f(y) .$$

By the p-harmonicity of $G_pf - G_{p\Omega}f$ on Ω and the lower semicontinuity of E,

$$E(G_pf) \le \underline{\lim}_{\Omega\to R} E(G_{p\Omega}f) \le G_p(|f|,|f|) < \infty .$$

In view of Lebesgue's convergence theorem,

$$\begin{aligned} E(G_pf) &= \lim_{\Omega\to R} E(G_{p\Omega}f) \\ &= \lim_{\Omega\to R} \int_{R\times R} g_{p\Omega}(x,y)*f(x)*f(y) \\ &= \int_{R\times R} g_p(x,y)*f(x)*f(y) \\ &= G_p(f,f) . \end{aligned}$$

<u>4.4.</u> <u>The</u> (p,q)-<u>biharmonic projection.</u> As a preparation for the (p,q)-biharmonic projection, we introduce a number of classes of functions on R. Let M_{p3} be the class of continuous functions with finite energy integrals; M_{p2} and M_{p4} the Wiener and the Royden p-algebras on R; and N_{pi} the potential p-subalgebra of M_{pi} for $i = 2,3,4$ (cf. Sario-Nakai [1]). For the sake of simplicity, we set $X_T = \{f \mid Tf \in X\}$ and $XY = X \cap Y$ for given classes of functions X, Y, and a given operator T. Furthermore, we write $M_{ij} = M_{pi}(M_{qj})_{\Delta_p}$ and $N_{ij} = N_{pi}(N_{qj})_{\Delta_p}$ for all i, j. Let P', B, and E be the classes of essentially positive functions, bounded functions, and functions with finite energy integrals, respectively. Set $H_{p1} = H_pP'$, $H_{p2} = H_pB$, $H_{p3} = H_pE$, and $H_{p4} = H_pK$, where $K = BE$. It is known that the direct sum decompositions $M_{pi} = H_{pi} \oplus N_{pi}$ are valid for all i. The p-harmonic part of a function $f \in M_{pi}$

is called the p-harmonic projection of M_{pi} and denoted by $\pi_{pi}f$. It is also known that the decompositions are orthogonal in the sense that $E(f) = E(\pi_{pi}f) + E(f - \pi_{pi}f)$ for $f \in M_{pi}$ and $i = 3,4$ (e.g., Sario-Nakai [1]). Let

$$F_{p2} = \{f \in F_{p1} \mid \sup_R |G_p f| < \infty\} ,$$

$$F_{p3} = \{f \in F_{p1} \mid G_p(|f|,|f|) < \infty\}$$

$$F_{p4} = F_{p2} \cap F_{p3} ,$$

and

$$\Phi_{ij} = M_{ij}(F_{pi})_{\pi_{qj}\Delta_p} , \qquad i,j = 1,2,3,4 .$$

THEOREM. *On an arbitrary Riemannian manifold* R *(hyperbolic if* $p \equiv 0$*), the functions in* Φ_{ij} *have unique decompositions into* (p,q)*-biharmonic functions and* (p,q)*-potentials:*

$$\Phi_{ij} = H^2_{pq}\Phi_{ij} \oplus N_{ij}\Phi_{ij} .$$

Proof. Let $f \in \Phi_{ij}$. By the decomposition theorem of M_{pi} and M_{qj}, $f = \pi_{pi}f + h_i$ with $\pi_{pi}f \in H_{pi}$ and $h_i \in N_{pi}$, $\Delta_p f = \pi_{qj}\Delta_p f + k_j$ with $\pi_{qj}\Delta_p f \in H_{qj}$ and $k_j \in N_{qj}$. Since $\pi_{qj}\Delta_p f \in F_{pi}$ and $F_{pi} \subset F_{p1}$, the function $w_{ij} = \pi_{pi}f + G_p\pi_{qj}\Delta_p f$ is well defined. By Lemmas 4.1 and 4.2, we see that $w_{ij} \in H^2_{pq}$ for all i, j, and $w_{ij} \in \Phi_{ij}$ for $i = 1,2$, and all j. In view of the second part of Lemma 4.3,

$$E(w_{ij}) \leq E(\pi_{pi}f) + E(G_p\pi_{qj}\Delta_p f)$$

$$= E(\pi_{pi}f) + G_p(\pi_{qj}\Delta_p f, \pi_{qj}\Delta_p f) < \infty$$

for $i = 3,4$. Therefore, $w_{ij} \in H^2_{pq}\Phi_{ij}$ for all i, j. It remains to show that $f - w_{ij} \in N_{ij}\Phi_{ij}$. Clearly, $\Delta_p(f - w_{ij}) = k_j \in N_{qj}$ and $\pi_{qj}\Delta_p(f - w_{ij}) = 0$. By Lemma 4.2, $f - w_{ij} = h_i - G_p\pi_{qj}\Delta_p f \in N_{pi}$. Therefore, $w_{ij} + (f - w_{ij})$ is the desired decomposition.

To prove the uniqueness, let $v \in H^2_{pq}\Phi_{ij} \cap N_{ij}\Phi_{ij}$. Since

$\Delta_p v \in H_{qj} \cap N_{qj} = \{0\}$, $v \in H_{pi} \cap N_{pi}$, and consequently, $v \equiv 0$ on R.

We call the function $w_{ij} \in H^2_{pq}\Phi_{ij}$ in the Theorem the (p,q)-biharmonic projection of $f \in \Phi_{ij}$. It is the solution of the (p,q)-biharmonic Dirichlet problem with

$$w_{ij}|\beta_i = f|\beta_i \quad \text{and} \quad \Delta_p w_{ij}|\beta_j = \Delta_p f|\beta_j \ ,$$

where β_i and β_j are the p- and q-harmonic boundaries corresponding to M_{pi} and M_q, respectively. From the uniqueness of the decomposition, we see that the solution is unique except for the cases $i = 1$ or $j = 1$. In these cases, there exist singular p-harmonic functions which vanish on the p-harmonic boundary.

4.5. (p,q)-quasiharmonic classification of Riemannian manifolds. The (p,q)-biharmonic projection was obtained in Theorem 4.4 for certain restricted classes of functions on arbitrary Riemannian manifolds. In order to relax the conditions on the functions, it is necessary to impose conditions on the manifold. We shall see that such conditions are intimately related to the (p,q)-quasiharmonic classification of manifolds.

We continue denoting by O^N_X the class of Riemannian N-manifolds on which there exist no nonconstant functions in a given class X, and we retain the notation in 4.1 and 4.4. Various (p,q)-quasiharmonic null-manifolds are determined by the p-harmonic Green's function and the q-elliptic measure:

THEOREM. On a q-hyperbolic Riemannian manifold R (hyperbolic if $p \equiv 0$),

(a) $R \in \tilde{O}^N_{Q_{pq}P} \Leftrightarrow G_p e_q < \infty$,

(b) $R \in \tilde{O}^N_{Q_{pq}B} \Leftrightarrow \sup_R G_p e_q < \infty$,

(c) $R \in \tilde{O}^N_{Q_{pq}E} \Leftrightarrow G_p(e_q, e_q) < \infty$,

(d) $R \in \tilde{O}^N_{Q_{pq}K} \Leftrightarrow \sup_R G_p e_q < \infty$ and $G_p(e_q, e_q) < \infty$.

Proof. For every $u \in Q^N_{pq}$ and every regular subregion $\Omega \subset R$,

$$u(x) = h_{p\Omega u}(x) + \int_R g_{p\Omega}(x,y) * e_q(y) .$$

Suppose $R \in \tilde{O}^N_{Q_{pq}P}$, that is, there exists a nonnegative $u \in Q_{pq}$. Clearly, u is p-superharmonic and bounded from below on R. Therefore, $h_{pu} = \lim_{\Omega \to R} h_{p\Omega u}$ exists. By the monotone convergence theorem,

$$G_p e_q = \lim_{\Omega \to R} \int_R g_{p\Omega}(\cdot, y) * e_q(y) = u - h_{pu} < \infty .$$

Conversely, $G_p e_q \in Q_{pq}P$, and (a) follows. Relation (b) is established in a similar manner.

Suppose $R \in \tilde{O}^N_{Q_{pq}E}$ and take a $v \in Q_{pq}E$. For every regular subregion $\Omega \subset R$,

$$v = h_{p\Omega v} + G_{p\Omega} e_q$$

and

$$E(v) = E(h_{p\Omega v}) + E(G_{p\Omega} e_q) .$$

As in the proof of the second part of Lemma 4.3,

$$E(G_{p\Omega} e_q) = G_{p\Omega}(e_q, e_q) .$$

The monotone convergence theorem yields

$$\lim_{\Omega \to R} G_{p\Omega}(e_q, e_q) = G_p(e_q, e_q) .$$

Since $G_p(e_q) - G_{p\Omega}(e_q)$ is p-harmonic on Ω, Lemma 4.3 implies

$$E(G_p e_q) \leq \underline{\lim}_{\Omega \to R} E(G_{p\Omega} e_q) \leq E(v) < \infty .$$

By Lebesgue's convergence theorem,

$$G_p(e_q, e_q) = E(G_p e_q) < \infty .$$

Conversely, if $G_p(e_q, e_q) < \infty$, then $G_p e_q < \infty$ and $\Delta_p G_p e_q = e_q$. By virtue of

$$E(G_p e_q) \leq \underline{\lim}_{\Omega \to R} E(G_{p\Omega} e_q) \leq G_p(e_q, e_q) < \infty ,$$

$G_p e_q \in Q_{pq}E$ and (c) follows. The last assertion of the Theorem is an immediate consequence of (b) and (c).

4.6. Decomposition. An important biproduct of the proof of Theorem 4.4 is that the (p,q)-biharmonic functions restricted to the class Φ_{ij} can be uniquely decomposed into the p-harmonic part and the potential part:

THEOREM. On an arbitrary Riemannian manifold R (hyperbolic if $p \equiv 0$), every function $w_{ij} \in H^2_{pq}\Phi_{ij}$ can be uniquely written as

$$w_{ij} = u_i + G_p v_j ,$$

with $u_i \in H_{pi}$ and $v_j \in H_{qj}$ for $i,j = 1,2,3,4$.

4.7. Nondegenerate manifolds. We shall show that, by imposing a suitable condition on the manifold R, the restrictions we have set on the functions which have (p,q)-biharmonic projections can be relaxed.

We write $X_1 = P$, $X_2 = B$, $X_3 = E$, $X_4 = K$, and let $H^2_{pq}X_1(X_j)_{\Delta_p}$ stand for $H^2_{pq}M_{P1}(X_j)_{\Delta_p}$.

THEOREM. On a Riemannian manifold which carries $Q_{pq}X_i$ functions,

$$M_{ij} = H^2_{pq}X_i(X_j)_{\Delta_p} \oplus N_{ij}$$

with $i = 1,2,3,4$, and $j = 2,4$.

On a Riemannian manifold R which carries nonnegative Q_{pq} functions,

$$H^2_{pq}X_i(X_j)_{\Delta_p} \subset H_{pi} \oplus G_p H_{qj}$$

with $i = 1,2,3,4$, and $j = 2,4$. Moreover,

$$H^2_{pq}X_i(X_j)_{\Delta_p} = H_{pi} \oplus G_p H_{qj}$$

if and only if $R \in \tilde{O}^N_{Q_{pq}X_i}$.

Proof. For the first part, it is sufficient to show that $f \in H^2_{pq}B_{\Delta_p}$ implies the p-harmonizability of f on $R \in \tilde{O}^N_{Q_{pq}P}$.

For every regular subregion Ω of R, and every $f \in H^2_{pq}B_{\Delta_p}$,

$$f = h_{p\Omega f} + \int_R g_{p\Omega}(\cdot,y)*\Delta_p f(y) .$$

Since $R \in \tilde{O}^N_{Q_{pq}P}$, $G_p e_q < \infty$ by Theorem 4.5. In view of $|g_{p\Omega}\Delta_p f| \leq k g_p e_q$ for some constant k, the Lebesgue convergence theorem implies the existence of $\lim_{\Omega\to R} \int_R g_{p\Omega}(\cdot,y)*\Delta_p f(y)$. Thus $h_{p\Omega f}$ converges, and f is p-harmonizable. The first part of the Theorem follows.

The rest of the Theorem is a consequence of Theorems 4.5 and 4.6 and the first part of the present Theorem.

On a manifold $R \in \tilde{O}^N_{Q_{pq}X_i}$, let φ, ψ be continuous functions on the harmonic boundaries β_i and β_j corresponding to M_{pi} and M_{qj}, respectively. The last assertion of the Theorem implies that if $h_{p\varphi}$ and $h_{q\psi}$ are solutions of the p-harmonic and q-harmonic boundary value problems with boundary values φ and ψ, respectively, then our (p,q)-biharmonic Dirichlet problem has a solution which is in $H^2_{pq}X_i(X_j)_{\Delta_p}$ and takes the form $h_{p\varphi} + G_p h_{q\psi}$.

4.8. Special density functions. In the case where the density function is bounded from below by a positive constant, we have more explicit results:

THEOREM. *If* $\inf_R p(x) > 0$ *on a q-hyperbolic Riemannian manifold, then*

$$M_{ij} = H^2_{pq}X_i(X_j)_{\Delta_p} \oplus N_{ij}$$

and

$$H^2_{pq}X_i(X_j)_{\Delta_p} = H_{pi} \oplus G_p H_{qj}$$

with $i = 1,2$ *and* $j = 2,4$. *Furthermore, if* $\int_R *p(x) < \infty$, *then the above assertions are true also for* $i = 3,4$.

Proof. To prove the first assertion, it is sufficient to show that $R \in \tilde{O}^N_{Q_{pq}B}$ for $\inf_R p(x) > 0$. On every regular subregion Ω, we have $1 = h_{p\Omega 1} + \int_R g_{p\Omega}(\cdot,y)*p(y)$, and consequently $G_p p \leq 1$ upon letting $\Omega \to R$. Therefore, $G_p e_q \leq G_p 1 \leq 1/m$ with $m = \inf_R p$. By Theorem 4.5, $R \in \tilde{O}^N_{Q_{pq}B}$. Suppose, furthermore, that $\int_R *p(x) < \infty$. Then the volume of R is $V(R) = \int_R *1 \leq m^{-1} \int_R *p(x) < \infty$ and

$$G_p(e_q, e_q) \leq \frac{1}{m^2} V(R) < \infty .$$

The second assertion ensues from Theorem 4.5, and the proof is complete.

4.9. Inclusion relations. By the fact that $g_p(x,y) \leq g_r(x,y)$ for $p \geq r$, and Theorem 4.5, we have:

THEOREM. *For q-hyperbolic Riemannian manifolds* R (*hyperbolic if* $p \equiv 0$),

(a) $O^N_{Q_{pq}P} \subset O^N_{Q_{pq}B} \subset O^N_{Q_{pq}K}$, and $O^N_{Q_{pq}P} \subset O^N_{Q_{pq}E} \subset O^N_{Q_{pq}K}$,
(b) $O^N_{Q_{pq}X} \subset O^N_{Q_{rq}X}$ for $p \geq r$,
(c) $O^N_{Q_{pq}X} \subset O^N_{Q_{ps}X}$ for $q \geq s$,
(d) $O^N_{Q_{pq}X} \subset O^N_{Q_{rs}X}$ for $p \geq r$ and $q \geq s$, with $X = P, B, E, K$.

We note that if R is q-parabolic, $Q_{pq} = H_p$ and $\emptyset = O^N_{H_pP} \subset O^N_{H_pB} \subset O^N_{H_pE} = O^N_{H_pK}$, that is, (a) is still true. However, (b)-(d) are no longer valid, for $\emptyset = O^N_{H_pP} = O^N_{H_rP}$, $O^N_{H_rB} \subset O^N_{H_pB}$, and $O^N_{H_rE} = O^N_{H_rK} \subset O^N_{H_pK} = O^N_{H_pE}$ if $p \geq r$.

From (d), we see that if the (r,s)-biharmonic Dirichlet problem is solvable by the decomposition method of the latter part of Theorem 4.7, then the (p,q)-biharmonic boundary value problem has a solution for $p \geq r$ and $q \geq s$. In particular, the (p,q)-biharmonic problem is solvable if the biharmonic problem is.

NOTES TO §4. The (p,q)-biharmonic functions were introduced and Theorems 4.4-4.9 established in Sario-Wang [1]. A corresponding theory of the biharmonic projection and decomposition was first developed in Sario-Wang-Range [1].

Numerous new null classes were introduced in the present section. An interesting problem would be to systematically fit them into the harmonic, quasiharmonic, and biharmonic classification schemes of Riemannian manifolds.

Several relations could be read off from the above considerations, whereas others would require construction of new counterexamples.

CHAPTER VIII

BIHARMONIC GREEN'S FUNCTION γ

The harmonic Green's function was originally introduced as the electrostatic potential of a point charge in a grounded system. Its characterization by the fundamental singularity and vanishing boundary values permitted its generalization to regular subregions Ω of an abstract Riemann surface and Riemannian manifold R. The Green's function g on R was then defined as the directed limit, if it exists, of the Green's function g_Ω on Ω as $\{\Omega\}$ exhausts R. The distinction of Riemann surfaces and Riemannian manifolds into hyperbolic and parabolic types according as g does or does not exist is still a cornerstone of the harmonic classification theory.

The biharmonic Green's function also has an important physical meaning: it is the deflection of a thin elastic plate under a point load. However, in sharp contrast with the harmonic case, nothing was known about its existence on noncompact spaces. The purpose of Chapters VIII-XI is to tackle this fundamental problem of biharmonic classification theory.

Biharmonicity being not meaningful on abstract Riemann surfaces, our aim is to generalize the definition of the biharmonic Green's function to Riemannian manifolds R and to explore its existence on them. On a regular subregion Ω of R, there exist two biharmonic Green's functions, to be denoted by β and γ, with a biharmonic fundamental singularity, and with boundary data $\beta = \partial\beta/\partial n = 0$ and $\gamma = \Delta\gamma = 0$. For dimension 2, both functions give the deflection under a point load of a thin plate which is clamped or simply supported at the edges, respectively.

In the present chapter, we shall deal exclusively with γ, which could be called the biharmonic Green's function of a <u>simply supported</u> body. In §1, we give a useful existence criterion for γ in terms of the harmonic measure. In §2, we introduce what we call the biharmonic measure; its nondegeneracy can also serve as a test for the existence of γ. In §§3-4, we relate the existence of γ to harmonic and quasiharmonic null classes.

§1. EXISTENCE CRITERION FOR γ

On a regular subregion Ω of a Riemannian manifold R, let γ_Ω be the biharmonic Green's function defined by a biharmonic fundamental singularity and the boundary data $\gamma_\Omega = \Delta\gamma_\Omega = 0$. The function increases with Ω, and we set $\gamma = \lim_{\Omega\to R} \gamma_\Omega$ on R if the limit exists.

We first study the existence of γ on the Euclidean N-space, which we shall denote in the present chapter by R^N (to distinguish the space R^N_α to be introduced from the space E^N_α discussed earlier). The result is fascinatingly simple: γ exists on R^N if and only if $N > 4$. By way of preparation, we recall the peculiar behavior of the biharmonic fundamental singularity at the origin: $r^2 \log r$ for $N = 2$, r for $N = 3$, $\log r$ for $N = 4$, and r^{4-N} for $N > 4$.

No parabolic Riemannian manifold carries γ. For a hyperbolic Riemannian manifold R, we deduce a useful criterion: If $\bar{R}_0 \subset \Omega \subset R$, and ω_Ω is harmonic on $\Omega - \bar{R}_0$ with boundary values 1 on ∂R_0, 0 on $\partial\Omega$, and we denote by $\omega = \lim_{\Omega\to R} \omega_\Omega$ the harmonic measure of ∂R_0 on $R - R_0$, <u>then</u> γ <u>exists on</u> R <u>if and only if</u> $\omega \in L^2(R - R_0)$. An essential step of the reasoning is the proof that the existence of γ is independent of the choice of the fundamental singularity. This property allows us to introduce the class O^N_γ of Riemannian N-manifolds which do not carry γ, in analogy with the class of parabolic manifolds.

As a simple illustration of our criterion, we generalize the above result that $R^N \in O^N_\gamma$ if and only if $N \leq 4$. We ask whether one could induce γ to exist even for these low dimensions by replacing the Euclidean metric $ds = |dx|$ by $ds = (1 + r^2)^\alpha |dx|$, with the constant α sufficiently large. The answer is intriguing: the resulting space is in O^N_γ for $N \leq 4$ regardless of what α is chosen. For $N > 4$, γ continues to exist if and only if $\alpha > -\frac{1}{2}$.

The usefulness of the criterion $\omega \in L^2$ for $R \in \tilde{O}^N_\gamma$ lies in the fact that it also applies if there is no way of obtaining an expression for the approximating Green's functions γ_Ω, and even if nothing is known about the metric in the complement of an arbitrarily small neighborhood of the ideal boundary of the Riemannian manifold.

1.1. Definition. Let Ω be a regular subregion of a Riemannian manifold R, carrying the biharmonic Green's function $\gamma_\Omega(x,y)$ on $\bar{\Omega}$, with the biharmonic fundamental singularity at $y \in \Omega$ and with the boundary data

$$\gamma_\Omega | \partial\Omega = 0, \qquad \Delta\gamma_\Omega | \partial\Omega = 0 .$$

In terms of the harmonic Green's function $g_\Omega(x,y)$ on Ω with singularity y, the function $\gamma_\Omega(x,y)$ has the integral representation

$$\gamma_\Omega(x,y) = \int_\Omega g_\Omega(x,z)*g_\Omega(z,y) .$$

To see this, it suffices to verify

$$\Delta \int_\Omega g_\Omega(x,z)*g_\Omega(z,y) = g_\Omega(x,y) ,$$

where Δ is taken with respect to x. For every $\varphi \in C_0^\infty$,

$$\begin{aligned}\int_\Omega \Delta\left(\int_\Omega g_\Omega(x,z)*g_\Omega(z,y)\right)*\varphi(x) &= \int_\Omega\left(\int_\Omega g_\Omega(x,z)*g_\Omega(z,y)\right)*\Delta\varphi(x)\\ &= \int_\Omega\left(\int_\Omega g_\Omega(x,z)*\Delta\varphi(x)\right)*g_\Omega(z,y)\\ &= \int_\Omega g_\Omega(z,y)*\varphi(z) .\end{aligned}$$

Therefore, our assertion is true in the sense of distributions and a fortiori in the conventional sense.

We introduce the biharmonic Green's function $\gamma(x,y)$ on a Riemannian manifold R by setting

$$\gamma(x,y) = \lim_{\Omega\to R} \gamma_\Omega(x,y) ,$$

provided the limit exists for some exhaustion $\{\Omega\}$. We shall later show that the existence is independent of the exhaustion $\{\Omega\}$ and the choice of the singularity y.

1.2. Existence on N-space. We start by examining the existence of $\gamma(x,y)$ in an illuminating special case, the Euclidean N-space R^N. Here the computation is elementary. A function $h(r)$ is harmonic if $-r^{-N+1}(r^{N-1}h'(r))' = 0$. We obtain

$$h(r) = \begin{cases} a \log r + b, & N = 2\ , \\ ar^{-N+2} + b, & N > 2\ , \end{cases}$$

where a, b are arbitrary constants. If $u(r)$ belongs to the class H^2 of nonharmonic biharmonic functions on R^N, then $\Delta u(r) = h(r)$. A straightforward integration yields the biharmonic, quasiharmonic, harmonic, and constant components of $u(r)$:

$$u(r) = \begin{cases} ar^2 \log r + br^2 + c \log r + d, & N = 2\ , \\ ar + br^2 + cr^{-1} + d, & N = 3\ , \\ a \log r + br^2 + cr^{-2} + d, & N = 4\ , \\ ar^{-N+4} + br^2 + cr^{-N+2} + d, & N > 4\ , \end{cases}$$

with a, b, c, d arbitrary constants.

Let B_ρ be a ball $\{r < \rho\}$ and take $x \in B_\rho$, $|x| = r$. In view of $\Delta r^2 = -2N$, the biharmonic Green's function on B_ρ with singularity 0 is

$$\gamma_\rho(x,0) = \begin{cases} r^2 \log \frac{r}{\rho} - (r^2 - \rho^2), & N = 2\ , \\ -r + \rho + \frac{1}{3}(r^2 - \rho^2)\rho^{-1}, & N = 3\ , \\ -\log \frac{r}{\rho} + \frac{1}{4}(r^2 - \rho^2)\rho^{-2}, & N = 4\ , \\ r^{-N+4} - \rho^{-N+4} + \frac{N-4}{N}(r^2 - \rho^2)\rho^{-N+2}, & N > 4\ , \end{cases}$$

where the signs have been so chosen that $\gamma_\rho > 0$. As $\rho \to \infty$, $\{B_\rho\}$ exhausts R^N, and we obtain

$$\gamma(x,0) = r^{-N+4}, \qquad N > 4 ,$$

whereas $\lim_{\rho\to\infty} \gamma_\rho = \infty$ for $N = 2, 3, 4$.

We have proved:

THEOREM. *The biharmonic Green's function* γ *exists on the Euclidean* N-*space if and only if* $N > 4$.

1.3. Biharmonic Dirichlet problem. For further illustration, we recall the simple computation showing the significance of the biharmonic Green's function in expressing values of a biharmonic function u in terms of the boundary values of u and Δu. Given a regular subregion Ω of a Riemannian manifold, let $y \in \Omega$, $u \in H^2(\bar{\Omega})$, and $\gamma \in H^2(\Omega - y) \cap C^3(\bar{\Omega} - y)$. Take a compact smooth hypersurface α enclosing y and suppose that as α shrinks to y,

$$\int_\alpha \Delta\gamma{*}du \to 0, \qquad \int_\alpha \Delta u{*}d\gamma \to 0, \qquad \int_\alpha \gamma{*}d\Delta u \to 0 .$$

Then

$$u(y) = \frac{1}{F(\Delta\gamma)} \int_{\partial\Omega} u{*}d\Delta\gamma - \Delta\gamma{*}du + \Delta u{*}d\gamma - \gamma{*}d\Delta u ,$$

with $F(\Delta\gamma) = \int_\alpha {*}d\Delta\gamma$ the flux of $\Delta\gamma$. In fact, by Stokes' formula, the integral on the right taken along $\partial\Omega - \alpha$ is

$$(du,d\Delta\gamma) - (u,\Delta^2\gamma) - (d\Delta\gamma,du) + (\Delta\gamma,\Delta u)$$

$$+ (d\Delta u,d\gamma) - (\Delta u,\Delta\gamma) - (d\gamma,d\Delta u) + (\gamma,\Delta^2 u) = 0 ,$$

where the inner products are taken over the region bounded by $\partial\Omega \cup \alpha$. If $\langle\varepsilon\rangle$ stands for a quantity which $\to 0$ as $\alpha \to y$, then

$$\int_\alpha u{*}d\Delta\gamma = \int_\alpha (u(y) + \langle\varepsilon\rangle){*}d\Delta\gamma \to u(y)F(\Delta\gamma) ,$$

and the assertion follows.

In the Euclidean case of 1.2, Ω is the ball $\{r < \rho\}$, y the origin, and

$\gamma = \gamma_\rho$. Accordingly,

$$u(0) = \frac{1}{F(\Delta\gamma_\rho)} \int_{r=\rho} u{*}d\Delta\gamma_\rho + \Delta u{*}d\gamma_\rho ,$$

provided the three integrals in the above hypothesis tend to 0.

Denote by dS the surface element on the sphere $S_\rho = \{r = \rho\}$ and let α be the sphere $\{r = \delta < \rho\}$. For $N = 2$,

$$*d\gamma_\rho = \left(2r \log \frac{r}{\rho} - r\right) dS, \qquad \Delta\gamma_\rho = -4 \log \frac{r}{\rho} ,$$

$$*d\Delta\gamma_\rho = -4r^{-1}dS, \qquad F(\Delta\gamma_\rho) = -4\int_\alpha r^{-1}dS = -\int_0^{2\pi} 4d\varphi = -8\pi .$$

On the other hand, $|u|$, $|\partial u/\partial n|$, $|\Delta u|$ are bounded on B_ρ, and as $\delta \to 0$,

$$\left|\int_\alpha \Delta\gamma_\rho {*}du\right| < M\left|\delta \log \frac{\delta}{\rho}\right| \to 0 ,$$

$$\left|\int_\alpha \Delta u{*}d\gamma_\rho\right| < M\left|2\delta \log \frac{\delta}{\rho} - \delta\right|\delta \to 0 ,$$

$$\left|\int_\alpha \gamma_\rho {*}d\Delta u\right| = c(\delta)F(\Delta u) = 0 .$$

The above representation formula for $u(0)$ follows, with $F(\Delta\gamma_\rho) = -8\pi$.

For $N = 3$, we have

$$*d\gamma_\rho = \left(-1 + \frac{2}{3} r\rho^{-1}\right) dS, \qquad \Delta\gamma_\rho = 2r^{-1} - 2\rho^{-1} ,$$

$$*d\Delta\gamma_\rho = -2r^{-2}dS, \qquad F(\Delta\gamma_\rho) = -2\int_\alpha r^{-2}r^2 d\omega = -8\pi ,$$

where $d\omega$ is the area element of the unit sphere. As $\delta \to 0$,

$$\left|\int_\alpha \Delta\gamma_\rho {*}du\right| < M(2\delta^{-1} - 2\rho^{-1}) \cdot 4\pi\delta^2 \to 0 ,$$

$$\left|\int_\alpha \Delta u * d\gamma_\rho\right| < M\left|-1 + \frac{2}{3}\delta\rho^{-1}\right| \cdot 4\pi\delta^2 \to 0 ,$$

$$\left|\int_\alpha \gamma_\rho * d\Delta u\right| = c(\delta)F(\Delta u) = 0 ,$$

and the formula for $u(0)$ follows, with $F(\Delta\gamma_\rho) = -8\pi$.

For $N = 4$,

$$*d\gamma_\rho = \left(-r^{-1} + \frac{1}{2}r\rho^{-2}\right)dS, \qquad \Delta\gamma_\rho = 2r^{-2} - 2\rho^{-2} ,$$

$$*d\Delta\gamma_\rho = -4r^{-3}dS, \qquad F(\Delta\gamma_\rho) = -4\int_\alpha r^{-3}r^3 d\omega = -4A_4 ,$$

where A_4 is the area of the unit sphere in 4-space. As $\delta \to 0$,

$$\left|\int_\alpha \Delta\gamma_\rho * du\right| < M(2\delta^{-2} - 2\rho^{-2})A_4\delta^3 \to 0 ,$$

$$\left|\int_\alpha \Delta u * d\gamma_\rho\right| < M\left|-\delta^{-1} + \frac{1}{2}\delta\rho^{-2}\right|A_4\delta^3 \to 0 ,$$

$$\left|\int_\alpha \gamma_\rho * d\Delta u\right| = c(\delta)F(\Delta u) = 0 ,$$

and the formula for $u(0)$ follows, with $F(\Delta\gamma_\rho) = -4A_4$.

Finally, for $N > 4$,

$$*d\gamma_\rho = (N - 4)r^{-N+3}\left(-1 + \frac{2}{N}\left(\frac{r}{\rho}\right)^{N-2}\right)dS ,$$

$$\Delta\gamma_\rho = (N - 4)\left[-(N - 3)\left(r^{-N+2} + \frac{2}{N(N - 3)}\rho^{-N+2}\right) - (N - 1)\left(-r^{-N+2} + \frac{2}{N}\rho^{-N+2}\right)\right] ,$$

$$*d\Delta\gamma_\rho = -2(N - 2)(N - 4)r^{-N+1}dS ,$$

$$F(\Delta\gamma_\rho) = -2(N - 2)(N - 4)A_N ,$$

where A_N is the area of the unit sphere in N-space. As $\delta \to 0$,

$$\left|\int_\alpha \Delta\gamma_\rho {*du}\right| < M|c_1\delta^{-N+2} + c_2|c_3\delta^{N-1} \to 0 ,$$

$$\left|\int_\alpha \Delta u {*d\gamma_\rho}\right| < M|c_1\delta^{-N+3} + c_2\delta|c_3\delta^{N-1} \to 0 ,$$

$$\left|\int_\alpha \gamma_\rho {*d\Delta u}\right| = c(\delta)F(\Delta u) = 0 ,$$

and the formula for $u(0)$ again follows, with $F(\Delta\gamma_\rho) = -2(N-2)(N-4)A_N$.

In summary, the formula for $u(0)$ is true for all N. For our present purpose of illustrating the use of γ this will suffice. The formula generalizes, however, in various directions. First, since the fundamental singularity is locally defined, the above reasoning continues to hold when $u(0)$ is replaced by $u(y)$ with $y \in B_\rho$, and $\gamma_\rho(x,0)$ by $\gamma_\rho(x,y)$. On a Riemannian manifold R, the geodesic distance d is, in a sufficiently small neighborhood of y, a constant multiple of the Euclidean r in a parametric ball, added by a function which together with its derivatives is bounded. Therefore, the above elementary reasoning, mutatis mutandis, remains valid and the representation formula holds for $u(y)$ when S_ρ is replaced by the boundary $\partial\Omega$ of a regular subregion of R, and γ_ρ by the biharmonic Green's function $\gamma_\Omega(x,y)$ on Ω.

1.4. Independence. We return to our existence problem. The proof of the following intermediate result is essential.

LEMMA. *The existence of a biharmonic Green's function* γ *on a Riemannian manifold is independent of the exhaustion and the choice of the singularity.*

Proof. Let $\{\Omega\}$ be some exhaustion of R. In view of $\gamma_\Omega(x,y) = \int_\Omega g_\Omega(x,z){*g_\Omega(z,y)}$ and Lebesgue's Dominated Convergence Theorem, if a biharmonic Green's function γ on R with singularity y exists, it can be written as

$$\gamma(x,y) = \int_R g(x,z){*g(z,y)} ,$$

where g is the harmonic Green's function on R, and the integral is the directed limit of $\int_\Omega g_\Omega(x,z)*g_\Omega(z,y)$ as $\Omega \to R$. Thus we have immediately the independence of the exhaustion. It remains to show that if γ exists for some (x_1,y_1), then it exists for any (x_2,y_2).

From the above representation, we see that γ can exist only on hyperbolic manifolds, and

$$\int_R g(x_1,z)*g(z,y_1) < \infty .$$

Choose a regular subregion R_0 of R containing x_1, x_2, y_1, and y_2, with boundary α. Clearly, the above relation is equivalent to

$$\int_{R-R_0} g(x_1,z)*g(z,y_1) < \infty ,$$

which in turn is equivalent to

$$\int_{R-R_0} g(x_1,z)*g(y_1,z) < \infty ,$$

by virtue of the symmetry of the harmonic Green's function. Since $g(x,z)$, $g(y,z) > 0$ on R_0 for each $z \in R - \bar{R}_0$, Harnack's inequality gives $g(y,z) < Kg(x,z)$ for all x, y in R_0, with the constant K depending only on R and R_0. Therefore,

$$K^{-1}\int_{R-R_0} *_z g(x,z)^2 < \int_{R-R_0} g(x,z)*_z g(y,z) < K\int_{R-R_0} *_z g(x,z)^2 ,$$

and we conclude that $\gamma(x_1,y_1)$ exists if and only if $\int_{R-R_0} *_z g(x_1,z)^2 < \infty$, that is,

$$g(\cdot,x_1) \in L^2(R - R_0) .$$

Next we shall show that $g(\cdot,x_1) \in L^2(R - R_0)$ if and only if the harmonic measure ω of α on $R - R_0$ belongs to $L^2(R - R_0)$. Recall that ω is the limit of the harmonic function ω_Ω on $\Omega - \bar{R}_0$ with boundary values 1 on α,

0 on $\partial\Omega$, for some and hence every exhaustion $\{\Omega\}$ of R with $\overline{R}_0 \subset \Omega$. The function ω always exists, and $\omega \equiv 1$ if and only if R is parabolic. Set

$$m_\Omega = \min_\alpha g_\Omega(\cdot, x_1), \qquad M_\Omega = \max_\alpha g_\Omega(\cdot, x_1) .$$

By the maximum principle for harmonic functions,

$$m_\Omega \omega_\Omega \le g_\Omega(\cdot, x_1) \le M_\Omega \omega_\Omega$$

on $\overline{\Omega} - R_0$. On letting $\Omega \to R$, we obtain

$$m\omega \le g(\cdot, x_1) \le M\omega$$

on $R - R_0$, with

$$m = \min_\alpha g(\cdot, x_1), \qquad M = \max_\alpha g(\cdot, x_1) .$$

Thus $g(\cdot, x_1) \in L^2(R - R_0)$ if and only if $\omega \in L^2(R - R_0)$. Since $g(\cdot, x_2) \in L^2(R - R_0)$ is characterized by the same condition, we have proved the existence of $\gamma(x_2, y_2)$ as a consequence of that of $\gamma(x_1, y_1)$.

We also conclude from the above reasoning that $\lim_{\Omega\to R} \gamma_\Omega$ is either identically infinite or a biharmonic function γ on $R - y$ with the fundamental singularity y.

1.5. Existence criterion. In view of Lemma 1.4, we may introduce the class of Riemannian N-manifolds R,

$$O_\gamma^N = \{R \mid \gamma \text{ does not exist}\} .$$

Let $\tilde{O}_\gamma^N$ be the class of manifolds which do carry γ. From the proof of the Lemma, we have the following criterion:

THEOREM. *Every parabolic Riemannian manifold R belongs to O_γ^N. A hyperbolic R belongs to $\tilde{O}_\gamma^N$ if and only if $\omega \in L^2(R - R_0)$ for some and hence every R_0.*

1.6. Illustration. Using Theorem 1.5, we can extend Theorem 1.2. We ask whether γ exists even for the low dimensions if R^N is generalized to the Riemannian manifold R_α^N with the same base manifold $\{r < \infty\}$ but with the metric

$$ds = (1 + r^2)^\alpha |dx| ,$$

α a constant. For a harmonic $h(r)$ we have by direct computation

$$h(r) = a \int_1^r r^{-N+1}(1 + r^2)^{-(N-2)\alpha} dr + b$$

as $r \to \infty$. If $\alpha \leq -\frac{1}{2}$, this is unbounded, hence $R_\alpha^N \in O_G^N \subset O_\gamma^N$. For $\alpha > -\frac{1}{2}$, the harmonic measure ω satisfies

$$\omega(r) \sim ar^{-(N-2)(2\alpha+1)} ,$$

and the L^2-norm of $\omega(r)$ over the annulus $(1, r)$ satisfies

$$\|\omega\|_2^2 \sim c \int_1^r r^{-2(N-2)(2\alpha+1)+N-1+2N\alpha} dr = c \int_1^r r^{-2(N-4)\alpha-(N-3)} dr .$$

This is bounded if and only if $N > 4$ and $\alpha > -\frac{1}{2}$. We have proved that R_α^N fails to carry γ for $N \leq 4$ regardless of how rapidly or slowly the metric $(1 + r^2)^\alpha |dx|$ grows:

THEOREM. $R_\alpha^N \in \tilde{O}_\gamma^N$ *if and only if* $N > 4$ *and* $\alpha > -\frac{1}{2}$.

NOTES TO §1. The biharmonic Green's function γ on a Riemannian manifold was introduced and Theorems 1.2, 1.5, and 1.6 established in Sario [11]. Applications of Theorem 1.5 to relations between O_γ^N and other null classes will be discussed in §§3-4.

§2. BIHARMONIC MEASURE

We recall from I.1.3 that the harmonic Green's function g exists on a Riemannian manifold R if and only if the harmonic measure ω of the ideal boundary δ does not reduce to a constant. This measure is a harmonic function on the complement $S_0 = R - \bar{R}_0$ of the closure of a regular region R_0, with essentially the boundary values 1 on ∂R_0 and 0 on δ. From 1.5 we know that the biharmonic Green's function γ on R, with "boundary values" $\gamma = \Delta\gamma = 0$ on δ, exists if and only if $\omega \in L^2(S_0)$. In this section we ask: Can a non-harmonic biharmonic function be introduced on S_0 with the property that its nondegeneracy characterizes the existence of γ, in analogy with the nondegeneracy of ω characterizing the existence of g? We shall show that this is possible. The function, which we will call the biharmonic measure σ of δ, is the limit of biharmonic functions σ_Ω on subregions $S_0 \cap \Omega$ of S_0, with the Ω regular subregions of R containing $\bar{R}_0$ and exhausting R. In contrast with the approximating harmonic measures ω_Ω with $\omega_\Omega = 1$ on ∂R_0, $\omega_\Omega = 0$ on $\partial\Omega$, the function σ_Ω, with $\Delta\sigma_\Omega = \omega_\Omega$, vanishes on the entire boundary of $S_0 \cap \Omega$, and spans $S_0 \cap \Omega$ like an arc shaped bridge. As Ω increases, the height of this arch σ_Ω increases, and its limit σ as Ω exhausts R is either an arch spanning S_0 or else the constant ∞. We shall show that the finiteness of σ is independent of the choice of S_0, and we can, therefore, introduce the class O_σ^N of Riemannian manifolds with boundaries of infinite biharmonic measure.

We first explore O_σ^N in its own right. For radial spaces, which we have seen to play an important role in biharmonic classification theory, we decompose σ into its biharmonic, harmonic, quasiharmonic, and constant components. The biharmonic type of R can then be easily tested. In particular, $\sigma < \infty$ if both the biharmonic and harmonic components of σ tend to zero as one approaches δ. We use this test to determine the type of a number of important manifolds. E.g., for the Poincaré N-ball, we obtain the following complete characterization:

$$B_\alpha^N \in O_\sigma^N \Leftrightarrow \begin{cases} \alpha \leq -3/2, & N = 2\ , \\ \alpha \notin (-3,1), & N = 3\ , \\ \alpha \geq (N-2)^{-1}, & N > 3\ . \end{cases}$$

After this study of σ we establish its characteristic property in our original problem:

$$O_\sigma^N = O_\gamma^N .$$

In particular, the above values of α exclude γ on the Poincaré N-ball B_α^N.

It should be stressed that, to determine whether or not a given Riemannian N-manifold belongs to O_γ^N, Theorem 1.5 offers a vastly more efficient means than the relation $O_\sigma^N = O_\gamma^N$. It is, however, of considerable methodological interest to explore the latter avenue. This is the purpose of much of the present section. No use of it will be made later, and the reader can, if he prefers, continue the reading from §3.

2.1. Definition. Let R_0, Ω be regular subregions of R, with $\bar{R}_0 \subset \Omega$, and set $S_0 = R - \bar{R}_0$, $\alpha = \partial R_0$, $\delta_\Omega = \partial\Omega$. Take $\omega_\Omega \in H(S_0 \cap \Omega) \cap C(\bar{S}_0 \cap \bar{\Omega})$, $\omega_\Omega|\alpha = 1$, $\omega_\Omega|\delta_\Omega = 0$. In the present section, we call the directed limit $\omega = \lim_{\Omega\to R} \omega_\Omega$ the harmonic measure on $\bar{S}_0$ of the ideal boundary δ of R.

We introduce:

DEFINITION. The directed limit

$$\sigma = \lim_{\Omega\to R} \sigma_\Omega ,$$

where

$$\sigma_\Omega \in H^2(S_0 \cap \Omega) \cap C(\bar{S}_0 \cap \bar{\Omega}), \qquad \Delta\sigma_\Omega = \omega_\Omega, \qquad \sigma_\Omega|\alpha = \sigma_\Omega|\delta_\Omega = 0 ,$$

is the biharmonic measure of the ideal boundary δ of the Riemannian manifold R.

The limit always exists. In fact, if $g_{S_0\cap\Omega}(x,y)$ is the harmonic Green's function on $S_0 \cap \Omega$ with pole y, then

$$\sigma_\Omega(x) = \int_{S_0\cap\Omega} g_{S_0\cap\Omega}(x,y) * \omega_\Omega(y) .$$

Since both $g_{S_0\cap\Omega}$ and ω_Ω increase with Ω, so does σ_Ω, and the limit σ exists, finite or infinite, at every $x \in S_0$. We shall show in 2.5 that the finiteness is independent of x and S_0.

2.2. Biharmonic measure on N-space. We first study the biharmonic measure on radial spaces, e.g., the N-space and the N-ball, each endowed with a radial metric $ds = \lambda(r)|dx|$, $r = |x|$. We choose $R_0 = \{r < r_0\}$, $S_0 = \{r > r_0\}$.

The space of all radial biharmonic functions $u(r)$, $\Delta^2 u(r) = 0$, is generated by four functions: any $u_0(r) \in H^2 - Q$, $\Delta u_0(r)$, any $q_0(r) \in Q$, and the constant 1. Thus every $u(r)$ has a decomposition

$$u(r) = au_0(r) + b\Delta u_0(r) + cq_0(r) + d ,$$

which depends on the choice of u_0 and q_0. In particular, if the biharmonic measure

$$\sigma(x) = \sigma(r) = \int_{S_0} g_{S_0}(x,y) * \omega(y)$$

is finite, it has such a decomposition, and $\sigma(x) \to 0$ as $x \to \delta$.

Testing of the finiteness of σ is facilitated by the following simple criterion:

LEMMA. *If every radial $u \in H^2$ on S_0 is unbounded, then the biharmonic measure σ is infinite. If there exists a function $u_0(r) \in H^2 - Q$ with $u_0(r) \to 0$, $\Delta u_0(r) \to 0$ as $x \to \delta$, then $\sigma < \infty$.*

Proof. The first part of the Lemma is clear, since $\Delta\sigma = \omega$ entails $\sigma \in H^2$. If $u_0(r) \to 0$, $\Delta u_0(r) \to 0$ as $x \to \delta$ for some $u_0 \in H^2 - Q$, then $a = 1/\Delta u_0(r_0)$ gives $\Delta(au_0) = \omega$, and $b = -au_0(r_0)/\Delta u_0(r_0)$ provides us with the function $\sigma = au_0 + b\Delta u_0$ which has all the properties required of the biharmonic measure.

For our first radial space, we take the Euclidean N-space R^N. The type distinction is here fascinatingly simple:

THEOREM. *The ideal boundary of R^N has a finite biharmonic measure if and only if* $N > 4$.

Proof. Let us first construct σ for $N > 4$ by means of the definition

$\sigma = \lim_{\Omega \to R} \sigma_\Omega$. Choose $r_0 = 1$, $\rho > 1$, $\Omega = \{r < \rho\}$, and write σ_ρ for σ_Ω. For the decomposition

$$\sigma_\rho = a_\rho u_0(r) + b_\rho \Delta u_0(r) + c_\rho q_0(r) + d_\rho ,$$

we take

$$\sigma_\rho = a_\rho r^{-N+4} + b_\rho r^{-N+2} + c_\rho r^2 + d_\rho ,$$

where we have absorbed in b_ρ the constant $[2(N-4)]^{-1}$ from $\Delta r^{-N+4} = 2(N-4)r^{-N+2}$, and in c_ρ the constant $-(2N)^{-1}$ from $\Delta r^2 = -2N$. By the definition of σ_ρ,

$$\begin{cases} \sigma_\rho(1) = a_\rho + b_\rho + c_\rho + d_\rho = 0 , \\ \sigma_\rho(\rho) = a_\rho \rho^{-N+4} + b_\rho \rho^{-N+2} + c_\rho \rho^2 + d_\rho = 0 , \\ \Delta\sigma_\rho(1) = 2(N-4)a_\rho - 2Nc_\rho = 1 , \\ \Delta\sigma_\rho(\rho) = 2(N-4)a_\rho \rho^{-N+2} - 2Nc_\rho = 0 . \end{cases}$$

From the third and fourth equations we obtain

$$a_\rho = [2(N-4)(1-\rho^{-N+2})]^{-1} ,$$

$$c_\rho = \rho^{-N+2}[2N(1-\rho^{-N+2})]^{-1} .$$

If a, b, c, d are the limits of $a_\rho, b_\rho, c_\rho, d_\rho$ as $\rho \to \infty$, then

$$a = [2(N-4)]^{-1}, \qquad c = 0 .$$

The first and second equations above then give

$$b = \lim_{\rho\to\infty} \{-(1-\rho^{-N+2})^{-1}[a_\rho(1-\rho^{-N+4}) + c_\rho(1-\rho^2)]\} = -[2(N-4)]^{-1} ,$$

$$d = \lim_{\rho\to\infty} [-(a_\rho + b_\rho + c_\rho)] = 0 .$$

Thus we have the explicit expression

$$\sigma(r) = [2(N-4)]^{-1}(r^{-N+4} - r^{-N+2})$$

for the biharmonic measure of the ideal boundary of R^N for $N > 4$.

We now deduce the same result for $N > 4$ using the Lemma and its proof. For $u_0(r) = r^{-N+4}$, $\Delta u_0(r) = 2(N-4)r^{-N+2}$, we have

$$a = [2(N-4)]^{-1} r_0^{N-2}, \qquad b = -[2(N-4)]^{-2} r_0^N ,$$

and for $r_0 = 1$, the function $\sigma = au_0 + b\Delta u_0$ is the biharmonic measure $[2(N-4)]^{-1}(r^{-N+4} - r^{-N+2})$.

To prove that $\sigma = \infty$ for $N \leq 4$, we use the representation for $\sigma < \infty$,

$$\sigma(r) = \begin{cases} ar^2 \log r + b \log r + cr^2 + d, & N = 2 , \\ ar + br^{-1} + cr^2 + d, & N = 3 , \\ a \log r + br^{-2} + cr^2 + d, & N = 4 . \end{cases}$$

For $N = 2$, this is unbounded unless $a = b = c = 0$, and σ is constant. For $N = 3$ or 4, it is unbounded unless $a = c = 0$ and σ is harmonic. However, by $\Delta\sigma = \omega$, σ cannot be harmonic, a contradiction.

2.3. Radial metric. Can the biharmonic measure of the ideal boundary of R^N be made finite for $N \leq 4$ if we "shrink" or "expand" the boundary by replacing the Euclidean metric $ds = |dx|$ by the metric $(1 + r^2)^{\alpha}|dx|$, with α a sufficiently small or large constant? Denote the resulting space by R_{α}^N.

THEOREM. The biharmonic measure σ of the ideal boundary of R_{α}^N is infinite for all α if $N \leq 4$. For $N > 4$, σ is infinite if and only if $\alpha \leq -\frac{1}{2}$.

Proof. An explicit construction of σ as the limit of σ_{Ω} as at the beginning of the proof of Theorem 2.2 is now not possible, and we make use of Lemma 2.2. We know that σ has the form

$$\sigma(r) = au_0(r) + b\Delta u_0(r) + cq_0(r) + d .$$

First we shall find bounded functions u_0, Δu_0 for $N > 4$. Choose again $S_0 = \{r > 1\}$. For $h(r) \in H(S_0)$,

$$\Delta h(r) = -r^{-N+1}(1 + r^2)^{-N\alpha}[r^{N-1}(1 + r^2)^{(N-2)\alpha}h'(r)]' = 0 ,$$

$$h(r) = c\int^r r^{-N+1}(1 + r^2)^{-(N-2)\alpha}dr \sim c\int^r r^{-N+1-2(N-2)\alpha}dr$$

$$= \begin{cases} a + br^{-(N-2)(1+2\alpha)} & \text{if } N > 2 \text{ and } \alpha \neq -\frac{1}{2} , \\ a + b \log r & \text{if } N = 2 \text{ or } \alpha = -\frac{1}{2} . \end{cases}$$

Thus $h(r)$ belongs to the family B of bounded functions if and only if $N > 2$, $\alpha > -\frac{1}{2}$, an assumption we shall make for the present. Here and later we disregard irrelevant multiplicative and additive constants, and we choose an h_0 with

$$h_0(r) \sim r^{-(N-2)(1+2\alpha)} ,$$

which $\to 0$ as $r \to \infty$ if $N > 2$, $\alpha > -\frac{1}{2}$.

For $\Delta u(r) = h_0(r)$, we obtain

$$[r^{N-1}(1 + r^2)^{(N-2)\alpha}u'(r)]' \sim r^{1+4\alpha} .$$

Accordingly, in view of $\alpha \neq -\frac{1}{2}$, we have $[\ ] \sim r^{2+4\alpha}$, and, again by virtue of $\alpha \neq -\frac{1}{2}$, we can take u_0 with

$$u_0(r) \sim r^{-(N-4)(1+2\alpha)} .$$

This $\to 0$ as $r \to \infty$ if $-2(N - 4)\alpha < N - 4$, which in turn, under our assumption $\alpha > -\frac{1}{2}$, holds if and only if $N > 4$. By Lemma 2.2, we conclude that $\sigma < \infty$ if $N > 4$, $\alpha > -\frac{1}{2}$.

In the discussion of the case $\sigma = \infty$, the nonuniqueness of the generators u_0, Δu_0, q_0 (and 1) makes it necessary to consider the unboundedness of all four components of $\sigma(r) \in H^2$. For $\Delta q(r) = 1$, we obtain

$$[r^{N-1}(1 + r^2)^{(N-2)\alpha}q'(r)]' \sim r^{N-1+2N\alpha} ,$$

and therefore,

$$q'(r) \sim \begin{cases} r^{1+4\alpha}, & \alpha \neq -\frac{1}{2}, \\ r^{-1} \log r, & \alpha = -\frac{1}{2}. \end{cases}$$

We choose

$$q_0(r) \sim \begin{cases} r^{2+4\alpha}, & \alpha \neq -\frac{1}{2}, \\ (\log r)^2, & \alpha = -\frac{1}{2}. \end{cases}$$

For $N = 2$ and any α, $h_0(r) \sim \log r$, and $u'(r)$ satisfies

$$[ru'(r)]' \sim r^{1+4\alpha} \log r, \quad [\ ] \sim r^{2+4\alpha} \log r, \quad u'(r) \sim r^{1+4\alpha} \log r ,$$

so that we can take

$$u_0(r) \sim \begin{cases} r^{2+4\alpha} \log r, & \alpha \neq -\frac{1}{2}, \\ (\log r)^2, & \alpha = -\frac{1}{2}. \end{cases}$$

Therefore,

$$\sigma(r) \sim \begin{cases} ar^{2+4\alpha} \log r + b \log r + cr^{2+4\alpha} + d, & \alpha \neq -\frac{1}{2}, \\ a(\log r)^2 + b \log r + d, & \alpha = -\frac{1}{2}. \end{cases}$$

By virtue of $\Delta\sigma = \omega$, we have $\sigma \notin B$, hence $\sigma = \infty$.

For $N = 3$, $\alpha \neq -\frac{1}{2}$,

$$h_0(r) \sim r^{-1-2\alpha}, \qquad u_0(r) \sim r^{1+2\alpha}, \qquad q_0(r) \sim r^{2+4\alpha} .$$

We have $\sigma \notin B$, hence $\sigma = \infty$.

For $N = 3$, $\alpha = -\frac{1}{2}$, $h_0(r) \sim \log r$ and

$$[r^2(1 + r^2)^{-1/2}u'(r)]' \sim r^2(1 + r^2)^{-3/2} \log r \sim r^{-1} \log r ,$$

$$[\ ] \sim (\log r)^2, \qquad u'(r) \sim r^{-1}(\log r)^2, \qquad u_0(r) \sim (\log r)^3 .$$

Since $q_0(r) \sim (\log r)^2$, we have $\sigma(r) \sim (\log r)^3 \notin B$, hence $\sigma = \infty$.

For $N = 4$, $\alpha \neq -\frac{1}{2}$, $h_0(r) \sim r^{-2-4\alpha}$,

$$[r^3(1 + r^2)^{2\alpha}u'(r)]' \sim r^{1+4\alpha}, \qquad u'(r) \sim r^{-1}, \qquad u_0(r) \sim \log r .$$

In view of $q_0(r) \sim r^{2+4\alpha}$, $\sigma(r)$ grows at least as rapidly as $\log r$, hence $\sigma = \infty$.

For $N = 4$, $\alpha = -\frac{1}{2}$, $h_0(r) \sim \log r$,

$$[r^3(1 + r^2)^{-1}u'(r)]' \sim r^{-1} \log r, \quad u'(r) \sim r^{-1}(\log r)^2, \quad u_0(r) \sim (\log r)^3 .$$

Since $q_0(r) \sim (\log r)^2$, we have $\sigma(r) \sim (\log r)^3 \notin B$, hence $\sigma = \infty$. We have proved that $\sigma = \infty$ for $N \leq 4$, all α.

For $N > 4$, $\alpha < -\frac{1}{2}$,

$$h_0(r) \sim r^{-(N-2)(1+2\alpha)}, \qquad u_0(r) \sim r^{-(N-4)(1+2\alpha)}, \qquad q_0(r) \sim r^{2+4\alpha} .$$

Therefore, $\sigma \sim r^{\beta}$, with

$$\beta \geq \min[-(N - 2)(1 + 2\alpha), -(N - 4)(1 + 2\alpha)] .$$

The two quantities are both positive for $\alpha < -\frac{1}{2}$, so that $\sigma \notin B$, hence $\sigma = \infty$.

For $N > 4$, $\alpha = -\frac{1}{2}$, $h_0(r) \sim \log r$,

$$[r^{N-1}(1 + r^2)^{-(N-2)/2}u'(r)]' \sim r^{-1} \log r, \quad u'(r) \sim r^{-1}(\log r)^2, \quad u_0(r) \sim (\log r)^3,$$

and $q_0(r) \sim (\log r)^2$. Consequently, $\sigma(r) \sim (\log r)^3 \notin B$, and $\sigma = \infty$.

We have shown that $\sigma = \infty$ for $N > 4$, $\alpha \leq -\frac{1}{2}$. The proof of the Theorem is complete.

2.4. Poincaré N-ball. Next we consider the Poincaré N-ball B^N_α, which we have seen to play an important role in general biharmonic classification theory. By definition,

$$B^N_\alpha = \{x = (x^1, \ldots, x^N) \mid |x| = r, \ r < 1, \ ds = (1 - r^2)^\alpha |dx|, \ \alpha \in R^1\}.$$

We shall give a complete characterization of the finiteness of σ:

THEOREM. *The biharmonic measure of the ideal boundary of the Poincaré N-ball B^N_α is finite if and only if*

$$\begin{cases} \alpha > -3/2, & N = 2, \\ \alpha \in (-3,1), & N = 3, \\ \alpha < (N-2)^{-1}, & N > 3. \end{cases}$$

<u>Proof</u>. For $h(r) \in H$,

$$\Delta h(r) = -r^{-N+1}(1-r^2)^{-N\alpha}[r^{N-1}(1-r^2)^{(N-2)\alpha}h'(r)]' = 0,$$

and we choose

$$h_0(r) \sim \begin{cases} \log r \sim 1-r, & N = 2, \quad \text{any } \alpha, \\ (1-r)^{-(N-2)\alpha+1}, & N > 2, \ \alpha \neq (N-2)^{-1}, \\ \log(1-r), & N > 2, \ \alpha = (N-2)^{-1}. \end{cases}$$

For $\Delta u(r) = h_0(r)$,

$$[r^{N-1}(1-r^2)^{(N-2)\alpha}u'(r)]' \sim (1-r)^{N\alpha}h_0(r),$$

$$u'(r) \sim (1-r)^{-(N-2)\alpha}\int^r (1-r)^{N\alpha}h_0(r)dr.$$

For $N = 2$, we take

$$u_0(r) \sim \int^r\int^r (1-s)^{2\alpha+1}dsdr \sim \begin{cases} \int^r (1-r)^{2\alpha+2}dr \sim (1-r)^{2\alpha+3}, & \alpha \neq -1, -3/2, \\ \int^r \log(1-r)dr \sim (1-r)\log(1-r), & \alpha = -1, \\ \int^r (1-r)^{-1}dr \sim \log(1-r), & \alpha = -3/2. \end{cases}$$

For $N = 3$, we obtain successively

$$u(r) \sim \begin{cases} \displaystyle\int^r (1-r)^{-\alpha} \int^r (1-s)^{2\alpha+1} ds\,dr, & \alpha \neq 1 , \\ \displaystyle\int^r (1-r)^{-1} \int^r (1-s)^3 \log(1-s) ds\,dr, & \alpha = 1 , \end{cases}$$

$$u(r) \sim \begin{cases} \displaystyle\int^r (1-r)^{\alpha+2} dr, & \alpha \neq 1, -1 , \\ \displaystyle\int^r (1-r)^3 \log(1-r) dr, & \alpha = 1 , \\ \displaystyle\int^r (1-r)\log(1-r) dr, & \alpha = -1 , \end{cases}$$

$$u_0(r) \sim \begin{cases} (1-r)^{\alpha+3}, & \alpha \neq 1, -1, -3 , \\ (1-r)^4 \log(1-r), & \alpha = 1 , \\ (1-r)^2 \log(1-r), & \alpha = -1 , \\ \log(1-r), & \alpha = -3 . \end{cases}$$

For $N = 4$,

$$u(r) \sim \begin{cases} \displaystyle\int^r (1-r)^{-2\alpha} \int^r (1-s)^{2\alpha+1} ds\,dr, & \alpha \neq \tfrac{1}{2} , \\ \displaystyle\int^r (1-r)^{-1} \int^r (1-s)^2 \log(1-s) ds\,dr, & \alpha = \tfrac{1}{2} , \end{cases}$$

$$u_0(r) \sim \begin{cases} (1-r)^3, & \alpha \neq \tfrac{1}{2}, -1 , \\ (1-r)^3 \log(1-r), & \alpha = \tfrac{1}{2} , \\ (1-r)^3 \log(1-r), & \alpha = -1 . \end{cases}$$

For $N > 4$,

$$u(r) \sim \begin{cases} \int^r (1-r)^{-(N-2)\alpha} \int^r (1-s)^{2\alpha+1} ds dr, & \alpha \neq (N-2)^{-1}, \\ \int^r (1-r)^{-1} \int^r (1-s)^{N/(N-2)} \log(1-s) ds dr, & \alpha = (N-2)^{-1}, \end{cases}$$

$$u_0(r) \sim \begin{cases} (1-r)^{-(N-4)\alpha+3}, & \alpha \neq 3(N-4)^{-1},\ (N-2)^{-1},\ -1, \\ \log(1-r), & \alpha = 3(N-4)^{-1}, \\ (1-r)^{(2N-2)/(N-2)} \log(1-r), & \alpha = (N-2)^{-1}, \\ (1-r)^{N-1} \log(1-r), & \alpha = -1. \end{cases}$$

For $\Delta q(r) = 1$,

$$[r^{N-1}(1-r^2)^{(N-2)\alpha} q'(r)]' \sim (1-r)^{N\alpha},$$

$$q'(r) \sim \begin{cases} (1-r)^{2\alpha+1}, & \alpha \neq -N^{-1}, \\ (1-r)^{(N-2)/N} \log(1-r), & \alpha = -N^{-1}. \end{cases}$$

For $N \geq 2$,

$$q_0(r) \sim \begin{cases} (1-r)^{2\alpha+2}, & \alpha \neq -N^{-1},\ -1, \\ (1-r)^{(2N-2)/N} \log(1-r), & \alpha = -N^{-1}, \\ \log(1-r), & \alpha = -1. \end{cases}$$

As $r \to 1$,

$$h_0(r) \to 0 \quad \text{if} \quad \begin{cases} N = 2, & \text{any } \alpha, \\ N > 2, & \alpha < (N-2)^{-1}, \end{cases}$$

$$u_0(r) \to 0 \quad \text{if} \quad \begin{cases} N = 2, & \alpha > -3/2 , \\ N = 3, & \alpha > -3 , \\ N = 4, & \text{any } \alpha , \\ N > 4, & \alpha < 3(N-4)^{-1} . \end{cases}$$

We conclude by Lemma 2.2 that

$$\sigma < \infty \quad \text{if} \quad \begin{cases} N = 2, & \alpha > -3/2 , \\ N = 3, & \alpha \in (-3,1) , \\ N > 3, & \alpha < (N-2)^{-1} , \end{cases}$$

as claimed.

In preparation for the case $\sigma = \infty$, we observe that

$$h_0 \notin B \Leftrightarrow \begin{cases} N = 2, & \text{no } \alpha , \\ N > 2, & \alpha \geq (N-2)^{-1} , \end{cases}$$

$$u_0 \notin B \Leftrightarrow \begin{cases} N = 2, & \alpha \leq -3/2 , \\ N = 3, & \alpha \leq -3 , \\ N = 4, & \text{no } \alpha , \\ N > 4, & \alpha \geq 3(N-4)^{-1} , \end{cases}$$

$$q_0 \notin B \Leftrightarrow N \geq 2,\ \alpha \leq -1 .$$

We have obtained

$$au_0 + b\Delta u_0 \notin B \Leftrightarrow \begin{cases} N = 2, & \alpha \leq -3/2 , \\ N = 3, & \alpha \notin (-3,1) , \\ N > 3, & \alpha \geq (N-2)^{-1} , \end{cases}$$

except that we shall return later to the case $N > 4$, $\alpha \geq 3(N-4)^{-1}$. Here for

$N = 2$, we have $h_0 \in B$, $u_0 \notin B$, with

$$u_0(r) \sim \begin{cases} (1 - r)^{2\alpha+3}, & \alpha < -3/2 , \\ \log(1 - r), & \alpha = -3/2 , \end{cases}$$

whereas

$$q_0(r) \sim (1 - r)^{2\alpha+2}, \qquad \alpha \le -3/2 .$$

Thus the rates of growth of u_0 and q_0 are different for $\alpha \le -3/2$, and we have $\sigma \notin B$, hence $\sigma = \infty$ as claimed.

For $N = 3$, $h_0 \notin B$, $u_0 \in B$ if $\alpha \ge 1$, with

$$h_0(r) \sim \begin{cases} (1 - r)^{-\alpha+1}, & \alpha > 1 , \\ \log(1 - r), & \alpha = 1 , \end{cases}$$

whereas

$$q_0(r) \sim (1 - r)^{2\alpha+2}, \qquad \alpha \ge 1 .$$

Thus the rates of growth are different for $\alpha \ge 1$, hence $\sigma = \infty$. Moreover, $h_0 \in B$, $u_0 \notin B$ for $\alpha \le -3$, with

$$u_0(r) \sim \begin{cases} (1 - r)^{\alpha+3}, & \alpha < -3 , \\ \log(1 - r), & \alpha = -3 , \end{cases}$$

whereas

$$q_0(r) \sim (1 - r)^{2\alpha+2}, \qquad \alpha \le -3 .$$

The rates of growth are different for $\alpha \le -3$, hence $\sigma = \infty$.

For $N = 4$, $h_0 \notin B$, $u_0 \in B$ for $\alpha \ge \frac{1}{2}$, with

$$h_0(r) \sim \begin{cases} (1 - r)^{-2\alpha+1}, & \alpha > \frac{1}{2} , \\ \log(1 - r), & \alpha = \frac{1}{2} , \end{cases}$$

whereas

$$q_0(r) \sim (1 - r)^{2\alpha+2}, \qquad \alpha \geq \tfrac{1}{2}.$$

The rates of growth are different for $\alpha \geq \frac{1}{2}$, hence $\sigma = \infty$.

For $N > 4$, $h_0 \notin B$, $u_0 \in B$ if $\alpha \in [(N - 2)^{-1}, 3(N - 4)^{-1})$, with

$$h_0(r) \sim \begin{cases} (1 - r)^{-(N-2)\alpha+1}, & \alpha \in ((N - 2)^{-1}, 3(N - 4)^{-1}), \\ \log(1 - r), & \alpha = (N - 2)^{-1}, \end{cases}$$

whereas

$$q_0(r) \sim (1 - r)^{2\alpha+2}, \qquad \alpha \in [(N - 2)^{-1}, 3(N - 4)^{-1}).$$

Thus the rates of growth are different for $\alpha \in [(N - 2)^{-1}, 3(N - 4)^{-1})$, hence $\sigma = \infty$. Moreover, $h_0 \notin B$, $u_0 \notin B$ for $\alpha \geq 3(N - 4)^{-1}$, with

$$h_0(r) \sim (1 - r)^{-(N-2)\alpha+1}, \qquad \alpha \geq 3(N - 4)^{-1},$$

$$u_0(r) \sim \begin{cases} (1 - r)^{-(N-4)\alpha+3}, & \alpha > 3(N - 4)^{-1}, \\ \log(1 - r), & \alpha = 3(N - 4)^{-1}, \end{cases}$$

whereas

$$q_0(r) \sim (1 - r)^{2\alpha+2}, \qquad \alpha \geq 3(N - 4)^{-1}.$$

The rates of growth are all different for $\alpha \geq 3(N - 4)^{-1}$, hence $\sigma = \infty$.

This completes the proof of the Theorem.

2.5. Independence. We proceed to the proof of the fundamental property of σ referred to at the end of 2.1. Let R be an arbitrary Riemannian manifold, R_0 its regular subregion, and $x \in S_0 = R - \bar{R}_0$.

THEOREM. *The finiteness of the biharmonic measure* $\sigma(x)$ *on* $\bar{S}_0$ *is independent of* R_0 *and of* x.

<u>Proof</u>. For any region G, let $g_G(x,y)$ be the harmonic Green's function on G, with pole y. Denote the harmonic measure on $\overline{S}_0$ by ω. The biharmonic measure on S_0 and the biharmonic Green's function on R, if they exist, are

$$\sigma_{S_0}(x) = \int_{S_0} g_{S_0}(x,y)*\omega(y) \ ,$$

$$\gamma(p,q) = \int_R g_R(p,y)*g_R(y,q) \ .$$

We are to prove:

I. <u>If</u> $\sigma_{S_0}(x) < \infty$ <u>for some</u> S_0, $x \in S_0$, <u>then</u> $\gamma(p,q) < \infty$ <u>for any</u> $p,q \in R$.

II. <u>If</u> $\gamma(p,q) < \infty$ <u>for some</u> $p,q \in R$, <u>then</u> $\sigma_{S_0}(x) < \infty$ <u>for any</u> S_0, $x \in S_0$.

<u>Proof of I</u>. Given $\sigma_{S_0}(x) < \infty$ for some S_0, $x \in S_0$, choose any $p,q \in R$ and regular subregions R_1, Ω of R with

$$\overline{R}_0 \cup x \cup p \cup q \subset R_1 \subset \overline{R}_1 \subset \Omega \ .$$

Set $\alpha_0 = \partial R_0$, $\alpha_1 = \partial R_1$, $\delta_\Omega = \partial\Omega$, $S_1 = R - \overline{R}_1$, and take

$$\omega_\Omega \in H(\Omega \cap S_0) \cap C(\overline{\Omega} \cap \overline{S}_0), \qquad \omega_\Omega | \alpha_0 = 1, \qquad \omega_\Omega | \delta_\Omega = 0 \ .$$

We shall use the following constants:

$$m_{1\Omega} = \min_{y\in\alpha_1} g_{S_0\cap\Omega}(y,x), \qquad M_{1\Omega} = \max_{y\in\alpha_1} g_{S_0\cap\Omega}(y,x) \ ,$$

$$m_{2\Omega} = \min_{\alpha_1} \omega_\Omega, \qquad M_{2\Omega} = \max_{\alpha_1} \omega_\Omega \ ,$$

$$m_{3\Omega} = \min_{y\in\alpha_1} g_\Omega(y,p), \qquad M_{3\Omega} = \max_{y\in\alpha_1} g_\Omega(y,p) \ ,$$

$$m_{4\Omega} = \min_{y\in\alpha_1} g_\Omega(y,q), \qquad M_{4\Omega} = \max_{y\in\alpha_1} g_\Omega(y,q) \ ,$$

$$m_i = \lim_{\Omega\to R} m_{i\Omega}, \quad M_i = \lim_{\Omega\to R} M_{i\Omega}, \quad i = 1,2,3,4 ,$$

$$k_1 = \frac{M_3 M_4}{m_1 m_2}, \qquad k_2 = \frac{M_1 M_2}{m_3 m_4} .$$

We obtain

$$g_\Omega(y,p) \le \frac{M_{3\Omega}}{m_{1\Omega}} g_{S_0\cap\Omega}(y,x) \quad \text{on} \quad \alpha_1 \cup \delta_\Omega, \quad \text{hence on} \quad \bar{\Omega} \cap \bar{S}_1 ,$$

$$g_R(y,p) \le \frac{M_3}{m_1} g_{S_0}(y,x) \quad \text{on} \quad \bar{S}_1 ,$$

$$g_\Omega(y,q) \le \frac{M_{4\Omega}}{m_{2\Omega}} \omega_\Omega(y) \quad \text{on} \quad \bar{\Omega} \cap \bar{S}_1 ,$$

$$g_R(y,q) \le \frac{M_4}{m_2} \omega(y) \quad \text{on} \quad \bar{S}_1 .$$

Therefore,

$$\begin{aligned}\int_{S_1} g_R(p,y)*g_R(y,q) &= \int_{S_1} g_R(y,p)*g_R(y,q) \\ &\le k_1 \int_{S_1} g_{S_0}(y,x)*\omega(y) \\ &= k_1 \int_{S_1} g_{S_0}(x,y)*\omega(y) \\ &< k_1 \int_{S_0} g_{S_0}(x,y)*\omega(y) < \infty ,\end{aligned}$$

and a fortiori,

$$\begin{aligned}\gamma(p,q) &= \int_R g_R(p,y)*g_R(y,q) \\ &= C_1 + \int_{S_1} g_R(p,y)*g_R(y,q) \\ &< C_1 + k_1 \sigma_{S_0}(x) < \infty .\end{aligned}$$

Proof of II. Suppose $\gamma(p,q) < \infty$ for some $p,q \in R$. Take any regular region R_0 and an $x \in S_0 = R - \bar{R}_0$. For R_1, Ω chosen as before,

$$g_{S_0 \cap \Omega}(y,x) \le \frac{M_{1\Omega}}{m_{3\Omega}} g_\Omega(y,p) \quad \text{on} \quad \alpha_1 \cup \delta_\Omega, \quad \text{hence on} \quad \bar{S}_1 \cap \bar{\Omega},$$

$$g_{S_0}(y,x) \le \frac{M_1}{m_3} g_R(y,p) \quad \text{on} \quad \bar{S}_1,$$

$$\omega_\Omega(y) \le \frac{M_{2\Omega}}{m_{4\Omega}} g_\Omega(y,q) \quad \text{on} \quad \alpha_1 \cup \delta_\Omega, \quad \text{hence on} \quad \bar{S}_1 \cap \bar{\Omega},$$

$$\omega(y) \le \frac{M_2}{m_4} g_R(y,q) \quad \text{on} \quad \bar{S}_1 .$$

Therefore,

$$\begin{aligned}
\sigma_{S_0}(x) &= \int_{S_0} g_{S_0}(x,y) * \omega(y) \\
&= C_2 + \int_{S_1} g_{S_0}(x,y) * \omega(y) \\
&\le C_2 + k_2 \int_{S_1} g_R(y,p) * g_R(y,q) \\
&= C_2 + k_2 \left(C_3 + \int_R g_R(p,y) * g_R(y,q) \right) \\
&= C_2 + k_2 (C_3 + \gamma(p,q)) < \infty .
\end{aligned}$$

2.6. Conclusion. In view of Theorem 2.5, we may introduce the class of Riemannian N-manifolds R with ideal boundaries of infinite biharmonic measure:

$$O_\sigma^N = \{R \mid \sigma = \infty\} .$$

The class of Riemannian N-manifolds which do not carry biharmonic Green's function γ has been denoted O_γ^N. Properties I and II of σ and γ provide us with our main result:

THEOREM. $O_\sigma^N = O_\gamma^N$.

NOTES TO §2. The biharmonic measure was introduced and Theorems 2.2-2.6 established in Sario [9]. The purpose of Theorems 2.2-2.4 is to study the degeneracy of the biharmonic measure as a problem in its own right. These theorems can, of course, also be deduced from Theorem 2.6 by means of the test $\omega \in L^2$ for the existence of γ.

§3. BIHARMONIC GREEN'S FUNCTION γ AND HARMONIC DEGENERACY

The harmonic and biharmonic classification theories of Riemannian manifolds have developed in somewhat opposite directions. In harmonic classification theory, the existence of the Green's function was first explored, and then its relations to various harmonic null classes established. In biharmonic classification theory, a rather complete array of relations for quasiharmonic and biharmonic null classes was first developed, without any reference to biharmonic Green's functions. The reason was that no explicit tests for the existence of these functions were known. Such tests, as presented in §§1-2, then opened the road to finding relations between the class O_γ^N of Riemannian N-manifolds which do not carry γ, and other null classes considered in classification theory.

The present section is devoted to harmonic null classes. The first question here is: Is there any relation between O_γ^N and the class O_G^N? We shall show that the strict inclusion

$$O_G^N < O_\gamma^N$$

holds for every dimension $N \geq 2$.

We know from Chapter I that $O_G^N < O_{HP}^N < O_{HB}^N < O_{HD}^N = O_{HC}^N$. Where does O_γ^N fit into this scheme? We shall prove that its behavior is quite different from that of the O_{HX}^N classes: O_γ^N neither contains nor is contained in any of the classes O_{HX}^N for any X = P, B, D, C, or any $N \geq 2$.

We also take up the class $O_{HL^p}^N$ of Riemannian N-manifolds which admit no harmonic functions of finite L^p norm, $p \geq 1$. We show that this class shares the property of the above O_{HX}^N: the classes

$$\tilde{O}_\gamma^N \cap \tilde{O}_{HL^p}^N, \quad O_\gamma^N \cap O_{HL^p}^N, \quad O_\gamma^N \cap \tilde{O}_{HL^p}^N, \quad \tilde{O}_\gamma^N \cap O_{HL^p}^N$$

are all nonvoid for every $p \geq 1$ and every $N \geq 2$.

3.1. Alternative proof of the test for O_{γ}^{N}. We showed in §1 that a Riemannian N-manifold R belongs to O_{γ}^{N} if and only if $\omega \notin L^2$. Here we first give a proof which is slightly different in that it does not make use of Harnack's inequality.

Given a fixed regular subregion R_0 of R, set $S_0 = R - \bar{R}_0$, $\alpha_0 = \partial R_0$, and choose a regular subregion Ω with $\bar{R}_0 \subset \Omega$, $\delta_\Omega = \partial\Omega$. On $\bar{\Omega} \cap \bar{S}_0$, take

$$\omega_\Omega \in H(\Omega \cap S_0) \cap C(\bar{\Omega} \cap \bar{S}_0), \qquad \omega_\Omega | \alpha_0 = 1, \qquad \omega_\Omega | \delta_\Omega = 0 .$$

First suppose $\gamma(x,y)$ exists for some $x, y \in R$,

$$\gamma(x,y) = \int_R g(x,z)*g(z,y) .$$

Since the existence of γ entails that of g, we assume henceforth that $R \in \tilde{O}_G^N$. We shall show that $\omega \in L^2(S_0)$, $S_0 = R - \bar{R}_0$, for any regular subregion R_0 of R. Take regular subregions R_1 and Ω of R with $\bar{R}_0 \cup x \cup y \subset R_1 \subset \bar{R}_1 \subset \Omega$ and set $\alpha_1 = \partial R_1$, $S_1 = R - \bar{R}_1$,

$$m_{1\Omega} = \min_{z \in \alpha_1} g_\Omega(z,x), \qquad M_{1\Omega} = \max_{z \in \alpha_1} g_\Omega(z,x) ,$$

$$m_{2\Omega} = \min_{z \in \alpha_1} g_\Omega(z,y), \qquad M_{2\Omega} = \max_{z \in \alpha_1} g_\Omega(z,y) ,$$

$$m_{3\Omega} = \min_{\alpha_1} \omega_\Omega, \qquad M_{3\Omega} = \max_{\alpha_1} \omega_\Omega ,$$

$$k_{1\Omega} = \frac{M_{3\Omega}^2}{m_{1\Omega} m_{2\Omega}}, \qquad k_{2\Omega} = \frac{M_{1\Omega} M_{2\Omega}}{m_{3\Omega}^2} .$$

Denote by $m_1, M_1, m_2, M_2, m_3, M_3, k_1, k_2$ the corresponding limits as $\Omega \to R$. Then

$$\omega_\Omega(z)^2 \leq k_{1\Omega} g_\Omega(z,x) g_\Omega(z,y)$$

on $\alpha_1 \cup \delta_\Omega$, hence on $\bar{\Omega} \cap \bar{S}_1$. A fortiori,

$$\omega(z)^2 \leq k_1 g(z,x) g(z,y) \quad \text{on} \quad \bar{S}_1 .$$

By the symmetry of g,

$$\|\omega\|_2^2 = \int_{S_0} *\omega(z)^2 = c + \int_{S_1} *\omega(z)^2$$

$$\leq c + k_1 \int_{S_1} g(x,z) * g(z,y)$$

$$< c_1 + k_1 \int_R g(x,z) * g(z,y)$$

$$= c_1 + k_1 \gamma(x,y) < \infty .$$

Therefore, $\gamma(x,y) < \infty$ for some (x,y) implies $\omega \in L^2(S_0)$ for any $S_0 = R - \bar{R}_0$.

Conversely, suppose $\omega \in L^2(S_0)$ for some $S_0 = R - \bar{R}_0$. Take any $x, y \in R$ and choose R_1, Ω as above. Then

$$g_\Omega(z,x) g_\Omega(z,y) \leq k_{2\Omega} \omega_\Omega(z)^2$$

on $\alpha_1 \cup \delta_\Omega$, hence on $\bar{\Omega} \cap \bar{S}_1$, and therefore,

$$g(z,x) g(z,y) \leq k_2 \omega(z)^2 \quad \text{on} \quad \bar{S}_1 .$$

It follows that

$$\gamma(x,y) = c + \int_{S_1} g(x,z) * g(z,y)$$

$$\leq c + k_2 \int_{S_1} *\omega(z)^2 < \infty .$$

We conclude that $\omega \in L^2(S_0)$ for some $S_0 = R - \bar{R}_0$ implies $\gamma(x,y) < \infty$ for any $x, y \in R$. This proves our criterion. As a consequence, the finiteness of $\gamma(x,y)$ and the nondegeneracy of ω are independent of x, y, and R_0.

<u>3.2. Harmonic and biharmonic Green's functions.</u> By means of our test, we now tackle our first problem, that of determining the relation between O_G^N

and O^N_γ.

<u>THEOREM</u>. <u>For</u> $N \geq 2$,

$$O^N_G < O^N_\gamma .$$

<u>Proof</u>. We already observed that $O^N_G \subset O^N_\gamma$, and we only have to prove the strictness of the inclusion. Take the N-cylinder

$$R = \{(x, y^1, \ldots, y^{N-1}) | x > 0, \ |y^i| \leq 1, \ i = 1, \ldots, N-1\}$$

with the metric

$$ds^2 = x^\alpha dx^2 + x^{\alpha/(N-1)} \sum_{i=1}^{N-1} dy^{i2} ,$$

α a constant. For $h(x) \in H$,

$$\Delta h(x) = -x^{-\alpha}(x^\alpha x^{-\alpha} h'(x))' = 0 ,$$

hence $h(x) = ax + b$. For $R_0 = \{1 < x < 2\}$, $S_0 = \{0 < x < 1\} \cup \{x > 2\}$, the harmonic measure on $\{0 < x < 1\}$ is $\omega(x) = x$, hence $R \in \tilde{O}^N_G$. On the other hand, for $\alpha \leq -3$,

$$\|\omega\|_2^2 \geq c \int_0^1 x^2 x^\alpha dx = \infty$$

and therefore, $R \in O^N_\gamma$.

3.3. <u>Relation to harmonic degeneracy</u>. We proceed to show that there are no inclusion relations between O^N_γ and the harmonic null classes O^N_{HX}, with $X = P, B, D, C$. We recall that $O^N_G < O^N_{HP} < O^N_{HB} < O^N_{HD} = O^N_{HC}$. To begin with, the Euclidean N-ball gives trivially

$$\tilde{O}^N_\gamma \cap \tilde{O}^N_{HX} \neq \emptyset, \quad X = P, B, D, C, \quad N \geq 2 .$$

To see that

$$O_\gamma^N \cap O_{HX}^N \neq \emptyset, \quad X = P, B, D, C, \quad N \geq 2 ,$$

consider the N-cylinder

$$R = \{|x| < \infty, \ |y^i| \leq 1, \ i = 1, \ldots, N - 1\} ,$$

with the Euclidean metric. Every $h(x) \in H$ has the form $h(x) = ax + b$, which is unbounded for $a \neq 0$. Therefore, $\omega(x) \equiv 1$, and $R \in O_G^N < O_{HX}^N$. Moreover, $\|\omega\|_2^2 > c \int^\infty 1 \, dx = \infty$, hence $R \in O_\gamma^N$.

It remains to show:

THEOREM. *For* $X = P, B, D, C,$ *and* $N \geq 2$,

$$O_\gamma^N \cap \tilde{O}_{HX}^N \neq \emptyset, \quad \tilde{O}_\gamma^N \cap O_{HX}^N \neq \emptyset .$$

Proof. To prove the first relation, take the N-cylinder

$$R = \{|x| < 1, \ |y^i| \leq 1, \ i = 1, \ldots, N - 1\}$$

with the metric

$$ds^2 = \lambda(x)^2 dx^2 + \lambda(x)^{2/(N-1)} \sum_{i=1}^{N-1} dy^{i2} ,$$

where $\lambda \in C^\infty(-1,1)$. For $h(x) \in H$,

$$\Delta h = -\lambda^{-2}(\lambda^2 \lambda^{-2} h')' = 0 ,$$

$h = ax + b$, and the Dirichlet integral is

$$D(h) = c \int_{-1}^{1} \lambda^{-2} \lambda^2 dx < \infty ,$$

hence $x \in HD$ and $R \in \tilde{O}_{HD}^N \supset \tilde{O}_{HX}^N$. On the other hand,

$$\|\omega\|_2^2 > c \int^1 (ax + b)^2 \lambda^2 dx .$$

For $\lambda = (1 - x^2)^\alpha$, $\alpha \leq -3/2$, this gives $\|\omega\|_2 = \infty$ and $R \in O_\gamma^N$.

Regarding the second relation of the Theorem, the case $N > 4$ offers no difficulty. In fact, on the Euclidean N-space R^N, we have for every $h \in HP(R^N)$, $x \in R^N$, $r = |x|$, $r < \rho < \infty$, the Harnack inequality

$$\left(\frac{\rho}{\rho + r}\right)^{N-2} \frac{\rho - r}{\rho + r} h(0) \leq h(x) \leq \left(\frac{\rho}{\rho - r}\right)^{N-2} \frac{\rho + r}{\rho - r} h(0) ,$$

which for $\rho \to \infty$ gives $h = \text{const}$, $R^N \in O^N_{HP} \subset O^N_{HX}$. The harmonic measure on $\{r > 1\}$ is $\omega(x) = r^{-N+2}$ and

$$\|\omega\|_2^2 = c \int_1^\infty r^{-2N+4} r^{N-1} dr < \infty ,$$

hence $R^N \in \tilde{O}^N_\gamma$.

The above argument fails if $N \leq 4$, as then $R^N \in O^N_\gamma$. However, we know from II.7.6 that there exist manifolds in $\tilde{O}^N_{QC} \cap O^N_{HP}$, $N > 2$, and we shall see in 4.2 that $O^N_\gamma \subset O^N_{QC}$. A fortiori, $\tilde{O}^N_\gamma \cap O^N_{HX} \neq \emptyset$ for $N > 2$.

For $N = 2$, we make use of the base surface constructed in I.1.6 to show the strictness of $O^2_G < O^2_{HP}$. For our present purpose, we turn the surface into a Riemannian manifold R which continues excluding all nonconstant HP functions, but which nevertheless admits γ.

The function

$$\omega(z(p)) = -\log|z(p)|/\log 2$$

is the harmonic measure on that part of R which lies above $\{\frac{1}{2} < |z| < 1\}$, its value and harmonicity being unaffected by the rotations about $z = 0$ of the partial regions combining into single or multiple disks. In view of $N = 2$, endowing R with a conformal metric does not alter the harmonicity of ω. Any conformal metric $ds_0(p) = \lambda(p)|dt(p)|$ on R can be reduced to another conformal metric $ds(p) = \mu(p)ds_0(p)$ for which, as p tends to the ideal boundary of R, $\mu(p) \to 0$ so rapidly that the volume of R is finite. In this metric, since $\omega \in B$, we have $\|\omega\|_2 < \infty$, and therefore, $R \in \tilde{O}^2_\gamma$. The introduction of the conformal metric has no bearing on the class HP, so that we continue having $R \in O^2_{HP} \subset O^2_{HX}$.

This completes the proof of the Theorem.

3.4. Neither γ nor HL^p functions. We now take up the class $O^N_{HL^p}$ of Riemannian N-manifolds which do not carry harmonic functions with a finite L^p norm, $1 \leq p < \infty$. In view of the Euclidean N-ball, we have

$$\tilde{O}^N_{\gamma} \cap \tilde{O}^N_{HL^p} \neq \emptyset ,$$

for $1 \leq p < \infty$, $N \geq 2$.

Next we exclude both γ and HL^p functions.

THEOREM. *For* $1 \leq p < \infty$, $N \geq 2$,

$$O^N_{\gamma} \cap O^N_{HL^p} \neq \emptyset .$$

Proof. On the N-cylinder

$$R = \{|x| < \infty,\ |y^i| \leq \pi,\ i = 1,\ldots,N-1\}$$

with the Euclidean metric, every $h(x) \in H$ has the form $h(x) = ax + b$, hence $R \in O^N_G \subset O^N_{\gamma}$. As in II.2.4, every $h \in H$ can be expanded into a series $h = \Sigma f_n G_n$, where $f_n G_n = f_n(x) G_n(y) \in H$, $y = (y^1,\ldots,y^{N-1})$, and G_n ranges over all products of the form

$$G_n(y) = \prod_{i=1}^{N-1} \begin{matrix}\cos\\ \sin\end{matrix}\ n_i y^i$$

with $n = (n_1,\ldots,n_{N-1})$, the n_i integers ≥ 0, and $f_0(x) = h_0(x) \in H$. Set $\eta^2 = \Sigma_{i=1}^{N-1} n_i^2$. In view of $\Delta(f_n G_n) = -(f_n'' G_n - \eta^2 f_n G_n) = 0$, we have

$$h = h_0(x) + \Sigma' (a_n e^{\eta x} + b_n e^{-\eta x}) G_n ,$$

where the sum Σ' is extended over all $n \neq (0,\ldots,0)$ and $n_i \geq 0$.

Assume there exists an $h \in HL^p$ and choose a continuous function $\rho(x) \geq 0$ with $\operatorname{supp} \rho \subset (0,1)$, $\int_0^1 \rho dx = 1$. Suppose $a_n \neq 0$ for some $n \neq (0,\ldots,0)$.

Then for $\rho_t(x) = \rho(x - t)$, $\varphi_t = \rho_t G_n$, with $t > 0$,

$$(h, \varphi_t) \sim ce^{\eta t} \int_t^{t+1} \rho_t dx = ce^{\eta t} ,$$

hence $|(h,\varphi_t)| \to \infty$ as $t \to \infty$. But $\|\varphi_t\|_\infty = \text{const} < \infty$ and for $p^{-1} + q^{-1} = 1$,

$$\|\varphi_t\|_q = c \left(\int_t^{t+1} \rho_t^q dx \right)^{1/q} = \text{const} < \infty$$

if $p > 1$. Thus, $|(h,\varphi_t)| \le \|h\|_p \|\varphi_t\|_q = \text{const} < \infty$ for all $t > 0$. It follows that $a_n = 0$ for all $n \ne (0,\ldots,0)$. An analogous argument with $t < 0$, $t \to -\infty$ shows that $b_n = 0$ for all $n \ne (0,\ldots,0)$. Therefore, $h = h_0(x) = ax + b$. But $\|h_0\|_p = \infty$ unless $a = b = 0$, and we have proved that $R \in O^N_{HL^p}$.

3.5. HL^p functions but no γ. Next we prove:

THEOREM. For $p \ge 1$, $N \ge 2$,

$$O^N_\gamma \cap \tilde{O}^N_{HL^p} \ne \emptyset .$$

Proof. On the N-cylinder

$$R = \{|x| < \infty,\ |y^i| \le 1,\ i = 1,\ldots,N-1\} ,$$

choose the metric

$$ds^2 = e^{-x^2} dx^2 + e^{-x^2/(N-1)} \sum_{i=1}^{N-1} dy^{i2} .$$

For $h(x) \in H$ we again have $h = ax + b$, and therefore, $R \in O^N_G \subset O^N_\gamma$. On the other hand, $x \in HL^p$, since

$$\|x\|_p^p = c \int_{-\infty}^{\infty} |x|^p e^{-x^2} dx < \infty .$$

3.6. γ but no HL^p functions. It remains to show:

<u>THEOREM.</u> <u>For</u> $p \geq 1$, $N \geq 2$,

$$\tilde{O}^N_\gamma \cap O^N_{HL^p} \neq \emptyset .$$

<u>Proof.</u> Consider the N-space E^N with the metric

$$ds^2 = \varphi(r)dr^2 + \psi(r)^{1/(N-1)} \sum_{i=1}^{N-1} \gamma_i(\theta)d\theta^{i2} ,$$

where $\varphi, \psi \in C^\infty[0,\infty)$,

$$\varphi(r) = \begin{cases} 1 & \text{for } r < \frac{1}{2} , \\ e^{-r} & \text{for } r > 1 , \end{cases} \qquad \psi(r) = \begin{cases} r^{2(N-1)} & \text{for } r < \frac{1}{2} , \\ e^r & \text{for } r > 1 , \end{cases}$$

and the γ_i are trigonometric functions of $\theta = (\theta^1,\ldots,\theta^{N-1})$ such that the metric is Euclidean on $\{r < \frac{1}{2}\}$. For $h(r) \in H(\{r > 1\})$, we have $-(e^r h'(r))' = 0$. The harmonic measure $\omega(r) = e^{1-r}$ on $\{r \geq 1\}$ gives

$$\|\omega\|_2^2 = c\int_1^\infty e^{2-2r}dr < \infty ,$$

hence $R \in \tilde{O}^N_\gamma$.

To see that $R \in O^N_{HL^p}$, expand $h \in HL^p$ into a series $h = \Sigma f_n(r)S_n(\theta)$, where $f_n S_n \in H$ and the S_n are spherical harmonics. If $f_{n_0} \not\equiv 0$ for some $n_0 \geq 0$, the maximum principle applied to $f_{n_0}S_{n_0}$ gives $|f_{n_0}| > c_0 > 0$ on $[1,\infty)$, and we may assume $f_{n_0} > 0$ on $[1,\infty)$.

In the case $p = 1$, take $g(r) \in C^\infty[0,\infty)$, $0 < g < 1$, with $g(r) = (2r)^{-1}$ for $r > 1$. Then for some c, c_1, c_2

$$\|h\|_1 \geq c\left| \int_R hg{*}S_{n_0}\right| = c_1 + c_2 \int_1^\infty f_{n_0}gdr$$

$$\geq c_1 + c_2\int_1^\infty gdr = \infty,$$

a contradiction. If $p > 1$, take q with $p^{-1} + q^{-1} = 1$. Then $gS_{n_0} \in L^q$, and $(\cdot, gS_{n_0})$ is a linear functional on L^p. Since

$$|(h, gS_{n_0})| = \left| c\int_0^\infty f_{n_0} g dr \right| = \infty ,$$

we have a contradiction with $h \in L^p$, and conclude that $R \in O^N_{HL^p}$ for all $p \geq 1$.

NOTES TO §3. Theorems 3.2-3.6 were established in Sario [10]. In the above proofs of Theorems 3.4-3.6, essential use was made of counterexamples in Sario-Wang [13], Chung-Sario-Wang [1], and Chung-Sario [2].

§4. BIHARMONIC GREEN'S FUNCTION γ AND QUASIHARMONIC DEGENERACY

Our next problem is to determine whether or not there are inclusion relations between O^N_γ and the quasiharmonic null classes. We shall show that, in interesting contrast with harmonic null classes, we have strict inclusions:

$$\begin{array}{ccccc} & & O^N_{QB} & & \\ O^N_G < O^N_\gamma < O^N_{QP} & \angle & & \urcorner & O^N_{QC} \; . \\ & \urcorner & & \angle & \\ & & O^N_{QD} & & \vee \\ & & \vee & & \\ & & O^N_{QL^1} & & O^N_{QL^p} \end{array}$$

The first inclusion having been established in 3.2, and the inclusions between quasiharmonic null classes in II.2.8, we only have to show that $O^N_\gamma < O^N_{QP}$ and that there are no inclusion relations between O^N_γ and $O^N_{QL^p}$, $p \geq 1$.

4.1. Existence test for QP functions. We first establish a useful test for the existence of positive quasiharmonic functions.

Given a Riemannian N-manifold R, fix a regular subregion R_0, and take regular subregions Ω with $\bar{R}_0 \subset \Omega$. As in §3, let $\omega = \lim_{\Omega \to R} \omega_\Omega$ be the harmonic measure of the ideal boundary of R relative to R_0. We know from I.1.3 that $\omega \equiv 1$ if and only if $R \in O^N_G$.

Consider the Poisson equation

$$\Delta u = f, \quad f \geq 0, \quad f \not\equiv 0 \text{ on } R .$$

If

$$Gf(x) = \int_R g(x,y)*f(y) < \infty ,$$

then Gf is a positive solution of this equation. Conversely, if this equation has a positive solution u, then the Riesz decomposition yields

$$u(x) = h_\Omega(x) + \int_\Omega g_\Omega(x,y)*f(y)$$

on Ω, where $h_\Omega \in H(\Omega) \cap C(\bar{\Omega})$, $h_\Omega | \partial\Omega = u | \partial\Omega$. Since u is positive superharmonic and $u \geq h_\Omega$ on Ω, the limit $h = \lim_{\Omega\to R} h_\Omega$ exists and

$$Gf(x) = \lim_{\Omega\to R} \int_\Omega g_\Omega(x,y)*f(y) < \infty .$$

Therefore, $\Delta u = f$ has a positive solution if and only if $Gf(x) < \infty$ on R.

For a given $x \in R$, choose a regular subregion R_0 containing x. Let

$$M_x = \max_{y\in\partial R_0} g(x,y), \qquad m_x = \min_{y\in\partial R_0} g(x,y) .$$

By means of an exhaustion $\Omega \to R$ we see that

$$m_x\omega(y) \leq g(x,y) \leq M_x\omega(y)$$

for $y \in R - R_0$, and therefore,

$$m_x\int_{R-R_0} \omega(y)*f(y) \leq \int_{R-R_0} g(x,y)*f(y) \leq M_x\int_{R-R_0} \omega(y)*f(y) .$$

We have proved that on a hyperbolic Riemannian manifold, the Poisson equation $\Delta u = f$ with $f \geq 0$, $f \not\equiv 0$, has a positive solution if and only if $\int_{R-R_0} \omega(y)*f(y) < \infty$.

In the special case $f \equiv 1$ on R, we have the following simple but useful criterion:

THEOREM. For $N \geq 2$,

$$R \in \tilde{O}^N_{QP} \Leftrightarrow \omega \in L^1(R - R_0)$$

for some and hence every regular subregion R_0.

This criterion greatly simplifies nonexistence proofs of QP functions on certain Riemannian manifolds.

4.2. Strict inclusion. We are ready to show:

THEOREM. For $X = P, B, D, C$, and $N \geq 2$,

$$O^N_\gamma < O^N_{QX} .$$

Proof. We know from 1.5 that $R \in \tilde{O}^N_\gamma$ if and only if $\omega \in L^2(R - R_0)$ for some and hence every regular subregion R_0. By

$$\int_{R-R_0} g(x,z)*g(z,y) \leq M\int_{R-R_0} *g(x,z) ,$$

where R_0 is a regular subregion containing y, and $M = \max_{z\in\partial R_0} g(z,y)$, we have $O^N_\gamma \subset O^N_{QX}$.

If the volume of R is finite, then

$$\left(\int_{R-R_0} *\omega(z)\right)^2 \leq \mathrm{Vol}(R - R_0)\int_{R-R_0} *\omega(z)^2 .$$

Thus $O^N_{QX} \subset O^N_\gamma$ and consequently $O^N_{QX} = O^N_\gamma$.

To prove the strictness of $O^N_\gamma \subset O^N_{QP}$, take the N-cylinder

$$R = \{|x| < 1, \ |y^i| \leq \pi, \ i = 1,\ldots,N - 1\} ,$$

with the metric

$$ds = (1 - x^2)^{-1}ds_E ,$$

where ds_E is the Euclidean metric.

For $h(x) \in H$,

$$\Delta h(x) = -(1 - x^2)^N[(1 - x^2)^{-(N-2)}h'(x)]' = 0 .$$

If $N = 2$, $h(x) = ax + b$. If $N > 2$,

$$h'(x) = c(1 - x^2)^{N-2} \sim c(1 - |x|)^{N-2} ,$$

as $|x| \to 1$, and

$$h(x) \sim a(1 - |x|)^{N-1} + b .$$

This holds, in particular, for the harmonic measure $\omega(x)$ on

$$\{-1 < x < -\tfrac{1}{2}\} \cup \{\tfrac{1}{2} < x < 1\} ,$$

say. Thus

$$\omega(x) \sim a(1 - |x|)^{N-1}$$

for all $N \geq 2$. Since

$$\int_{\frac{1}{2}<|x|<1} *\omega^2 \sim c\int_{\frac{1}{2}}^{1} (1 - x)^{2N-2}(1 - x)^{-N}dx < \infty ,$$

we have $R \in \tilde{O}^N_\gamma$. On the other hand,

$$\int_{\frac{1}{2}<|x|<1} *\omega \sim c\int_{\frac{1}{2}}^{1} (1 - x)^{N-1}(1 - x)^{-N}dx = \infty ,$$

and we obtain $R \in O^N_{QP}$.

4.3. Relation to QL^p degeneracy. It is of interest that the inclusion relation 4.2 does not hold for $X = L^p$. From the proof of Theorem II.2.4 we know that $O^N_G \cap O^N_{QL^p} \neq \emptyset$, and from the proof of Theorem II.2.3 that $O^N_G \cap \tilde{O}^N_{QL^p} \neq \emptyset$. On the other hand, $\tilde{O}^N_{QP} \cap O^N_{QL^p} \neq \emptyset$ by Theorem II.2.5, and $\tilde{O}^N_{QP} \cap \tilde{O}^N_{QL^p} \neq \emptyset$ by Theorem II.2.4. In view of the inclusion $O^N_G < O^N_\gamma < O^N_{QP}$, we conclude that the classes

$$O_\gamma^N \cap O_{QL^p}^N, \quad O_\gamma^N \cap \tilde{O}_{QL^p}^N, \quad \tilde{O}_\gamma^N \cap O_{QL^p}^N, \quad \tilde{O}_\gamma^N \cap \tilde{O}_{QL^p}^N$$

are all disjoint and nonvoid. Thus, despite the close relationship between the existence of the biharmonic Green's function γ and of QX functions for $X = P, B, D, C$, <u>there are no relations whatever between the existence of</u> γ <u>and of</u> QL^p <u>functions</u>.

<u>NOTES TO §4.</u> Theorems 4.1-4.3 are due to Wang [1].

We have not considered the problem, of some interest, of relations between O_γ^N and $O_{H^2X}^N$.

CHAPTER IX

BIHARMONIC GREEN'S FUNCTION β: DEFINITION AND EXISTENCE

The remainder of the present monograph, Chapters IX-XI, will be devoted to the biharmonic Green's function β. On a regular subregion of a Riemannian manifold, it is defined by the boundary data $\beta = \partial\beta/\partial n = 0$. In the case of dimension 2, it gives the deflection, under a point load, of a clamped thin elastic plane plate. For higher dimensions, β could be called the biharmonic Green's function of a clamped body.

In the present chapter, we give the definition and discuss the existence of β. In Chapter X, we consider the role played by the nonexistence of β in the classification of Riemannian N-manifolds. Chapter XI is a comprehensive discussion of an interesting property of β: even on quite simple regular regions, β can be of nonconstant sign, in violation of Hadamard's famous conjecture on clamped plates.

In contrast with the harmonic Green's function g, and the biharmonic Green's function γ of a simply supported body, virtually nothing is known about β on noncompact carriers. The difficulty seems to lie in the fact that, whereas the monotonic increase of g and γ with exhausting subregions make their definitions on noncompact carriers quite simple, the possibility of a nonconstant sign of β on regular subregions causes complications in the convergence problem and in the independence of the convergence of the exhaustion and of the location of the pole.

The purpose of the present chapter is to define β on an arbitrary Riemannian manifold of any dimension, to give a test for the existence of β, and to show that the existence is independent of the exhaustion and of the pole. To reach these goals, we reverse the customary procedure. First we give a definition of β that applies directly to an abstract noncompact Riemannian manifold R, and give a condition assuring the existence of β which is jointly continuous on $R \times R$. Then we prove, a posteriori, that β on R

is the directed limit of the corresponding conventionally defined functions β_Ω on regular subregions Ω exhausting R.

§1. INTRODUCTION: DEFINITION AND MAIN RESULT

1.1. Conventional definition. We return to our earlier notation E^N for the Euclidean N-space, $N \geq 2$. Let R be a regular subregion of E^N. The conventional definiton of the biharmonic Green's function $\beta(x,y)$ of the clamped body R is as follows: $\beta(\cdot,y)$ is of class C^4 on R - y and

$$\Delta^2\beta(\cdot,y) = \Delta(\Delta\beta(\cdot,y)) = 0$$

on R - y, where Δ is the Laplacian $-\Sigma_{i,j=1}^N \delta^{ij}\partial^2/\partial x^i \partial x^j$;

$$\Delta\beta(\cdot,y) - g(\cdot,y) \in H(R) ,$$

where $g(\cdot,y)$ is the harmonic Green's function on R normalized by the flux $\int_\alpha (\partial g(x,y)/\partial n)dS_x = 1$ across a sphere $|x - y| < \varepsilon$ about y, with $\partial/\partial n$ the inner normal derivative, and H(R) is the class of harmonic functions on R; $\beta(\cdot,y)$ is of class C^1 on $\bar{R}$ - y and has the boundary values

$$\beta(\cdot,y) = \frac{\partial}{\partial n}\beta(\cdot,y) = 0$$

on $\partial R = \bar{R} - R$. The unique existence and fundamental properties of $\beta(x,y)$ have been deduced by several authors; a good reference is Miranda [1].

In contrast, complications set in for a general R, such as an unbounded R, or an R with a nonsmooth ∂R, or an unbounded R with a nonsmooth ∂R. The first two properties above are meaningful even for a general R but the third property is not. One natural way is to make use of the potential-theoretic ideal boundary theory in order to make the third property meaningful. However, this approach often ends up as a mere formality not supplying any applicable results. We prefer to reformulate the third property in such a manner that it becomes meaningful also for a not necessarily smooth ∂R. To motivate our new definition, we make the following observation concerning $\beta(x,y)$ on a regular R.

We call

$$H(\cdot,y) = \Delta\beta(\cdot,y)$$

the <u>density</u> of $\beta(\cdot,y)$ since, by $\Delta^2\beta = 0$ on $R - y$ and $\beta = 0$ on ∂R, $H(,y)$ is actually the density of the Green potential

$$\beta(x,y) = \int_R g(x,\xi)*H(\xi,y) ,$$

where $*1 = d\xi^1 \ldots d\xi^N$. It is a well known property of β (e.g., John [1]) that $H(\cdot,y)$ is continuous and, in fact, of class C^1, on $\bar{R} - y$. Therefore, in view of a basic property of Green's potentials (e.g., Miranda [1]),

$$\frac{\partial}{\partial n_x} \beta(x,y) = \int_R \frac{\partial}{\partial n_x} g(x,\xi)*H(\xi,y)$$

on ∂R. On multiplying both sides by an h in $H(R) \cap C(\bar{R})$ and then integrating with respect to the surface measure dS_x on ∂R, we infer, by the Fubini theorem and the Poisson type representation of harmonic functions, that

$$\int_{\partial R} h(x) \frac{\partial}{\partial n_x} \beta(x,y)dS_x = \int_R h(\xi)*H(\xi,y) .$$

Thus $\partial\beta/\partial n = 0$ on ∂R is equivalent to

$$\int_R h(\xi)*H(\xi,y) = 0$$

for every $h \in H(R) \cap C(\bar{R})$. Since $H(R) \cap C(\bar{R})$ is dense, e.g., in $H_2(R) = H(R) \cap L^2(R)$, we can replace $H(R) \cap C(\bar{R})$ by $H_2(R)$ in the above statement. The function $h = g(\cdot,x) - H(\cdot,x)$ belongs to $H_2(R)$, and it follows that

$$\beta(x,y) = \int_R H(\xi,x)*H(\xi,y) .$$

Suppose that there exists an $H'(\cdot,y)$ such that $H'(\cdot,y) - g(\cdot,y)$ belongs to $H_2(R)$ and $(h(\cdot), H'(\cdot,y)) = 0$ for every h in $H_2(R)$. Then $H'(\cdot,y) - H(\cdot,y) \in H_2(R)$, and in view of $(h,H) = (h,H') = 0$ with $h = H'(\cdot,y) - H(\cdot,y)$, we obtain

$$\int_R *(H'(\xi,y) - H(\xi,y))^2 = 0 ,$$

that is, $H' = H$.

1.2. New definition. We can now define the biharmonic Green's function $\beta(x,y)$ on a regular R as follows: $\beta(x,y)$ is the function given by $(H(\cdot,x), H(\cdot,y))$, with $H(\cdot,y)$ characterized by three properties: $H(\cdot,y) \in H(R - y)$ has the harmonic fundamental singularity at y, $H(\cdot,y)$ is square integrable on R off any neighborhood of y, and $H(\cdot,y)$ satisfies $(h,H) = 0$ for every h in $H_2(R)$. The merit of this definition lies in the fact that no explicit reference is made to the boundary ∂R, and the regularity of R is not needed. Thus this definition is meaningful on a general R as well. The purpose of this chapter is to construct, adopting the above definition, the biharmonic Green's function $\beta(x,y)$ on an arbitrary noncompact Riemannian manifold R, under a certain condition which is implied by the existence of a harmonic Green's function or Evans function square integrable off the pole, and as consequences of the construction, to establish the joint continuity of $\beta(x,y)$ on $R \times R$ and the consistency relation $\beta = \lim_{\Omega \to R} \beta_\Omega$.

Henceforth we consider a noncompact C^∞ Riemannian manifold R of dimension $N \geq 2$, with $x = (x^1, \ldots, x^N)$ a local parameter of R. As before, let $H(R) = \{h \in C^2(R) | \Delta h = 0\}$ be the class of harmonic functions on R. We continue denoting by $g_\Omega(x,y)$ the harmonic Green's function on a regular subregion Ω of R. The fundamental function space in our discussion is

$$H_2(R) = H(R) \cap L^2(R) ,$$

where $L^2(R)$ is taken with respect to the volume element $*1 = dx = g^{1/2} dx^1 \ldots dx^N$. We shall show in §2 that $H_2(R)$ is a locally bounded Hilbert space.

In view of the above observation, we introduce:

DEFINITION. The biharmonic Green's function $\beta(x,y)$ on R is defined as

$$\beta(x,y) = \int_R H(\xi,x)*H(\xi,y) ,$$

where the density $H(\cdot,y)$ of $\beta(\cdot,y)$ satisfies

$$H(\cdot,y) - g_\Omega(\cdot,y) \in H(\Omega)$$

and

$$H(\cdot,y) \in H_2(R - \Omega)$$

for any regular subregion Ω of R containing y, and $H(\cdot,y)$ is orthogonal to $H_2(R)$,

$$\int_R h(\xi)*_\xi H(\xi,y) = 0$$

for every h in $H_2(R)$.

1.3. Main Theorem. The following condition will be instrumental in our study:

CONDITION [*]. There exists a positive harmonic function h on an open subset S of R with a compact complement $R - S$ such that

$$\int_S *h(x)^2 < \infty$$

and the flux of h across the ideal boundary of R does not vanish, that is,

$$\int_{\partial\Omega} *dh \neq 0$$

for any regular subregion Ω *of* R *with* $\Omega \supset R - S$.

In this chapter, we shall prove:

MAIN THEOREM. *If* R *satisfies condition* [*], *then there exists a biharmonic Green's function* $\beta(x,y)$ *on* R *such that* $\beta(x,y)$ *is continuous on* $R \times R$ *in the extended sense and* $\{\beta(x,y) - \beta_\Omega(x,y)\}$ *converges to zero uniformly on each compact subset of* $R \times R$ *as the regular subregions* Ω *of* R *with* $y \in \Omega$ *exhaust* R.

Any regular subregion Ω of R satisfies condition [*] and thus the above result contains the classical existence theorem. If R carries a harmonic Green's function square integrable off its pole, or an Evans potential square integrable off its pole, then [*] is fulfilled by R.

1.4. *Plan of this chapter.* We start by showing, in §2, that the space $H_p(R) = H(R) \cap L^p(R)$, with $L^p = \{f \mid \|f\|_p < \infty\}$, is a locally bounded Banach space. In §3, we introduce the concept of harmonic fundamental kernel $K(x,y)$ defined, in essence, by continuity properties and square integrability off the pole. The corresponding functional k_y on $H_2(R)$ is shown to belong to $H_2(R)^*$, so that $y \to k_y$ can be viewed as a mapping from R into $H_2(R)$. Using the local boundedness of $H_2(R)$ established in §2 one sees that the mapping $y \to k_y$ is continuous.

§4 will be devoted to the result, central in our reasoning, that if a fundamental kernel K exists, then so does β; this is shown by choosing $K(x,y) - k_y(x)$ for the density $H(x,y)$ of $\beta(x,y)$. Moreover, β is continuous on $R \times R$ and, as will be shown in §5, consistent on R, that is, β is the direct limit of the β_Ω on regular subregions Ω exhausting R.

In §§6 and 7 we discuss the cases of hyperbolic and parabolic manifolds, respectively. We show that, in both cases, condition [*] implies the existence of a fundamental kernel K, and the Main Theorem follows. Condition [*] in turn is satisfied if, for a hyperbolic R, the harmonic Green's function g of R is square integrable off its pole, or if, for a parabolic R, some positive $h \in H_2(S)$ is unbounded on every ideal boundary neighborhood.

Much of §7 is devoted to proving the existence and joint continuity of a generalized Evans kernel, to be called the h-kernel, on every parabolic R. By definition, the h-kernel differs from a given $h \in H(S)$ by a bounded function. For an h satisfying condition [*], the h-kernel provides us with a fundamental kernel, hence with the existence of β.

In §8, we illustrate the use of our test by several examples.

NOTES TO §1. Since the square integrability of a harmonic Green's function off its pole entails condition [*], the Main Theorem contains the result, first proved in Ralston-Sario [1], that the existence of γ implies that of β. Similarly, since the square integrability of an Evans potential e entails condition [*], the Main Theorem contains the result, established in Nakai-Sario [12], that $\|e(\cdot,y)\|_{R-B_y} < \infty$ for every pole y and a geodesic ball B_y about y guarantees the existence of β. We shall return to these consequences of the Main Theorem in Chapter X.

All results in the present chapter were established in Nakai-Sario [20].

§2. LOCAL BOUNDEDNESS

2.1. An auxiliary result. At a point $\xi \in R$, fix a (relatively compact) geodesic ball $B = \{|x - \xi| < \delta\}$, sufficiently small to justify the operations to be performed on it, and denote by B_r the concentric geodesic ball $|x - \xi| < r$ for $r \in (0,\delta)$. Fix a $p \in [1,\infty)$.

LEMMA. *There exists a positive constant* q *for any fixed* a *in* $(0,\delta)$ *such that*

$$\sup_{x \in B_a} |u(x)| \leq q\left(\int_{B-B_a} *|u(x)|^p\right)^{1/p}$$

for every u *in* H(B).

For the proof, suppose such a q does not exist. Then we can choose a sequence $\{u_n\}$ in $H(B)$ such that

$$\lim_{n\to\infty} \int_{B-B_a} *|u_n(x)|^p = 0$$

and

$$\sup_{x\in B_a} |u_n(x)| = 1 .$$

For each $r \in (a,\delta)$, set

$$f_n(r) = \int_{\partial B_r} |u_n(x)|^p \, dS_x ,$$

with dS the Riemannian surface element on ∂B_r. In geodesic coordinates,

$$\int_{B-B_a} *|u_n(x)|^p = \int_a^\delta f_n(r)dr ,$$

and we see that $\{f_n(r)\}$ is a zero sequence in $L^1(a,\delta)$. By choosing a subsequence, if necessary, we may, therefore, assume that $\{f_n(r)\}$ converges to zero almost everywhere on (a,δ), and thus there exists a $b \in (a,\delta)$ such that $\lim_{n\to\infty} f_n(b) = 0$, that is,

$$\lim_{n\to\infty} \int_{\partial B_b} |u_n(x)|^p \, dS_x = 0 .$$

Let $g(x,y) = g_{B_b}(x,y)$ be the harmonic Green's function on B_b. We fix a $t \in (a,b)$ and let k be the Harnack constant of the compact set $\overline{B}_a$ in the region B_t. Then

$$g(x,y) \leq kg(\xi,y)$$

for every $x,\xi \in \overline{B}_a$ and every $y \in \overline{B}_b - \overline{B}_t$. Since $g(x,y) = kg(\xi,y) = 0$ for every $y \in \partial B_b$, we see that

$$0 \leq \frac{\partial}{\partial n_y} g(x,y) \leq k \frac{\partial}{\partial n_y} g(\xi,y)$$

for every $y \in \partial B_b$, with $\partial/\partial n$ the Riemannian inner normal derivative on ∂B_b. Since $k\partial g(\xi,y)/\partial n_y$ is continuous on ∂B_b, it is bounded by a constant, A, say. From

$$u_n(x) = \int_{\partial B_b} u_n(y) \frac{\partial}{\partial n_y} g(x,y)dS_y ,$$

it follows that

$$\begin{aligned} |u_n(x)| &\leq \int_{\partial B_b} |u_n(y)| \frac{\partial}{\partial n_y} g(x,y)dS_y \\ &\leq k \int_{\partial B_b} |u_n(y)| \frac{\partial}{\partial n_y} g(\xi,y)dS_y \\ &\leq A \int_{\partial B_b} |u_n(y)|\,dS_y \\ &\leq A\left(\int_{\partial B_b} dS_y\right)^{1/p^*}\left(\int_{\partial B_b} |u_n(y)|^p\, dS_y\right)^{1/p} , \end{aligned}$$

where $1/p^* + 1/p = 1$, with $p^* = \infty$ if $p = 1$. By $\sup_{B_a} |u_n(x)| = 1$, we arrive at the following violation of $\lim_{n\to\infty} \int_{\partial B_b} |u_n(x)|^p\, dS_x = 0$:

$$1 = \sup_{x\in B_a} |u_n(x)| \leq A\left(\int_{\partial B_b} dS_y\right)^{1/p^*}\left(\int_{\partial B_b} |u_n(y)|^p\, dS_y\right)^{1/p}$$

for every $n = 1,2,\dots$.

2.2. Locally bounded Banach space. We denote by $L^p(R)$ the Banach space $L^p(R,{*1})$, $p \geq 1$, with the norm $\|f\|_p = (\int_R *|f(x)|^p)^{1/p}$ and consider the space

$$H_p(R) = H(R) \cap L^p(R) .$$

As a consequence of Lemma 2.1, we obtain:

THEOREM. *The space $H_p(R)$ is a Banach space with the norm $\|u\|_p$. It is locally bounded, that is, for any compact subset K of R, there exists a constant q_K such that*

$$\max_{x\in K} |u(x)| \leq q_K \|u\|_p$$

for every u *in* $H_p(R)$.

By the compactness of K, there exists a set $\{\xi_i\}$, $i = 1,\ldots,j$, of points in K such that the union of B_a^i, $i = 1,\ldots,j$, contains K, where the B^i are geodesic balls $|x - \xi_i| < \delta_i$, $i = 1,\ldots,j$, and $B_a^i = \{|x - \xi_i| < a_i\}$ with $a_i \in (0,\delta_i)$. Let q_i be the constant in Lemma 2.1 for B^i and B_a^i. Then the required q_K is $\max_{1\leq i\leq j} q_i$. In fact,

$$\begin{aligned}\max_{x\in K} |u(x)| &\leq \max_{1\leq i\leq j}\left(\sup_{x\in B_a^i} |u(x)|\right)\\ &\leq \max_{1\leq i\leq j}\left(q_i\left(\int_{B^i-B_a^i} *|u(x)|^p\right)^{1/p}\right)\\ &\leq \left(\max_{1\leq i\leq j} q_i\right)\left(\int_R *|u(x)|^p\right)^{1/p}\\ &= q_K\|u\|_p\ .\end{aligned}$$

Let $\{u_n\}$ be a Cauchy sequence in $H_p(R)$. Since it is also Cauchy in $L^p(R)$, there exists a unique u in $L^p(R)$ such that $\{\|u_n - u\|_p\}$ is a zero sequence, and a fortiori, a subsequence of $\{u_n\}$ converges to u almost everywhere on R. On the other hand, by the second part of the Theorem, $\{u_n\}$ is also a Cauchy sequence with respect to the supremum norm on each compact subset of R. Therefore, $\{u_n\}$ converges to u uniformly on each compact subset of R, and we have $u \in H(R)$, that is, $u \in H_p(R)$.

2.3. Locally bounded Hilbert space. We shall make use of Theorem 2.2 only in the case $p = 2$. Then $H_2(R)$ is a locally bounded Hilbert space with the inner product $(f,g) = \int_R f(x)*g(x)$ related to the norm $\|f\|_2$ by $(f,f) = \|f\|_2^2$. Henceforth, we simply write $\|f\|$ instead of $\|f\|_2$. We also use $(f,g)_S$ and $\|f\|_S$ for the inner product and the norm in $L^2(S)$ for a subset S of R. We always understand (f,g) and $\|f\|$ as $(f,g)_R$ and

$\|f\|_R$. Since $H_2(R)$ is locally bounded, there exists, by the general theory of Aronszajn, a <u>reproducing kernel</u> $\Phi(\cdot,y) \in H_2(R)$ such that

$$u(y) = (u, \Phi(\cdot,y))$$

for every $u \in H_2(R)$. This kernel is interesting from the view point of harmonic degeneracy of the ideal boundary of R. We shall not, however, discuss it in this book.

We denote by $H_2(R)^*$ the dual space of $H_2(R)$, that is, the space of bounded linear functionals on $H_2(R)$. By the Riesz theorem,

$$H_2(R)^* = H_2(R)$$

in the sense that $v \in H_2(R)^*$ can be identified with an element $v \in H_2(R)$ by the relation $v(u) = (u,v)$ for every $u \in H_2(R)$.

<u>NOTES TO §2</u>. In the context of 2.3, we pose the following problem: prove or disprove $H_p(R)^* = H_{p^*}(R)$ with $1/p + 1/p^* = 1$.

§3. FUNDAMENTAL KERNEL

<u>3.1</u>. <u>Harmonic Green's function</u>. Fix a regular subregion Ω of R. The harmonic Green's function $g_\Omega(x,y)$ on Ω for each fixed $y \in \Omega$ is, by definition, an element in the class $H(\Omega - y) \cap C(\bar{\Omega} - y)$ with $g_\Omega(\cdot,y)|\partial\Omega = 0$ and $\Delta g_\Omega(\cdot,y) = \delta_y$ in the sense of distributions, where δ_y is the Dirac δ at y. As a kernel function on $\Omega \times \Omega$, $g_\Omega(x,y)$ is symmetric, that is, $g_\Omega(x,y) = g_\Omega(y,x)$, and finitely continuous on $\Omega \times \Omega$ off the diagonal set. Moreover, $\lim_{x,y\to\xi} g_\Omega(x,y) = \infty$ for every $\xi \in \Omega$. Fix a parametric ball $B = \{|x - \xi| < \frac{1}{2} - \varepsilon\}$, $\varepsilon \in (0,\frac{1}{2})$, at $\xi \in \bar{\Omega}$. Then there exists a constant c such that

$$g_\Omega(x,y) \le cS(x,y) = \begin{cases} c \log|x - y|^{-1}, & N = 2, \\ c|x - y|^{-N+2}, & N \ge 3, \end{cases}$$

for every pair of points x and y in $B \cap \bar{\Omega}$. We extend $g_\Omega(x,y)$ to all of $R \times R$ by setting $g_\Omega(x,y) = 0$ whenever x or y belongs to $R - \Omega$. For the existence and the above properties of $g_\Omega(x,y)$, we refer to, e.g., Miranda [1].

3.2. Fundamental kernel. We call a kernel function $K(x,y)$ on R a fundamental kernel if the following five conditions are satisfied:

[a] $K(\cdot,y) \in H(R - y)$ *for every fixed* $y \in R$.

[b] $K(x,y)$ *is symmetric, that is,* $K(x,y) = K(y,x)$ *for all* x *and* y *in* R.

[c] *The difference* $K(x,y) - g_\Omega(x,y)$ *is finitely continuous on all of* $\Omega \times \Omega$ *for any regular subregion* Ω *of* R.

[d] $K(\cdot,y)$ *is square integrable on* R *off the pole* y, *that is, there exists a regular subregion* Ω_y *of* R *containing an arbitrarily fixed* $y \in R$ *such that*

$$\int_{R-\Omega_y} *K(x,y)^2 < \infty .$$

[e] *For any point* $\eta \in R$ *and any regular subregion* Ω *of* R *containing* η,

$$\lim_{y\to\eta} \int_{R-\Omega} *|K(x,y) - K(x,\eta)|^2 = 0 .$$

3.3. Corresponding functional. Here and hereafter in §§3-5 we assume that a fundamental kernel $K(x,y)$ exists on R. We then consider the functional k_ξ on $H_2(R)$ given by

$$k_\xi(u) = \int_R *u(x)K(x,\xi) ,$$

for every $u \in H_2(R)$ and for any fixed $\xi \in R$.

LEMMA. *The functional* k_ξ *belongs to* $H_2(R)^* = H_2(R)$.

Fix a parametric ball $B = \{|x - \xi| < 1\}$ about ξ and an $a \in (0,1)$. For $B_a = \{|x - \xi| < a\}$, there exists, by Theorem 2.2, a constant q such that

$$\sup_{x \in B_a} |u(x)| \leq q\|u\|$$

for every $u \in H_2(R)$. We estimate

$$\int_R |u(x)| *|K(x,\xi)| = \int_{R-B_a} |u(x)| *|K(x,\xi)| + \int_{B_a} |u(x)| *|K(x,\xi)| \ .$$

The first term on the right is dominated by

$$\left(\int_{R-B_a} *u(x)^2\right)^{1/2} \cdot \left(\int_{R-B_a} *K(x,\xi)^2\right)^{1/2} ,$$

and the second, by $(\sup_{B_a} |u|) \cdot \int_{B_a} *|K(x,\xi)|$. A fortiori, on setting

$$c = \left(\int_{R-B_a} *K(x,\xi)^2\right)^{1/2} + q\int_{B_a} *|K(x,\xi)| ,$$

we deduce

$$\int_R |u(x)| *|K(x,\xi)| \leq c\|u\| \ .$$

The first term of c is finite by [d], and the second term by [c] and 3.1, so that $c < \infty$. This means that k_ξ is well defined on $H_2(R)$ and $|k_\xi(u)| \leq c\|u\|$ for every $u \in H_2(R)$. Since k_ξ is clearly linear, we conclude that $k_\xi \in H_2(R)^*$.

3.4. Continuity. We can view $\xi \to k_\xi$ as a mapping from R into $H_2(R)$. We next show that this mapping is continuous:

LEMMA. *For any fixed* $\xi \in R$,

$$\lim_{\eta \to \xi} \|k_\eta - k_\xi\| = 0 \ .$$

Given an arbitrary $\varepsilon > 0$, fix a parametric ball $B = \{|x - \xi| < 1\}$ and set $B_t = \{|x - \xi| < t\}$ for $t \in (0,1)$. Let $b \in (0,1/4)$. There exists, by [c] and 3.1, a constant c such that

$$*|K(x,\eta)| \leq c *_E S(x,\eta)$$

for all x and η in B_b, with $*_E 1 = dx^1 \cdots dx^N$ the Euclidean volume element. Let $a \in (0,b/2)$. Then for every $\eta \in B_a$,

$$\begin{aligned}\int_{B_a} *|K(x,\xi) - K(x,\eta)| &\leq \int_{B_a} *|K(x,\xi)| + \int_{B_a} *|K(x,\eta)| \\ &\leq c\int_{B_a} *_E S(x,\xi) + c\int_{B_a} *_E S(x,\eta) \\ &\leq c\int_{B_a} *_E S(x,\xi) + c\int_{|x-\eta|<2a} *_E S(x,\eta) \\ &\leq 2c\int_{B_{2a}} *_E S(x,\xi) .\end{aligned}$$

The last term tends to zero as $a \to 0$. Hence we can fix an $a \in (0,b/2)$ such that

$$\int_{B_a} *|K(x,\xi) - K(x,\eta)| \leq \varepsilon/2q$$

for every $\eta \in B_a$, where q is the constant in Theorem 2.2 for $\bar{B}_b$ and $p = 2$. By [e], we can find a $\delta \in (0,a)$ such that

$$\int_{R-B_a} *|K(x,\xi) - K(x,\eta)|^2 \leq (\varepsilon/2)^2$$

for every $\eta \in B_\delta$. Fix an arbitrary $u \in H_2(R)$. Then

$$|(u,k_\xi - k_\eta)| = |k_\xi(u) - k_\eta(u)| \leq \int_R *|u(x)(K(x,\xi) - K(x,\eta))|$$

$$\leq \int_{R-B_a} |u(x)|*|K(x,\xi) - K(x,\eta)| + \int_{B_a} |u(x)|*|K(x,\xi) - K(x,\eta)| .$$

Here the first term is dominated by

$$\left(\int_{R-B_a} *|K(x,\xi) - K(x,\eta)|^2\right)^{1/2}\left(\int_{R-B_a} *u(x)^2\right)^{1/2} ,$$

which in turn is dominated by $\frac{1}{2}\varepsilon\|u\|$. The second term is dominated by

$$\left(\sup_{x\in B_a} |u(x)|\right)\int_{B_a} *|K(x,\xi) - K(x,\eta)| \leq q\|u\|\int_{B_a} *|K(x,\xi) - K(x,\eta)| ,$$

which is again dominated by $\frac{1}{2}\varepsilon\|u\|$. A fortiori,

$$|(u, k_\xi - k_\eta)| \leq \varepsilon\|u\|$$

for every $\eta \in B_\delta$. On choosing $u = k_\xi - k_\eta$, we conclude that

$$\|k_\xi - k_\eta\| \leq \varepsilon$$

whenever $\eta \in B_\delta$. The Lemma follows.

3.5. Auxiliary function. We consider the auxiliary function

$$\kappa(x,y) = \int_R K(x,\xi)*K(y,\xi) .$$

In view of [b]-[d] and 3.1, $\kappa(x,y)$ is well defined on $R \times R$ with $|\kappa(x,y)| < \infty$ for $x \neq y$, $|\kappa(x,x)| < \infty$ for dimensions $N = 2,3$, and $\kappa(x,x) = \infty$ for $N > 3$. We fix a regular subregion Ω of R. By [c], we have

$$\kappa(x,y) = \gamma_\Omega(x,y) + w_\Omega(x,y)$$

on $\Omega \times \Omega$, where

$$\gamma_\Omega(x,y) = \int_\Omega g_\Omega(x,\xi)*g_\Omega(y,\xi) ,$$

and $w_\Omega(x,y)$ is the sum of four functions $I_j(x,y)$, $1 \leq j \leq 4$, on $\Omega \times \Omega$

given as follows:

$$I_1(x,y) = \int_{R-\Omega} K(x,\xi)*K(y,\xi) ,$$

which, by [a], is harmonic on Ω as a function of x for any fixed $y \in \Omega$ and, by [e], continuous on $\Omega \times \Omega$; for $u_\Omega(y,\xi) = K(y,\xi) - g_\Omega(y,\xi)$,

$$I_2(x,y) = \int_\Omega g_\Omega(x,\xi)*u_\Omega(y,\xi) ,$$

which is biharmonic on Ω as a function of x for any fixed $y \in \Omega$ and continuous on $\Omega \times \Omega$;

$$I_3(x,y) = \int_\Omega u_\Omega(x,\xi)*g_\Omega(y,\xi) ,$$

which is identical with $I_2(y,x)$, harmonic on Ω as a function of x for any fixed $y \in \Omega$, and continuous on $\Omega \times \Omega$;

$$I_4(x,y) = \int_\Omega u_\Omega(x,\xi)*u_\Omega(y,\xi) ,$$

which is harmonic on Ω as a function of x for any fixed $y \in \Omega$ and continuous on $\Omega \times \Omega$. We conclude that $w_\Omega(x,y) = w_\Omega(y,x)$ is continuous on $\Omega \times \Omega$ and $w_\Omega(\cdot,y)$ is biharmonic on Ω for any fixed $y \in \Omega$.

Properties of $\gamma_\Omega(\cdot,y)$ can be readily deduced from 3.1. In particular, $\gamma_\Omega(\cdot,y)$ is biharmonic on $\Omega - y$ for any fixed $y \in \Omega$ and $\gamma_\Omega(x,y) = \gamma_\Omega(y,x)$ is finitely continuous on $\Omega \times \Omega$ off the diagonal set. Moreover, $\lim_{x,y\to\xi} \gamma_\Omega(x,y) < \infty$ if $N = 2,3$, and ∞ if $N > 3$, for every $\xi \in \Omega$. As a consequence, $\kappa(x,y)$ has these same properties:

<u>LEMMA.</u> <u>The kernel</u> $\kappa(x,y)$ <u>is symmetric on</u> R, $\kappa(\cdot,y)$ <u>is biharmonic on</u> $\Omega - y$ <u>for any</u> $y \in \Omega$ <u>and</u> $\kappa(x,y)$ <u>is continuous on</u> $R \times R$, <u>viz.</u>, $\kappa(x,y)$ <u>is finitely continuous off the diagonal set, and</u> $\lim_{x,y\to\xi} \kappa(x,y) = \kappa(\xi,\xi)$

for any $\xi \in R$.

NOTES TO §3. The functional k_ξ and the function κ will serve to give, in §4, to β a simple expression which will yield the joint continuity of β.

§4. EXISTENCE OF β

4.1. Fundamental kernel and β. After the preparations in §§2-3, we are ready to establish the following result, which is central in our reasoning:

THEOREM. Suppose there exists a fundamental kernel $K(x,y)$ on R. Then the biharmonic Green's function $\beta(x,y)$ with the following two properties exists on R: $\beta(x,y)$ is continuous on $R \times R$ in the extended sense, that is, $\beta(x,y)$ is finitely continuous on $R \times R$ off the diagonal set and $\lim_{x,y\to\xi} \beta(x,y) = \beta(\xi,\xi)$; $\beta(x,y)$ has the consistency property, that is, $\{\beta(x,y) - \beta_\Omega(x,y)\}$ converges to zero uniformly on each compact subset of $R \times R$ as regular subregions Ω of R with $y \in \Omega$ exhaust R.

The proof will be given in 4.2-5.3.

4.2. Existence and uniqueness. The existence causes no difficulty. The functional k_ξ defined in 3.3 is in $H_2(R)$ by Lemma 3.3. Set

$$H(x,y) = K(x,y) - k_y(x)$$

for x and y in R. Since $k_y \in H(R)$, [a] and [c] of 3.2 assure that $H(\cdot,y)$ satisfies the relation $H(\cdot,y) - g_\Omega(\cdot,y) \in H(\Omega)$ of 1.2 for every regular subregion Ω of R. Condition $H(\cdot,y) \in H_2(R - \Omega)$ is clearly satisfied by virtue of [d] and $k_y \in H_2(R)$. Fix an arbitrary $u \in H_2(R)$. Then

$$\int_R u(\xi) *_\xi H(\xi,y) = \int_R u(\xi) *_\xi K(\xi,y) - (u,k_y)$$

$$= k_y(u) - (u,k_y) = 0 ,$$

so that the final condition in 1.2 is met.

The uniqueness of such an $H(\cdot,y)$ is also clear. Suppose $H'(\cdot,y)$ satisfies the last three conditions in 1.2. Then $h = H(\cdot,y) - H'(\cdot,y)$ belongs to $H_2(R)$. The last condition in 1.2 for $H(\cdot,y)$ and $H'(\cdot,y)$ with this particular h implies that $(h,h) = \|h\|^2 = 0$, hence $H(\cdot,y) \equiv H'(\cdot,y)$.

We have proved that

$$\beta(x,y) = \int_R H(\xi,x)*H(\xi,y)$$

is the required biharmonic Green's function on R.

4.3. Joint continuity. To prove the joint continuity of $\beta(x,y)$, we observe that, since $k_x = K(\cdot,x) - H(\cdot,x) \in H_2(R)$, k_x is orthogonal to $H(\cdot,y) = K(\cdot,y) - k_y$ and, a fortiori,

$$\beta(x,y) = \kappa(x,y) - \int_R K(\xi,x)*k_y(\xi) .$$

By the definition of k_x,

$$\beta(x,y) = \kappa(x,y) - (k_x,k_y) .$$

In view of Lemma 3.5, we only have to show the joint continuity of (k_x,k_y) in order to prove that of $\beta(x,y)$. We obtain

$$|(k_x,k_y) - (k_\xi,k_\eta)| = |(k_x - k_\xi,k_y) + (k_\xi,k_y - k_\eta)|$$

$$\leq \|k_x - k_\xi\| \cdot \|k_y\| + (\|k_x\| + \|k_x - k_\xi\|) \cdot \|k_y - k_\eta\| .$$

By Lemma 3.4, $\lim_{(x,y)\to(\xi,\eta)} (k_x,k_y) = (k_\xi,k_\eta)$ for every (ξ,η) in $R \times R$.

4.4. Existence on regular subregions. Before proceeding to the consistency relation, we insert here the following comment on the existence of β in the special case of a regular subregion Ω. Consider the boundary conditions

$$v(x) = \frac{\partial}{\partial n_x} v(x) = 0$$

on $\partial\Omega$.

COROLLARY. <u>The biharmonic Green's function</u> $\beta_\Omega(x,y)$ <u>with the following two properties exists on any regular subregion</u> Ω <u>of</u> R: $\beta_\Omega(x,y)$ <u>is continuous on</u> $\Omega \times \Omega$ <u>in the extended sense, that is,</u> $\beta_\Omega(x,y)$ <u>is finitely continuous on</u> $\Omega \times \Omega$ <u>off the diagonal set and</u> $\lim_{x,y\to\xi} \beta_\Omega(x,y) = \beta_\Omega(\xi,\xi)$; $\beta_\Omega(\cdot,y)$ <u>is of class</u> C^1 <u>on</u> $\bar{\Omega} - y$ <u>and satisfies the above boundary conditions.</u>

It is easily verified that $g_\Omega(x,y)$ is a fundamental kernel on Ω. In view of 4.2-4.3, we thus have the existence of the biharmonic Green's function

$$\beta_\Omega(x,y) = \int_\Omega H_\Omega(\xi,x)*H_\Omega(\xi,y) ,$$

which is jointly continuous. As in §1, we see that the biharmonic fundamental solution $\bar{\beta}_\Omega(x,y)$ with the above boundary data, if it exists, is identical with $\beta_\Omega(x,y)$, given by the above integral, where $\Delta\bar{\beta}_\Omega(\cdot,y) = H_\Omega(\cdot,y)$. However, to show that this β_Ω implies the existence of $\bar{\beta}_\Omega$ or, equivalently, to show that this β_Ω actually satisfies the above boundary conditions, requires deep analysis and is <u>not</u> in the plan of the present book. Here we choose the following easy approach. It is a classical result that $\bar{\beta}_\Omega(x,y)$ exists on any regular subregion Ω (e.g., Miranda [1]). Let $\bar{H}_\Omega(\cdot,y) = \Delta\bar{\beta}_\Omega(\cdot,y)$. We have seen in §1 that $\bar{H}_\Omega(\cdot,y)$ satisfies the conditions in Definition 1.2; in particular,

$$\bar{\beta}_\Omega(x,y) = \int_\Omega \bar{H}_\Omega(\xi,x)*\bar{H}_\Omega(\xi,y) .$$

By the uniqueness of H_Ω, we have $\bar{H}_\Omega \equiv H_\Omega$, hence $\bar{\beta}_\Omega \equiv \beta_\Omega$, and β_Ω satisfies the above boundary conditions.

We reword the conclusion:

<u>The biharmonic Green's function</u> $\beta_\Omega(x,y)$ <u>on any regular subregion</u> Ω <u>of</u> R <u>in the sense of the definition in §1 is identical with the biharmonic</u>

<u>fundamental function which satisfies boundary conditions</u> $\beta_\Omega = \partial\beta_\Omega/\partial n = 0$ <u>on</u> $\partial\Omega$.

For convenience, we extend $\beta_\Omega(\cdot,y)$ and $H_\Omega(\cdot,y)$ to all of R by the convention $\beta_\Omega(\cdot,y) = H_\Omega(\cdot,y) = 0$ on $R - \Omega$.

<u>NOTES TO §4</u>. We will devote a separate section, §5, to the somewhat delicate proof of the last part of Theorem 4.1, the consistency property of β.

§5. β AS A DIRECTED LIMIT

<u>5.1</u>. <u>Consistency</u>. We proceed to the consistency relation:

$$\lim_{\Omega\to R} (\beta(x,y) - \beta_\Omega(x,y)) = 0 ,$$

uniformly on each compact subset of $R \times R$, where $\{\Omega\}$ is a directed set of regular subregions Ω of R with $y \in \Omega$. We start from

$$\beta(x,y) - \beta_\Omega(x,y) = \int_R (H(\xi,x)*H(\xi,y) - H_\Omega(\xi,x)*H_\Omega(\xi,y)) .$$

Since $H(\xi,x) - H_\Omega(\xi,x) \in H_2(\Omega)$ and $H_\Omega(\xi,y)$ is orthogonal to $H_2(\Omega)$,

$$\int_R H(\xi,x)*_\xi H_\Omega(\xi,y) = \int_R H_\Omega(\xi,x)*_\xi H_\Omega(\xi,y)$$

and similarly,

$$\int_R H_\Omega(\xi,x)*_\xi H(\xi,y) = \int_R H_\Omega(\xi,x)*_\xi H_\Omega(\xi,y) .$$

Therefore,

$$\beta(x,y) - \beta_\Omega(x,y) = \int_R (H(\xi,x) - H_\Omega(\xi,x))*(H(\xi,y) - H_\Omega(\xi,y))$$

on $\Omega \times \Omega$. On setting

$$d_\Omega(x) = \int_R *(H(\xi,x) - H_\Omega(\xi,x))^2$$

we obtain, by the Schwarz inequality,

$$|\beta(x,y) - \beta_\Omega(x,y)| \le (d_\Omega(x)d_\Omega(y))^{1/2}$$

on $\Omega \times \Omega$.

5.2. Continuity. Next we show that $d_\Omega(x)$ is continuous on Ω. Let

$$H_\Omega(\xi,x) = g_\Omega(\xi,x) - g_x(\xi) ,$$

with the convention $g_\Omega(\cdot,x) = g_x = 0$ on $R - \Omega$ for any fixed $x \in \Omega$. For $h_\Omega(\xi,x) = H(\xi,x) - H_\Omega(\xi,x)$,

$$\left|\int_R h_\Omega(\xi,x)*_\xi(h_\Omega(\xi,y) - h_\Omega(\xi,x))\right|$$
$$\le \left(\int_R *_\xi h_\Omega(\xi,x)^2\right)^{1/2}\left(\int_R *_\xi(h_\Omega(\xi,y) - h_\Omega(\xi,x))^2\right)^{1/2} .$$

The second factor is dominated by the sum of

$$J_1(y) = \left(\int_R *[(K(\xi,y) - g_\Omega(\xi,y)) - (K(\xi,x) - g_\Omega(\xi,x))]^2\right)^{1/2}$$

and

$$J_2(y) = \|k_y - k_x\| + \|g_y - g_x\|_\Omega .$$

For $u_\Omega(\xi,y) = K(\xi,y) - g_\Omega(\xi,y)$ on $\Omega \times \Omega$ with $y \in \Omega$, $J_1(y)$ is dominated by the sum of

$$\left(\int_\Omega *_\xi(u_\Omega(\xi,y) - u_\Omega(\xi,x))^2\right)^{1/2} \quad \text{and} \quad \left(\int_{R-\Omega} *_\xi(K(\xi,y) - K(\xi,x))^2\right)^{1/2} .$$

By [c], the former tends to zero as $y \to x$, and by [e], so does the latter. Therefore, $J_1(y) \to 0$ as $y \to x$. Since g_Ω is a fundamental kernel on Ω, the relation $H_\Omega(\xi,x) = g_\Omega(\xi,x) - g_x(\xi)$ shows that g_x is the associated functional, and Lemma 3.4 applies both to k_x and g_x. Therefore, $J_2(y) \to 0$

as $y \to x$. We have shown that

$$d_\Omega(x) = \lim_{y\to x} (\beta(x,y) - \beta_\Omega(x,y))$$

for every $x \in \Omega$. By relations $\beta(x,y) = \kappa(x,y) - (k_x,k_y)$, $\beta_\Omega(x,y) = (g_\Omega(\cdot,x),H_\Omega(\cdot,y)) = \gamma_\Omega(x,y) - (g_x,g_y)_\Omega$, and $\kappa(x,y) = \gamma_\Omega(x,y) + w_\Omega(x,y)$, we deduce

$$\beta(x,y) - \beta_\Omega(x,y) = w_\Omega(x,y) - (k_x,k_y) + (g_x,g_y)_\Omega .$$

It follows that

$$d_\Omega(x) = w_\Omega(x,x) - \|k_x\|^2 + \|g_x\|_\Omega^2 .$$

Since $w_\Omega \in C(\Omega \times \Omega)$ (cf. 3.5), $w_\Omega(x,x)$ is continuous on Ω. By Lemma 3.4, $\|k_x\|^2$ and $\|g_x\|_\Omega^2$ are also continuous in x on Ω. A fortiori, $d_\Omega \in C(\Omega)$.

<u>5.3</u>. <u>Convergence to zero</u>. We proceed to show that $\{d_\Omega\}$ converges to zero decreasingly on R as regular subregions Ω of R exhaust R. Let Ω' be a regular subregion of R with $\Omega' \supset \Omega$. Since $h_{\Omega'}(\xi,x)$ belongs to $H_2(\Omega') \subset H_2(\Omega)$ and $H_{\Omega'}(\xi,x)$ ($H_\Omega(\xi,x)$, resp.) is orthogonal to $H_2(\Omega')$ ($H_2(\Omega)$, resp.), we see that

$$(h_{\Omega'}(\cdot,x),h_\Omega(\cdot,x)) = \|h_{\Omega'}(\cdot,x)\|^2$$

and therefore,

$$\|h_{\Omega'}(\cdot,x) - h_\Omega(\cdot,x)\|^2 = \|h_\Omega(\cdot,x)\|^2 - \|h_{\Omega'}(\cdot,x)\|^2 ,$$

that is,

$$d_\Omega(x) - d_{\Omega'}(x) = \|h_{\Omega'}(\cdot,x) - h_\Omega(\cdot,x)\|^2 \geq 0$$

for every $x \in \Omega$. This means that $\{d_\Omega\}$ is a decreasing net and that there exists an $h \in H_2(R)$ such that

$$\lim_{\Omega\to R} \|h - h_\Omega(\cdot,x)\| = 0 .$$

We set $F(\cdot,x) = H(\cdot,x) - h$. Then

$$F(\cdot,x) - H_\Omega(\cdot,x) = h_\Omega(\cdot,x) - h$$

shows that $F(\cdot,x) - H_\Omega(\cdot,x) \in H_2(\Omega)$ and

$$\lim_{\Omega\to R} \int_R *_\xi(F(\xi,x) - H_\Omega(\xi,x))^2 = 0 .$$

Thus the second and third relation in Definition 1.2 are satisfied by F. Let $u \in H_2(R) \subset H_2(\Omega)$. Since

$$|(u,F(\cdot,x) - H_\Omega(\cdot,x))| \leq \|u\|\cdot\|F(\cdot,x) - H_\Omega(\cdot,x)\| \to 0$$

as $\Omega \to R$, and $(u,H_\Omega(\cdot,x)) = 0$, we obtain $(u,F(\cdot,x)) = 0$, that is, the last relation in 1.2 is also satisfied by F. By the uniqueness of H, we conclude that $F(\cdot,x) = H(\cdot,x)$, that is, $h \equiv 0$. In view of $\lim_{\Omega\to R} \|h - h_\Omega(\cdot,x)\| = 0$ and $d_\Omega(x) = \|h_\Omega\|^2$, it follows that

$$\lim_{\Omega\to R} d_\Omega(x) = 0$$

for every fixed $x \in R$. Since $\{d_\Omega\} \subset C(K)$, with K a compact subset of R and $\Omega \supset K$, is a decreasing net converging to zero, Dini's theorem yields the uniform convergence of $\{d_\Omega\}$ to zero on K.

In view of the last relation in 5.1, we infer that

$$\lim_{\Omega\to R} |\beta(x,y) - \beta_\Omega(x,y)| = 0 ,$$

uniformly on each compact subset of $R \times R$.

The proof of Theorem 4.1 is herewith complete.

5.4. Existence only. It is obvious that for the mere existence of β, it is sufficient to assume the existence of a $K(x,y)$ with properties [a], [d], and a part of [c], viz., $K(\cdot,y) - g_\Omega(\cdot,y)$ is harmonic at y. The other conditions in the theorem are to assure the joint continuity and the consistency property. Since H, if it exists, satisfies [a], [d], and the above part of [c], these conditions are also necessary:

COROLLARY. For the mere existence of β, it is necessary and sufficient that there exists a kernel K with properties [a], [d], and $K(\cdot,y) - g_\Omega(\cdot,y) \in H(\Omega)$.

NOTES TO §5. Corollary 5.4, as the rest of the results in this chapter, were obtained in Nakai-Sario [20].

§6. EXISTENCE OF β ON HYPERBOLIC MANIFOLDS

6.1. Hyperbolicity. In view of Theorem 4.1, it is essential to determine whether or not a fundamental kernel exists on R. We shall now prove that an R satisfying condition [*] in 1.3 admits a fundamental kernel. This result together with Theorem 4.1 implies the Main Theorem in 1.3.

To prove that [*] entails the existence of a fundamental kernel, it is convenient to consider separately the cases of hyperbolic and parabolic manifolds. We recall the definitions. For $y \in R$, let $\{\Omega\}$ be the directed set of regular subregions Ω of R with $y \in \Omega$. Since $\{g_\Omega(x,y)\}$ is an increasing net, Harnack's principle assures that either $\lim_\Omega g_\Omega(x,y) < \infty$ on $R \times R$ off the diagonal or $\lim_\Omega g_\Omega(x,y) = \infty$ on $R \times R$. In the former case, R is called hyperbolic and

$$g(x,y) = \lim_{\Omega\to R} g_\Omega(x,y)$$

is referred to as the harmonic Green's function on R. By 3.1, we see that $g(\cdot,y) \in H(R - y)$, $\Delta g(\cdot,y) = \delta_y$, and $g(x,y) = g(y,x)$.

In the latter case, we say that R is called parabolic. Let Ω_0 and Ω be regular subregions of R with $\bar{\Omega}_0 \subset \Omega$, and denote by w_Ω the harmonic function on $\Omega - \bar{\Omega}_0$ with boundary values 1 on $\partial\Omega_0$ and 0 on $\partial\Omega$. Since $\{w_\Omega\}$ is an increasing net as Ω exhausts R, $w = \lim_{\Omega\to R} w_\Omega$ is harmonic on $R - \bar{\Omega}_0$, with boundary values 1 on $\partial\Omega_0$. We set $w = 1$ on $\bar{\Omega}_0$ and have

$0 < w \leq 1$ on R. By Stokes' formula,

$$\int_{\partial(\Omega-\overline{\Omega}_0)} (g_\Omega(\cdot,y)*dw_\Omega - w_\Omega*dg_\Omega(\cdot,y)) = 0$$

for $y \in \Omega_0$, and a fortiori,

$$\int_{\partial\Omega_0} g_\Omega(\cdot,y)*dw_\Omega = -1.$$

Therefore, R is parabolic if and only if

$$\int_{\partial\Omega_0} *dw = \lim_{\Omega\to R} \int_{\partial\Omega_0} *dw_\Omega = 0.$$

Since $*dw \leq 0$ on $-\partial\Omega_0$, this is equivalent to

$$\lim_{\Omega\to R} w_\Omega = w \equiv 1 .$$

6.2. Existence of fundamental kernel. We now prove that, on a hyperbolic R, condition [*] implies the existence of a fundamental kernel. We actually show:

The following three statements are equivalent by pairs for a hyperbolic R: (i) R satisfies condition [*]; (ii) the harmonic Green's function $g(x,y)$ on R is a fundamental kernel on R; (iii) $g(\cdot,y)$ is square integrable on R off any neighborhood of y.

The implications (ii) → (iii) → (i) are trivial, and we only have to prove that (i) implies (ii). Assume that condition [*] is satisfied by R. Fix an arbitrary $y \in R$ and choose a regular subregion Ω of R such that $\Omega \supset (R - S) \cup \{y\}$. Set

$$c = \left(\max_{x\in\partial\Omega} g(x,y)\right) \Big/ \left(\min_{x\in\partial\Omega} h(x)\right) .$$

Then by $g = \lim_{\Omega\to R} g_\Omega$, we have $g(\cdot,y) \leq ch$ on $R - \Omega$, and, in view of

$\int_S {}^*h(x)^2 < \infty$,

$$\int_{R-\Omega} {}^*_\xi g(\xi,y)^2 < \infty .$$

If this is valid for some regular subregion Ω with $y \in \Omega$, then it holds for every such Ω, and condition [d] in 3.2 for a fundamental kernel is satisfied by $g(x,y)$. Conditions [a] and [b] are trivially valid. To see that [c] is satisfied, note that the function $u(x,y) = g(x,y) - g_\Omega(x,y)$ is positive and symmetric on $\Omega \times \Omega$, and $u(\cdot,y) \in H(\Omega)$. As a consequence of the Harnack inequality, $u \in C(\Omega \times \Omega)$, and g satisfies [c].

It remains to show that g satisfies [e]. Fix a parametric ball $B = \{|y - \eta| < 1\}$ with $\bar{B} \subset \Omega$ and let $c \geq 1$ be the Harnack constant for $\bar{B}$, that is,

$$c^{-1}g(x,\eta) \leq g(x,y) \leq cg(x,\eta)$$

for every $y \in \bar{B}$ and $x \in R - \Omega$. Then

$$|g(x,y) - g(x,\eta)|^2 \leq (c-1)^2 g(x,\eta)^2$$

for every $x \in R - \Omega$ with $y \in \bar{B}$. For each fixed $x \in R - \Omega$, the left-hand side tends to zero as $y \to \eta$. Therefore, the Lebesgue theorem yields

$$\lim_{y\to\eta} \int_{R-\Omega} {}^*_x|g(x,y) - g(x,\eta)|^2 = 0 ,$$

and g satisfies [e]. Thus g is a fundamental kernel on R.

6.3. Existence of β. On combining the results in 6.2 and Theorem 4.1 we obtain:

COROLLARY. *If R admits a harmonic Green's function square integrable on R off the pole, then the biharmonic Green's function $\beta(x,y)$ with the following properties exists on R: $\beta(x,y)$ is continuous on $R \times R$ in the extended sense and $\{\beta(x,y) - \beta_\Omega(x,y)\}$ converges to zero uniformly on each* compact subset of $R \times R$ as the regular subregions Ω with $y \in \Omega$ exhaust R.

To existence and the uniform convergence of $\{\beta(x,y) - \beta_\Omega(\cdot,y)\}$ to zero on each compact subset of R for each fixed $y \in R$ were proved in Ralston-Sario [1].

NOTES TO §6. We know from VIII.1.4 that the square integrability of the harmonic Green's function off its pole is identical with the square integrability of the harmonic measure ω that takes the value 1 on a compact set. On the other hand, by VIII.1.5, $\omega \in L^2$ characterizes the existence of γ. We shall see later that there exist manifolds which carry β but no γ. Thus the condition in Corollary 6.3 is sufficient but not necessary for the existence of β. We shall return to this question in §8, with a view not only to the existence of β but also its joint continuity and consistency.

§7. EXISTENCE OF β ON PARABOLIC MANIFOLDS

7.1. Imitation problem. We shall now prove that condition [*] in 1.3 for R implies the existence of a fundamental kernel on R even if R is parabolic. We start by showing the existence of a function $p(\cdot,y) \in H(R - y)$ for each $y \in R$ for which the difference $p(\cdot,y) - g_B(\cdot,y)$ belongs to $H(B)$, with B any parametric ball $|x - y| < 1$, and which imitates the behavior of $-h$ in the sense that

$$\sup_{x \in R-\Omega} |p(x,y) + h(x)| < \infty$$

for any regular subregion $\Omega \supset (R - S) \cup \bar{B}$.

By taking S smaller, if necessary, we may assume in condition [*] that $R - \bar{S}$ is a regular subregion and h is harmonic on an open set containing $\bar{S}$. Let Ω be a regular subregion with $R - S \subset \Omega$ and h_Ω the harmonic function on $\Omega \cap S$ with boundary values $h_\Omega = h$ on ∂S and $h_\Omega = 0$ on $\partial\Omega$. Since $0 \le h_\Omega \le h$ on $\Omega \cap S$, $\{h_\Omega\}$ is an increasing net as Ω exhausts R, and a fortiori $h_0 = \lim_{\Omega \to R} h_\Omega$ is harmonic on S, with $0 \le h_0 \le h$ on S. By Stokes' formula,

$$\int_{\partial S} (w_\Omega {*dh_\Omega} - h_\Omega {*dw_\Omega}) = 0,$$

where w_Ω is as in 6.1 with $\Omega_0 = R - \overline{S}$. Hence,

$$\left| \int_{\partial S} {*dh_\Omega} \right| \leq \left(\sup_{\partial S} h \right) \cdot \left| \int_{\partial S} {*dw_\Omega} \right|,$$

and by 6.1,

$$\int_{\partial S} {*dh_0} = \lim_{\Omega \to R} \int_{\partial S} {*dh_\Omega} = 0.$$

Therefore,

$$\int_{\partial S} {*dh} = \int_{\partial S} {*d(h - h_0)} > 0,$$

and we may assume that

$$\int_{\partial S} {*dh} = 1.$$

To see that h id unbounded, it suffices to show that $h_1 = h - h_0$ is unbounded. By Stokes' formula,

$$\int_{\partial(\Omega \cap S)} (h_1 {*dw_\Omega} - w_\Omega {*dh_1}) = 0$$

and thus, if $\sup_S h_1 = c < \infty$,

$$\left| \int_{\partial S} {*dh_1} \right| = \left| \int_{\partial \Omega} h_1 {*dw_\Omega} \right| \leq c \left| \int_{\partial \Omega} {*dw_\Omega} \right| = c \left| \int_{\partial S} {*dw_\Omega} \right|.$$

By 6.1, we arrive at the desired contradiction $\int_{\partial S} {*dh_1} = 0$.

7.2. Principal functions. We proceed to the construction of $p(\cdot, y)$.

Fix a regular subregion $\Omega \supset (R - S) \cup \overline{B}$ and set $A_0 = R - \overline{\Omega}$, $A = A_0 \cup (B - y)$, $\alpha = \partial A$. Define $s \in H(A)$ by $s|A_0 = -h$ and $s\epsilon(B - y) = g_B(\cdot, y)$. By $\int_{\partial S} {*dh} = 1$,

$$\int_\alpha {*ds} = 0.$$

Let $\{R_n\}$ be an increasing sequence of regular subregions of R, $n = 1,2,\ldots,$ with $R_1 \supset \overline{\Omega}$, and $\{B_n\}$ the sequence of concentric balls $|x - y| < 1/(n + 1)$, $n = 1,2,\ldots,$ in B. For each $f \in C(\alpha)$, let $L_n f$ be the harmonic function on $A \cap (R_n - \overline{B}_n)$ with boundary values $L_n f = f$ on α and $L_n f = \text{const}$ on $\partial(R_n - \overline{B}_n)$ such that $\int_\beta {*dL_n f} = 0$, where $\beta \subset A$ is a smooth hypersurface homotopic to α. It is easy to see that

$$Lf = \lim_{n\to\infty} L_n f$$

exists, $L : C(\alpha) \to H(A) \cap C(\overline{A})$ is a positive linear operator with $Lf = f$ on α, and $\int_\beta {*dLf} = 0$. Thus L is a <u>normal operator</u> on A and s is the <u>singularity function</u> in the terminology of principal functions (cf. I.1.2) and, by the Main Existence Theorem and $\int_\alpha {*ds} = 0$, there exists a $p(\cdot,y) \in H(R - y)$, the (L,s)-<u>principal function</u>, such that

$$L(p(\cdot,y) - s) = p(\cdot,y) - s$$

on A, and in particular $\sup_A |p(\cdot,y) - s| < \infty$. This implies $p(\cdot,y) - g_B(\cdot,y) \in H(B)$ and $\sup_{x\in R-\Omega} |p(x,y) + h(x)| < \infty$.

<u>7.3</u>. <u>Maximum principle</u>. Let F be a subregion of R with a smooth $\partial F \neq \emptyset$, compact or noncompact. We shall make essential use of the following <u>maximum principle</u>: For a $u \in H(F) \cap C(\overline{F})$ with $\sup_F u < \infty$,

$$\sup_F u = \sup_{\partial F} u .$$

To verify this, choose Ω_0 in $R - \overline{F}$ and consider w_Ω as in 6.1. Take the harmonic function ω_Ω on $\Omega \cap F$ with boundary values 1 on $(\partial F) \cap \Omega$ and zero on $(\partial\Omega) \cap F$. Then $w_\Omega \le \omega_\Omega \le 1$ on $\Omega \cap F$ and by the last relation in 6.1,

$$\lim_{\Omega \to R} \omega_\Omega = 1$$

on F. Let $\sup_F u = k$ and $\sup_{\partial F} u = k'$. Then $k' \le k$ and

$$u \le k' + k(1 - \omega_\Omega)$$

on $\partial(\Omega \cap F)$ and thus on $\Omega \cap F$. On letting $\Omega \to R$, we deduce the inequality $u \le k'$ on F and thus $k \le k'$. The assertion $k = k'$ follows.

7.4. Generalization of Evans kernel. We call a kernel function $E(x,y)$ on R an h-kernel on R if the following conditions are satisfied: $E(\cdot,y) \in H(R - y)$ for every $y \in R$; $E(x,y) = E(y,x)$ for every pair of points x and y in R; $E(\cdot,y) - g_\Omega(\cdot,y) \in H(\Omega)$ for every regular subregion Ω of R with $y \in \Omega$; $\sup_{R-\Omega} |E(\cdot,y) + h| < \infty$ for every regular subregion Ω of R such that h is defined on $R - \Omega$. If $\lim h = \infty$ at the ideal boundary of R, then the h-kernel is called an Evans kernel. The existence of an Evans kernel and its joint continuity were proved in Nakai [6] (see also Sario-Nakai [1]). As a generalization, we shall establish here the existence of an h-kernel, and its joint continuity.

For the existence proof we use the function $p(\cdot,y)$ constructed in 7.2. We fix a point $y_0 \in R$ and choose, using the Sard theorem, an increasing divergent sequence $\{\rho_n\}$ of positive numbers ρ_n, $n = 1,2,\ldots$, such that the level surface $p(\cdot,y_0) = -\rho_n$ is smooth for each $n = 1,2,\ldots$. The set

$$R_n = \{x \in R | p(x,y_0) > -\rho_n\}$$

may or may not be relatively compact. Since $y_0 \in R_n$, R_n is nonempty for every n, and $\{R_n\}$ is an increasing sequence exhausting R. Moreover, R_n is connected. In fact, if there is a component R_n' of R_n with $y_0 \notin R_n'$, then, since $p(\cdot,y_0)$ is bounded from below on R_n', the maximum principle in 7.3

implies that $p(\cdot, y_0) \equiv -\rho_n$ on R'_n, a contradiction. Thus each R_n is a subregion of R with a smooth, not necessarily compact, relative boundary ∂R_n. In view of $\partial R_n \neq \emptyset$ and, for regular regions Ω exhausting R,

$$g_n(\cdot, y) = g_{R_n}(\cdot, y) = \lim_{\Omega \to R} g_{\Omega \cap R_n}(\cdot, y) ,$$

R_n is hyperbolic. Set

$$E_n(x,y) = g_n(x,y) - \rho_n$$

on $R_n \times R_n$ and let $n(y)$ be the smallest integer > 0 such that y_0 and y belong to $R_{n(y)}$. By $\sup_{x \in R - \Omega} |p(x,y) + h(x)| < \infty$,

$$c(y) = \sup_{x \in R - R_{n(y)}} |p(x,y) - p(x,y_0)| < \infty .$$

Since $p(\cdot, y) - E_n(\cdot, y)$ is harmonic and bounded on R_n, $n \geq n(y)$, the maximum principle in 7.3 implies that

$$\sup_{x \in R_n} |E_n(x,y) - p(x,y)| = \sup_{x \in \partial R_n} |E_n(x,y) - p(x,y)| = \sup_{x \in \partial R_n} |-\rho_n - p(x,y)|$$

$$= \sup_{x \in \partial R_n} |p(x,y_0) - p(x,y)| = c(y)$$

and, therefore,

$$|E_n(x,y) - p(x,y)| \leq c(y)$$

on R_n. It follows that

$$|E_{n+k}(x,y) - E_n(x,y)| \leq 2c(y)$$

for every $x \in R_n$, $n \geq n(y)$, and every $k = 1,2,\ldots$. Consequently, any subfamily of $\{E_n(\cdot, y)\}$ contains a subsequence uniformly convergent on each compact subset of R.

Let D be a countable dense subset of R. Using the Cantor diagonal process, we can find a subsequence $\{n_k\}$ of positive integers such that

$\{E_{n_k}(\cdot,y)\}$ converges uniformly on each compact subset of R for every $y \in D$. For any $x \in R$, since $E_{n_k}(\cdot,x) = E_{n_k}(x,\cdot)$, $\{E_{n_k}(\cdot,x)\}$ is convergent on D. Again by the above inequality, it is uniformly convergent on any compact subset of R. Thus,

$$E(x,y) = \lim_{k\to\infty} E_{n_k}(x,y)$$

exists for every $(x,y) \in R \times R$, $E(x,y) = E(y,x)$, $E(\cdot,y) \in H(R - y)$ and $|E(\cdot,y) - p(\cdot,y)| \leq c(y)$ on R_n, $n \geq n(y)$, and hence on R.

Suppose $\{m_k\}$ is another subsequence of positive integers of the same kind and set

$$E'(x,y) = \lim_{k\to\infty} E_{m_k}(x,y) .$$

Then $a(\cdot,y) = E(\cdot,y) - E'(\cdot,y)$ is bounded on R, since

$$|a(\cdot,y)| \leq |E(\cdot,y) - p(\cdot,y)| + |E'(\cdot,y) - p(\cdot,y)| \leq 2c(y) .$$

Thus $a(\cdot,y)$ is a bounded harmonic function on R and the maximum principle in 7.3 shows that $a(\cdot,y)$ is a constant $a(y)$ on R, that is, for any $x \in R$,

$$a(y) = E(y,x) - E'(y,x) .$$

By the same reasoning, $a(y)$ is a bounded harmonic function on R and $a(y)$ is a constant a on R. Thus,

$$E(x,y) - E'(x,y) = a$$

on $R \times R$. Observe that $E_n(x,y_0) = p(x,y_0)$ and therefore,

$$a = E(x,y_0) - E'(x,y_0) = p(x,y_0) - p(x,y_0) = 0 .$$

We have established the existence of

$$E(x,y) = \lim_{n\to\infty} E_n(x,y)$$

on $R \times R$. It is easily verified that $E(x,y)$ is an h-kernel.

7.5. Continuity. Next we prove that $E(x,y)$ is continuous on $R \times R$ in the extended sense, that is, $E(x,y)$ is finitely continuous on $R \times R$ off the diagonal set and $\lim_{x,y\to\xi} E(x,y) = E(\xi,\xi)$.

Fix a regular subregion Ω of R and a point $\eta \in \Omega$. We consider

$$\rho(y) = \max_{x\in\partial\Omega} |E(x,y) - E(x,\eta)|$$

for $y \in \Omega$. By the maximum principle in 7.3,

$$|E(x,y) - E(x,\eta)| \le \rho(y)$$

for every $x \in R - \Omega$ with a fixed $y \in \Omega$. We now maintain:

$$\lim_{y\to\eta} \rho(y) = 0 .$$

Suppose this is not the case. Then there exists a sequence $\{y_n\}$ of points in Ω such that $\{y_n\} \to \eta$, $\rho(y_n) > 0$, and $\lim_{n\to\infty} \rho(y_n) > 0$. Choose a regular subregion Ω' such that $\{y_n\} \cup \{\eta\} \subset \Omega' \subset \bar{\Omega}' \subset \Omega$. The function $v_n(x) = (E(x,y_n) - E(x,\eta))/\rho(y_n)$ is harmonic on $R - \Omega'$, and $|v_n(x)| \le 1$ on $R - \Omega$. For $u_n(x) = (g_\Omega(x,y_n) - g_\Omega(x,\eta))/\rho(y_n)$, the functions $\pm u_n + 1 \mp v_n$ are harmonic and nonnegative on Ω. Thus

$$-1 + \min_{\partial\Omega'} u_n \le v_n \le 1 + \max_{\partial\Omega'} u_n$$

on $\Omega - \Omega'$ and, consequently, on $R - \Omega'$. Since g_Ω is continuous on $\Omega \times \Omega$ and $\{1/\rho(y_n)\}$ is bounded, the above bounds of v_n tend to finite limits as $n \to \infty$. Thus $\{v_n\}$ is uniformly bounded on $R - \Omega'$. Let x be an arbitrary but fixed point in $R - \Omega'$. By the definition of $E_n(x,y)$, it is symmetric and so is, by $E = \lim E_n$, $E(x,y)$. Since, therefore, $E(x,y_n) \to E(x,\eta)$ as $n \to \infty$, and $\{1/\rho(y_n)\}$ is bounded, $\lim_{n\to\infty} v_n(x) = 0$. Thus $\{v_n\}$ converges to zero uniformly on each compact subset of $R - \Omega'$ and, in particular, on $\partial\Omega$. This is, however, impossible since $\max_{x\in\partial\Omega} |v_n(x)| = \rho(y_n)/\rho(y_n) = 1$.

Let $(\xi,\eta) \in R \times R$ with $\xi \neq \eta$, and choose a regular subregion Ω of R with $(\xi,\eta) \in (R - \bar{\Omega}) \times \Omega$. For $(x,y) \in (R - \bar{\Omega}) \times \Omega$, $|E(x,y) - E(x,\eta)| \leq \rho(y)$ yields

$$|E(x,y) - E(\xi,\eta)| \leq |E(x,\eta) - E(\xi,\eta)| + \rho(y) .$$

By $\lim_{y\to\eta} \rho(y) = 0$, we conclude that

$$\lim_{(x,y)\to(\xi,\eta)} E(x,y) = E(\xi,\eta) ,$$

that is, $E(x,y)$ is finitely continuous on $R \times R$ off the diagonal set.

For an arbitrary regular subregion Ω of R, let

$$v(x,y) = E(x,y) - g_\Omega(x,y)$$

on $\Omega \times \Omega$. To deduce the continuity of $E(x,y)$ on $R \times R$ in the extended sense, it suffices to show that $v \in C(\Omega \times \Omega)$. We have already seen that v is continuous on $\Omega \times \Omega$ off the diagonal set. Fix an arbitrary $\xi \in \Omega$ and let Ω' be a regular subregion with $\xi \in \Omega' \subset \bar{\Omega}' \subset \Omega$. Given an $\varepsilon > 0$, we can find an open neighborhood U of ξ such that $\bar{U} \subset \Omega'$ and

$$v(x,\xi) - \varepsilon < v(x,y) < v(x,\xi) + \varepsilon$$

for every $(x,y) \in (\partial\Omega') \times U$. By the maximum principle, this remains true on $\bar{\Omega}' \times U$ and, in particular, on $U \times U$. Therefore,

$$|v(x,y) - v(\xi,\xi)| \leq |v(x,\xi) - v(\xi,\xi)| + \varepsilon .$$

Since, trivially, $|v(x,\xi) - v(\xi,\xi)| \to 0$ as $x \to \xi$, we conclude that

$$\lim_{(x,y)\to(\xi,\xi)} v(x,y) = v(\xi,\xi) .$$

7.6. Fundamental kernel. Let h be the function in condition [*] on a parabolic R. We shall prove that any h-kernel $E(x,y)$ on R is a fundamental kernel in the sense of 3.2. Conditions [a] and [b] are trivially satisfied by E. By 7.5, E also satisfies [c]. To prove [d] for every $y \in R$, let Ω be a regular subregion of R such that $\Omega \supset (R - S) \cup \{y\}$. The constant

$$c = \left(\sup_{R-\Omega} |E(\cdot,y) + h|\right)/\inf_{\partial\Omega} h$$

is finite and, by the maximum principle in 7.3,

$$-ch \le E(\cdot,y) + h \le ch$$

on $R - \Omega$. Thus there exists a constant c_1 such that

$$E(\cdot,y)^2 \le c_1 h^2$$

on $R - \Omega$, hence

$$\int_{R-\Omega} *_\xi E(\xi,y)^2 \le c_1 \int_{R-\Omega} *h(\xi)^2 < \infty ,$$

and E satisfies [d].

To prove [e] for E, choose an arbitrary $\eta \in R$ and take any regular subregion Ω of R containing η. Let U be a neighborhood of η with $\overline{U} \subset \Omega$. By the finite continuity of E on $(\partial\Omega) \times \overline{U}$,

$$c = \left(\sup_{(x,y)\in(\partial\Omega)\times\overline{U}} |E(x,y) - E(x,\eta)|\right)/\inf_{\partial\Omega} h < \infty .$$

Thus,

$$-ch(x) \le E(x,y) - E(x,\eta) \le ch(x)$$

for $(x,y) \in (\partial\Omega) \times \overline{U}$, and since $|E(x,y) - E(x,\eta)|$ is bounded on $R - \Omega$ and $h > 0$, the maximum principle in 7.3 assures that these inequalities remain valid for every $(x,y) \in (R - \Omega) \times \overline{U}$. A fortiori, there exists a constant c_1 such that

$$|E(x,y) - E(x,\eta)|^2 \le c_1 h(x)^2$$

for $(x,y) \in (R - \Omega) \times \overline{U}$. For each $x \in R - \Omega$, $E(x,y) - E(x,\eta) \to 0$ as $y \to \eta$. By Lebesgue's theorem,

$$\lim_{y\to\eta} \int_{R-\Omega} *|E(x,y) - E(x,\eta)|^2 = 0 ,$$

and E satisfies [e]. We have proved that any h-kernel with h in [*] is a fundamental kernel on R.

7.7. Unboundedness of h. We know from 7.1 that h in [*] is unbounded. Conversely, assume that $h > 0$ is harmonic on a neighborhood of the ideal boundary of R and unbounded on any neighborhood of the ideal boundary of R. Let Ω be a regular subregion of R such that h is defined on $R - \Omega$. Take the normal operator L for $A = R - \Omega$ (cf. 7.2). If $\int_{\partial\Omega} *dh = 0$, then there exists a harmonic function p on R such that $L(p - h) = p - h$. By adding a constant, we may assume $p > 0$. Since R is parabolic, p must be constant (e.g., Sario-Nakai [1]). Thus h is bounded, a contradiction. We have seen:

The last condition in [*] is equivalent to the unboundedness of h on any neighborhood of the ideal boundary of R.

7.8. Existence of β. By virtue of the results in 7.1-7.4 and Theorem 4.1, we have:

COROLLARY. If R is parabolic and there exists a square integrable positive harmonic function on a neighborhood of the ideal boundary of R, unbounded on any neighborhood of the ideal boundary of R, then the biharmonic Green's function $\beta(x,y)$ with the following properties exists on R: $\beta(x,y)$ is continuous on $R \times R$ in the extended sense and $\{\beta(x,y) - \beta_\Omega(x,y)\}$ converges to zero uniformly on each compact subset of $R \times R$ as the regular subregions Ω with $y \in \Omega$ exhaust R.

NOTES TO §7. The existence of $\beta(x,y)$ and the uniform convergence of $\{\beta(x,y) - \beta_\Omega(x,y)\}$ to zero on each compact subset of R for each fixed $y \in R$ were proved in Nakai-Sario [12] under the additional requirement that $\lim h = \infty$ at the ideal boundary of R. In Corollary 7.8, we only assume that the cluster set of h at the ideal boundary contains ∞, and the condition $\lim h = \infty$ at the ideal boundary of R is not needed.

§8. EXAMPLES

8.1. Euclidean N-space. We saw in Corollary 6.3 that if

$$\int_{R-\Omega} *g(x,y)^2 < \infty ,$$

with Ω any regular subregion of R containing y, then the biharmonic Green's function β exists on R and is continuous (in the extended sense) on $R \times R$, and $\lim_{\Omega\to R} \beta_\Omega = \beta$ uniformly on each compact subset of $R \times R$. We referred to this last relation as the consistency of β on $R \times R$. We shall consider the Euclidean space E^N of dimension $N \geq 2$ from the view point of the above assertion, in particular the condition $\int_{R-\Omega} *g^2 < \infty$. We recall that E^N is hyperbolic if and only if $N \geq 3$, and thus we exclude the case $N = 2$ for the present. Since

$$g_{E^N}(x,y) = |x - y|^{2-N}, \qquad N \geq 3 ,$$

we see that

$$\int_{a<|x-y|<\infty} *g_{E^N}(x,y)^2 = cA_N \int_a^\infty r^{3-N} dr, \qquad a > 0 ,$$

with $*1 = dx^1 \cdots dx^N$, and A_N the area of the unit sphere in E^N; here and later we suppress some multiplicative constants irrelevant in our reasoning. The above integral is finite if and only if $N \geq 5$. Thus we have:

The biharmonic Green's function β_{E^N} on the Euclidean space E^N exists for $N \geq 5$, and in this case β_{E^N} is continuous on $E^N \times E^N$ and consistent on $E^N \times E^N$, that is, $\lim_{\Omega\to E^N} \beta_\Omega = \beta_{E^N}$ uniformly on each compact subset of $E^N \times E^N$.

8.2. Dimensions 3 and 4. We had to exclude E^3 and E^4 in the above assertion since our discussion was based on $\int_{R-\Omega} {*g^2} < \infty$. This does not yet mean that there exists no β_{E^N} for $N = 3$ and 4. However, the nonexistence can be seen by contradiction as follows. Suppose β_{E^N} exists for $N = 3$ or 4. Then the density $H_{E^N}(x,y)$ exists. Since E^N is translation invariant, we have

$$H_{E^N}(x,y) = H_{E^N}(x - y, 0),$$

and the existence or nonexistence of $H_{E^N}(\cdot,y)$ is completely determined by that of $H_{E^N}(\cdot,0)$. This remark applies, of course, to every dimension. In the cases $N = 3$ and 4, we consider

$$h_N(r) = \int_S H_{E^N}((r,\theta),0)d\theta ,$$

where $(r,\theta) = (r,\theta^1,\ldots,\theta^{N-1})$ are the polar coordinates on E^N, and $d\theta$ is the surface element on the unit sphere S in E^N. The function $h_N(x) = h_N(|x|)$ is harmonic on $E^N - 0$. Since $H_{E^N}(x,0) - |x|^{2-N}$ is bounded in the vicinity of $x = 0$, the same is true of $h_N(r) - r^{2-N}$ in the vicinity of $r = 0$. On the other hand, by the Schwarz inequality,

$$h_N(r)^2 \leq A_N \int_S H_{E^N}((r,\theta),0)^2 d\theta$$

and therefore,

$$\int_a^\infty h_N(r)^2 r^{N-1} dr \leq A_N \int_{|x|>a} {*H_{E^N}(x,0)^2} < \infty, \qquad a > 0 .$$

Since $\int_a^\infty r^{N-1}dr = \infty$, $\liminf_{r\to\infty} h_N(r)^2 = 0$. By the subharmonicity of $h_N(x)^2 = h_N(|x|)^2$ and the maximum principle, we conclude that $\lim_{r\to\infty} h_N(r) = 0$. Thus $h_N(x) - |x|^{2-N}$ is a bounded harmonic function on E^N tending to zero as $|x| \to \infty$. Therefore, $h_N(r) = r^{2-N}$, but this is impossible in view of the above inequality. We have proved:

<u>There exists no biharmonic Green's function</u> β_{E^N} <u>on</u> E^N <u>for</u> $N = 3$ <u>and</u> 4.

8.3. Complement of unit ball. In contrast with the conclusions in 8.1 and 8.2, the following observation may be of interest. We consider the complement $\Sigma^N = \{|x| > 1\}$ of the closed unit ball in E^N, $N \geq 2$. We shall prove in 8.4-8.6:

The biharmonic Green's function β_{Σ^N} exists on Σ^N for every dimension $N \geq 2$, is continuous on $\Sigma^N \times \Sigma^N$ and consistent on $\Sigma^N \times \Sigma^N$, that is, $\lim_{\Omega \to \Sigma^N} \beta_\Omega = \beta_{\Sigma^N}$ uniformly on each compact subset of $\Sigma^N \times \Sigma^N$.

The assertion is of no interest if $N \geq 5$, because $g_{\Sigma^N} \leq g_{E^N}$ on Σ^N, so that $\int_{R-\Omega} *g^2 < \infty$ follows for g_{Σ^N} from that for g_{E^N}, and the above statement contains nothing beyond the result in 8.1 for $N \geq 5$. The relation $g_{\Sigma^N} \leq g_{E^N}$ itself is true also for $N = 3$ and 4, but this is of no use inasmuch as g_{E^N} is not square integrable near the point at infinity. The very interesting aspect here is that the situation is no better for g_{Σ^N}, that is,

$$\int_{\Sigma^N - \Omega} *g_{\Sigma^N}(x,y)^2 = \infty, \qquad N = 2,3,4 ,$$

where Ω is any regular subregion of Σ^N containing y. This is seen at once by

$$g_{\Sigma^N}(x,y) \sim g_{E^N}(x,y)$$

as $|x| \to \infty$, $N = 3,4$, and by $g_{\Sigma^2}(\infty, y) = \log|y| > 0$. In passing, we state the conclusion:

The condition in Corollary 6.3 is only sufficient but in general not necessary.

Thus we cannot have our discussion of the existence of β_{Σ^N}, $N = 2,3,4$, based on $\int_{R-\Omega} *g^2 < \infty$. Nevertheless, we can follow the procedure in 3.2-5.4. We consider the function

$$K(\xi,y) = g_{E^N}(\xi,y) - g_{E^N}(\xi,0) = |\xi - y|^{2-N} - |\xi|^{2-N}$$

on Σ^N for $N = 3,4$; the case $N = 2$ will be treated slightly differently in 8.6. The function $K(\xi,y)$ is <u>not</u> a fundamental kernel on Σ^N, but it shares most of its important properties and the result in §5 will apply. Since we must obtain more than the mere existence of β_{Σ^N}, we next describe the entire procedure.

8.4. Properties of K. We list those properties of the above function $K(\xi,y)$ which will be needed to carry out the construction, almost identical with that in 3.2-5.4:

[α] $K(\cdot,y) \in H(\Sigma^N - y)$, $K(\xi,\cdot) \in H(\Sigma^N - \xi)$.

[β] $K(x,y) - g_\Omega(x,y) \in C(\Omega \times \Omega)$ for every regular subregion Ω of Σ^N.

[γ] $K(\cdot,y)$ is square integrable on Σ^N off any neighborhood of y,

$$\int_{\Sigma^N-\Omega} *K(\xi,y)^2 < \infty ,$$

where Ω is any regular subregion of Σ^N containing y.

[δ] For any point $\eta \in \Sigma^N$ and any regular subregion Ω of Σ^N containing η,

$$\lim_{y\to\eta} \int_{\Sigma^N-\Omega} *|K(\xi,y) - K(\xi,\eta)|^2 = 0 .$$

Properties [α] and [β] are easily verified. To prove [γ] we only have to show that

$$\int_{|\xi|>a} *K(\xi,y)^2 < \infty$$

for $a = |y| + 1$, say. But this is implied by the following readily verified inequalities:

$$K(\xi,y) \le \begin{cases} c|\xi|^{-1}, & N = 3\ , \\ c|\xi|^{-2}, & N = 4\ , \end{cases}$$

with $c = c(y)$ a suitable constant for $|\xi| > a$. Property [δ] is an immediate consequence of the first equation in 8.3 and the Lebesgue convergence theorem.

8.5. Functional k_ξ. We consider

$$k_\xi(u) = \int_{\Sigma^N} u(x){*}K(x,\xi)$$

for $u \in H_2(\Sigma^N)$. By the proof in 3.3 we see that $k_\xi \in H_2(\Sigma^N)$. The proof in 3.4 applies verbatim and gives

$$\lim_{\xi\to\eta} \|k_\xi - k_\eta\| = 0\ .$$

It is readily seen that

$$H_{\Sigma^N}(x,y) = K(x,y) - k_y(x)$$

is the density in Definition 1.2. Thus the biharmonic Green's function exists and is given by

$$\beta_{\Sigma^N}(x,y) = \int_{\Sigma^N} H_{\Sigma^N}(\xi,x){*}H_{\Sigma^N}(\xi,y)\ .$$

By [δ] and $\lim_{\xi\to\eta} \|k_\xi - k_\eta\| = 0$, we easily see that

$$\lim_{x\to y} \int_{\Sigma^N - \Omega} {*}H_{\Sigma^N}(\xi,x)^2 = \int_{\Sigma^N - \Omega} {*}H_{\Sigma^N}(\xi,y)^2$$

for any fixed regular subregion Ω of Σ^N containing y. From this, the continuity (in the genuine sense) of β_{Σ^N} on $\Sigma^N \times \Sigma^N$ off the diagonal set follows. Again by the above equality and the definition of K in 8.3, we see that

$$\beta_{\Sigma^N}(x,y) - \int_\Omega |\xi - x|^{2-N} *_\xi |\xi - y|^{2-N} \in C(\Omega \times \Omega)$$

for any regular subregion Ω of Σ^N, and therefore,

$$\beta_{\Sigma^N}(x,y) - \beta_\Omega(x,y) \in C(\Omega \times \Omega) .$$

We have deduced the continuity of β_{Σ^N} on $\Sigma^N \times \Sigma^N$. The entire discussion in 5.1-5.3 applies verbatim and shows that $\lim_{\Omega \to \Sigma^N} \beta_\Omega = \beta_{\Sigma^N}$ uniformly on each compact subset of $\Sigma^N \times \Sigma^N$.

8.6. Dimension 2. It remains to discuss the case $N = 2$. It will be convenient to view E^2 as the complex plane. We have (cf. 8.1)

$$g_{\Sigma^2}(z,w) = \log\left|\frac{1 - \bar{w}z}{z - w}\right|$$

$$= \log|w| + \sum_{n=1}^{\infty} r^{-n}(a_n(w) \cos n\theta + b_n(w) \sin n\theta)$$

for $z = re^{i\theta}$ with $r \in [1 + |w|, \infty)$, where

$$a_n(w) = (t^n - t^{-n}) \cos ns, \qquad b_n(w) = (t^n - t^{-n}) \sin ns$$

for $w = te^{is}$ and $n = 1,2,\ldots$. We set

$$g_1(re^{i\theta},w) = \log|w| + (a_1(w) \cos \theta + b_1(w) \sin \theta)r^{-1}$$

$$= \log|w| + (t - t^{-1})r^{-1} \cos(\theta - s)$$

and consider

$$K(z,w) = g_{\Sigma^2}(z,w) - g_1(z,w).$$

By means of the above explicit expressions of g_{Σ^2} and g_1, it is easy to see that K satisfies $[\alpha]$-$[\delta]$ in 8.4 and thus the discussion in 8.5, with trivial modifications, such as the replacement of $|\xi - x|^{2-N}$ and $|\xi - y|^{2-N}$ by $\log|\xi - x|^{-1}$ and $\log|\xi - y|^{-1}$ in the next to the last relation in 8.5, can be repeated in the present case.

<u>The biharmonic Green's function</u> β_{Σ^N} <u>exists on</u> Σ^N, $N = 2,3,4$, <u>and is continuous and consistent on</u> $\Sigma^N \times \Sigma^N$, <u>despite the relation</u>

$$\int_{\Sigma^N - \Omega} *g_{\Sigma^N}(\xi, y)^2 = \infty, \qquad N = 2,3,4\ ,$$

<u>where</u> Ω <u>is any regular subregion of</u> Σ^N <u>containing</u> y.

<u>NOTES TO §8</u>. Since for $N = 2,3,4$, g_{E^N} on E^N is not square integrable off its pole, the harmonic measure ω_{E^N} of the ideal boundary of E^N is not square integrable (cf. VIII.1.4) and the same is, therefore, true of the harmonic measure ω_{Σ^N} of the ideal boundary of Σ^N. Thus, by Theorem VIII.1.5, Σ^N carries no γ for $N = 2,3,4$, whereas it does carry β. Hence there exist Riemannian manifolds which carry β and g but no γ. This anticipates the more general result, to be proved in X.§5, that, in the notation there, $O_\gamma^N \cap \tilde{O}_\beta^N \cap \tilde{O}_G^N \neq \emptyset$ for all $N \geq 2$.

CHAPTER X

RELATION OF O_β^N TO OTHER NULL CLASSES

After the discussion of the existence of γ in Chapter VIII and that of β in Chapter IX, the natural question arises: How are the corresponding null classes of Riemannian N-manifolds related? We shall show in §1 that $O_\beta^N \subset O_\gamma^N$ for all $N \geq 2$. In terms of the Evans kernel, we then deduce a sufficient condition for a manifold to carry β. We use this test to prove the relation $O_G^N \cap \tilde{O}_\beta^N \neq \emptyset$, which gives the strict inclusion $O_\beta^N < O_\gamma^N$ for $N > 2$. The case $N = 2$ will be discussed in §4 in its natural context.

In §2, we first consider relations between O_β^N and the harmonic null classes O_{HX}^N, $X = P, B, D, C$. In view of $O_{HP}^N < O_{HB}^N < O_{HD}^N = O_{HC}^N$, we start with O_{HD}^N. We establish a test, which seems to have interest in its own right, for a manifold to belong to O_β^N, and illustrate it by showing that $O_\beta^2 \cap \tilde{O}_{HD}^2 \neq \emptyset$. Using a different technique, viz., that of doubling of a Riemannian manifold, we prove in §3 that $O_\beta^N \cap \tilde{O}_{HD}^N \neq \emptyset$ for every $N \geq 2$.

Since $O_G^N < O_{HP}^N$, the next step in discussing relations between O_β^N and O_{HX}^N is to consider O_G^N and O_β^N. Again we start with the interesting special case $N = 2$ and give, in §4, a complete characterization for a plane with a radial conformal metric to belong to O_β^2. As a corollary, $O_G^2 \cap O_\beta^2 \neq \emptyset$ and $O_G^2 \cap \tilde{O}_\beta^2 \neq \emptyset$. Thus the relation $O_G^N \cap \tilde{O}_\beta^N \neq \emptyset$ and its consequence $O_\beta^N < O_\gamma^N$, proved in §1 for $N > 2$, have been established for every $N \geq 2$.

All remaining relations between O_G^N, O_{HX}^N, O_β^N, and O_γ^N are settled in §5, where we show that the classes $O_G^N \cap O_\beta^N$, $\tilde{O}_G^N \cap O_\beta^N$, and trivially, $\tilde{O}_G^N \cap \tilde{O}_\beta^N$ are nonvoid for every N and that, moreover, we have the strict inclusion

$$O_G^N \cup O_\beta^N < O_\gamma^N.$$

The question as to how O_β^N and O_γ^N are related to the quasiharmonic null classes is immediately answered by

$$O_\beta^N < O_\gamma^N < O_{QP}^N$$

and will not be discussed further. As to relations of O_β^N and O_γ^N to the biharmonic null classes, we believe that there are no inclusions, but we have not carried out the construction of counterexamples.

§1. INCLUSION $O_\beta^N < O_\gamma^N$

In view of the fundamental importance of the relation $O_\beta^N \subset O_\gamma^N$, we present two proofs for it. The first one, in 1.1 - 1.7, is that originally given and uses techniques from the theory of partial differential equation. The second one, in 1.8, is of later vintage and is quite elementary.

We then deduce, in 1.9, a useful sufficient condition for a parabolic manifold to carry β: the square integrability of the Evans kernel off its pole. As a fascinating consequence of this test we note in passing that, for $N = 2,3$, every compact Riemannian manifold punctured at a point carries β. In 1.10, we apply the Evans kernel test to obtain $O_G^N \cap \tilde{O}_\beta^N \neq \emptyset$, hence $O_\beta^N < O_\gamma^N$ for $N > 2$. The case $N = 2$ will be discussed in §4.

<u>1.1. Definitions of</u> O_β^N <u>and</u> $\tilde{O}_\beta^N$. It is time to introduce notation for the classes of Riemannian manifolds which do, or do not, carry β. It would be logical to define $\tilde{O}_\beta^N$ as the class of Riemannian N-manifolds R which carry a β that is jointly continuous on $R \times R$ and has the consistency property $\beta = \lim_{\Omega \to R} \beta_\Omega$. In some cases, however, in which the proofs of joint continuity and consistency, though not difficult, would entail dull and nonconstructive reasoning, we choose to not carry them out. Accordingly, we adopt the following convenient definitions:

$$O_\beta^N = \{R \mid \beta(x,y) \text{ does not exist for some } y \text{ and some exhaustion}\},$$

$$\tilde{O}_\beta^N = \{R \mid \beta(x,y) \text{ exists for some } y \text{ and some exhaustion}\}.$$

Whenever the joint continuity and consistency are verified for some class Z^N, then $(O_\beta^N \cup \tilde{O}_\beta^N) \cap Z^N = Z^N$.

As concrete examples, we consider the Euclidean N-spaces E^N for $N \geq 2$. On exhausting balls $\Omega = \{r < \rho\}$, $\rho \to \infty$, the functions γ_Ω, β_Ω with poles at the origin can be explicitly constructed. We shall prove that $\gamma_\Omega \to \infty$ and $\beta_\Omega \to \infty$ at every point if $N \leq 4$, whereas both $\gamma = \lim_\Omega \gamma_\Omega$ and $\beta = \lim_\Omega \beta_\Omega$ exist on E^N if $N > 4$. This simultaneous existence of γ and β raises the question: Is there some relationship between the existence of γ and β in the general case (and, in particular, on a plate of arbitrary shape)? We shall show that the answer is in the affirmative: <u>on an arbitrarily given Riemannian manifold, if</u> γ <u>exists, so does</u> β.

We recall that the biharmonic fundamental singularity σ at $r = 0$ is $r^2 \log r$ for $N = 2$, r for $N = 3$, $\log r$ for $N = 4$, and r^{-N+4} for $N > 4$, the singularity manifesting itself in that $\sigma \notin C^4$ at the origin. We normalize σ to vanish for $r = \rho$ and then subtract a multiple of the regular biharmonic function $r^2 - \rho^2$ such that the Laplace-Beltrami operator $\Delta = d\delta + \delta d$ acting on γ_Ω gives zero values for $r = \rho$. The resulting γ_Ω on $\bar{\Omega}$ with pole 0 is for $N = 2,3,4,>4$, respectively,

$$\gamma_\Omega = \begin{cases} r^2 \log \frac{r}{\rho} - (r^2 - \rho^2), \\ -r + \rho + \frac{1}{3}\rho^{-1}(r^2 - \rho^2), \\ -\log \frac{r}{\rho} + \frac{1}{4}\rho^{-2}(r^2 - \rho^2), \\ r^{-N+4} - \rho^{-N+4} + N^{-1}(N-4)\rho^{-N+2}(r^2 - \rho^2). \end{cases}$$

We conclude that $\gamma = \lim_{\rho\to\infty} \gamma_\Omega$ does not exist on E^N for $N = 2,3,4$, whereas for $N > 4$ it is

$$\gamma = r^{-N+4}.$$

For β_Ω on Ω the construction is the same except that now a multiple of $r^2 - \rho^2$ is to be subtracted which makes $\partial\beta_\Omega/\partial n = 0$ for $r = \rho$. We obtain for $N = 2,3,4,>4$,

$$\beta_\Omega = \begin{cases} r^2 \log \frac{r}{\rho} - \frac{1}{2}(r^2 - \rho^2), \\ -r + \rho + \frac{1}{2}\rho^{-1}(r^2 - \rho^2), \\ -\log \frac{r}{\rho} + \frac{1}{2}\rho^{-1}(r^2 - \rho^2), \\ r^{-N+4} - \rho^{-N+4} + \frac{1}{2}(N - 4)\rho^{-N+2}(r^2 - \rho^2). \end{cases}$$

As $\rho \to \infty$, we again deduce that $\beta = \lim_{\rho\to\infty} \beta_\Omega$ does not exist on E^N for $N \leq 4$, whereas for $N > 4$ it is

$$\beta = r^{-N+4}.$$

We know from VIII.1.4, IX.8.1, and IX.8.2 that the existence of $\gamma = \gamma(x,y)$ and $\beta = \beta(x,y)$ is independent of (x,y) and the exhaustion.

The biharmonic Green's functions γ and β exist on E^N for precisely the same dimensions: $N > 4$.

This observation suggests an inquiry into a relationship between the existence of γ and β on every Riemannian manifold.

1.2. Operators β_Ω and γ_Ω. Given an arbitrary Riemannian manifold R, let Ω be a regular subregion of R. Take a point $y \in \Omega$ and a geodesic ball B_y about y, $\alpha_y = \partial B_y$, $\alpha_y \cap \partial\Omega = \emptyset$, and orient both α_y and $\partial\Omega$ positively in regard to B_y and Ω, respectively. Let $\beta_\Omega = \beta_\Omega(x,y)$, $\gamma_\Omega = \gamma_\Omega(x,y)$ be the biharmonic Green's functions on $\overline{\Omega}$, with the biharmonic fundamental singularity at y normalized by $\int_{\alpha_y} *d\Delta\beta_\Omega = \int_{\alpha_y} *d\Delta\gamma_\Omega = -1$, and with boundary data

$$\beta_\Omega = \frac{\partial\beta_\Omega}{\partial n} = \gamma_\Omega = \Delta\gamma_\Omega = 0 \quad \text{on} \quad \partial\Omega.$$

For $f \in C_0^\infty(\Omega)$, set

$$(\beta_\Omega f)(x) = \int_\Omega \beta_\Omega(x,y) * f(y), \qquad (\gamma_\Omega f)(x) = \int_\Omega \gamma_\Omega(x,y) * f(y).$$

By general results on elliptic boundary value problems (e.g., Hörmander [1, Ch. X]), there are unique functions $u_\beta, u_\gamma \in C^\infty(\overline{\Omega})$ satisfying

$$\Delta^2 u_\beta = \Delta^2 u_\gamma = f \quad \text{on} \quad \Omega,$$

$$u_\beta = \frac{\partial u_\beta}{\partial n} = u_\gamma = \Delta u_\gamma = 0 \quad \text{on} \quad \partial\Omega.$$

For every u_β, u_γ with these properties,

$$u_\beta = \beta_\Omega f, \qquad u_\gamma = \gamma_\Omega f.$$

In fact, by Stokes' formula,

$$\int_{\partial\Omega - \alpha_y} u_\beta * d\Delta\beta_\Omega - \Delta\beta_\Omega * du_\beta - \beta_\Omega * d\Delta u_\beta + \Delta u_\beta * d\beta_\Omega$$

$$= -\int_{\Omega - B_y} u_\beta * \Delta^2\beta_\Omega - \Delta\beta_\Omega * \Delta u_\beta - \beta_\Omega * f + \Delta u_\beta * \Delta\beta_\Omega.$$

As the geodesic radius of α_y tends to 0, $\int_{-\alpha_y} \to u_\beta(y)$, and we obtain $u_\beta = \beta_\Omega f$. Similarly, the equality

$$\int_{\partial\Omega - \alpha_y} u_\gamma * d\Delta\gamma_\Omega - \Delta\gamma_\Omega * du_\gamma - \gamma_\Omega * d\Delta u_\gamma + \Delta u_\gamma * d\gamma_\Omega$$

$$= -\int_{\Omega - B_y} u_\gamma * \Delta^2\gamma_\Omega - \Delta\gamma_\Omega * \Delta u_\gamma - \gamma_\Omega * f + \Delta u_\gamma * \Delta\gamma_\Omega$$

gives $u_\gamma = \gamma_\Omega f$.

<u>1.3.</u> <u>Monotonicity</u>. Let Ω, Ω' with $\Omega \subset \Omega'$ be regular subregions of R, and set $(f,g) = \int_R f * g$.

<u>LEMMA</u>. <u>For</u> $f \in C_0^\infty(\Omega)$,

$$(f, \beta_\Omega f) \leq (f, \beta_{\Omega'} f).$$

<u>Proof</u>. On Ω, consider the Sobolev space

$$H_{2,\beta}(\Omega) = \{u \mid D^\alpha u \in L^2(\Omega),\ |\alpha| \leq 2,\ u = \frac{\partial u}{\partial n} = 0 \quad \text{on} \quad \partial\Omega\}$$

and the functional on $H_{2,\beta}(\Omega)$,

$$J(u) = \int_\Omega \tfrac{1}{2}\,\Delta u*\Delta u - u*f.$$

For $u \in H_{2,\beta}$, we have the elliptic estimate for $|\alpha| \leq 2$, with C_α a constant,

$$\|D^\alpha u\|_\Omega \leq C_\alpha \|\Delta u\|_\Omega,$$

where $\|\cdot\|_\Omega$ stands for the L^2 norm over Ω. Thus $J(u)$ is bounded from below. Let $\{u_n\}$ be a sequence such that $J(u_n) \to \inf_{H_{2,\beta}} J(u)$. Then $\{\Delta u_n\}$ is a bounded sequence in $L^2(\Omega)$. Using the above elliptic estimate and the Rellich compactness theorem, we may choose a subsequence $\{u_{n_k}\}$ converging to $\bar{u}$ in $L^2(\Omega)$ and converging weakly to $\bar{u}$ in the Sobolev space

$$H_2(\Omega) = \{u \mid D^\alpha u \in L^2(\Omega),\ |\alpha| \leq 2\}.$$

Since $H_{2,\beta}$ is a closed subspace of $H_2(\Omega)$, it follows that $\bar{u} \in H_{2,\beta}$. Now, $J(u_n) \to \inf_{H_{2,\beta}} J(u)$ implies $\lim_{k\to\infty} \|\Delta u_{n_k}\| = \|\Delta \bar{u}\|$. Hence, $\{\Delta u_{n_k}\}$ converges to $\Delta\bar{u}$ in $L^2(\Omega)$ and

$$\inf_{H_{2,\beta}} J(u) = J(\bar{u}).$$

Thus $\bar{u}$ satisfies

$$0 = \int_\Omega \Delta\bar{u}*\Delta\varphi - \varphi*f$$

for all $\varphi \in H_{2,\beta}$. By direct computation, we see that u_β also satisfies this equation. A fortiori,

$$0 = \int_\Omega \Delta(\bar{u} - u_\beta)*\Delta\varphi$$

for all $\varphi \in H_{2,\beta}$. In particular, this holds for $\varphi = \bar{u} - u_\beta$ and thus the above elliptic estimate implies $\bar{u} = u_\beta$. As a consequence,

$$\inf_{u\in H_{2,\beta}} J(u) = J(u_\beta) = \int_\Omega \tfrac{1}{2}\,\Delta u_\beta*\Delta u_\beta - u_\beta*\Delta^2 u_\beta.$$

An integration by parts gives

$$J(u_\beta) = -\tfrac{1}{2}\int_\Omega \Delta u_\beta * \Delta u_\beta .$$

Every $u \in H_{2,\beta}(\Omega)$ can be extended to a function $u \in H_{2,\beta}(\Omega')$ by setting $u = 0$ on $\Omega' - \Omega$. Therefore,

$$\min_{H_{2,\beta}(\Omega)} J(u) \geq \min_{H_{2,\beta}(\Omega')} J(u),$$

that is,

$$\int_\Omega \Delta(\beta_\Omega f) * \Delta(\beta_\Omega f) \leq \int_{\Omega'} \Delta(\beta_{\Omega'} f) * \Delta(\beta_{\Omega'} f),$$

which gives

$$\int_\Omega \Delta^2(\beta_\Omega f) * \beta_\Omega f \leq \int_\Omega \Delta^2(\beta_{\Omega'} f) * \beta_{\Omega'} f.$$

The Lemma follows.

1.4. Comparison. We now compare β_Ω and γ_Ω on the same regular subregion $\Omega \subset R$.

LEMMA. *For* $f \in C_0^\infty(\Omega)$,

$$(f, \beta_\Omega f) \leq (f, \gamma_\Omega f).$$

Proof. For the Sobolev spaces

$$H_{2,\gamma}(\Omega) = \{u \mid D^\alpha u \in L^2(\Omega),\ |\alpha| \leq 2,\ u|\partial\Omega = 0\},$$

$$H_2(\Omega) = \{u \mid D^\alpha u \in L^2(\Omega),\ |\alpha| \leq 2\},$$

we have $H_{2,\beta} \subset H_{2,\gamma} \subset H_2$, and therefore,

$$\min_{H_{2,\beta}(\Omega)} J(u) \geq \min_{H_{2,\gamma}(\Omega)} J(u).$$

An argument completely analogous to that in 1.3, with $H_{2,\beta}$ and u_β replaced by $H_{2,\gamma}$ and u_γ, gives

$$J(u_\gamma) = \inf_{H_{2,\gamma}(\Omega)} J(u).$$

Thus,

$$-\tfrac{1}{2}\int_\Omega \Delta u_\beta * \Delta u_\beta = J(u_\beta) \geq J(u_\gamma) = -\tfrac{1}{2}\int_\Omega \Delta u_\gamma * \Delta u_\gamma.$$

As a consequence,

$$\int_\Omega \Delta u_\beta * \Delta u_\beta \leq \int_\Omega \Delta u_\gamma * \Delta u_\gamma.$$

Here,

$$\int_\Omega \Delta u_\gamma * \Delta u_\gamma = \int_\Omega u_\gamma * f,$$

and we infer that

$$\int_\Omega u_\beta * f \leq \int_\Omega u_\gamma * f,$$

hence the Lemma.

<u>1.5.</u> <u>Exhaustion</u>. Thus far Ω has been a fixed regular subregion of R. We now let Ω exhaust R, and indicate inner products taken over Ω by the subscript Ω. We know from VIII.1.4 that either $\lim_\Omega \gamma_\Omega(x,y) = \infty$ for every (x,y), or else the biharmonic Green's function $\gamma(x,y) = \lim_\Omega \gamma_\Omega(x,y)$ on R exists for every (x,y), with the convergence uniform on compact sets of R. Let $L_0^2(R)$ be the space of L^2 functions on R with compact supports.

<u>LEMMA</u>. <u>If</u> γ <u>exists on</u> R, <u>then as</u> $\Omega \to R$, $(g, \beta_\Omega f)_\Omega$ <u>converges for all</u> $f, g \in L_0^2(R)$.

<u>Proof</u>. Since C_0^∞ is dense in L^2, Lemmas 1.3 and 1.4 remain valid for

$f \in L_0^2(R)$, $\operatorname{supp} f \subset \Omega$. If γ exists, the uniform convergence $\gamma_\Omega \to \gamma$ on compact sets entails the existence of a constant M_f such that

$$(f,\gamma_\Omega f)_\Omega < M_f$$

for $f \in C_0^\infty(R)$ and hence for $f \in L_0^2(R)$. By Lemmas 1.3 and 1.4 for $f \in L_0^2(R)$, $(f,\beta_\Omega f)_\Omega$ converges. For $f,g \in L_0^2(R)$,

$$2(g,\beta_\Omega f)_\Omega = (f + g,\beta_\Omega(f + g))_\Omega - (f,\beta_\Omega f)_\Omega - (g,\beta_\Omega g)_\Omega$$

and therefore, $(g,\beta_\Omega f)_\Omega$ converges.

1.6. Convergence of β_Ω. To prove the convergence of β_Ω in every compact set K of an R which carries γ, we may assume that $y \notin K$. In fact, if $\rho(x,y)$ is the geodesic distance between x and y, we may choose, for sufficiently small constants $\rho_1,\rho_2 > 0$,

$$K = \{x | \rho_1 \le \rho(x,y) \le \rho_2\}.$$

The uniform convergence of β_Ω on K will imply that of the harmonic function $\Delta\beta_\Omega$ on K and hence on $K_0 = \{x | 0 \le \rho(x,y) \le \rho_2\}$. A fortiori, the potential part in the Riesz decomposition of β_Ω converges uniformly on K_0 and so does, by the convergence of β_Ω on K, the harmonic part on K and consequently on K_0.

Having chosen an arbitrary but then fixed compact set K of R, $y \notin K$, we take a compact set K_y, $y \in K_y$, $K \cap K_y = \emptyset$, and a function $\varphi \in C_0^\infty(R)$, $\varphi \ge 0$, $\varphi | K_y = 1$, $K_\varphi = \operatorname{supp} \varphi$ a regular region, $K_\varphi \cap K = \emptyset$, and let Ω henceforth contain $K_\varphi \cup K$. We denote by β_φ the biharmonic Green's function on K_φ and set

$$\tilde{\beta}_\Omega = \beta_\Omega - \varphi\beta_\varphi, \qquad \Delta^2\tilde{\beta}_\Omega = f,$$

where f is independent of Ω. Since $\Delta^2(\tilde{\beta}_\Omega - \beta_\Omega f) = 0$ and $\tilde{\beta}_\Omega - \beta_\Omega f = \partial(\tilde{\beta}_\Omega - \beta_\Omega f)/\partial n = 0$ on $\partial\Omega$, we have

$$\tilde{\beta}_\Omega = \beta_\Omega f.$$

Denote the L^2 norm by $\|\cdot\|$ and choose compact sets K_1, K_2 with

$K \subset \operatorname{int} K_1 \subset K_1 \subset \operatorname{int} K_2 \subset K_2$ and $K_2 \cap K_\varphi = \emptyset$, $K_2 \subset \Omega$.

LEMMA. The L^2 norm $\|\tilde{\beta}_\Omega\|_{K_2}$ is bounded in Ω.

Proof. For the functions $g \in L_0^2(R)$, the restrictions $g|K_2$ form a Banach space B, on which $\tilde{\beta}_\Omega$ gives a bounded linear functional $(g,\tilde{\beta}_\Omega)_{K_2}$,

$$|(g,\tilde{\beta}_\Omega)_{K_2}| \le k_\Omega \|g\|_{K_2}, \qquad k_\Omega = \|\tilde{\beta}_\Omega\|_{K_2}.$$

By the Uniform Boundedness Principle, either there exists a constant k independent of Ω, with

$$|(g,\tilde{\beta}_\Omega)_{K_2}| \le k\|g\|_{K_2}$$

for all $g \in B$, or else

$$|(g,\tilde{\beta}_\Omega)_{K_2}| \to \infty$$

for some g (in fact for g in a dense set in B) as $\Omega \to R$. But we know from Lemma 1.5 that $(g,\tilde{\beta}_\Omega) = (g,\beta_\Omega f)$ converges for every g. Therefore, the first alternative occurs. In particular, for $g = \tilde{\beta}_\Omega|K_2$,

$$(\|\tilde{\beta}_\Omega\|_{K_2})^2 \le k\|\tilde{\beta}_\Omega\|_{K_2},$$

hence the Lemma.

1.7. Inclusion $O_\beta^N \subset O_\gamma^N$. We are ready to state:

THEOREM. On an arbitrarily given Riemannian N-manifold R, $N \ge 2$, if the biharmonic Green's function of the simply supported body,

$$\gamma(x,y) = \lim_{\Omega \to R} \gamma_\Omega(x,y)$$

exists for some, and hence every, (x,y), then so does the biharmonic Green's function of the clamped body,

$$\beta(x,y) = \lim_{\Omega \to R} \beta_\Omega(x,y),$$

<u>and the convergence is uniform on compact sets of</u> R.

<u>Proof</u>. By virtue of the interior regularity estimate (cf. Agmon [1, Sec. 5]) in Sobolev norms,

$$\|\tilde{\beta}_\Omega\|_{s+4,K_2} \leq C(\|f\|_{s,K_2} + \|\tilde{\beta}_\Omega\|_{K_2}),$$

Lemma 1.6 implies that

$$\|D^\alpha \tilde{\beta}_\Omega\|_{K_1} \quad \text{is bounded in } \Omega$$

for every $m \geq 0$, $|\alpha| \leq m$. Therefore, by the Rellich compactness theorem (cf. [loc. cit., Sec. 3]), there exists a sequence $\{\tilde{\beta}_n = \tilde{\beta}_{\Omega_n}\}$ such that, for $|\alpha| \leq m$,

$$\|D^\alpha(\tilde{\beta}_{n+p} - \tilde{\beta}_n)\|_{K_1} \to 0$$

as $n \to \infty$, $p \geq 0$. This in turn implies by Sobolev's Inequality (cf. [loc. cit., Sec. 3]) that, for $|\alpha| \leq m$,

$$|D^\alpha(\tilde{\beta}_{n+p} - \tilde{\beta}_n)| \to 0.$$

In particular, for $\alpha = 0$ we have by $\tilde{\beta}_n|K = \beta_n|K$ the desired uniform convergence of β_n on K. The uniqueness of the limiting function β on R follows from

$$(g,\beta_n g) \nearrow \sup_{\{\Omega\}}(g,\beta_\Omega g),$$

which implies

$$(g,\beta g) = \sup_{\{\Omega\}}(g,\beta_\Omega g)$$

for all $g \in L_0^2(R)$ with $\operatorname{supp} g \subset K$.

This completes the proof of the Theorem.

<u>1.8.</u> <u>Elementary proof</u>. In view of the significance of the relation $O_\beta^N \subset O_\gamma^N$, we give a second proof. Take a manifold R in $\tilde{O}_\gamma^N$, $N \geq 2$. Choose a regular subregion Ω of R and continue denoting by $g_\Omega(x,y)$ the harmonic Green's function on $\bar{\Omega}$, and by $\beta_\Omega(x,y)$ the biharmonic Green's function of the clamped

body on $\bar{\Omega}$. Write

$$H_\Omega(x,y) = \Delta_x \beta_\Omega(x,y)$$

and set $H_\Omega(\cdot,y) = g_\Omega(\cdot,y) = 0$ on $R - \bar{\Omega}$. Then

$$\beta_\Omega(x,y) = (g_\Omega(x,\cdot), H_\Omega(\cdot,y)).$$

Let $h \in H(\Omega) \cap C^1(\bar{\Omega})$. In view of

$$\int_{\partial\Omega} h(x) *_x d\beta_\Omega(x,y) = 0$$

and

$$(dh, d\beta_\Omega)_\Omega = \int_{\partial\Omega} \beta_\Omega * dh = 0,$$

we have

$$(h(\cdot), H_\Omega(\cdot,y))_\Omega = 0$$

for all $h \in H(\Omega) \cap C^1(\bar{\Omega})$.

Fix $x,y \in R$ and take regular subregions Ω_0, Ω_1 with $\bar{\Omega}_0 \subset \Omega_1$ and $x,y \in \Omega_0$. We recall from VIII.1.4 that the existence of the biharmonic Green's function of a simply supported body on R,

$$\gamma(x,y) = (g(\cdot,x), g(\cdot,y)),$$

is equivalent to

$$\|g(\cdot,x)\|_{R-\Omega_1} < \infty$$

for every $x \in \Omega_0$. Since $\beta_\Omega(x,y) = (g_\Omega(\cdot,x), H_\Omega(\cdot,y))_\Omega$, we obtain

$$\beta_{\Omega'}(x,y) - \beta_\Omega(x,y) = (g(\cdot,x) - H_{\Omega_1}(\cdot,x), H_{\Omega'}(\cdot,y) - H_\Omega(\cdot,y))_{\Omega'}$$

for $\Omega \subset \Omega'$ with $\Omega_1 \subset \Omega$ and for any $x \in \Omega_0$. The quantity

$$K = \sup_{x\in\Omega_0} \|g(\cdot,x) - H_{\Omega_1}(\cdot,x)\|$$

is finite by virtue of the continuity of $g(z,x) - H_{\Omega_1}(z,x)$ on $\Omega_1 \times \Omega_1$. The Schwarz inequality yields

$$|\beta_{\Omega'}(x,y) - \beta_{\Omega}(x,y)|^2 \leq K^2 \|H_{\Omega'}(\cdot,y) - H_{\Omega}(\cdot,y)\|^2_{\Omega'}$$

for $\Omega' \supset \Omega \supset \Omega_1$ and $x \in \Omega_0$. Here,

$$\|H_{\Omega'}(\cdot,y) - H_{\Omega}(\cdot,y)\|^2_{\Omega'} = \|H_{\Omega'}(\cdot,y) - H_{\Omega_1}(\cdot,y)\|^2_{\Omega'} - \|H_{\Omega}(\cdot,y) - H_{\Omega_1}(\cdot,y)\|^2_{\Omega}.$$

Since

$$(g(\cdot,y) - H_{\Omega}(\cdot,y),\ H_{\Omega}(\cdot,y) - H_{\Omega_1}(\cdot,y))_{\Omega} = 0,$$

we obtain

$$(g(\cdot,y) - H_{\Omega_1}(\cdot,y),\ H_{\Omega}(\cdot,y) - H_{\Omega_1}(\cdot,y))_{\Omega} = \|H_{\Omega}(\cdot,y) - H_{\Omega_1}(\cdot,y)\|^2_{\Omega}.$$

The Schwarz inequality gives

$$\|H_{\Omega}(\cdot,y) - H_{\Omega_1}(\cdot,y)\|_{\Omega} \leq \|g(\cdot,y) - H_{\Omega_1}(\cdot,y)\| \leq K$$

for every Ω. Therefore,

$$\lim_{\Omega' \supset \Omega \nearrow R} \|H_{\Omega'}(\cdot,y) - H_{\Omega}(\cdot,y)\|^2_{\Omega'} = 0$$

and

$$\lim_{\Omega' \supset \Omega \nearrow R} |\beta_{\Omega'}(x,y) - \beta_{\Omega}(x,y)| = 0,$$

uniformly for $x \in \Omega_0$. Thus,

$$\beta(x,y) = \lim_{\Omega \nearrow R} \beta_{\Omega}(x,y)$$

exists on R for any fixed y, and the convergence is uniform for x in any compact subset of R.

The proof of $O^N_{\beta} \subset O^N_{\gamma}$ is complete.

<u>1.9. A criterion for the existence of</u> β. Suppose $R \in O^N_G$, and let $e(x,y)$ be an Evans kernel in the sense of Nakai [6]. For the definition and properties

of $e(x,y)$ to be used below, we refer to Sario-Nakai [1, pp. 354-361]; the discussion there is for Riemann surfaces, but it applies verbatim to Riemannian manifolds. Let B_y be a geodesic ball $|x - y| < \varepsilon$ about y.

THEOREM. *If an Evans kernel* e *on* $R \in O_G^N$ *satisfies*

$$\|e(\cdot,y)\|_{R-B_y} < \infty$$

for every y, *then* $R \in \tilde{O}_\beta^N$, *with* $N \geq 2$.

Proof. As we pointed out in the Notes to IX.§1, this is a consequence of the Main Theorem in IX.1.3. Here we give an independent proof of the existence of β. Using $h(\cdot) = e(\cdot,y) - H_\Omega(\cdot,y)$, we have, by the convention $H_{\Omega'}(\cdot,y) = 0$ on $R - \bar{\Omega}'$,

$$(e(\cdot,y) - H_\Omega(\cdot,y),\ H_{\Omega'}(\cdot,y))_\Omega = 0$$

for $\Omega' = \Omega \supset \bar{B}_y$ and $\Omega \supset \Omega' = B_y$. We set $f(\cdot) = e(\cdot,y) - H_{B_y}(\cdot,y)$ and $t_\Omega(\cdot) = H_\Omega(\cdot,y) - H_{B_y}(\cdot,y)$ and obtain

$$(f(\cdot) - t_\Omega(\cdot),\ t_\Omega(\cdot))_\Omega = 0.$$

By the Schwarz inequality,

$$\|t_\Omega(\cdot)\|_\Omega^2 = (f(\cdot),\ t_\Omega(\cdot))_\Omega \leq \|f(\cdot)\|_\Omega \cdot \|t_\Omega(\cdot)\|_\Omega.$$

In view of the assumption of the Theorem, and the joint continuity of $e(x,y)$ on $R \times R$,

$$\|H_\Omega(\cdot,y) - H_{B_y}(\cdot,y)\|_\Omega^2 \leq \|e(\cdot,y) - H_{B_y}(\cdot,y)\|^2 = K(y) < K(R_0) < \infty$$

for every Ω and for all y in an arbitrarily chosen compact subset R_0 of R. By the orthogonality $h \perp H_\Omega$ in 1.8,

$$\beta_\Omega(x,y) = (H_\Omega(\cdot,x),\ H_\Omega(\cdot,y))_\Omega$$

$$\beta_{\Omega_0}(x,y) = (H_\Omega(\cdot,x),\ H_{\Omega_0}(\cdot,y))_\Omega,$$

where we again use the convention $H_{\Omega'}(\cdot,x) = 0$ on $R - \bar{\Omega}'$. It follows that

$$\beta_{\Omega}(x,y) - \beta_{\Omega_0}(x,y) = (H_{\Omega}(\cdot,x),\, H_{\Omega}(\cdot,y) - H_{\Omega_0}(\cdot,y))_{\Omega}$$

$$= (H_{\Omega}(\cdot,x) - H_{\Omega_0}(\cdot,x),\, H_{\Omega}(\cdot,y) - H_{\Omega_0}(\cdot,y))_{\Omega}.$$

By the Schwarz inequality,

$$|\beta_{\Omega}(x,y) - \beta_{\Omega_0}(x,y)|^2 \le \|H_{\Omega}(\cdot,x) - H_{\Omega_0}(\cdot,x)\|_{\Omega}^2 \cdot \|H_{\Omega}(\cdot,y) - H_{\Omega_0}(\cdot,y)\|_{\Omega}^2$$

$$= I_1(x)^2 \cdot I_2(y)^2,$$

where

$$I_1(x) = \|H_{\Omega}(\cdot,x) - H_{\Omega_0}(\cdot,x)\|_{\Omega}$$

$$\le \|H_{\Omega}(\cdot,x) - H_{B_x}(\cdot,x)\|_{\Omega} + \|H_{\Omega_0}(\cdot,x) - H_{B_x}(\cdot,x)\|_{\Omega}$$

$$\le 2K(R_0)^{1/2} < \infty$$

for all $x \in R_0$. Since

$$\|H_{\Omega}(\cdot,y) - H_{B_y}(\cdot,y)\| \le K(y)^{1/2} < \infty$$

for all Ω,

$$I_2(y) = \|(H_{\Omega}(\cdot,y) - H_{B_y}(\cdot,y)) - (H_{\Omega_0}(\cdot,y) - H_{B_y}(\cdot,y))\|_{\Omega} \to 0$$

as $\Omega \supset \Omega_0 \nearrow R$. We conclude that

$$\beta(x,y) = \lim_{\Omega\to R} \beta_{\Omega}(x,y) = \lim_{\Omega\to R} \beta_{\Omega}(y,x)$$

exists and the convergence is uniform on every compact subset of R for any fixed $y \in R$. The proof of the Theorem is complete.

We note in passing the following immediate consequence of the Theorem:

COROLLARY. For $N = 2,3$, every compact Riemannian manifold punctured at a point carries β.

In fact, the punctured manifold is parabolic and, therefore, $e(x,y)$ tends to $-\infty$ asymptotically as $\log \rho$ or $-1/\rho$ for $N = 2,3$, respectively, where ρ is the geodesic distance from x to the punctured point.

1.10. Strictness of the inclusion. We are ready to prove:

THEOREM. For $N \geq 2$,

$$O_\beta^N < O_\gamma^N.$$

More precisely,

$$O_G^N \cap \tilde{O}_\beta^N \neq \emptyset.$$

Proof. For $N = 2$, the proof will be given in §4 by means of a necessary and sufficient condition for the complex plane with a conformal radial metric to carry β. For $N > 2$, consider the N-ball $R = \{r < 1, ds\}$ with the metric $ds = \lambda(x)^{1/2}|dx|$, where $r = |x|$, $x = (x^1,\ldots,x^N)$, $\lambda \in C^\infty(R)$, $\lambda > 0$, and on $\{\frac{1}{2} < r < 1\}$,

$$\lambda(x) = \lambda(r) = r^{(2-2N)/(N-2)}(1 - r)^{4/(N-2)}.$$

Since the function $h(r) = 1/(1 - r)$ satisfies on $\{\frac{1}{2} < r < 1\}$ the harmonic equation

$$\Delta h(r) = -g^{-1/2}(g^{1/2}g^{rr}h'(r))' = 0,$$

R is parabolic. Therefore, there exists an Evans kernel $e(x,y)$ on R such that

$$e(x,y) \sim \frac{1}{1 - r} \quad \text{as} \quad r \to 1.$$

By virtue of

$$g(x)^{1/2} = \lambda(r)^{N/2} \sim [(1 - r)^{4/(N-2)}]^{N/2} = (1 - r)^{2N/(N-2)},$$

we obtain for $\rho \in (|y|, 1)$,

$$\|e(\cdot,y)\|^2_{r>\rho} \sim \int_\rho^1 \left(\frac{1}{1-r}\right)^2 (1-r)^{2N/(N-2)}\, r^{N-1}\, dr$$

$$\sim \int_\rho^1 (1-r)^{-2+2N/(N-2)}\, dr$$

$$= \int_\rho^1 (1-r)^{4/(N-2)}\, dr < \infty.$$

By Theorem 1.9, β exists on R, hence $O_G^N \cap \tilde{O}_\beta^N \neq \emptyset$.

NOTES TO §1. Theorem 1.7 and its Lemmas 1.3 - 1.6 were proved in Ralston-Sario [1]. The elementary proof in 1.8 was later given in Nakai-Sario [12], where also Theorem 1.9 with its Corollary, and Theorem 1.10 for $N > 2$ were established.

§2. A NONEXISTENCE TEST FOR β

The purpose of this section is to introduce a convenient test for the nonexistence of the biharmonic Green's function β on a Riemannian manifold R. As an application, we exhibit a Riemannian 2-manifold R whose boundary is harmonically so strong that $R \in \tilde{O}^2_{HD}$ but which nevertheless carries no β.

2.1. The class O_β^N. Let $\{\Omega\}$ be the directed set of regular subregions of a noncompact Riemannian manifold R of dimension $N \geq 2$. By definition,

(β.1) $\beta_\Omega(\cdot,y) \in C^2(\Omega - y)$ and $\Delta\beta_\Omega(\cdot,y) - g_\Omega(\cdot,y) \in H(\Omega)$ for every y in Ω;

(β.2) $\beta_\Omega(\cdot,y) \in C^1(\bar{\Omega} - y)$ and $\beta_\Omega(\cdot,y) = *d\beta_\Omega(\cdot,y) = 0$ on $\partial\Omega$.

We recall that $\beta_{\Omega'} - \beta_\Omega \in C(\Omega \times \Omega)$ for $\Omega \subset \Omega'$ and thus we can define $\beta_{\Omega'}(y,y) - \beta_\Omega(y,y)$ as $\lim_{x\to y}(\beta_{\Omega'}(x,y) - \beta_\Omega(x,y))$. A function $\beta(x,y)$ on $R \times R$ with values in $(-\infty,\infty]$, finite on $R \times R$ off the diagonal and such that

$$\lim_{\Omega\to R} (\beta(x,y) - \beta_\Omega(x,y)) = 0$$

on $R \times R$, is, if it exists, the biharmonic Green's function of the clamped body on R. We understand the above relation for (y,y) as the existence of a finite $\lim_{\Omega'\to R}(\beta_{\Omega'}(y,y) - \beta_\Omega(y,y))$ for one and hence for every Ω. As before, we denote by O^N_β the class of noncompact Riemannian N-manifolds R on which there exists no β. We observe that $R \notin O^N_\beta$ is equivalent to

(β.3) $\lim_{\Omega' \to R}(\beta_{\Omega'}(x,y) - \beta_\Omega(x,y))$ _exists and is finite on_ $\Omega \times \Omega$ _for one and hence for every_ Ω.

2.2. _The class_ $O^N_{SH_2}$. Consider the class $H_2(R) = H(R) \cap L^2(R,*1)$, where $*1$ is the volume element of R. We know from IX.§2 that $H_2(R)$ is a locally bounded Hilbert space, and the norm convergence implies the uniform convergence on each compact subset of R. It is easy to show that $H(\Omega) \cap C(\overline{\Omega})$ is dense in $H_2(\Omega)$. Denote by $O^N_{H_2}$ the class of Riemannian N-manifolds R with $H_2(R) = \{0\}$ and by $O^N_{SH_2}$ the class of Riemannian N-manifolds R such that there exists a subregion $S \neq \emptyset$ of R with $R - \overline{S} \neq \emptyset$ and $S \in O^N_{H_2}$, that is, $H_2(S) = \{0\}$. We have the strict inclusion relation

$$O^N_{SH_2} < O^N_{H_2}.$$

The mere inclusion is trivial and the strictness is seen as follows, by means of the Euclidean space E^N of dimension $N \geq 2$. First we prove that

$$E^N \in O^N_{H_2}.$$

Take any $h \in H_2(E^N)$. Let $(r,\theta) = (r,\theta^1,\ldots,\theta^{N-1})$ be the polar coordinates, and $d\omega$ the surface element on $\omega = \{|x| = 1\}$. Then

$$f(x) = f(|x|) = \int_\omega h(|x|,\theta)^2 d\omega \geq 0$$

is subharmonic on E^N and, by the maximum principle, $f(r)$ is an increasing function on $[0,\infty)$. If $f(r) \not\equiv 0$, then there exist constants $c > 0$ and $\sigma > 0$ such that $f(r) \geq c$ on $[\sigma,\infty)$. Thus

$$\begin{aligned}\infty = c\int_\sigma^\infty r^{N-1}\,dr &\leq \int_\sigma^\infty f(r)r^{N-1}dr \\ &= \int_\sigma^\infty \int_\omega h(r,\theta)^2 r^{N-1} d\omega dr \\ &= \int_{|x|>\sigma} h(x)^2 dx^1 \cdots dx^N < \infty,\end{aligned}$$

a contradiction. Therefore, $f \equiv 0$ and a fortiori, $h \equiv 0$ on every $|x| = \rho > 0$, hence $h \equiv 0$, and we conclude that $H_2(E^N) = \{0\}$.

Next we show that

$$E^N \notin O^N_{SH_2}.$$

Suppose there exists a subregion $S \neq \emptyset$ of E^N with $E^N - \bar{S} \neq \emptyset$ and $H_2(S) = \{0\}$. Let $x_0 \in E^N - \bar{S}$ and $\{|x - x_0| < \rho\} \subset E^N - \bar{S}$. By a parallel translation, if necessary, we obtain $S \subset S_\rho = \{|x| > \rho\}$. Since $H_2(S_\rho) \subset H_2(S) = \{0\}$, we have $H_2(S_\rho) = \{0\}$, but this is impossible because

$$h(x) = r^{-(n+N-2)} S_n(\theta) \in H_2(S_\rho),$$

$(r,\theta) = x$, $n \geq 2$, with $S_n(\theta)$ any nonzero spherical harmonic of degree n.

The main purpose of the present section is to prove:

<u>THEOREM</u>. <u>For</u> $N \geq 2$,

$$O^N_{SH_2} \subset O^N_\beta.$$

This will give a convenient test for $R \in O^N_\beta$. We only have to find a subregion $S \neq \emptyset$ of R with $R - \bar{S} \neq \emptyset$ and $H_2(S) = \{0\}$ to conclude that $R \in O^N_\beta$. Note that this is not a characterization of O^N_β, that is, the above inclusion is not an equality in general. In fact, by 1.1,

$$E^N \in O^N_\beta, \quad N = 2,3,4.$$

From this and $E^N \notin O^N_{SH_2}$ we see that the equality does not hold in $O^N_{SH_2} \subset O^N_\beta$ for $N = 2$, 3, and 4.

The proof of the Theorem will be given in 2.7 after we have estabalished, in 2.3 - 2.6, three <u>complete characterizations</u> of O^N_β, instead of merely an inclusion as in the above Theorem. The significance of the Theorem lies in its applicability to concrete cases to show the nonexistence of β.

<u>2.3</u>. <u>The</u> β-<u>density</u> $H_\Omega(\cdot,y)$. As a consequence of (β.1) and (β.2) in 2.1, $\beta_\Omega(\cdot,y)$ is a Green's potential with the β-density $H_\Omega(\cdot,y) = \Delta\beta_\Omega(\cdot,y)$,

$$\beta_\Omega(\cdot,y) = \int_\Omega g_\Omega(\cdot,\xi) * H_\Omega(\xi,y).$$

Since $H_\Omega(\cdot,y) \in C(\overline{\Omega} - y)$, a property of the Green's kernel (e.g., Miranda [1]) gives

$$*d\beta_\Omega(\cdot,y) = \int_\Omega *dg_\Omega(\cdot,\xi) *_\xi H_\Omega(\xi,y) = 0$$

on $\partial\Omega$. Again by (β.2),

$$\int_\Omega *_x d_x g_\Omega(x,\xi) *_\xi H_\Omega(\xi,y) = 0$$

for every $x \in \partial\Omega$. On multiplying both sides by an arbitrary $h \in H(\Omega) \cap C(\overline{\Omega})$ and integrating over $\partial\Omega$, we obtain by Fubini's theorem

$$\int_\Omega \int_{\partial\Omega} h(x) *_x d_x g_\Omega(x,\xi) *_\xi H_\Omega(\xi,y) = 0.$$

By the reproducing property of g_Ω, we conclude that

$$\int_\Omega h(\xi) *_\xi H_\Omega(\xi,y) = 0$$

for every $h \in H(\Omega) \cap C(\overline{\Omega})$, and since $H(\Omega) \cap C(\overline{\Omega})$ is dense in $H_2(\Omega)$, for every $h \in H_2(\Omega)$. We have reviewed for the β-density the following orthogonality property which plays an important role in the study of O_β^N:

$$H_\Omega(\cdot,y) \perp H_2(\Omega).$$

2.4. The β-span S_β. We denote by $(\cdot,\cdot)$ and $\|\cdot\|$ the inner product and the norm in $L^2(R,*1)$. We consider $\beta_\Omega(x,y)$ and $H_\Omega(x,y)$ as defined on all of $R \times R$ by giving values zero outside of their original domains of definition. First observe that, by the above orthogonality and $H_\Omega(\cdot,y) - g_\Omega(\cdot,y) \in H_2(\Omega)$,

$$\beta_\Omega(x,y) = \int_\Omega H_\Omega(\xi,x) * H_\Omega(\xi,y).$$

Similarly by $H_\Omega(\cdot,y) \perp H_2(\Omega)$ and $H_{\Omega'}(\cdot,y) - H_\Omega(\cdot,y) \in H_2(\Omega)$, we have for $\Omega \subset \Omega'$,

$$\beta_{\Omega'}(x,y) - \beta_{\Omega}(x,y) = \int_{\Omega'} (H_{\Omega'}(\xi,x) - H_{\Omega}(\xi,x)) * (H_{\Omega'}(\xi,y) - H_{\Omega}(\xi,y)).$$

In particular,

$$\beta_{\Omega'}(y,y) - \beta_{\Omega}(y,y) = \|H_{\Omega'}(\cdot,y) - H_{\Omega}(\cdot,y)\|^2.$$

Again by $H_{\Omega}(\cdot,y) \perp H_2(\Omega)$,

$$\|H_{\Omega''}(\cdot,y) - H_{\Omega'}(\cdot,y)\|^2$$

$$= \|H_{\Omega''}(\cdot,y) - H_{\Omega}(\cdot,y)\|^2 - \|H_{\Omega'}(\cdot,y) - H_{\Omega}(\cdot,y)\|^2$$

for $\Omega \subset \Omega' \subset \Omega''$. It follows that $\{\beta_{\Omega'}(y,y) - \beta_{\Omega}(y,y)\}$, $\Omega' \supset \Omega$, is an increasing net. Therefore, we can define for $y \in R$ and Ω with $y \in \Omega$,

$$S_\beta(y) = S_\beta(y;R) = S_\beta(y;\Omega,R)$$

$$= \lim_{\Omega' \to R} (\beta_{\Omega'}(y,y) - \beta_{\Omega}(y,y))$$

$$= \lim_{\Omega' \to R} \|H_{\Omega'}(\cdot,y) - H_{\Omega}(\cdot,y)\|^2 \in (0,\infty],$$

which we will call the β-<u>span</u> of R at $y \in R$ with respect to Ω. The property $S_\beta(y) < \infty$ is clearly independent of the choice of Ω and is thus a property of (R,y). We maintain:

<u>THEOREM</u>. <u>For</u> $N \geq 2$,

$$R \notin O_\beta^N$$

<u>if and only if the</u> β-<u>span</u> $S_\beta(y)$ <u>of</u> R <u>is finite at every point</u> $y \in R$.

If $R \notin O_\beta^N$, or $(\beta.3)$ is valid, then we trivially have $S_\beta(y) < \infty$ for every $y \in R$. Conversely, assume that $S_\beta(y) < \infty$ for every $y \in R$. Then, by the above expressions for $\beta_{\Omega'}(x,y) - \beta_{\Omega}(x,y)$, the Schwarz inequality implies that

$$|\beta_{\Omega''}(x,y) - \beta_{\Omega'}(x,y)| \leq \|H_{\Omega''}(\cdot,x) - H_{\Omega'}(\cdot,x)\| \cdot \|H_{\Omega''}(\cdot,y) - H_{\Omega'}(\cdot,y)\|$$

on $\Omega' \times \Omega'$ for $\Omega' \subset \Omega''$. By the above expressions for $\|H_{\Omega''}(\cdot,y) - H_{\Omega'}(\cdot,y)\|^2$ and $S_\beta(y)$, and by $S_\beta(x) < \infty$ and $S_\beta(y) < \infty$, we see that the right-hand side converges to zero on $\Omega \times \Omega$ as $\Omega' \to R$ for any $\Omega \subset \Omega' \subset \Omega''$, and, since Ω is arbitrary,

$$\lim_{\Omega' \subset \Omega'', \Omega' \to R} (\beta_{\Omega''}(x,y) - \beta_{\Omega'}(x,y)) = 0$$

on $R \times R$,

$$\lim_{\Omega'' \to R} |\beta_{\Omega''}(x,y) - \beta_\Omega(x,y)| \le S_\beta(x)^{1/2}\, S_\beta(y)^{1/2},$$

and $S_\beta(y) = \lim_{\Omega'' \to R}(\beta_{\Omega''}(y,y) - \beta_\Omega(y,y)) < \infty$. Thus ($\beta$.3) is fulfilled for every Ω and, therefore, $R \notin O_\beta^N$.

<u>2.5.</u> <u>The</u> β-<u>density</u> $H(\cdot,y)$. Assume $R \notin O_\beta^N$. By Theorem 2.4 and the last two relations preceding it, we conclude that $\{H_{\Omega'}(\cdot,y) - H_\Omega(\cdot,y)\}$, $\Omega' \supset \Omega$, is a Cauchy net in $L^2(R,*1)$ and has a limit $H_{R\Omega}(\cdot,y) \in L^2(R,*1)$. Set

$$H(\cdot,y) = H_{R\Omega}(\cdot,y) + H_\Omega(\cdot,y).$$

Then since

$$H(\cdot,y) - H_{\Omega'}(\cdot,y) = H_{R\Omega}(\cdot,y) - (H_{\Omega'}(\cdot,y) - H_\Omega(\cdot,y)),$$

the net $\{H(\cdot,y) - H_{\Omega'}(\cdot,y)\}$ is convergent to zero in $L^2(R,*1)$. Fix an arbitrary $\Omega_0 \subset \Omega'$ with $y \notin \Omega_0$. Then $\{H_{\Omega'}(\cdot,y)\}$ is a Cauchy net in $H_2(\Omega_0)$, $H(\cdot,y)$ is its limit, and a fortiori $H(\cdot,y) \in H_2(\Omega_0)$. Therefore,

$$H(\cdot,y) \in H(R - y).$$

Fix an arbitrary Ω with $y \in \Omega$. Observe that $\{H_{\Omega'}(\cdot,y) - g_\Omega(\cdot,y)\}$, $\Omega' \supset \Omega$, is also a Cauchy net in $H_2(\Omega)$, convergent to $H(\cdot,y) - g_\Omega(\cdot,y)$, which is again in $H_2(\Omega)$. Thus we have

$$H(\cdot,y) - g_\Omega(\cdot,y) \in H(\Omega)$$

for one and hence for every Ω with $y \in \Omega$. It is also clear that

$$H(\cdot,y) \in H_2(R - \Omega)$$

for any Ω with $y \in \Omega$. Besides the above three properties of $H(\cdot,y)$, we have the orthogonality relation

$$H(\cdot,y) \perp H_2(R),$$

or, equivalently,

$$\int_R h(\xi) *_\xi H(\xi,y) = 0$$

for every $h \in H_2(R)$. Here the integral is well defined because of the second and third properties of $H(\cdot,y)$. For the proof, observe that the inequality

$$|(h,H(\cdot,y) - H_\Omega(\cdot,y))| \leq \|h\| \cdot \|H(\cdot,y) - H_\Omega(\cdot,y)\|$$

implies

$$\int_R h(\xi) *_\xi H(\xi,y) = \lim_{\Omega \to R} \int_\Omega h(\xi) *_\xi H_\Omega(\xi,y).$$

Since $h \in H_2(R) \subset H_2(\Omega)$, equality $\int_\Omega h(\xi) *_\xi H_\Omega(\xi,y) = 0$ yields $\int_R h(\xi) *_\xi H(\xi,y) = 0$.

As in IX.§1, we call a function $H(\cdot,y)$ on $R - y$ with the above four properties the β-density on R for $y \in R$. It is unique. In fact, if $K(\cdot,y)$ has these four properties, then $h = H(\cdot,y) - K(\cdot,y) \in H_2(R)$, and by the fourth property for $H(\cdot,y)$ and $K(\cdot,y)$ we obtain $(h,h) = \|h\|^2 = 0$, hence $h \equiv 0$ on R.

We claim:

<u>THEOREM</u>. <u>For</u> $N \geq 2$,

$$R \notin O_\beta^N$$

<u>if and only if the</u> β-<u>density</u> $H(\cdot,y)$ <u>exists on</u> R <u>for every</u> $y \in R$.

We only have to show that the existence of the β-density $H(\cdot,y)$ on R for every $y \in R$ implies $R \notin O_\beta^N$. Let $\Omega \subset \Omega'$. Since $H(\cdot,y) - H_{\Omega'}(\cdot,y) \in H_2(\Omega') \subset H_2(\Omega)$, we have

$$((H(\cdot,y) - H_\Omega(\cdot,y)) - (H_{\Omega'}(\cdot,y) - H_\Omega(\cdot,y)), H_{\Omega'}(\cdot,y) - H_\Omega(\cdot,y))$$

$$= (H(\cdot,y) - H_{\Omega'}(\cdot,y), H_{\Omega'}(\cdot,y)) - (H(\cdot,y) - H_{\Omega'}(\cdot,y), H_\Omega(\cdot,y)) = 0$$

and a fortiori,

$$(H(\cdot,y) - H_\Omega(\cdot,y), H_{\Omega'}(\cdot,y) - H_\Omega(\cdot,y)) = \|H_{\Omega'}(\cdot,y) - H_\Omega(\cdot,y)\|^2.$$

By the Schwarz inequality,

$$\|H_{\Omega'}(\cdot,y) - H_\Omega(\cdot,y)\| \leq \|H(\cdot,y) - H_\Omega(\cdot,y)\|.$$

In view of the relation for $S_\beta(y)$ in 2.4, it follows that $S_\beta(y) < \infty$ for every $y \in R$, that is, $R \notin O_\beta^N$.

COROLLARY. The β-span $S_\beta(y)$ is finite if and only if the β-density $H(\cdot,y)$ exists on R at y, and in this case,

$$S_\beta(y;\Omega,R) = \|H(\cdot,y) - H_\Omega(\cdot,y)\|^2.$$

2.6. An extremum property of $H(\cdot,y)$. Assume the existence of β. Then by the relations for $\lim_\Omega(\beta - \beta_\Omega)$ in 2.1, for $\beta_{\Omega'} - \beta_\Omega$ in 2.4, and $\lim_\Omega\|H(\cdot,y) - H_\Omega(\cdot,y)\| = 0$, we have

$$\beta(x,y) - \beta_\Omega(x,y) = (H(\cdot,x) - H_\Omega(\cdot,x), H(\cdot,y) - H_\Omega(\cdot,y))$$

on $\Omega \times \Omega$. By the orthogonality property in 2.3 and the expression for $\beta_\Omega(x,y)$ in 2.4,

$$\beta(x,y) = \int_R H(\xi,x)*H(\xi,y)$$

on $\Omega \times \Omega$ for every Ω and a fortiori on $R \times R$. Instead of $\lim_\Omega(\beta - \beta_\Omega) = 0$, we can take, as in Chapter IX, this relation as the definition of β, starting from β-densities $H(\cdot,y)$ for all $y \in R$.

In this connection, we consider the family $F(R,y)$ of functions $K(\cdot,y)$ on $R - y$ which have the first three properties in 2.5, with K replacing H. If $H(\cdot,y)$ exists, then it is in the class $F(R,y)$ and thus

$$F(R,y) \neq \emptyset.$$

Since $H(\cdot,y) - K(\cdot,y) \in H_2(R)$, we have

$$(H(\cdot,y) - K(\cdot,y),\ H(\cdot,y) - H_\Omega(\cdot,y)) = 0$$

for every Ω. By the Schwarz inequality applied to

$$\|H(\cdot,y) - H_\Omega(\cdot,y)\|^2 = (K(\cdot,y) - H_\Omega(\cdot,y),\ H(\cdot,y) - H_\Omega(\cdot,y)),$$

we obtain the following extremum property of $H(\cdot,y)$:

$$\|H(\cdot,y) - H_\Omega(\cdot,y)\| = \min_{K \in F(R,y)} \|K(\cdot,y) - H_\Omega(\cdot,y)\|$$

for any Ω with $y \in \Omega$.

This property actually characterizes $H(\cdot,y)$ in the class $F(R,y)$ if $F(R,y) \neq \emptyset$. In fact, in this case, if we fix an arbitrary Ω, then the family

$$X_\Omega = \{K(\cdot,y) - H_\Omega(\cdot,y) \mid K(\cdot,y) \in F(R,y)\}$$

is clearly a nonempty convex set in $L^2(R,*1)$. It is also closed. To see this, let $\{K_n(\cdot,y) - H_\Omega(\cdot,y)\}$, $n = 1,2,\ldots,$ be a sequence in X_Ω converging to a $\bar{K} \in L^2(R,*1)$. Set $K = \bar{K} + H_\Omega(\cdot,y)$. Then $\{K_n - K\}$ is a Cauchy sequence in $L^2(R,*1)$, and therefore, $\{K_n\}$ is Cauchy in $H_2(\Omega)$ for every Ω with $y \notin \Omega$, and $\{K_n - g_\Omega(\cdot,y)\}$ is Cauchy in $H_2(\Omega)$ for every Ω with $y \in \Omega$. Thus K enjoys the first three properties of H in 2.5, that is, $K - H_\Omega(\ ,y) = \lim_n(K_n(\cdot,y) - H_\Omega(\cdot,y)) \in X_\Omega$, and X_Ω is closed.

Since any nonempty closed convex subset of a Hilbert space contains a unique element of minimum norm, there exists a unique element $K_0 - H_\Omega(\cdot,y) \in X_\Omega$ such that

$$\|K_0 - H_\Omega(\cdot,y)\| = \min_{K \in F(R,y)} \|K(\cdot,y) - H_\Omega(\cdot,y)\|.$$

Let h be any element in $H_2(\Omega)$, and $t > 0$. In view of $K_0 + th \in F(R,y)$, we have

$$\|K_0 - H_\Omega(\cdot,y) + th\|^2 \geq \|K_0 - H_\Omega(\cdot,y)\|^2$$

or

$$2t(K_0 - H_\Omega(\cdot,y),h) + t^2\|h\|^2 \geq 0.$$

Since this is true for every $t > 0$,

$$(K_0 - H_\Omega(\cdot,y),h) = 0$$

for every $h \in H_2(R)$. From this and $(h,H_\Omega(\cdot,y))_\Omega = 0$, we deduce $K_0 \perp H_2(R)$. Thus K_0 has the four properties of H in 2.5, that is, K_0 is the β-density $H(\cdot,y)$ on R for y. We have shown:

The β-density $H(\cdot,y)$ on R for $y \in R$ exists if and only if $F(R,y) \neq \emptyset$.

We restate this in the following form:

THEOREM. The manifold R does not belong to O_β^N if and only if there exists a harmonic function $K(\cdot,y)$ on $R - y$ which has the harmonic fundamental singularity at y and is square integrable on R off any neighborhood of y.

2.7. Proof of Theorem 2.2. Inclusion $O_{SH_2}^N \subset O_\beta^N$ can now be established using Theorem 2.5 or 2.6. For example, let $R \in O_{SH_2}^N$. Then there exists a subregion $S \neq \emptyset$ of R with $R - \overline{S} \neq \emptyset$ and $H_2(S) = \{0\}$. If $R \notin O_\beta^N$, choose a point $y \in R - \overline{S}$. By Theorem 2.5, the β-density $H(\cdot,y)$ exists on R for y, and by taking Ω with $y \in \Omega$ and $\overline{\Omega} \subset R - \overline{S}$, we infer by $H(\cdot,y) \in H_2(R - \Omega)$ that $H(\cdot,y)|S \in H_2(S) = \{0\}$. Thus $H(\cdot,y) \equiv 0$ on S. By the unique continuation property of harmonic functions, $H_\Omega(\cdot,y) \equiv 0$ on $R - y$, which contradicts $H(\cdot,y) - g_\Omega(\cdot,y) \in H(\Omega)$. Therefore, $R \in O_\beta^N$, and we have proved Theorem 2.2.

2.8. Plane with HD functions but no β. As an illustration of the use of our test $O_{SH_2}^N \subset O_\beta^N$, we exhibit a manifold R which shows that

$$O^2_\beta \cap \tilde{O}^2_{HD} \neq \emptyset.$$

Let E^2_λ be the plane with the metric

$$ds = \lambda(x)^{1/2}|dx|, \qquad \lambda(x) = e^{r^2},$$

with $r = |x|$. Choose $R = \{r > 1\}$ in E^2_λ. Since $HD(R,ds) = HD(R,|dx|)$, we clearly have $(R,ds) \in \tilde{O}^2_{HD}$. Let $S_\rho = \{r > \rho\}$, $1 < \rho < \infty$. We assert that $H_2(S_\rho,ds) = \{0\}$. It suffices to show that if h is harmonic on $r > \rho$ and

$$A = \int_{S_\rho} h(r,\theta)^2\lambda(r)rdrd\theta < \infty,$$

then $h \equiv 0$. The expansion

$$h(r,\theta) = a \log r + \sum_{n=-\infty}^{\infty} r^n(a_n\cos n\theta + b_n\sin n\theta)$$

on S_ρ gives

$$L(r) = \int_0^{2\pi} h(r,\theta)^2d\theta = 2\pi(a_0 + a \log r)^2 + \pi \sum_{\substack{n=-\infty \\ n\neq 0}}^{\infty} (a_n^2 + b_n^2)r^{2n},$$

and we have

$$A = \int_\rho^\infty L(r)\lambda(r)rdr$$

$$= 2\pi\int_\rho^\infty (a_0 + a \log r)^2 e^{r^2} rdr + \pi \sum_{\substack{n=-\infty \\ n\neq 0}}^{\infty} (a_n^2 + b_n^2) \int_\rho^\infty r^{2n+1}e^{r^2}dr.$$

From $A < \infty$ we infer that $a = a_n = b_n = 0$ for $n = 0, \pm 1, \ldots,$ that is, $h \equiv 0$ on S_ρ. Therefore,

$$(R,ds) \in O^2_{SH_2},$$

and Theorem 2.2 yields $(R,ds) \in O^2_\beta$.

NOTES TO §2. Theorems 2.2, 2.4, 2.5, and 2.6 were proved in Nakai-Sario [19]. We shall return to the relation $O^N_\beta \cap \tilde{O}^N_{HD} \neq \emptyset$ for $N > 2$ in §3.

§3. MANIFOLDS WITH STRONG HARMONIC BOUNDARIES BUT WITHOUT β

In this section, we discuss systematically the question whether or not a harmonic nondegeneracy of the ideal boundary of a Riemannian manifold R of dimension N is sufficient to entail the existence of β. We know from Chapter I that $O_G^N < O_{HP}^N < O_{HB}^N < O_{HD}^N = O_{HC}^N$ for $N \geq 2$. Thus the strongest harmonic nondegeneracy is $R \in \tilde{O}_{HD}^N$, and we ask explicitly: Does $R \in \tilde{O}_{HD}^N$ assure that $R \in \tilde{O}_\beta^N$? By §2, the answer is in the negative for $N = 2$, and we shall now show that this continues to hold for <u>every</u> N, that is,

$$O_\beta^N \cap \tilde{O}_{HD}^N \neq \emptyset, \quad N \geq 2.$$

Since any subregion of the Euclidean N-space E^N for $N > 4$, and any subregion of E^N with an exterior point for $N = 2,3,4$ carries β (cf. IX.§8), a manifold in $O_\beta^N \cap \tilde{O}_{HD}^N$ will be, by necessity, somewhat intricate.

3.1. Double of a Riemannian manifold. Consider the sets

$$\Sigma^N = \{r > 1\}, \quad \Gamma^N = \{r = 1\}$$

in the Euclidean space E^N of dimension $N \geq 2$. Denote by $\hat{\Sigma}^N$ the <u>double</u> of Σ^N with respect to Γ^N, that is, the topological manifold $(\Sigma^N)_1 \cup (\Sigma^N)_2 \cup \Gamma^N$ defined in an obvious manner, with the $(\Sigma^N)_i$, $i = 1,2$, duplicates of Σ^N.

The Euclidean metric (line element) $|dx|$ in $\Sigma^N \cup \Gamma^N$ is given by

$$|dx|^2 = dr^2 + r^2 \sum_{j=1}^{N-1} \gamma_j(\theta) d\theta^{j2}.$$

We introduce the Riemannian metric

$$ds^2 = \phi(r)^2 dr^2 + \psi(r)^2 r^2 \sum_{j=1}^{N-1} \gamma_j(\theta) d\theta^{j2}$$

on $\Sigma^N \cup \Gamma^N$, with ϕ and ψ strictly positive C^∞ functions on $[1,\infty)$ such that ds^2 defines a C^∞ metric on $\hat{\Sigma}^N$ by symmetry. Denote by $\hat{\Sigma}^N_{\phi,\psi}$ the manifold

$\hat{\Sigma}^N$ with this metric. We could actually allow ds to be any smooth metric on $\hat{\Sigma}^N_{\phi,\psi}$ as long as it coincides with the metric given by the above ds^2 in a neighborhood of each "point at infinity" δ_i, $i = 1,2$, that is, duplicate in $(\Sigma^N)_i$ of the point at infinity δ of Σ^N and hence of E^N. The manifold $\hat{\Sigma}^N_{\phi,\psi}$ can also be viewed as the double of the Riemannian manifold $\Sigma^N_{\phi,\psi} = (\Sigma^N, ds)$ with respect to Γ^N.

3.2. Manifolds with HD functions but without β. We consider the following two conditions on the functions ϕ and ψ:

$$\int_1^\infty \frac{\phi(r)}{r^{N-1}\,\psi(r)^{N-1}}\,dr < \infty,$$

$$\int_1^\infty \left(\int_r^\infty \frac{\phi(\rho)}{\rho^{N-1}\psi(\rho)^{N-1}}\,d\rho\right)^2 \phi(r)\psi(r)^{N-1}r^{N-1}dr = \infty.$$

The existence of functions satisfying these conditions is obvious, the choice $\phi(r) = r^2$ and $\psi(r) = r^{-(N-5)/(N-1)}$ for large r being an example. The purpose of the present section is to prove:

THEOREM. For $N \geq 2$,

$$\hat{\Sigma}^N_{\phi,\psi} \in O^N_\beta \cap \tilde{O}^N_{HD}$$

if and only if the functions ϕ and ψ satisfy the above two conditions.

We shall prove, in 3.3 and 3.4, two lemmas from which the Theorem will follow at once.

3.3. Existence of HD functions. Denote by $W = W(\sigma)$, $\sigma > 1$, the region $\{r > \sigma\}$ in Σ^N, and by $W_i = W(\sigma)_i$ its duplicates in $(\Sigma^N)_i$, $i = 1,2$. On each $W(\sigma)_i$, the Laplace-Beltrami operator for $\hat{\Sigma}^N_{\phi,\psi}$ takes the form

$$\Delta u = -\frac{1}{\phi(r)\psi(r)^{N-1}r^{N-1}}\frac{\partial}{\partial r}\left(\frac{r^{N-1}\psi(r)^{N-1}}{\phi(r)}\frac{\partial}{\partial r}u\right) - \frac{1}{r^2\psi(r)^2\gamma(\theta)}\sum_{j=1}^{N-1}\frac{\partial}{\partial\theta^j}\left(\gamma(\theta)\gamma_j(\theta)^{-1}\frac{\partial}{\partial\theta^j}u\right),$$

with $\gamma(\theta) = (\Pi_1^{N-1} \gamma_j(\theta))^{1/2}$. If the first condition in 3.2 is satisfied, then

$$w(x;\sigma) = \left(\int_r^\infty \frac{\phi(\rho)}{\rho^{N-1}\psi(\rho)^{N-1}} d\rho\right) \Big/ \left(\int_1^\infty \frac{\phi(\rho)}{\rho^{N-1}\psi(\rho)^{N-1}} d\rho\right)$$

is a harmonic function on $W(\sigma)_i$. This means that the harmonic measure $1 - w(x;\sigma)$ of δ_i on $W(\sigma)_i$ is positive and, therefore,

$$(W(\sigma)_i, \partial W(\sigma)_i) \notin SO_{HD},$$

with SO_{HD} the class of those Riemannian manifolds R with compact or noncompact smooth boundaries α which carry no nonconstant Dirichlet finite harmonic functions with boundary values zero on α. By the two region criterion (e.g., Sario-Nakai [1, p. 242]), the above relation implies that $\hat{\Sigma}^N_{\phi,\psi} \in \tilde{O}^N_{HD}$ if the first condition in 3.2 is assumed.

Suppose now that the first condition in 3.2 is not satisfied. The function

$$e(x;\sigma) = \int_1^r \frac{\phi(\rho)}{\rho^{N-1}\psi(\rho)^{N-1}} d\rho$$

is also harmonic on $W(\sigma)$, but $\lim_{x\to\delta_i} e(x;\sigma) = \infty$, $i = 1,2$. Hence, $\hat{\Sigma}^N_{\phi,\psi} \in O^N_G$ and, a fortiori, $\hat{\Sigma}^N_{\phi,\psi} \in O^N_{HD}$.

We have shown:

LEMMA. *For* $N \geq 2$,

$$\hat{\Sigma}^N_{\phi,\psi} \in \tilde{O}^N_{HD}$$

if and only if the functions ϕ *and* ψ *satisfy the first condition in* 3.2.

3.4. Nonexistence of β. The surface element of Γ^N considered in E^N is $d\theta = \gamma(\theta)d\theta^1 \cdots d\theta^{N-1}$, with finite total measure

$$A_N = \int_{\Gamma^N} d\theta = 2\pi^{N/2}/\Gamma(N/2).$$

Therefore, the surface element of $\{r > \sigma\}$ in $\hat{\Sigma}^N_{\phi,\psi}$ is $r^{N-1}\psi(r)^{N-1}d\theta$, and the volume element of $W(\sigma)_i$, in $\hat{\Sigma}^N_{\phi,\psi}$ is

$$*1 = \phi(r)\psi(r)^{N-1}r^{N-1}drd\theta.$$

Under the first assumption in 3.2, the second assumption is thus equivalent to

$$\int_{W(\sigma)_1 \cup W(\sigma)_2} *w(x;\sigma)^2 = \infty.$$

If we assume that this condition does not hold, then by Theorem 1.7, we conclude that $\hat{\Sigma}^N_{\phi,\psi} \in \tilde{O}^N_\beta$.

Conversely suppose, under the first assumption in 3.2, that $\hat{\Sigma}^N_{\phi,\psi} \in \tilde{O}^N_\beta$. We wish to deduce the invalidity of the second assumption in 3.2, or equivalently, of the above relation $w \notin L^2$. The proof is by contradiction. Suppose that this relation holds. Fix an arbitrary $y \in \hat{\Sigma}^N_{\phi,\psi}$ and then a $\sigma > 1$ such that $y \notin \overline{W(\sigma)_1 \cup W(\sigma)_2}$. Let $H(\cdot,y) = \Delta\beta(\cdot,y)$ be the β-density. Then by §§1-2,

$$\int_{W(\sigma)_1 \cup W(\sigma)_2} *_x H(x,y)^2 < \infty.$$

The function $h = H(\cdot,y)$ has, for any fixed $r \geq \sigma$, the "Fourier expansion"

$$h(r,\theta) = A_N^{-1/2}\, h_0(r) + \sum_{n=1}^{\infty} \sum_{m=1}^{m_n} h_{nm}(r)\, S_{nm}(\theta),$$

where $\{S_{nm}(\theta)\}$, $m = 1,\ldots,m_n$, is a complete orthonormal system of spherical harmonics of degree $n \geq 1$, and thus $\{A_N^{-1/2}\} \cup \{S_{nm}(\theta)\}$, $n = 1,2,\ldots,$ $m = 1,\ldots,m_n$, is a complete orthonormal system in $L^2(\Gamma^N;d\theta)$. By the Parseval identity,

$$f(r) \equiv \int_{\Gamma^N} h(r,\theta)^2 d\theta = h_0(r)^2 + \sum_{n=1}^{\infty} \sum_{m=1}^{m_n} h_{nm}(r)^2$$

for every $r \in [\sigma,\infty)$. Since $h_0(r)$ satisfies $\Delta h_0 = 0$ and $h_{nm}(r)$ satisfies the "P-harmonic equation"

$$\Delta u = Pu, \quad P = n(n + N - 2)r^{-2}\psi(r)^{-2} > 0,$$

we see that h_0^2 and h_{nm}^2 are all subharmonic on $W(\sigma)_1 \cup W(\sigma)_2$ and consequently

$f(r)$ is subharmonic on $W(\sigma)_1 \cup W(\sigma)_2$. Therefore, $f(r)$ takes on its maximum in $[a,b] \subset [\sigma,\infty)$ at a or b.

By the above relation $H \in L^2$,

$$\int_\sigma^\infty f(r)\phi(r)\psi(r)^{N-1}r^{N-1}dr = \int_\sigma^\infty \left(\int_{\Gamma^N} h(r,\theta)^2 d\theta\right)\phi(r)\psi(r)^{N-1}r^{N-1}dr$$

$$= \int_\sigma^\infty *h(x)^2$$

$$= \frac{1}{2}\int_{W(\sigma)_1 \cup W(\sigma)_2} *_x H(x,y)^2 < \infty.$$

As a direct consequence of the second assumption in 3.2,

$$\int_\sigma^\infty \phi(r)\psi(r)^{N-1}r^{N-1}dr = \infty.$$

In view of these two relations, we have $\liminf_{r\to\infty} f(r) = 0$, that is, there exists an increasing divergent sequence $\{r_n\}$ in (σ,∞) such that $f(r_n) \to 0$ as $n \to \infty$. Since

$$\max_{r_n \le r \le r_{n+1}} f(r) = \max\,(f(r_n),\, f(r_{n+1}))$$

for every $n = 1,2,\ldots,$ we have $\lim_{r\to\infty} f(r) = 0$, that is,

$$\lim_{x\to\sigma_1} H(x,y) = 0$$

for $i = 1,2$. This with $\lim_{x\to y} H(x,y) = \infty$ and the maximum principle yields

$$H(\cdot,y) > 0$$

on $\hat{\Sigma}^N_{\phi,\psi}$. Again by the maximum principle, $w(\cdot,\sigma) \le cH(\cdot,y)$ on $W(\sigma)_1 \cup W(\sigma)_2$, with $c = (\inf_{|x|=\sigma} H(x,y))^{-1} < \infty$. Hence,

$$\int_{W(\sigma)_1 \cup W(\sigma)_2} *w(x;\sigma)^2 \le c^2 \int_{W(\sigma)_1 \cup W(\sigma)_2} *_x H(x,y)^2.$$

However, this is impossible in view of the above relations $w \notin L^2$ and $H \in L^2$.

Thus $\hat{\Sigma}^N_{\phi,\psi} \in \tilde{O}^N_\beta$ implies that the second condition in 3.2 is not satisfied.

LEMMA. Suppose the first condition in 3.2 is satisfied. Then for $N \geq 2$,

$$\hat{\Sigma}^N_{\phi,\psi} \in O^N_\beta$$

if and only if the second condition in 3.2 is satisfied.

NOTES TO §3. Theorem 3.2 and Lemmas 3.3 and 3.4 were established in Nakai-Sario [13]. The above discussion on relations between O^N_β and harmonic degeneracy will be completed in §§4-5.

§4. PARABOLIC RIEMANNIAN PLANES CARRYING β

In the proof of Theorem 1.10, we established the relation $O^N_G \cap \tilde{O}^N_\beta \neq \emptyset$ only for $N > 2$, leaving the case $N = 2$ to the present section. We shall now show that $O^2_G \cap \tilde{O}^2_\beta \neq \emptyset$ as well.

Our counterexample is the complex plane C_λ endowed with a suitable radial conformal metric $ds = \lambda(|z|)^{1/2}|dz|$. In terms of λ, we shall give a necessary and sufficient condition for C_λ to carry β: the function $\log r$ must be square integrable over the Riemannian plane C_λ.

By way of preparation, we deduce, in 4.1-4.6, an extremum property of $\Delta\beta$ and give a number of applications, somewhat beyond our present need. We then discuss, in 4.7-4.8, convergence of functions β constructed on regular subregions exhausting the manifold. This discussion is a specialization to $N = 2$ of that in Chapter IX and may offer some interest. The main part of the present section is 4.9-4.14, where we give the above characterization of planes carrying β.

4.1. Density. Let R be a noncompact Riemannian 2-manifold, and Ω a regular subregion of R. We retain the notation $\beta_\Omega(x,y)$ for the biharmonic Green's function of the clamped plate on $\bar{\Omega}$, normalized by $\int_{\partial U_\varepsilon} *d\Delta_x\beta_\Omega = -1$, where U_ε is a geodesic ball $|x - y| < \varepsilon$ about y. For

$$H_\Omega(x,y) = \Delta_x \beta_\Omega(x,y)$$

and the harmonic Green's function $g_\Omega(x,y)$ on $\bar{\Omega}$ with $\int_{\partial U_\varepsilon} *dg_\Omega = -1$, set

$$h_\Omega(\cdot,y) = H_\Omega(\cdot,y) - g_\Omega(\cdot,y) \in H(\Omega) \cap C^\infty(\bar{\Omega}).$$

Here and later it is understood that we only consider such regular subregions Ω for which the membership of our functions in the class under consideration, here $C^\infty(\bar{\Omega})$, is assured. In the representation of β_Ω as a potential,

$$\beta_\Omega(x,y) = \int_\Omega g_\Omega(x,z) * H_\Omega(z,y),$$

we continue referring to H_Ω as the density of β_Ω. We shall give a characterization of this density in terms of an extremum property.

4.2. Potentials. Denote by $F(\Omega,y)$ the class of functions $f(x,y)$ on Ω such that $f(\cdot,y) - g_\Omega(\cdot,y) \in H(\Omega) \cap C^\infty(\bar{\Omega})$ and consider the potentials

$$p(x,y) = \int_\Omega g_\Omega(x,z) * f(z,y).$$

We start by proving that

$$D_\Omega(k,p(\cdot,y)) = \lim_{\varepsilon\to 0} D_{\Omega-\bar{U}_\varepsilon}(k,p(z,y)) = 0$$

for every $k \in H(\Omega) \cap C^\infty(\bar{\Omega})$. By Stokes' formula,

$$\begin{aligned} D_{\Omega-\bar{U}_\varepsilon}(k,p(\cdot,y)) &= \int_{\Omega-\bar{U}_\varepsilon} dp \wedge *dk \\ &= \int_{\partial(\Omega-\bar{U}_\varepsilon)} p*dk - \int_{\Omega-\bar{U}_\varepsilon} pd*dk \\ &= -\int_{\partial U_\varepsilon} p*dk. \end{aligned}$$

On letting $\varepsilon \to 0$ we obtain the asserted relation.

Next, we show that

$$\int_{\partial\Omega} k{*}dp = -\int_{\Omega} k(x){*}_x f(x,y)$$

for every $k \in H(\Omega) \cap C^{\infty}(\overline{\Omega})$. In fact, the relation $\Delta_x p(x,y) = f(x,y)$, Stokes' formula, and the relation just proved give

$$\int_{\Omega} k(x){*}_x f(x,y) = -\int_{\Omega} kd{*}dp$$

$$= D_{\Omega}(k,p) - \int_{\partial\Omega} k{*}dp = -\int_{\partial\Omega} k{*}dp.$$

4.3. Extremum property. We use $(\cdot,\cdot)_{\Omega}$ and $\|\cdot\|_{\Omega}$ to mean the inner product and the norm in $L^2(\Omega)$. We shall characterize $H_{\Omega}(\cdot,y)$ as the unique solution of the problem of finding $\inf_{f\in F(\Omega,y)} \|f(\cdot,y)\|_{\Omega}$.

THEOREM. A function $\overline{f}(\cdot,y) \in F(\Omega,y)$ coincides with $H_{\Omega}(\cdot,y)$ if and only if

$$(k,\overline{f}(\cdot,y))_{\Omega} = 0$$

for every $k \in H(\Omega) \cap C^{\infty}(\overline{\Omega})$, or, equivalently, if and only if

$$\|\overline{f}(\cdot,y)\|_{\Omega} = \min_{f\in F(\Omega,y)} \|f(\cdot,y)\|_{\Omega}.$$

Proof. Let $\overline{p} = \int_{\Omega} g_{\Omega}(\cdot,z){*}\overline{f}(z,y)$. By $\int_{\partial\Omega} k{*}dp = -\int_{\Omega} k{*}f$, we have

$$\int_{\partial\Omega} k{*}d\overline{p} = -(k,\overline{f}(\cdot,y))_{\Omega}.$$

If $f(\cdot,y) = H_{\Omega}(\cdot,y)$, then $\overline{p} = \beta_{\Omega}(\cdot,y)$ and ${*}d\overline{p} = {*}d\beta_{\Omega} = 0$ on $\partial\Omega$. Therefore, $(k,H_{\Omega}(\cdot,y))_{\Omega} = 0$. Conversely, if this equality is valid, then

$$\int_{\partial\Omega} k{*}d\overline{p} = 0$$

for every $k \in H(\Omega) \cap C^{\infty}(\overline{\Omega})$ and, a fortiori, for every $k \in C(\partial\Omega)$. From this, we immediately see that $\overline{p} - \beta_{\Omega} \equiv 0$ on $\overline{\Omega}$ and $\overline{f}(\cdot,y) = \Delta\overline{p}(\cdot,y) = \Delta\beta_{\Omega}(\cdot,y) = H_{\Omega}(\cdot,y)$

To see that the first condition of the Theorem implies the second condition, observe that $\overline{f}(\cdot,y) - f(\cdot,y) \in H(\Omega) \cap C^\infty(\overline{\Omega})$ and thus, $(\overline{f} - f,\overline{f})_\Omega = 0$ for every $f \in F(\Omega,y)$. By the Schwarz inequality,

$$\|\overline{f}\|_\Omega^2 = (f,\overline{f})_\Omega \leq \|f\|_\Omega \cdot \|\overline{f}\|_\Omega,$$

and the second condition of the Theorem follows. Conversely, assume the latter. Then, since $\overline{f} + \varepsilon k \in F(\Omega,y)$ for every $k \in H(\Omega) \cap C^\infty(\overline{\Omega})$ and every ε,

$$\|\overline{f} + \varepsilon k\|_\Omega \geq \|\overline{f}\|_\Omega,$$

or, equivalently,

$$2\varepsilon(k,\overline{f})_\Omega + \varepsilon^2\|k\|_\Omega^2 \geq 0$$

for every ε. Hence the first condition of the Theorem is true.

4.4. Consequences. We deduce some consequences of Theorem 4.3. Set $H_\Omega(\cdot,y) =$ on $R - \overline{\Omega}$ and use $(\cdot,\cdot)$ and $\|\cdot\|$ in reference to $L^2(R)$.

LEMMA. *For regular subregions* Ω *and* Ω' *of* R *with* $\Omega \subset \Omega'$,

$$\|H_\Omega(\cdot,y)\| \leq \|H_{\Omega'}(\cdot,y)\|.$$

In fact, $\|H_\Omega\| = \|H_\Omega\|_\Omega \leq \|H_{\Omega'}\|_\Omega \leq \|H_{\Omega'}\|_{\Omega'} = \|H_{\Omega'}\|$.

Another consequence of Theorem 4.3 is obtained by observing that by 4.1 and the first condition in Theorem 4.3,

$$\begin{aligned}\beta_\Omega(x,y) &= (g_\Omega(x,\cdot),\ H_\Omega(\cdot,y)) = (g_\Omega(\cdot,x),\ H_\Omega(\cdot,y)) \\ &= (g_\Omega(\cdot,x) + h_\Omega(\cdot,x),\ H_\Omega(\cdot,y)).\end{aligned}$$

Therefore,

$$\beta_\Omega(x,y) = (H_\Omega(\cdot,x),\ H_\Omega(\cdot,y)) = \int_\Omega H_\Omega(z,x) * H_\Omega(z,y).$$

In particular,

$$\beta_\Omega(x,x) = \|H_\Omega(\cdot,x)\|^2 = \int_\Omega *H_\Omega(z,x)^2,$$

and by the Schwarz inequality,

$$\beta_\Omega(x,y)^2 \leq \beta_\Omega(x,x)\beta_\Omega(y,y).$$

Note that the finiteness of $\beta_\Omega(y,y)$, say, is entailed by our restriction to dimension 2. In fact, in terms of the geodesic distance $\rho = |x - y|$, $\beta_\Omega(x,y) \sim \rho^2 \log \rho + \text{const}$.

4.5. Green's function of the simply supported plate. In passing, we also consider the biharmonic Green's function $\gamma_\Omega(x,y)$ of the simply supported plate:

$$\begin{cases} \Delta^2\gamma_\Omega = 0 & \text{on} \quad \Omega - y, \\ \gamma_\Omega = \Delta\gamma_\Omega = 0 & \text{on} \quad \partial\Omega. \end{cases}$$

The density of $\gamma_\Omega(x,y)$ is nothing but $g_\Omega(x,y)$,

$$\gamma_\Omega(x,y) = \int_\Omega g_\Omega(x,z)*g_\Omega(z,y).$$

We consider the difference

$$\delta_\Omega(x,y) = \beta_\Omega(x,y) - \gamma_\Omega(x,y).$$

By 4.1,

$$\delta_\Omega(x,y) = \int_\Omega g_\Omega(x,z)*h_\Omega(z,y).$$

As to counterparts of the last three relations in 4.4, we observe that by the first condition in Theorem 4.3,

$$0 = (h_\Omega(\cdot,y),\, H_\Omega(\cdot,x))_\Omega = (h_\Omega(\cdot,y),\, g_\Omega(\cdot,x))_\Omega + (h_\Omega(\cdot,y),\, h_\Omega(\cdot,x))_\Omega.$$

Thus,

$$\delta_\Omega(x,y) = -(h_\Omega(\cdot,x),\, h_\Omega(\cdot,y))_\Omega$$

and, in particular,

$$\delta_\Omega(x,x) = -\|h_\Omega(\cdot,x)\|_\Omega^2.$$

By the Schwarz inequality,

$$\delta_\Omega(x,y)^2 \le \delta_\Omega(x,x)\delta_\Omega(y,y).$$

Since $\delta_\Omega(x,x) < 0$, we have proved that

$$\beta_\Omega(x,x) < \gamma_\Omega(x,x).$$

4.6. Kernels. Besides the representation of β_Ω as a Green's potential, we have the other representation $\beta_\Omega(x,y) = (H_\Omega(\cdot,x), H_\Omega(\cdot,y))$. The kernel can actually be more general:

$$\beta_\Omega(x,y) = \int_\Omega \sigma(x,z)*H_\Omega(z,y)$$

for any $\sigma(\cdot,z) \in F(\Omega,z)$. If $R \in \tilde{O}_G^N$, then a good choice is $\sigma = g$,

$$\beta_\Omega(x,y) = \int_\Omega g(x,z)*H_\Omega(z,y).$$

If $R \in O_G^N$, then a good choice for σ is the Evans kernel $e(x,y)$ on R:

$$\beta_\Omega(x,y) = \int_\Omega e(x,z)*H_\Omega(z,y).$$

By definition, $e(x,z)$ is harmonic on $R - z$, has a positive fundamental singularity at z, and tends to $-\infty$ as x tends to the Alexandroff point a_∞ of R. Again we normalize by $\int_{\partial U_\varepsilon} *de = -1$.

4.7. Strong limits. If $\Omega \subset \Omega'$, then

$$(H_{\Omega'}(\cdot,x) - H_\Omega(\cdot,x), H_\Omega(\cdot,y))_\Omega = 0$$

and therefore,

$$(H_{\Omega'}(\cdot,x), H_\Omega(\cdot,y))_\Omega = (H_\Omega(\cdot,x), H_\Omega(\cdot,y))_\Omega.$$

For $x = y$, we have

$$\|H_{\Omega'}(\cdot,y) - H_\Omega(\cdot,y)\|_\Omega^2 = \|H_{\Omega'}(\cdot,y)\|_\Omega^2 - \|H_\Omega(\cdot,y)\|_\Omega^2.$$

If we again let $H_\Omega(\cdot,y) = 0$ on $R - \bar{\Omega}$ and write $\|\cdot\| = \|\cdot\|_R$ and $(\cdot,\cdot) = (\cdot,\cdot)_R$, then the above reasoning with $(\cdot,\cdot)_\Omega$ replaced by $(\cdot,\cdot)$ gives

$$\|H_{\Omega'}(\cdot,y)\|^2 - \|H_\Omega(\cdot,y)\|^2 = \|H_{\Omega'}(\cdot,y) - H_\Omega(\cdot,y)\|^2$$

for every $\Omega' \supset \Omega$. In view of 4.4, we set

$$c_y = \sup_{\Omega \subset R} \beta_\Omega(y,y) = \sup_{\Omega \subset R} \|H_\Omega(\cdot,y)\|^2.$$

We say that $u \in L^2$ is the strong limit of a sequence $\{u_n\} \subset L^2$,

$$u = \operatorname*{s-lim}_{n\to\infty} u_n,$$

if $\lim_n \|u - u_n\| = 0$.

LEMMA. *The strong limit*

$$H(\cdot,y) = \operatorname*{s-lim}_{\Omega\to R} H_\Omega(\cdot,y)$$

exists in $H(R - y) \cap L^2(R)$ *if and only if* $c_y < \infty$. *In this case,* $H(\cdot,y) - H_\Omega(\cdot,y) \in H(\Omega)$ *for every* Ω.

Proof. If $H = \text{s-lim}\, H_\Omega$ exists, then $\lim_\Omega \|H_\Omega\| = \|H\| < \infty$ and thus $c_y < \infty$. Conversely, if $c_y < \infty$, then by $\|H_{\Omega'}\|^2 - \|H_\Omega\|^2 = \|H_{\Omega'} - H_\Omega\|^2$, $H \equiv \text{s-lim}_\Omega H_\Omega$ exists in $L^2(R)$.

For $\varphi \in C_0^\infty(R)$, $(H_{\Omega'} - H_\Omega, \Delta\varphi) = 0$ for sufficiently large $\Omega \subset \Omega'$. On letting $\Omega' \nearrow R$, we obtain

$$(H(\cdot,y) - H_\Omega(\cdot,y), \Delta\varphi) = 0$$

for every $\varphi \in C_0^\infty(R)$ with $\operatorname{supp} \varphi \subset \Omega$. This means that $H(\cdot,y) - H_\Omega(\cdot,y) \in H(\Omega)$ for every Ω.

COROLLARY. *If* $c_x < \infty$ *and* $c_y < \infty$, *then* $\beta(x,y) = \lim_\Omega \beta_\Omega(x,y)$ *exists and*

$$\beta(x,y) = (H(\cdot,x),\ H(\cdot,y)).$$

Proof. By the second relation in 4.7 and $\beta_\Omega(x,y) = (H_\Omega(\cdot,x),\ H_\Omega(\cdot,y))$, we have $\beta_\Omega(x,y) = (H_{\Omega'}(\cdot,x),\ H_\Omega(\cdot,y))$ for every $\Omega' \supset \Omega$. Since $c_x < \infty$, $H(\cdot,x) = \text{s-lim}_{\Omega'} H_{\Omega'}(\cdot,x)$ exists and thus,

$$\beta_\Omega(x,y) = \lim_{\Omega'\to R} (H_{\Omega'}(\cdot,x),\ H_\Omega(\cdot,y)) = (H(\cdot,x),\ H_\Omega(\cdot,y)).$$

Again by $c_y < \infty$, $H(\cdot,y) = \text{s-lim}_\Omega H_\Omega(\cdot,y)$ exists and therefore,

$$\beta(x,y) = \lim_{\Omega\to R} \beta_\Omega(x,y) = \lim_{\Omega\to R} (H(\cdot,x),\ H_\Omega(\cdot,y)) = (H(\cdot,x),\ H(\cdot,y))$$

exists.

4.8. Convergence of β_Ω. For an $S \subset R$, we set

$$c_S = \sup_{x\in S} c_x.$$

THEOREM. If $c_S < \infty$ for every compact subset S of R, then $\beta(\cdot,y) = \lim_{\Omega\to R} \beta_\Omega(\cdot,y)$ exists, and the covergence is uniform on each compact subset of R for every fixed $y \in R$.

Proof. Let S be a compact subset of R, and x a point of S. By 4.7, $\beta(x,y) = (H(\cdot,x),\ H(\cdot,y))$ and $\beta_\Omega(x,y) = (H(\cdot,x),\ H_\Omega(\cdot,y))$. Therefore,

$$\begin{aligned}|\beta(x,y) - \beta_\Omega(x,y)| &= |(H(\cdot,x),\ H(\cdot,y) - H_\Omega(\cdot,y))| \\ &\le \|H(\cdot,x)\| \cdot \|H(\cdot,y) - H_\Omega(\cdot,y)\| \\ &\le c_S^{1/2} \|H(\cdot,y) - H_\Omega(\cdot,y)\|\end{aligned}$$

for every $x \in S$. Since $\|H(\cdot,y) - H_\Omega(\cdot,y)\| \to 0$ as $\Omega \to R$, $\beta_\Omega(x,y) \to \beta(x,y)$ uniformly on S for each fixed y.

The above proof also gives:

Let S be a subset of R and assume that $c_S < \infty$ and $c_y < \infty$. Then $\beta(x,y) = \lim_{\Omega\to R} \beta_\Omega(x,y)$, uniformly for $x \in S$.

4.9. Existence of β on C_λ. We endow the finite complex plane $C = \{r < \infty \mid r = |x|\}$ with the conformal metric

$$ds^2 = \lambda(x)((dx^1)^2 + (dx^2)^2),$$

where $\lambda(x) = \lambda(r)$ and $\lambda \in C^\infty[0,\infty)$ with $\lambda > 0$. The resulting Riemannian manifold is denoted by C_λ. The limit

$$\beta(x,y) = \lim_{\Omega \to C_\lambda} \beta_\Omega(x,y),$$

if it exists on $C_\lambda \times C_\lambda$, is the biharmonic Green's function of the clamped plate C_λ.

The main purpose of the present section is to prove:

THEOREM. The function β exists on C_λ if and only if $\log r \in L^2(C_\lambda)$, that is, λ satisfies

$$\int_0^\infty (\log r)^2 \lambda(r) r dr < \infty.$$

The proof will be given in 4.10-4.14.

4.10. Necessity. For $R \in (0,\infty)$, set $\Omega_R = \{|x| < R\}$ and $\beta_R = \beta_{\Omega_R}$. Let $H_\Omega(x,y) = \Delta_x \beta_\Omega(x,y)$, denote H_{Ω_R} by H_R, and set $H_\Omega(\cdot,y) = 0$ on $C_\lambda - \bar{\Omega}$. Then by 4.4,

$$\beta_R(0,0) = \|H_R(\cdot,0)\|^2.$$

First we assume that $\lim_\Omega \beta_\Omega$ exists. Then since

$$\|H_R(\cdot,0)\|^2 \le \|H_{R'}(\cdot,0)\|^2$$

for $0 < R < R'$, we have

$$\|H_R(\cdot,0)\|^2 \le \beta(0,0) < \infty.$$

For every regular subregion Ω, there exists an $\Omega_R \supset \Omega$. Therefore, $c_0 = \sup_\Omega \|H_\Omega(\cdot,0)\|^2 < \infty$ and by Lemma 4.7, the strong limit

$$H(\cdot,0) = \underset{\Omega\to C_\lambda}{\text{s-lim}}\, H_\Omega(\cdot,0)$$

exists in $L^2(C_\lambda)$. Clearly, $\beta_R(x,0)$ is radial, that is, $\beta_R(x,0) = \beta_R(r,0)$, and so is $H_R(\cdot,0) = \Delta\beta_R(\cdot,0)$. Hence,

$$h(re^{i\theta}) = h(r) = H(re^{i\theta},0) - \frac{1}{2\pi}\log\frac{1}{r} \in H(C_\lambda) = H(C),$$

and consequently, $h(r) = h_0$, a constant. We set

$$t(x) = \frac{1}{2\pi}\log\frac{1}{r},$$

so that

$$H(x,0) = t(x) + h_0.$$

There exist constants c and r_0 with $t(r)^2 \leq cH(r,0)^2$ for $r \geq r_0$. Hence,

$$\int_0^{2\pi}\int_{r_0}^{\infty} t(r)^2\lambda(r)r\,dr\,d\theta \leq c\int_{r>r_0} *H(x,0)^2 < c\|H(\cdot,0)\|^2,$$

which gives

$$\int_{r_0}^{\infty}(\log r)^2\lambda(r)r\,dr < \infty.$$

Clearly,

$$\int_0^{r_0}(\log r)^2\lambda(r)r\,dr \leq (\max_{0\leq r\leq r_0}\lambda(r))\int_0^{r_0}(\log r)^2 r\,dr < \infty,$$

and we obtain the inequality in Theorem 4.9.

<u>4.11.</u> <u>Sufficiency</u>. Conversely, suppose the inequality in Theorem 4.9 holds. Then

$$\Lambda = \int_0^{\infty}\lambda(r)r\,dr < \infty.$$

The Evans kernel on C_λ is

$$e(x,y) = \frac{1}{2\pi} \log \frac{1}{|x - y|} .$$

By 4.6,

$$\beta_R(x,y) = (e(x,\cdot), H_R(\cdot,y)).$$

We set on C_λ,

$$H_R(x,y) = e(x,y) + k_R(x,y),$$

with $H_R(\cdot,y) = 0$ on $C_\lambda - \overline{\Omega}_R$. Then $k_R \in H(\Omega_R) \cap C^\infty(\overline{\Omega}_R)$ and by Theorem 4.3, H_R is characterized by $(H_R(\cdot,y),u) = 0$ for every $u \in H(\Omega_R) \cap C^\infty(\overline{\Omega}_R)$,

$$(e(\cdot,y) + k_R(\cdot,y),u)_{\Omega_R} = 0.$$

This relation will be used to determine $k_R(\cdot,y)$.

We denote the inner product $(\ ,\ )_{\Omega_R}$ by $(\cdot,\cdot)_R$. In the expression

$$\beta_R(x,y) = (e(x,\cdot), e(\cdot,y))_R + (e(x,\cdot), k_R(\cdot,y))_R,$$

the term $(e(x,\cdot), e(\cdot,y))_R$ converges to $(e(x,\cdot), e(\cdot,y))$ uniformly in x on each compact subset S of C_λ for a fixed $y \in C_\lambda$. Thus we only have to discuss the convergence of $(e(x,\cdot), k_R(\cdot,y))_R$. Since C_λ is radial with respect to the origin 0 of C_λ, it suffices to consider, for $a \in [0,\infty)$,

$$(e(x,\cdot), k_R(\cdot,a))_R.$$

<u>4.12.</u> <u>Auxiliary formulas.</u> The basic formulas for our computation are

$$\int_0^{2\pi} \log \frac{1}{|re^{i\theta} - a|}\, d\theta = \begin{cases} 2\pi \log \frac{1}{a}, & r \in [0,a], \\ 2\pi \log \frac{1}{r}, & r \in [a,\infty), \end{cases}$$

and

$$\int_0^{2\pi} \left(\log \frac{1}{|re^{i\theta} - a|}\right) \cos n\theta\, d\theta = \begin{cases} \frac{\pi}{n} \cdot \frac{r^n}{a^n}, & r \in [0,a], \\ \frac{\pi}{n} \cdot \frac{a^n}{r^n}, & r \in [a,\infty) \end{cases}$$

for $n = 1,2,\dots,$ and $a > 0$. The first of these four formulas is an immediate consequence of the mean value property of harmonic functions, and the second formula for $r > a$ follows by $|re^{i\theta} - a| = |ae^{-i\theta} - r|$. Continuity gives the case $r = a$.

To prove the third formula, suppose $r \in [0,a]$. Then for $z = re^{i\theta}$ and any branch of $\log(z - a)^{-1}$,

$$\int_0^{2\pi} \left(\log \frac{1}{|re^{i\theta} - a|}\right) \cos n\theta \, d\theta = \operatorname{Re} \int_0^{2\pi} \left(\log \frac{1}{re^{i\theta} - a}\right) \cos n\theta \, d\theta$$

$$= \operatorname{Re} \int_{|z|=r} \left(\log \frac{1}{z - a}\right) \cdot \frac{1}{2}\left(\frac{z^n}{r^n} + \frac{r^n}{z^n}\right)\frac{dz}{iz}$$

$$= \operatorname{Re} \frac{1}{2i} \int_{|z|=r} \left(\log \frac{1}{z - a}\right)\left(\frac{z^{n-1}}{r^n} + \frac{r^n}{z^{n+1}}\right) dz.$$

Denote the integrand by $w(z)$. Then the integral is

$$2\pi i \operatorname*{Res}_{z=0} w(z) = 2\pi i \cdot \frac{1}{n!} \left(z^{n+1} w(z)\right)^{(n)}_{z=0} ,$$

where $z^{n+1} w = w_1 + w_2$ with

$$w_1(z) = \left(\log \frac{1}{z - a}\right) \frac{z^{2n}}{r^n}, \qquad w_2(z) = r^n \log \frac{1}{z - a}.$$

We obtain

$$w_1^{(n)}(0) = 0, \; w_2^{(n)}(0) = (n - 1)! \; r^n a^{-n},$$

and the third asserted formula follows for $r < a$.

The fourth formula is obtained for $r > a$ as before, and the case $r = a$ is again by continuity.

<u>4.13.</u> <u>Case</u> $y \neq 0$. We expand $e(re^{i\theta},a)$ into its Fourier series,

$$e(re^{i\theta},a) = \sum_{n=0}^{\infty} e_n(r)\cos n\theta .$$

Since $e(re^{-i\theta},a) = e(re^{i\theta},a)$, there are no sine terms, and

$$e_0(r) = \begin{cases} \frac{1}{2\pi} \log \frac{1}{a}, & r \in [0,a], \\ \frac{1}{2\pi} \log \frac{1}{r}, & r \in [a,\infty), \end{cases}$$

$$e_n(r) = \begin{cases} \frac{1}{2\pi n} \frac{r^n}{a^n}, & r \in [0,a], \\ \frac{1}{2\pi n} \frac{a^n}{r^n}, & r \in [a,\infty) \end{cases}$$

for $n = 1,2,\ldots$. By $(e + k_R, u)_R = 0$,

$$(e(re^{i\theta},a) + k_R(re^{i\theta},a),\ r^n \sin n\theta)_R = 0$$

for $n = 1,2,\ldots$. Since $(e(re^{i\theta},a),\ r^n \sin n\theta)_R = 0$, it follows that $(k_R(re^{i\theta},a),\ r^n \sin n\theta)_R = 0$ and

$$k_R(re^{i\theta},a) = \sum_{n=0}^{\infty} c_n r^n \cos n\theta.$$

To determine c_n, we again use $(e + k_R, u)_R = 0$, with $u = r^n \cos n\theta$:

$$\int_0^R \left(\int_0^{2\pi} \left(\sum_{m=0}^{\infty} (e_m(r) + c_m r^m) \cos m\theta \right) r^n \cos n\theta \, d\theta \right) \lambda(r) r \, dr = 0,$$

that is,

$$\int_0^R (e_n(r) r^n + c_n r^{2n}) \lambda(r) r \, dr = 0$$

and thus,

$$c_n = - \frac{\int_0^R e_n(r) r^n \lambda(r) r \, dr}{\int_0^R r^{2n} \lambda(r) r \, dr}$$

for $n = 0,1,2,\ldots$.

By virtue of the Fourier expansions of e and k_R,

$$(e(a,\cdot),\ k_R(\cdot,a))_R = \int_0^R (2\pi c_0 e_0(r) + \pi \sum_{n=1}^{\infty} c_n r^n e_n(r))\lambda(r)r\ dr$$

$$= 2\pi c_0 \int_0^R e_0(r)\lambda(r)r\ dr + \pi \sum_{n=1}^{\infty} c_n \int_0^R e_n(r)r^n\lambda(r)r\ dr.$$

By the above expression for c_n, we have

$$(e(a,\cdot),\ k_R(\cdot,a))_R = -2\pi \frac{\left(\int_0^R e_0(r)\lambda(r)r\ dr\right)^2}{\int_0^R \lambda(r)r\ dr}$$

$$- \pi \sum_{n=1}^{\infty} \frac{\left(\int_0^R e_n(r)r^n\lambda(r)r\ dr\right)^2}{\int_0^R r^{2n}\lambda(r)r\ dr}$$

$$= -\pi \sum_{n=0}^{\infty} \alpha_n.$$

For $n = 1,2,\ldots,$ we have by the above expression for e_n,

$$e_n(r)r^n = \frac{1}{2\pi n}\frac{r^n}{a^n}\, r^n \leq \frac{1}{2\pi n}\, r^n \qquad \text{for } r \in [0,a],$$

$$e_n(r)r^n = \frac{1}{2\pi n}\frac{a^n}{r^n}\, r^n \leq \frac{1}{2\pi n}\, r^n \qquad \text{for } r \in [a,\infty),$$

so that

$$e_n(r)r^n \leq \frac{1}{2\pi n}\, r^n \qquad \text{for } r \in [0,\infty).$$

It follows that, for $n = 1,2,\ldots,$

$$\left(\int_0^R e_n(r)r^n\lambda(r)r\ dr\right)^2 \leq \left(\int_0^R \frac{1}{2\pi n}\, r^n\lambda(r)r\ dr\right)^2$$

$$\leq \frac{1}{4\pi^2 n^2}\left(\int_0^R r^{2n}\lambda(r)r\ dr\right)\left(\int_0^R \lambda(r)r\ dr\right)$$

$$\leq \frac{\Lambda}{4\pi^2 n^2}\int_0^R r^{2n}\lambda(r)r\ dr.$$

Therefore,

$$\alpha_n \le \frac{\Lambda}{4\pi^2 n^2},$$

$n = 1,2,\ldots$. Using the Schwarz inequality and then letting $R \to \infty$ we obtain by the above expression for $(e,k_R)_R$,

$$\alpha_0 \le 2\left(\left(\frac{1}{2\pi}\log\frac{1}{a}\right)^2 \Lambda + \int_0^\infty \left(\frac{1}{2\pi}\log\frac{1}{r}\right)^2 \lambda(r) r\, dr\right) = A_0 < \infty,$$

and a fortiori,

$$|(e(a,\cdot),\ k_R(\cdot,a))_R| \le \pi\left(A_0 + \frac{\Lambda}{4\pi^2}\sum_{n=1}^\infty \frac{1}{n^2}\right) = A < \infty.$$

In view of the expression for β_R at the end of 4.11,

$$\beta_R(y,y) = \beta_R(a,a) \le (e(a,\cdot),\ e(\cdot,a))_R + |(e(a,\cdot),\ k_R(\cdot,a))_R|$$

with $a = |y|$. Since $\log|x|^{-1} - \log|x - a|^{-1}$ is bounded for x outside some compact set, we obtain by the inequality in Theorem 4.9 and the relation $\Lambda < \infty$,

$$\|e(\cdot,a)\|^2 = B < \infty$$

and thus

$$\beta_R(y,y) \le A + B$$

for every $R > 0$. For any regular subregion Ω of C_λ, there exists an $R > 0$ with $\Omega_R \supset \Omega$, and

$$\beta_\Omega(y,y) = \|H_\alpha(\cdot,y)\|^2 \le \|H_{\alpha_R}(\cdot,y)\|^2 = \beta_R(y,y).$$

Hence,

$$c_y = \sup_{\Omega\subset C_\lambda} \beta_\Omega(y,y) \le A + B < \infty.$$

Thus $c_y < \infty$ for every $y \in C_\lambda - \{0\}$, and the existence of $\lim_\Omega \beta_\Omega(x,y)$ follows for $y \neq 0$.

4.14. Case $y = 0$. The case $y = 0$ is simple. Since $k_R(x,0)$ is radial,

$$e(re^{i\theta},0) = \frac{1}{2\pi}\log\frac{1}{r},\quad k_R(re^{i\theta},0) = c_0 = \text{const.}$$

By $(e + k_R, u)_R = 0$ with $u = \text{const}$,

$$\int_0^R \left(\frac{1}{2\pi}\log\frac{1}{r} + c_0\right)\lambda(r)r\,dr = 0,$$

that is,

$$c_0 = -\frac{\int_0^R \left(\frac{1}{2\pi}\log\frac{1}{r}\right)\lambda(r)r\,dr}{\int_0^R \lambda(r)r\,dr}\,.$$

Therefore,

$$\beta_R(0,0) \le \|e(\cdot,0)\|^2 + 2\pi\,\frac{\left(\int_0^R \left(\frac{1}{2\pi}\log\frac{1}{r}\right)\lambda(r)r\,dr\right)^2}{\int_0^R \lambda(r)r\,dr}$$

$$\le 2\pi\int_0^\infty \left(\frac{1}{2\pi}\log\frac{1}{r}\right)^2\lambda(r)r\,dr + 2\pi\int_0^\infty \left(\frac{1}{2\pi}\log\frac{1}{r}\right)^2\lambda(r)r\,dr < \infty.$$

As in 4.13, the existence of $\lim_\Omega \beta_\Omega(x,y)$ follows for $y = 0$.

This completes the proof of Theorem 4.9.

We can actually say somewhat more: If λ satisfies

$$\int_0^\infty (\log r)^2\lambda(r)r\,dr < \infty,$$

then the convergence in

$$\beta(x,y) = \lim_{\Omega\to C_\lambda} \beta_\Omega(x,y)$$

is uniform in x on every compact subset of C_λ, for each fixed $y \in C_\lambda$. The proof is based on extending the integrals in 4.13 to ∞, obtaining for c_y an upper estimate φ_y which is continuous on C_λ, and invoking Theorem 4.8.

NOTES TO §4. Theorems 4.3, 4.8, and 4.9 were proved in Nakai-Sario [11]. We wished to present the discussion in 4.7-4.8 independently of Chapter IX. However, an inspection of the reasoning in 4.11-4.14 shows that the Main Theorem IX.1.3 and its consequence, Theorem 1.9, apply, and we have $C_\lambda \in \tilde{O}_\beta^N$ if and only if the condition in Theorem 4.9 is met. The resulting relation $O_G^N \cap \tilde{O}_\beta^N \neq \emptyset$ for $N = 2$ complements this relation proved in §1 for $N > 2$. Another consequence of Theorem 4.9, viz., $O_G^N \cap O_\beta^N \neq \emptyset$ for $N = 2$, will be extended in §5 to hold for $N > 2$ as well.

An interesting question, not discussed in the present book, is what relation, if any, Theorem 4.9 has with the sagging problem of circular elastic plates under a point load as the radius of the plate increases.

§5. FURTHER EXISTENCE RELATIONS BETWEEN HARMONIC AND BIHARMONIC GREEN'S FUNCTIONS

By §1, $O_\beta^N \subset O_\gamma^N$ and, in fact, $O_\beta^N < O_\gamma^N$ for every $N \geq 2$. On the other hand, by VIII.3.2, $O_G^N < O_\gamma^N$, so that $O_\beta^N \cup O_G^N \subset O_\gamma^N$. We ask: Is this inclusion strict? The main result of the present section is that the answer is in the affirmative: For $N \geq 2$,

$$O_\beta^N \cup O_G^N < O_\gamma^N,$$

that is, there exist Riemannian manifolds of any dimension which carry both β and g but nevertheless fail to carry γ.

As to relations between O_β^N and O_G^N, we know from 1.10 and 4.9 that $\tilde{O}_\beta^N \cap O_G^N \neq \emptyset$. We shall show that, for $N \geq 2$,

$$O_\beta^N \cap O_G^N \neq \emptyset, \qquad O_\beta^N \cap \tilde{O}_G^N \neq \emptyset$$

as well. That there also exist Riemannian N-manifolds carrying both β and g is trivial in view of the Euclidean N-ball.

5.1. Parabolic manifolds without β. We claim:

THEOREM. For $N \geq 2$,

$$O_\beta^N \cap O_G^N \neq \emptyset.$$

Proof. Let R be the N-space $\{0 \leq |x| = r < \infty\}$ with the metric

$$ds^2 = \varphi(r)^2 dr^2 + \psi(r)^{2/(N-1)} \sum_{i=1}^{N-1} \gamma_i(\theta) d\theta^{i2},$$

where φ, ψ are strictly positive functions in $C^\infty[0,\infty)$ with $\varphi^2 = 1$, $\psi^{2/(N-1)} = r^2$ on $\{r < \frac{1}{2}\}$ and the γ_i are the trigonometric functions of $\theta = (\theta^1, \ldots, \theta^{N-1})$ which make the metric Euclidean on $\{r < \frac{1}{2}\}$. Set

$$\sigma = \varphi\psi, \quad \tau = \varphi\psi^{-1}, \quad \gamma(\theta) = \left(\prod_{i=1}^{N-1} \gamma_i(\theta)\right)^{1/2}.$$

In terms of the metric tensor, we have $g^{1/2} = \sigma\gamma$ and $g^{1/2} g^{rr} = \tau^{-1}\gamma$. The Laplace-Beltrami operator is

$$\Delta = -g^{-1/2}\left[\frac{\partial}{\partial r}\left(g^{1/2} g^{rr} \frac{\partial}{\partial r}\right) + \sum_{i=1}^{N-1} \frac{\partial}{\partial \theta^i}\left(g^{1/2} g^{ii} \frac{\partial}{\partial \theta^i}\right)\right].$$

The function

$$h(r) = \int_1^r \tau(s) ds$$

satisfies the harmonic equation

$$\Delta h(r) = -\sigma^{-1}(\tau^{-1} h')' = 0.$$

For a fixed $\rho \in (0,\infty)$, the function

$$q_\rho(r) = \int_r^\rho \tau \int_1^t \sigma \, ds \, dt$$

satisfies the quasiharmonic equation

$$\Delta q(r) = -\sigma^{-1}(\tau^{-1} q')' = 1,$$

and the function

$$u_\rho(r) = \int_r^\rho \tau \int_1^v \sigma \int_1^t \tau \, ds \, dt \, dv$$

the equation

$$\Delta u(r) = -\sigma^{-1}(\tau^{-1}u')' = \int_1^r \tau(s)ds.$$

The function

$$\beta_\rho(r) = -u_\rho(r) + c_\rho q_\rho(r), \quad c_\rho = \frac{u'_\rho(\rho)}{q'_\rho(\rho)},$$

is biharmonic and meets the boundary conditions

$$\beta_\rho(\rho) = \beta'_\rho(\rho) = 0.$$

We write in extenso

$$\beta_\rho(r) = -\int_r^\rho \tau \int_1^v \sigma \int_1^t \tau \, ds \, dt \, dv + \frac{\int_1^\rho \sigma \int_1^t \tau \, ds \, dt}{\int_1^\rho \sigma \, ds} \int_r^\rho \tau \int_1^t \sigma \, ds \, dt.$$

On $\{r > 1\}$, choose $\sigma = \tau = 1$, that is, take the metric

$$ds^2 = dr^2 + \sum_{i=1}^{N-1} \gamma_i(\theta) d\theta^{i2}.$$

Then

$$h(r) = \int_1^r ds$$

is unbounded, hence the harmonic measure of the ideal boundary $r = \infty$ of R on $\{r \geq 1\}$ is $\omega = ah + b = b = \text{const}$, and we have $R \in O_G^N$. On the other hand,

$$\beta_\rho(r) = -\int_r^\rho \int_1^v (t - 1)dt \, dv + \frac{\int_1^\rho (t - 1)dt}{\rho - 1} \int_r^\rho (t - 1)dt$$

$$= -\frac{1}{6}\left[(\rho - 1)^3 - (r - 1)^3\right] + \frac{1}{4}(\rho - 1)\left[(\rho - 1)^2 - (r - 1)^2\right]$$

is unbounded in ρ, so that

$$\lim_{\rho\to\infty} \beta_\rho \equiv \infty.$$

Since the existence of β on R is independent of the pole and the exhaustion, we conclude that $R \in O_\beta^N$.

5.2. Hyperbolic manifolds without β. We know from 1.10 and 4.9 that there exist parabolic manifolds which carry β. We proceed to show:

THEOREM. For $N \geq 2$,

$$O_\beta^N \cap \tilde{O}_G^N \neq \emptyset.$$

Proof. This is a corollary of $O_\beta^N \cap \tilde{O}_{HD}^N \neq \emptyset$ established in §3. Here we give an independent proof. Consider again the N-space with the metric

$$ds^2 = \varphi(r)^2 dr^2 + \psi(r)^{2/(N-1)} \sum_{i=1}^{N-1} \gamma_i(\theta) d\theta^{i2}$$

but now take

$$h(r) = \int_r^\infty \tau(s)ds$$

and

$$\beta_\rho(r) = \int_r^\rho \tau \int_1^v \sigma \int_t^\infty \tau \, ds\, dt\, dv - \frac{\int_1^\rho \sigma \int_t^\infty \tau \, ds\, dt}{\int_1^\rho \sigma \, ds} \int_r^\rho \tau \int_1^t \sigma \, ds\, dt.$$

On $\{r > 1\}$, choose $\sigma = r$, $\tau = r^{-2}$, that is, consider the metric

$$ds^2 = r^{-1} dr^2 + r^{3/(N-1)} \sum_{i=1}^{N-1} \gamma_i(\theta) d\theta^{i2}.$$

Then

$$h(r) = \int_r^\infty s^{-2} ds$$

is bounded, hence $R \in \tilde{O}_G^N$, whereas

$$\beta_\rho(r) = \int_r^\rho v^{-2}\int_1^v t\cdot t^{-1}\,dt\,dv - \frac{\rho - 1}{(1/2)(\rho^2 - 1)}\int_r^\rho t^{-2}\,\tfrac{1}{2}(t^2 - 1)dt$$

$$= \log\frac{\rho}{r} + (\rho^{-1} - r^{-1}) - \frac{1}{\rho + 1}(\rho - r + \rho^{-1} - r^{-1})$$

is unbounded in ρ, and we have $R \in O_\beta^N$.

5.3. Hyperbolic manifolds with β but without γ. The goal of the remainder of the present section is to prove, for $N \geq 2$, the strictness of the inclusion

$$O_\beta^N \cup O_G^N < O_\gamma^N.$$

For $N = 2$, the Euclidean half-plane gives the desired counterexample, as was first shown by J. Ralston. The authors are pleased to acknowledge their gratitude to Professor Ralston for communicating this unpublished result to them. Ralston's elegant proof is based on an explicit formula for β in the case $N = 1$ and a further development of the technique in Ralston-Sario [1]. Here we give an alternate proof which utilizes results in Chapter IX. Let Π^2 be the half-plane $x^1 > 0$ in E^2. The harmonic Green's function

$$g(z,\zeta) = \frac{1}{2\pi}\log\left|\frac{z + \bar\zeta}{z - \zeta}\right|$$

on Π^2 gives trivially $\Pi^2 \in \tilde O_G^2$. Let Ω_ζ be a neighborhood of ζ with $\bar\Omega_\zeta \subset \Pi^2$, and set $G = \Pi^2 - \bar\Omega_\zeta$, $G_\rho = \{r > \rho \,\big|\, |\arg z| < \pi/4\} \cap G$. As $\rho \to \infty$,

$$\|g\|_G^2 = c\int_G \left|\log\left|\frac{z + \bar\zeta}{z - \zeta}\right|\right|^2 r\,dr\,d\theta \geq c\int_{G_\rho}\left|\log\left|\frac{z - \bar\zeta}{z - \zeta}\right|\right|^2 r\,dr\,d\theta$$

$$\sim c\int_\rho^\infty \frac{1}{r^2}\cdot r\,dr = \infty,$$

hence $\Pi^2 \in O_\gamma^N$. On the other hand, we know from IX.8.6 that the subregion $\Sigma^2 = \{|z| > 1\}$ of E^2 carries β. Since the existence of β on a Riemannian manifold entails that on a subregion, the relation $\Pi^2 + 1 \subset \Sigma^2 \in \tilde O_\beta^2$ gives $\Pi^2 \in \tilde O_\beta^2$.

Actually, for $\Sigma^N = \{r > 1\}$ in E^N, we have from IX.8.6 that

$$\Sigma^N \in O_\gamma^N \cap \tilde{O}_\beta^N \cap \tilde{O}_G^N \quad \text{for} \quad N = 2,3,4.$$

This example has the virtue of being simple and natural. However, since $E^N \notin O_\gamma^N \cup O_\beta^N \cup O_G^N$ for $N \geq 5$, every subregion of E^N has the same property. As a consequence, there do not exist "simple and natural" Riemannian manifolds in $O_\gamma^N \cap \tilde{O}_\beta^N \cap \tilde{O}_G^N$ for $N \geq 5$. That this class is, nevertheless, nonvoid for $N \geq 5$ as well is the main result of the present section.

5.4. A test for $O_\gamma^N \cap \tilde{O}_\beta^N \cap \tilde{O}_G^N$. Our construction will be guided by the following test, a direct consequence of our results in Chapter IX. On a Riemannian N-manifold $R \in \tilde{O}_\beta^N$ with exhausting regular subregions Ω, we continue referring to the uniform convergence of β_Ω to β on compact subsets of $R \times R$ as the consistency of β on $R \times R$.

THEOREM. *Let R be a hyperbolic Riemannian N-manifold, $N \geq 2$, with the harmonic Green's function $g(x,y)$. For a parametric ball B with center η, suppose*

$$\int_{R-\bar{B}} *_x g(x,\eta)^2 = \infty$$

but

$$\int_{R-\bar{\Omega}} *_x (g(x,y) - g(x,\eta))^2 < \infty$$

for any $y \in R - \bar{B}$ and any regular subregion Ω of R with $\Omega \supset \bar{B} \cup y$. Then

$$R - \bar{B} \in O_\gamma^N \cap \tilde{O}_\beta^N \cap \tilde{O}_G^N,$$

and $\beta_{R-\bar{B}}$ is continuous and consistent on $(R - \bar{B}) \times (R - \bar{B})$.

The relations $R - \bar{B} \in \tilde{O}_G^N$ and $R - \bar{B} \in O_\gamma^N$ are again immediate. The function $g(x,y) - g(x,\eta)$ is a fundamental kernel on $R - \bar{B}$ in the sense of IX.3.2, and a fortiori, $R - \bar{B} \in \tilde{O}_\beta^N$. If, in the definition of O_β^N, we disregard the continuity and consistency of β on the product space, then, $g(x,y) - g(x,\eta)$ being square integrable off its pole y, the characterization of O_β^N given in 2.6 makes the relation $R - \overline{B} \in \tilde{O}^N_\beta$ as trivial as $R - \overline{B} \in O^N_\gamma$.

5.5. Comparison principle. We insert here a general statement which will be used later. Let $0 \leq \alpha < \beta < \infty$ and $a \in C^1(\alpha,\beta)$. Consider the ordinary differential operator

$$Lu = (au')' - pu$$

with $p \in C(\alpha,\beta)$. If a function u satisfies

$$Lu \leq 0$$

on (α,β), then it is called a supersolution on (α,β). If a supersolution u satisfies

$$\liminf_{r \to \alpha} u(r) \geq 0, \quad \liminf_{r \to \beta} u(r) \geq 0,$$

then $u \geq 0$ on (α,β). This result was obtained, and called the comparison principle, in Nakai [8] (the proof for the above operator is the same as for the elliptic operator). We shall use this principle in the following form:

LEMMA. Let u be a solution of $Lu = 0$ in (α,β) with boundary values $u(\alpha)$ and $u(\beta)$, and let v be a supersolution, $Lv \leq 0$, on (α,β) with boundary values $v(\alpha) = u(\alpha)$ and $v(\beta) = u(\beta)$. Then $u \leq v$ on (α,β).

5.6. Expansions in spherical harmonics. For convenience, we summarize here some basics on spherical harmonics. At a point $x = (x^1,\ldots,x^N)$ of E^N, $N \geq 2$, $r = |x|$, the line element $|dx|^2 = \sum_1^N dx^{i2}$ reads in polar coordinates $(r,\theta) = (r,\theta^1,\ldots,\theta^{N-1})$,

$$|dx|^2 = dr^2 + r^2 \sum_{i=1}^{N-1} \gamma_i(\theta) d\theta^{i2},$$

and the area of the unit sphere $\omega = \{|x| = r = 1\}$ is

$$A_N = \int_\omega d\omega = 2\pi^{N/2}\,\Gamma(N/2)^{-1}.$$

For $\gamma(\theta) = (\Pi_1^{N-1}\gamma_i(\theta))^{1/2}$, the Euclidean Laplace-Beltrami operator

$$\Delta_E = -r^{-(N-1)} \frac{\partial}{\partial r}\left(r^{N-1}\frac{\partial}{\partial r}\right) - (\gamma(\theta)r^2)^{-1} \sum_{i=1}^{N-1} \frac{\partial}{\partial\theta^i}\left(\gamma(\theta)\gamma_i(\theta)^{-1}\frac{\partial}{\partial\theta^i}\right)$$

acting on a spherical harmonic $S_n(\theta)$ of degree $n \geq 1$ gives

$$\Delta_E S_n = n(n + N - 2)r^{-2}S_n.$$

The complete orthogonal system $\{S_{nm}\}$ of spherical harmonics of degree $n \geq 1$ gives a complete orthonormal system in $L^2(\omega, d\omega)$,

$$\left\{A_N^{-1/2}, S_{nm} \;\middle|\; m = 1,\ldots,m_n, \quad n = 1,2,\ldots\right\}.$$

Moreover, if $\varphi \in C^1(\omega)$, then

$$\varphi = c_0 + \sum_{n=1}^{\infty} \sum_{m=1}^{m_n} c_{nm} S_{nm}$$

with $c_0 = (\varphi,1)/A_N$, $c_{nm} = (\varphi, S_{nm})$, the inner product being in $L^2(\omega,d\omega)$, and the series is absolutely and uniformly convergent on ω; if φ depends on a parameter $r \in [r_1,r_2]$ and $\varphi \in C^1([r_1,r_2] \times \omega)$, then the convergence is uniform on $[r_1,r_2] \times \omega$.

<u>5.7.</u> <u>Main result</u>. We endow $\Sigma^N = \{r > 1\}$ with the metric

$$ds^2 = r^4 dr^2 + r^{8/(N-1)} \sum_{i=1}^{N-1} \gamma_i(\theta)d\theta^{i2}$$

and denote by Σ^N_{ds} the resulting Riemannian manifold.

<u>MAIN THEOREM</u>. <u>For</u> $N \geq 5$, <u>the manifold</u> Σ^N_{ds} <u>is hyperbolic, carries no</u> γ, <u>but carries a</u> β <u>which is continuous and consistent on</u> $\Sigma^N_{ds} \times \Sigma^N_{ds}$:

$$\Sigma^N_{ds} \in O^N_\gamma \cap \tilde{O}^N_\beta \cap \tilde{O}^N_G .$$

The proof will be given in 5.7-5.11.

Choose strictly positive C^∞ functions $R_1(r)$, $R_2(r)$ on $[0,\infty)$ such that $R_1(r) = r^4$, $R_2(r) = r^{8/(N-1)}$ on $[1,\infty)$ and $R_1(r) = 1$, $R_2(r) = r^2$ on $[0,\frac{1}{2}]$.

The metric

$$d\tilde{s}^2 = R_1(r)dr^2 + R_2(r) \sum_{i=1}^{N-1} \gamma_i(\theta)d\theta^{i2}$$

is C^∞ on E^N, $d\tilde{s} = ds$ on Σ^N, and $d\tilde{s} = |dx|$ on $|x| < \frac{1}{2}$. Accordingly, we may and will henceforth view ds as <u>a metric on</u> E^N, with $ds = |dx|$ on $|x| < \frac{1}{2}$.

<u>5.8.</u> <u>Hyperbolicity</u>. The metric ds on E^N gives the volume element on Σ^N

$$*1 = r^6 \, dr \, d\omega,$$

the surface element on $|x| = r \geq 1$ is

$$dS = r^4 \, d\omega,$$

and the interior normal derivative $\partial\varphi/\partial n$ on $|x| = r > 1$ is

$$\frac{\partial\varphi}{\partial n} = -r^{-2} \frac{\partial\varphi}{\partial r} \, .$$

The Laplace-Beltrami operator Δ with respect to ds takes the form

$$\Delta = -r^{-6} \frac{\partial}{\partial r}\left(r^2 \frac{\partial}{\partial r}\right) - \left(\gamma(\theta)r^{8/(N-1)}\right)^{-1} \sum_{i=1}^{N-1} \frac{\partial}{\partial\theta^i}\left(\gamma(\theta)\gamma_i(\theta)^{-1} \frac{\partial}{\partial\theta^i}\right).$$

For a function $\psi(\theta)$, the expression for Δ_E in 5.6 gives

$$\Delta\psi = r^{2-8/(N-1)}\Delta_E\psi.$$

In view of the expression for $\Delta_E S_n$ in 5.6,

$$\Delta S_n = n(n + N - 2)r^{-8/(N-1)}S_n$$

for every spherical harmonic S_n of degree $n \geq 1$.

For a function $\psi(r)$,

$$\Delta\psi = -r^{-6} \frac{d}{dr}(r^2 \frac{d}{dr}\psi).$$

In terms of the ordinary differential operator

$$L\psi = \frac{d}{dr}(r^2 \frac{d}{dr}\psi),$$

$\psi(x) = \psi(r)$ belongs to the class $H(r > 1)$ of harmonic functions relative to ds if and only if $L\psi = 0$, that is,

$$\psi = c_0 + c_1 r^{-1},$$

with c_0, c_1 constants. Thus $1 - \rho/r$ is the harmonic measure of the ideal boundary ∞ of Σ^N_{ds} (and of E^N_{ds}) on $r > \rho \geq 1$. Therefore, $E^N \in \tilde{O}^N_G$ and

$$\Sigma^N \in \tilde{O}^N_G.$$

<u>5.9.</u> <u>An inequality</u>. The constant $\rho_N = N^{(N-1)/(6N-14)}$ dominates 1 in our case $N \geq 5$. This ρ_N is so chosen that

$$n + N - 3 \leq nr^{(6N-14)/(N-1)}$$

for every $n = 1,2,\ldots$ and every $r \in [\rho_N,\infty)$. Consider the ordinary differential operator

$$L_n\psi = \frac{d}{dr}(r^2 \frac{d}{dr}\psi) - n(n + N - 2)r^{(6N-14)/(N-1)}\psi$$

for each $n = 0,1,2\cdots$ on $[1,\infty)$. Observe that

$$L_n r^{-1} = -n(n + N - 2)r^{(6N-14)/(N-1)}r^{-1} \leq 0,$$

that is, r^{-1} is a supersolution of $L_n\psi = 0$ on $[1,\infty)$. Since 0 is a solution of $L_n\psi = 0$, the Perron method assures the unique existence of a solution u of $L_n\psi = 0$ on $[1,\infty)$ with boundary values $u(1) = 1$ and $u(\infty) = 0$. Hence, there exists a unique solution

$$e_n(r;\rho)$$

of $L_n\psi = 0$ on $[\rho,\infty)$, $\rho \geq 1$, with boundary values $e_n(\rho;\rho) = 1$ and $e_n(\infty;\rho) = 0$. The key relation in our reasoning will be

$$0 < e_n(r;\rho) \leq \rho^{n+N-2}/r^{n+N-2}$$

for every $n = 1,2,\ldots$ and $r \in [\rho,\infty)$, with $\rho \geq \rho_N$. For the proof, $n + N - 3 \leq nr^{(6N-14)/(N-1)}$ gives

$$L_n r^{-n-N+2} = (n + N - 2)\left(n + N - 3 - nr^{(6N-14)/(N-1)}\right)r^{-n-N+2} \leq 0.$$

By the comparison principle, the above relation follows.

5.10. Fourier expansion. Let $h \in H(r > \rho)$, $\rho \geq \rho_N$. We consider its Fourier expansion

$$h(r,\theta) = h_0(r) + \sum_{n=1}^{\infty} \sum_{m=1}^{m_n} h_{nm}(r)\, S_{nm}(\theta)$$

for $r \in (\rho,\infty)$, with $\Delta h_0 = \Delta(h_{nm}\, S_{nm}) = 0$ for every n and m. The expansion converges absolutely and uniformly on compact sets of $\{r > \rho\}$. By 5.8,

$$L_0 h_0 = L_n h_{nm} = 0$$

for every n and m on (ρ,∞). We assume that $h(r,\theta)$ is continuous on $\rho \leq r \leq \infty$ and of class C^1 on $\rho \leq r < \infty$, and

$$\lim_{r\to\infty} h(r,\theta) = 0.$$

Then h_0 and h_{nm} also are continuous on $[\rho,\infty]$ and

$$h_0 = c_0 e_0(\cdot;\rho), \quad h_{nm} = c_{nm} e_n(\cdot;\rho),$$

with $c_0 = h_0(\rho)$, $c_{nm} = h_{nm}(\rho)$, and

$$|c_0| + \sum_{n=1}^{\infty} \sum_{m=1}^{m_n} |c_{nm}| < \infty, \quad |c_0|^2 + \sum_{n=1}^{\infty} \sum_{m=1}^{m_n} |c_{nm}|^2 < \infty.$$

The above expression for $h(r,\theta)$ thus takes the form

$$h(r,\theta) = c_0 e_0(r;\rho) + \sum_{n=1}^{\infty}\left(\sum_{m=1}^{m_n} c_{nm}\, S_{nm}(\theta)\right) e_n(r;\rho)$$

on $[\rho,\infty]$.

5.11. Conclusion. Let $g(x,y) = g(r,\theta;y)$ be the harmonic Green's function on E^N_{ds}, normalized by the flux $\int_\alpha *dg = -1$ across a hypersphere α about y. In particular,

$$g(r,\theta;0) = ce_0(r;1) = cr^{-1},$$

with $c = g(1,\theta;0)$, on Σ^N_{ds}. By $*1 = r^6 dr\, d\omega$,

$$\int_{\Sigma^N_{ds}} *g(x,0)^2 = cA_N \int_1^\infty r^{-2} \cdot r^6\, dr = \infty.$$

We claim that

$$\int_{r>\rho} *(g(x,y) - g(x,0))^2 < \infty$$

for $\rho = \rho_N + |y|$ and $y \in \Sigma^N_{ds}$. To see this, we apply the above expression for $h(r,\theta)$ to $g(\cdot,z)$ with $|z| \leq |y|$:

$$g(r,\theta;z) = c_0(z)e_0(r;\rho) + \sum_{n=1}^{\infty}\left(\sum_{m=1}^{m_n} c_{nm}(z)\, S_{nm}(\theta)\right) e_n(r;\rho)$$

for $r \geq \rho$. Since

$$c_0(z)e_0(r;\rho) = \frac{1}{A_N}\int_\omega g(r,\theta;z)d\omega,$$

we have

$$c_0(z)\, \frac{d}{dr}e_0(r;\rho) = \frac{1}{A_N}\int_\omega \frac{\partial}{\partial r}\, g(r,\theta;z)d\omega,$$

and, by the expressions for dS and $\partial\varphi/\partial n$ in 5.8, obtain

$$-c_0(z)r^2\, \frac{d}{dr}e_0(r;\rho) = \frac{1}{A_N}\int_{|x|=r} \frac{\partial}{\partial n_x}\, g(x;z)dS.$$

Here the integral is the value at z of the harmonic function on $|x| < r$ with boundary values 1 on $|x| = r$, that is, of the constant function 1. Since

$e_0(r;\rho) = \rho/r,$

$$c_0(z) = (A_N\rho)^{-1},$$

hence $c_0(z)$ is constant for all $|z| \le |y|$. Therefore,

$$g(r,\theta;y) - g(r,\theta;0) = \sum_{n=1}^{\infty}\left(\sum_{m=1}^{m_n} d_{nm} S_{nm}(\theta)\right) e_n(r;\rho),$$

where $d_{nm} = c_{nm}(y) - c_{nm}(0) = c_{nm}(y)$. Thus,

$$\int_\omega (g(r,\theta;y) - g(r,\theta;0))^2 d\omega = \sum_{n=1}^{\infty}\left(\sum_{m-1}^{m_n} d_{nm}^2\right) e_n(r;\rho)^2$$

and consequently,

$$\int_{|x|>\rho} *(g(x,y) - g(x,0))^2 = \sum_{n=1}^{\infty} k_n \int_\rho^\infty e_n(r;\rho)^2 r^6 dr$$

with $k_n = \sum_{m=1}^{m_n} d_{nm}^2$. Here, by the next to the last relation in 5.10,

$$\sum_{n=1}^{\infty} k_n < \infty.$$

Now we make use of $e_n(r;\rho) \le \rho^{n+N-2}/r^{n+N-2}$:

$$\int_\rho^\infty e_n(r;\rho)^2 r^6 dr \le \rho^{2(n+N-2)} \int_\rho^\infty r^{-2n-2N+4} \cdot r^6\, dr$$

$$\le \rho^{2(n+N-2)} \int_\rho^\infty r^{-2n} dr \le \frac{1}{2n-1}\rho^{2N-3},$$

since $N \ge 5$. It follows that

$$\int_{|x|>\rho} *(g(x,y) - g(x,0))^2 \le \sum_{n=1}^{\infty} \rho^{2N-3} \cdot \frac{k_n}{2n-1}.$$

This with $\sum k_n < \infty$ implies the asserted square integrability of $g(x,y) - g(x,0)$ over $|x| > \rho$.

In view of $g(x,0) \notin L^2$ and Theorem 5.4, the proof of the Main Theorem is complete.

<u>NOTES TO §5</u>. Theorems 5.1, 5.2, 5.4, and the Main Theorem 5.7 were proved in Nakai-Sario [18]. The Main Theorem completes the array of relations between O_G^N, O_β^N, O_γ^N, and the harmonic null classes.

CHAPTER XI

HADAMARD'S CONJECTURE ON THE GREEN'S FUNCTION OF A CLAMPED PLATE

The prize problem of the Paris Academy of integrating the equation of the clamped thin elastic plate was solved by Hadamard, in his monumental 1908 memoir [1]. He also made in it the famous conjecture that if such a plate is subject to a point load, the resulting deflection, given by the biharmonic Green's function of the clamped plate, is always of constant sign.

The first to challenge Hadamard's conjecture was Duffin [1], in 1949. He showed that if a uniform load is applied to a cross-section of an infinite strip clamped along the edges, the deflection will not be of constant sign but will oscillate as an exponentially damped sine wave.

Finite clamped plates with deflections of nonconstant sign under a point load were then exhibited by Garabedian [1] in 1951 and by Loewner [1] and Szegö [1] in 1953. Recent interest in the problem was stirred by the invited address of Duffin [2] before the annual meeting of American Mathematical Society in San Francisco in 1974.

In §1, we shall give a new simple counterexample to Hadamard's conjecture. We then show in §2 that the conjecture is untrue also for higher dimensions N: there exist Riemannian manifolds of any $N \geq 2$ on which the biharmonic Green's function of the clamped body is of nonconstant sign for some location of the pole. In §3, we show that any nested sequence of regular subregions exhausting Duffin's infinite strip gives eventually regions with a Green's function of the clamped plate of nonconstant sign. As special cases we obtain Duffin's strip example, sharpened to the case of a point load, and Garabedian's ellipse, without computations. A new example, a sufficiently elongated rectangle, also ensues.

§1. GREEN'S FUNCTIONS OF THE CLAMPED PUNCTURED DISK

If a thin elastic circular plate $B = \{|z| < 1\}$ is clamped (simply supported, resp.) along its edge $|z| = 1$, its deflection at $z \in B$ under a point load at $\zeta \in B$, measured positively in the direction of the gravitational pull, is the biharmonic Green's function $\beta(z,\zeta)$ of the clamped plate ($\gamma(z,\zeta)$ of the simply supported plate, resp.) (e.g., Muskhelishvili [1], Timoshenko [1]). We ask: How do $\beta(z,\zeta)$ and $\gamma(z,\zeta)$ compare with the corresponding deflections $\beta_0(z,\zeta)$ and $\gamma_0(z,\zeta)$ of the punctured circular plate $B_0 = \{0 < |z| < 1\}$ that is "clamped" or "simply supported", resp., also at the origin? We shall show that $\gamma(z,\zeta)$ is not affected by the puncturing, that is, $\gamma(\cdot,\zeta) = \gamma_0(\cdot,\zeta)$, whereas $\beta(\cdot,\zeta)$ is:

$$\beta_0(z,\zeta) = \beta(z,\zeta) - 16\pi\beta(z,0)\beta(\zeta,0)$$

on $B_0 \times B_0$. Moreover, while $\beta(\cdot,\zeta)$ is of constant sign, $\beta_0(\cdot,\zeta)$ is not. This gives a simple counterexample to Hadamard's conjecture:

<u>The biharmonic Green's function of a clamped concentric circular annulus is not of constant sign if the radius of the inner boundary circle is sufficiently small</u>.

<u>1.1</u>. <u>Clamping and simple supporting</u>. First we make precise what we mean by clamping and simple supporting at the isolated point 0. Denote by B_s the annulus $s < |z| < 1$ for $s \in (0,1)$. The corresponding biharmonic Green's function $\beta_s(z,\zeta)$ ($\gamma_s(z,\zeta)$, resp.) of the clamped (simply supported, resp.) annulus B_s is characterized by

$$\Delta^2\beta_s(\cdot,\zeta) = \delta_\zeta \qquad (\Delta^2\gamma_s(\cdot,\zeta) = \delta_\zeta)$$

on B_s, and

$$\beta_s(\cdot,\zeta) = \frac{\partial}{\partial n}\beta_s(\cdot,\zeta) = 0 \qquad (\gamma_s(\cdot,\zeta) = \Delta\gamma_s(\cdot,\zeta) = 0)$$

on the boundary ∂B_s of B_s. Here Δ is the Laplace-Beltrami operator $-(\partial^2/\partial x^2 + \partial^2/\partial y^2)$, δ_ζ is the Dirac delta at $\zeta \in B_s$, and $\partial/\partial n$ denotes the inner normal derivative. We shall define $\beta_0(\cdot,\zeta)$ and $\gamma_0(\cdot,\zeta)$ as the limits of $\beta_s(\cdot,\zeta)$ and $\gamma_s(\cdot,\zeta)$ as $s \to 0$.

1.2. Simply supported punctured disk. Denote by $g_s(\cdot,\zeta)$ the harmonic Green's function of B_s with pole $\zeta \in B_s$, and by $g(\cdot,\zeta)$ that of B. By the maximum principle and the Riemann removability theorem, $\{g(\cdot,\zeta) - g_s(\cdot,\zeta)\}$ converges decreasingly and uniformly to zero on each compact subset of $\overline{B} - 0$ as $s \to 0$, and

$$g(z,\zeta) = \frac{1}{2\pi} \log \left| \frac{1 - \overline{\zeta} z}{z - \zeta} \right| .$$

In view of the boundary conditions for γ_s, we have

$$\gamma_s(z,\zeta) = \int_{B_s} g_s(w,z) g_s(w,\zeta) du dv,$$

$w = u + iv$, on $B_s \times B_s$. On letting $s \to 0$, we see that

$$\gamma_0(z,\zeta) = \lim_{s \to 0} \gamma_s(z,\zeta)$$

exists uniformly on each compact subset of $B_0 \times B_0$, and

$$\gamma_0(z,\zeta) = \int_{B_0} g(w,z) g(w,\zeta) du dv.$$

On the other hand, since $\Delta^2 \gamma(\cdot,\zeta) = \delta_\zeta$ on B and $\gamma(\cdot,\zeta) = \Delta\gamma(\cdot,\zeta) = 0$ on ∂B,

$$\gamma(z,\zeta) = \int_B g(w,z) g(w,\zeta) du dv.$$

On comparing the right-hand sides we conclude that

$$\gamma_0(z,\zeta) = \gamma(z,\zeta)$$

on $B_0 \times B_0$. Thus simple supporting at a single point does not have any effect on the deflection of a simply supported disk. This result agrees with physical intuition: if we place the tip of a needle under a very thin plate that is simply supported along its periphery, and then put a sufficient point load on the plate, the plate will be pierced by the needle.

1.3. Clamped punctured disk. In contrast with the above, what happens to $\beta_0(\cdot,\zeta)$ is somewhat surprising. Denote by $H_s(\cdot,\zeta) = \Delta\beta_s(\cdot,\zeta)$ the β-density of $\beta_s(\cdot,\zeta)$. It is again readily deduced from Stokes' formula that $H_s(\cdot,\zeta) \perp H_2(B_s)$,

$$\int_{B_s} h(w)H_s(w,\zeta)dudv = 0,$$

for any h in the class $H_2(B_s)$ of square integrable harmonic functions on B_s (cf. IX.§1). As consequences, we easily obtain

$$\begin{cases} \beta_s(z,\zeta) = \int_{B_s} H_s(w,z)H_s(w,\zeta)dudv, \\ \|H_t(\cdot,\zeta) - H_s(\cdot,\zeta)\|^2 = \|H_t(\cdot,\zeta)\|^2 - \|H_s(\cdot,\zeta)\|^2, \quad 0 < t < s < 1, \\ |\beta_t(z,\zeta) - \beta_s(z,\zeta)| \le \|H_t(\cdot,z) - H_s(\cdot,z)\|\cdot\|H_t(\cdot,\zeta) - H_s(\cdot,\zeta)\|, \\ \|H_s(\cdot,\zeta)\| \le \|g(\cdot,\zeta)\|, \end{cases}$$

where $\|\cdot\|$ is the L^2-norm on B and functions here and hereafter are defined to be zero outside their genuine domains of definition. It follows that

$$\beta_0(z,\zeta) = \lim_{s\to 0} \beta_s(z,\zeta)$$

exists uniformly on each compact subset of $\overline{B} - 0$. If we denote by $H_0(\cdot,\zeta) = \Delta\beta_0(\cdot,\zeta)$ the β-density of $\beta_0(\cdot,\zeta)$, then

$$\begin{cases} H_0(\cdot,\zeta) \perp H_2(B_0), \\ \lim_{s\to 0} \|H_0(\cdot,\zeta) - H_s(\cdot,\zeta)\| = 0, \\ \beta_0(z,\zeta) = \int_{B_0} H_0(w,z)H_0(w,\zeta)dudv = \int_{B_0} H_0(w,z)K(w,\zeta)dudv, \end{cases}$$

where $K(\cdot,\zeta)$ is any square integrable function on B_0 with $\Delta K(\cdot,\zeta) = \delta_\zeta$ on B.

1.4. Clamped disk. The function $\beta(\cdot,\zeta)$ is defined by $\Delta^2\beta(\cdot,\zeta) = \delta_\zeta$ on B and $\beta(\cdot,\zeta) = \partial\beta(\cdot,\zeta)/\partial n = 0$ on ∂B. An explicit expression for $\beta(\cdot,\zeta)$ is known (e.g., Garabedian [2]):

$$\beta(z,\zeta) = \frac{1}{8\pi}\left[\,|z-\zeta|^2 \log\left|\frac{z-\zeta}{1-\bar{\zeta}z}\right| + \tfrac{1}{2}(|z|^2-1)(|\zeta|^2-1)\right]$$

on $B\times B$. Our immediate aim is to express $\beta_0(z,\zeta)$ in terms of $\beta(z,\zeta)$. The basis of our computation is $H_0(\cdot,\zeta) \perp H_2(B_0)$ and its counterpart $H(\cdot,\zeta) \perp H_2(B)$, where $H(\cdot,\zeta) = \Delta\beta(\cdot,\zeta)$ is the β-density of $\beta(\cdot,\zeta)$. The latter orthogonality relation implies that

$$\beta(z,\zeta) = \int_B H(w,z)H(w,\zeta)dudv = \int_B H(w,z)g(w,\zeta)dudv$$

on $B\times B$. Since $H(\cdot,\zeta) - H_0(\cdot,\zeta)$ is harmonic on B_0 and square integrable over B_0, we have

$$H(re^{i\theta},\zeta) - H_0(re^{i\theta},\zeta) = ag(r) + b + \sum_{n=1}^{\infty}\Big(\sum_{m=1}^{2} c_{nm}S_{nm}(\theta)\Big)r^n,$$

with uniform convergence on each compact subset of $\bar{B}-0$. Here a, b, and c_{nm} are constants, $g(r) = g(r,0) = -(1/2\pi)\log r$, $S_{n1}(\theta) = \cos n\theta$, and $S_{n2}(\theta) = \sin n\theta$ for $n = 1,2,\ldots$. We denote by $(\cdot,\cdot)$ the inner product on $L^2(B)$ and by $\|\cdot\|_1$ the norm on $L^1(B)$. Since $h_{nm}(re^{i\theta}) = S_{nm}(\theta)r^n$ is in the class $H_2(B) \subset H_2(B_0)$, and $\|h_{nm}\| \neq 0$,

$$c_{nm}\|h_{nm}\|^2 = (H(\cdot,\zeta) - H_0(\cdot,\zeta), h_{nm}) = 0$$

and $c_{nm} = 0$ for every n and m. Observe that

$$\begin{cases} (H(\cdot,\zeta) - H_0(\cdot,\zeta),1) = a\|g\|_1 + b\pi, \\ (H(\cdot,\zeta) - H_0(\cdot,\zeta),g) = a\|g\|^2 + b\|g\|_1. \end{cases}$$

By virtue of $1 \in H_2(B) \subset H_2(B_0)$ and $g \in H_2(B_0)$ (but $g \notin H_2(B)$), these equations take the form

$$\begin{cases} \|g\|_1 a + \pi b = 0, \\ \|g\|^2 a + \|g\|_1 b = \beta(\zeta,0). \end{cases}$$

In view of $\|g\|_1 = 1/4$ and $\|g\|^2 = 1/8\pi$, we obtain $a = 16\pi\beta(\zeta,0)$ and $b = -4\beta(\zeta,0)$, and conclude that

$$H_0(\cdot,\zeta) = H(\cdot,\zeta) - 16\pi\beta(\zeta,0)g(\cdot,0) + 4\beta(\zeta,0)$$

on B_0. We take the inner product of each side with $H(\cdot,z)$ and obtain the following main identity of the present section:

$$\beta_0(z,\zeta) = \beta(z,\zeta) - 16\pi\beta(z,0)\beta(\zeta,0)$$

on $B_0 \times B_0$. This is the desired representation of β_0 in terms of β.

By the explicit expression of β, we have $\beta(z,\zeta) > 0$ on $B\times B$, and a fortiori,

$$\beta_0(z,\zeta) < \beta(z,\zeta)$$

on $B_0 \times B_0$. Thus adding to the clamping at the periphery the clamping at a single point 0 does have a noticeable effect on the resulting deflection. Compared with the case of γ_0, this result is quite intriguing.

We now analyze the boundary behavior of β_0 in some more detail, with a view on our main identity.

1.5. Boundary behavior. The relations $\Delta^2\beta_0(\cdot,\zeta) = \delta_\zeta$ on B_0 and $\beta_0(\cdot,\zeta) = \partial\beta_0(\cdot,\zeta)/\partial n = 0$ on $\partial B = \{|z| = 1\}$ are immediately verified. Thus both clamping conditions are satisfied at the outer boundary ∂B. Since $\beta(0,0) = 1/16\pi$, our main identity together with the symmetry of β give

$$\beta_0(0,\zeta) = \lim_{z\to 0} \beta_0(z,\zeta) = \beta(0,\zeta) - 16\pi\beta(0,0)\beta(\zeta,0) = 0,$$

so that the first clamping condition is satisfied at the inner boundary $z = 0$.

We proceed to examine the second condition. Denote by $\partial/\partial n_\theta$ the directional derivative in the direction $e^{i\theta}$,

$$\frac{\partial}{\partial n_\theta}\beta_0(0,\zeta) = \lim_{t\to +0} \frac{\beta_0(te^{i\theta},\zeta) - \beta_0(0,\zeta)}{t}.$$

Again by the main identity,

$$\frac{\partial}{\partial n_\theta}\beta_0(0,\zeta) = \frac{\partial}{\partial n_\theta}\beta(0,\zeta) - 16\pi\beta(\zeta,0)\frac{\partial}{\partial n_\theta}\beta(0,0).$$

Since $\beta(te^{i\theta},0) = (8\pi)^{-1}[t^2\log t - \frac{1}{2}(t^2-1)]$ for $t > 0$, we have

$$\frac{\partial}{\partial n_\theta}\beta(0,0) = \lim_{t\to+0}\frac{\beta(te^{i\theta},0)-\beta(0,0)}{t} = 0$$

and therefore,

$$\frac{\partial}{\partial n_\theta}\beta_0(0,\zeta) = \frac{\partial}{\partial n_\theta}\beta(0,\zeta).$$

Since $\beta(z,\zeta)$ is real analytic on the neighborhood $|z| < |\zeta|$ of $z = 0$,

$$\frac{\partial}{\partial n_\theta}\beta(0,\zeta) = \left[\frac{\partial}{\partial t}\beta(te^{i\theta},\zeta)\right]_{t=0}.$$

Using the explicit representation for $\beta(z,\zeta)$, we obtain by direct calculation,

$$\frac{\partial}{\partial n_\theta}\beta_0(0,\zeta) = \frac{1}{8\pi}|\zeta|(|\zeta|^2 - 2\log|\zeta| - 1)\cos(\theta - \arg\zeta).$$

Thus the "normal derivative" of $\beta_0(z,\zeta)$ at $z = 0$ does not vanish identically, and the second clamping condition is not satisfied. However, this failed "clamping" will conveniently serve to disprove Hadamard's conjecture, as we shall now see.

1.6. Hadamard's conjecture. Hadamard [1] conjectured that the Green's function of a clamped thin elastic plate cannot take on negative values. We give here a simple counterexample based on $\beta_0(z,\zeta)$. Observe that for any $\zeta \in B_0$,

$$|\zeta|^2 - 2\log|\zeta| - 1 > 0.$$

From this we see that

$$\text{sign}\left.\frac{\partial\beta_0(0,\zeta)}{\partial n_\theta}\right|_{\theta=\pi+\arg\zeta} \neq \text{sign}\left.\frac{\partial\beta_0(0,\zeta)}{\partial n_\theta}\right|_{\theta=\arg\zeta}.$$

This means that $\beta_0(0,\zeta)$ takes on values of opposite sign on line segments $\{z|0 < |z| < \sigma, \arg z = \pi + \arg\zeta\}$ and $\{z|0 < |z| < \sigma, \arg z = \arg\zeta\}$ for a sufficiently small $\sigma \in (0,1)$. This rather agrees with our intuition provided $\beta_0(0,\zeta) = 0$, a fact which, however, is not clear a priori.

By 1.3 we see that $\beta_s(\cdot,\zeta)$ converges to $\beta_0(\cdot,\zeta)$ uniformly on each compact subset of B_0 and, therefore, $\beta_s(\cdot,\zeta)$ takes on values of nonconstant sign along with $\beta_0(\cdot,\zeta)$ if $s > 0$ is sufficiently small. Thus we have the following

COUNTEREXAMPLE TO HADAMARD'S CONJECTURE. The Green's function of a clamped concentric circular annulus is of nonconstant sign if the radius of the inner boundary circle is sufficiently small.

NOTES TO §1. The above counterexample was given in Nakai-Sario [17]. The drawback of the example compared with those of Duffin [1], Garabedian [1], Loewner [1], and Szegö [1] is that, whereas their examples are simply connected, ours is not. In the simplicity of the proof, however, there is no comparison.

Recently it was shown in Nakai-Sario [22] that clamping of B_0 at $z = 0$ is effective whether B_0 is clamped or simply supported along $\{|z| = 1\}$, whereas simple supporting at $z = 0$ is not effective in either case.

§2. HADAMARD'S PROBLEM FOR HIGHER DIMENSIONS

Consider a locally Euclidean compact bordered manifold R of dimension $N \geq 2$ with a smooth (C^∞) border ∂R. The biharmonic Green's function $\beta(p,q)$ of a clamped body is characterized by the equation

$$\Delta_p^2 \beta(p,q) = \Delta_p(\Delta_p \beta(p,q)) = \delta_q$$

on R, and the boundary conditions

$$\beta(\cdot,q) = \frac{\partial}{\partial n}\beta(\cdot,q) = 0$$

on ∂R. Here $\Delta_p = -\Sigma_{j=1}^{N} \partial^2/\partial p^{j2}$ is the Laplace-Beltrami operator, δ_q the Dirac delta at $q \in R$, and $\partial/\partial n$ the inner normal derivative on ∂R with respect to R. Set

$$\begin{cases} \bar{s}(q) = \sup\limits_{p \in R} \beta(p,q) \\ \underline{s}(q) = \inf\limits_{p \in R} \beta(p,q). \end{cases}$$

Hadamard's problem can be reworded as follows: Does there exist a plane region R such that

$$\underline{s}(q) < 0 < \bar{s}(q) \quad \text{for some} \quad q \in R?$$

Although the physical significance is lost for $N \geq 3$, it is natural and interesting from a purely mathematical view point to ask whether or not there exists a higher dimensional R with the above property. The purpose of this section is to remark that there does exist such an R for every dimension $N \geq 2$. Instead of constructing a concrete example we shall take the following indirect road to prove the above assertion. Let R_0 be a locally Euclidean compact bordered manifold of dimension $N_0 \geq 2$ with a smooth border ∂R_0, and denote by T a locally Euclidean torus of dimension $N(T) \geq 1$. The Cartesian product

$$R = R_0 \times T$$

is a locally Euclidean compact bordered manifold of dimension $N_0 + N(T)$ with the smooth border $(\partial R_0) \times T$. We shall show:

<u>If</u> R_0 <u>satisfies</u> $\underline{s}(q) < 0 < \overline{s}(q)$ <u>for some</u> $q \in R$, <u>then so does</u> $R = R_0 \times T$.

Our problem for higher dimensions thus reduces to that for $N = 2$: given an $N \geq 3$, choose any 2-dimensional R_0 with the above property, e.g., the one constructed in §1, and take T with $N(T) = N - 2$. Then, by the above result, the manifold $R = R_0 \times T$, which is of dimension N, has the desired property. However, if we seek an R which is a simply connected subregion of E^N, or even any subregion of E^N, and has the above property, then the problem is still wide open.

<u>2.1</u>. <u>The manifold</u>. Let x be a generic point of R_0, and $(x^1, \ldots, x^{N_0})$ its local coordinates such that $\Sigma_1^{N_0} dx^{j2}$ gives the metric ds^2 on R_0. We represent T by the points $y = (y^1, \ldots, y^{N(T)})$ in $E^{N(T)}$ with $|y^j| \leq t^j$, $j = 1, \ldots, N(T)$, where $t^j \in (0, \infty)$, and $y^j = t^j$ and $y^j = -t^j$ are identified for every $j = 1, \ldots, N(T)$. The points p in $R = R_0 \times T$ are of the form $p = (x,y)$, and their local coordinates $(x^1, \ldots, x^{N_0}, y^1, \ldots, y^{N(T)})$ give the locally Euclidean metric

$$ds^2 = \sum_{j=1}^{N_0} dx^{j2} + \sum_{j=1}^{N(T)} dy^{j2}$$

on R. Thus R is a locally Euclidean compact bordered manifold of dimension $N_0 + N(T)$ with the compact smooth border $\partial R = (\partial R_0) \times T$. If we denote by

$\Delta_x = -\Sigma_1^{N_0} \partial^2/\partial x^{j2}$ the Laplace-Beltrami operator on R_0 and by $\Delta_y = -\Sigma_1^{N(T)} \partial^2/\partial x^{j2}$ that on T, then on $R = R_0 \times T$ it is $\Delta_p = \Delta_x + \Delta_y$. If a function $\psi(p) = \psi(x,y)$ on R depends on x only, then $\partial\psi/\partial n_p$ on $\partial R = (\partial R_0) \times T$ is given by

$$\frac{\partial}{\partial n_p} \psi(p) = \frac{\partial}{\partial n_x} \psi(x)$$

on ∂R_0 with $x \in \partial R_0$ for $p = (x,y) \in \partial R$. Let $\beta_0(x,\xi)$ and $\beta(p,q)$ be the biharmonic Green's functions of a clamped body on R_0 and R, respectively. We assume that the functions $\underline{s}$ and $\overline{s}$ for R_0 satisfy

$$\underline{s}(q) < 0 < \overline{s}(q) \quad \text{for some} \quad q \in R_0,$$

and we have to show that the functions $\underline{s}$ and $\overline{s}$ for R satisfy $\underline{s}(q) < 0 < \overline{s}(q)$ for some $q \in R$.

2.2. Sign of β_0. From the assumption on $\underline{s}$ for R_0 it follows that there exists a pair (x_0,ξ_0) of points in R_0 such that $\beta_0(x_0,\xi_0) < 0$. Denote by Γ the set $\xi_0 \times T$, a compact subset of R, and let $\hat{\beta}_0(p) = \beta_0(x,\xi_0)$ for $p = (x,y)$. In view of the expression for ds^2 we see that $\Delta_p^2\hat{\beta}_0(p) = \Delta_p^2\beta_0(x,\xi_0) = \Delta_x^2\beta_0(x,\xi_0) = 0$ on $R - \Gamma$, and $\hat{\beta}_0$ satisfies the clamping conditions on ∂R. Therefore, $\hat{\beta}_0(p)$ is in some sense close to $\beta(p,q)$, and it is to be expected that $\inf \beta < 0$ along with $\inf \hat{\beta}_0 < 0$. This is our basic motivation for taking R as $R_0 \times T$. However, the rate of change of $\hat{\beta}_0(p)$ as $r = |p - q| \to 0$, $q = (\xi_0,y) \in \Gamma$, is identical with that of $\beta_0(x,\xi_0)$ as $r = |x - \xi_0| \to 0$, $p = (x,y)$, and this is significantly different from that of $\beta(p,q)$ as $r = |p - q| \to 0$. Therefore, at first sight it appears to be impossible to conveniently relate $\hat{\beta}_0(p)$ or $\beta_0(x,\xi_0)$ to $\beta(p,q)$. This difficulty, arising from the wide gap between the rates of change, can nevertheless be very easily overcome by the following simple device, which is the essence of our proof.

2.3. Biharmonic Poisson equation. Recall that $\beta_0 : R_0 \times R_0 \to (-\infty,\infty]$ is a continuous map (IX. §4) and, therefore, we can find a disk B_0 about ξ_0 such that $B_0 \cup \partial B_0 \subset R_0$ and

$$a = \sup_{\xi \in B_0} \beta_0(x_0,\xi) < 0.$$

Choosing a nonnegative C^∞ function $\rho_0(\xi)$ on R_0 such that $\rho_0(\xi_0) = 1$ and $\rho_0(\xi) = 0$ on $R_0 - B_0$, we form a "potential"

$$u_0(x) = \int_{R_0} \beta_0(x,\xi) {*}\rho_0(\xi)$$

on R_0, with $*1 = d\xi^1 \ldots d\xi^{N_0}$ the volume element of R_0. The function u_0 is the unique solution of the following boundary value problem for the biharmonic Poisson equation:

$$\begin{cases} \Delta_x^2 u_0(x) = \rho_0(x), & x \in R_0, \\ u_0(x) = \dfrac{\partial}{\partial n_x} u_0(x) = 0, & x \in \partial R_0. \end{cases}$$

In view of $a < 0$ and the definition of ρ_0, we have

$$u_0(x_0) \leq a \int_{B_0} {*}\rho_0(\xi) < 0.$$

2.4. First inequality. Consider functions u and ρ on R defined by $u(p) = u_0(x)$ and $\rho(p) = \rho_0(x)$ for $p = (x,y)$. The set $B = B_0 \times T$ is an open neighborhood of Γ with $B \cup \partial B \subset R$. Clearly, ρ is a nonnegative C^∞ function on R such that $\rho = 1$ on Γ and $\rho = 0$ on $R - B$. By the above biharmonic Poisson equation and the definition of ds^2,

$$\begin{cases} \Delta_p^2 u(p) = \rho(p), & p \in R = R_0 \times T, \\ u(p) = \dfrac{\partial}{\partial n_p} u(p) = 0, & p \in \partial R = (\partial R_0) \times T. \end{cases}$$

Since any solution of this Poisson equation satisfying the clamping conditions at the boundary can be represented as a potential of the Green's function $\beta(p,q)$ with density ρ, we have

$$u(p) = \int_R \beta(p,q) {*}\rho(q)$$

on R, with $*1 = d\xi^1 \ldots d\xi^{N_0} d\eta^1 \ldots d\eta^{N(T)}$ the volume element of R for $q = (\xi^1,\ldots,\xi^{N_0},\eta^1,\ldots,\eta^{N(T)})$. To prove the first inequality, $\underline{s}(q) < 0$, we maintain that the function $\underline{s}$ for R satisfies

$$\inf_{q\in R} \underline{s}(q) \leq b = \inf_{q\in B} \beta(p_0,q) < 0,$$

where $p_0 = (x_0,y_0)$ for a fixed $y_0 \in T$. Suppose that, contrary to the assertion, $b \geq 0$. Then

$$u(p_0) = \int_B \beta(p_0,q)*\rho(q) \geq b \int_B *\rho(q) \geq 0,$$

but this is impossible since $u(p_0) = u_0(x_0) < 0$ by 2.3. We have proved the first inequality, $\underline{s}(q) < 0$, for some $q \in R$.

2.5. Second inequality. To prove the second inequality, $0 < \overline{s}(q)$, for the same q, we shall show that

$$c_N = \beta(q,q) = \lim_{p\to q} \beta(p,q) > 0$$

for every $q \in R$. This is trivial for $N \geq 4$, since $\beta(p,q) \sim -c \log r \to +\infty$ for $N = 4$ and $\sim cr^{-N+4} \to +\infty$ for $N \geq 5$ as $r = |p-q| \to 0$, and thus $c_N = +\infty$ for $N \geq 4$. The essential cases are $N = 2,3$, since $c_N < +\infty$. But in these cases, the β-density $H(p,q) = \Delta_p\beta(p,q)$ is in the class $L^2(R) \cap H(R-q)$ and by Stokes' formula, orthogonal to $H_2(R) = L^2(R) \cap H(R)$,

$$(h,H(\cdot,q)) = 0$$

for every $h \in H_2(R)$ (IX. §1). Here $H(R)$ is the class of harmonic functions on R, and $(\cdot,\cdot)$ and $\|\cdot\|$ stand for the inner product and the norm in $L^2(R)$. It is readily seen that, as a consequence,

$$\beta(p,q) = (H(\cdot,p),H(\cdot,q))$$

on $R \times R$, and the Schwarz inequality gives

$$|\beta(p,q)|^2 \leq \beta(p,p)\cdot\beta(q,q)$$

for every $p \in R$. Here $\beta(q,q) = (H(\cdot,q),H(\cdot,q)) = \|H(\cdot,q)\|^2 \geq 0$. If $\beta(q,q) = 0$, then $\beta(\cdot,q) \equiv 0$ on R, in violation of $\Delta_p^2\beta(p,q) = \delta_q$. Therefore, $c_N > 0$, and

$\overline{s}(q) > 0$ for all $q \in R$ and all $N \geq 2$.

We have established the following result:

THEOREM. *If the biharmonic Green's function* $\beta(p,q)$ *of the clamped body* R_0 *is of nonconstant sign for some* $q \in R_0$, *then the same is true of the biharmonic Green's function* $\beta(p,q)$ *of the clamped body* $R = R_0 \times T$ *for some* $q \in R$.

COROLLARY. *Hadamard's conjecture does not hold for any dimension* $N \geq 2$.

NOTES TO §2. Theorem 2.5 and its Corollary were proved in Nakai-Sario [15].

Instead of taking our counterexample of §1 for R_0, we could have chosen, e.g., Garabedian's [1] ellipse. Thus the manifold

$$R = \{(x^1,x^2) \mid (\frac{x^1}{5})^2 + (\frac{x^2}{3})^2 \leq \frac{1}{16}\} \times \prod_{j=1}^{N-2} \{y^j \in (E^1/\text{mod } 1)\},$$

$$ds^2 = (dx^1)^2 + (dx^2)^2 + \sum_{j=1}^{N-2} dy^{j2},$$

satisfies $\underline{s}(q) < 0 < \overline{s}(q)$ for some $q \in R$, with E^N the Euclidean N-space.

§3. DUFFIN'S FUNCTION AND HADAMARD'S CONJECTURE

The purpose of the present section is to systematically apply our "beta densities" to Hadamard's problem. In particular, we shall examine in detail Duffin's [1] function w from our view point of beta densities and show that w is a potential of $\Delta^2 w \geq 0$ with respect to the biharmonic Green's kernel of a clamped plate. As a consequence, the biharmonic Green's function of the clamped infinite strip is of nonconstant sign along with w. On the other hand, we show using beta densities that the biharmonic Green's function of any clamped bounded subregion exhausting the strip tends to that of the clamped strip and, therefore, takes on both positive and negative values. Since the infinite strip can be exhausted by ellipses, we have at once, without carrying out any numerical computations, the Garabedian result: a sufficiently eccentric ellipse is a counterexample to Hadamard's conjecture. Since the strip can also be exhausted by rectangles, we can add a sufficiently long *rectangle* to counterexamples to Hadamard's conjecture. If this may be called a new example, then countless "new" examples can be produced by exhausting the strip by "new" subregions. Therefore, the main conclusion of the present section is that the negative solutions to Hadamard's problem began with Duffin's pioneering work [1] and, after all, end within its framework.

We give here a rough description of the contents of the present section. First we briefly review the definition and properties of beta densities in the present setting of simply connected plane regions. We then consider, in particular, the case of an infinite strip S and discuss the space $H_2(S)$ of square integrable harmonic functions on it. For this space, the ideal boundary of S is negligible. We show that, as a consequence, Duffin's function is a biharmonic Green's potential. Using this result we discuss, in the final part of this section, Hadamard's conjecture.

3.1. Beta densities. Since we shall make essential use of beta densities, we start by summarizing those fundamentals of their theory that are pertinent in our present setting. We denote by C the finite complex plane $|z| < \infty$, $z = x + iy$, and by R a simply connected subregion, to be called a plate, of C. For convenience, we say that a plate R is smooth (or piecewise smooth) if R is relatively compact and the relative boundary ∂R is a smooth, that is, C^∞, (or piecewise smooth) Jordan curve. Assume that R is a smooth plate and set $\bar{R} = R \cup \partial R$. There exists a unique function $\beta(z,\zeta)$ on $R \times R$ such that

$$\begin{cases} \Delta_z^2\beta(z,\zeta) = \Delta_z(\Delta_z\beta(z,\zeta)) = \delta_\zeta, & z \in R, \\ \beta(z,\zeta) = \dfrac{\partial}{\partial n_z}\beta(z,\zeta) = 0, & z \in \partial R, \end{cases}$$

where $\Delta_z = -(\partial^2/\partial x^2 + \partial^2/\partial y^2)$ is the Laplace-Beltrami operator, δ_ζ the Dirac delta at $\zeta \in R$, and $\partial/\partial n$ the inner normal derivative at ∂R with respect to R. The function $\beta(\cdot,\zeta)$, which is of class C^∞ on $\bar{R} - \zeta$, is the biharmonic Green's function of the clamped plate R with pole ζ.

On a smooth plate R, we call $H(\cdot,\zeta) \equiv \Delta\beta(\cdot,\zeta)$ the beta density with pole ζ. Let $g(\cdot,\zeta)$ be the harmonic Green's function on R with the singularity

$-(1/2\pi)\log|z-\zeta|$ at ζ. Since $\Delta H(\cdot,\zeta) = \Delta^2\beta(\cdot,\zeta) = \delta_\zeta$, the function $H(\cdot,\zeta) - g(\cdot,\zeta)$ belongs to the class $H(R)$ of harmonic functions on R. By the first boundary condition,

$$\beta(z,\zeta) = \int_R g(s,z)H(s,\zeta)dpdq,$$

$s = p + iq$. If $\beta(\cdot,\zeta)$ is viewed as a potential with respect to the harmonic Green's function, then $H(\cdot,\zeta)$ is the density of $\beta(\cdot,\zeta)$. Since $H(\cdot,\zeta)$ is of class C^2 on $\overline{R}-\zeta$, we have

$$\frac{\partial}{\partial n_z}\beta(z,\zeta) = \int_R \frac{\partial}{\partial n_z} g(s,z)H(s,\zeta)dpdq.$$

Multiply both sides by an $h \in H(R) \cap C(\overline{R})$ and integrate with respect to the line element $|dz|$ on ∂R. By the Fubini theorem and the Poisson type representation of harmonic functions,

$$\int_{\partial R} h(z)\frac{\partial}{\partial n_z}\beta(z,\zeta)|dz| = \int_R h(s)H(s,\zeta)dpdq.$$

Therefore, the second boundary condition is equivalent to

$$\int_R h(s)H(s,\zeta)dpdq = 0.$$

This relation is true for every $h \in H(R) \cap C(\overline{R})$ if and only if it is true for every $h \in H_2(R) \equiv H(R) \cap L^2(R)$, since $H(R) \cap C(\overline{R})$ is dense in $H_2(R)$ with respect to the L^2 norm $\|\cdot\|$ on R. In terms of the inner product $(\cdot,\cdot)$ on $L^2(R)$, we write the orthogonality relation simply as $H(\cdot,\zeta) \perp H_2(R)$. Since $g(\cdot,\zeta) - H(\cdot,\zeta)$ belongs to $H_2(R)$,

$$\beta(z,\zeta) = (H(\cdot,z),H(\cdot,\zeta)) = \int_R H(s,z)H(s,\zeta)dpdq.$$

We claim that the beta density $H(\cdot,\zeta)$ is characterized by the following properties:

$$\begin{cases} \Delta H(\cdot,\zeta) = \delta_\zeta, \\ H(\cdot,\zeta) \in L^2(R), \\ H(\cdot,\zeta) \perp H_2(R). \end{cases}$$

That $H(\cdot,\zeta)$ satisfies the first and third of these relations was explicitly shown above. On setting $z = \zeta$ in the expression for $\beta(z,\zeta)$ and observing that $\beta(\zeta,\zeta) = \lim_{z\to\zeta}\beta(z,\zeta) < \infty$, we conclude that the second relation is satisfied. Conversely, suppose a function $\bar{H}$ on R satisfies the above three conditions. Then, since $h = \bar{H} - H(\cdot,\zeta) \in H_2(R)$, we have $(h,\bar{H}) = 0$ and $(h,H(\cdot,\zeta)) = 0$ and a fortiori, $(h,\bar{H} - H(\cdot,\zeta)) = \|h\|^2 = 0$. Hence $h \equiv 0$, and $\bar{H}$ is the beta density on R.

3.2. Fundamental kernel. The importance of the above three conditions lies in the fact that they contain no reference to the boundary ∂R of the plate R. Therefore, we can define the beta density $H(\cdot,\zeta)$, if it exists, even for an arbitrary plate R by these conditions. Reversing the usual process, we subsequently define the biharmonic Green's function $\beta(z,\zeta)$ of a general clamped plate by

$$\beta(z,\zeta) = \int_R H(s,z)H(s,\zeta)dpdq$$

on $R \times R$.

At this point the biharmonic classification theory must come in: We classify plates into two categories, according as the beta density does or does not exist, in analogy with Riemann's classification of plates into hyperbolic and parabolic types. It would not be difficult to carry out this classification; however, what we really need is not the mere existence but detailed information on properties of $\beta(z,\zeta)$. To this end, we consider what we call a fundamental kernel $K(z,\zeta)$ on R characterized by

$$\begin{cases} K(\cdot,\zeta),\ K(\zeta,\cdot) \in H(R - \zeta), \\ K(z,\zeta) + \frac{1}{2\pi}\log|z - \zeta| \in H(R), \\ K(\cdot,\zeta) \in L^2(R), \\ \lim_{\zeta\to\zeta_0} \|K(\cdot,\zeta) - K(\cdot,\zeta_0)\| = 0. \end{cases}$$

Suppose there exists a fundamental kernel $K(z,\zeta)$ on R. We claim that there then exists a beta density $H(\cdot,\zeta)$ for every $\zeta \in R$ and a Green's kernel $\beta(z,\zeta)$

of the clamped plate R with the following properties: $\Delta^2\beta(\cdot,\zeta) = \delta_\zeta$; $\beta \in C(R\times R)$ (joint continuity); $\lim_\iota \sup_{F\times F}|\beta_\iota - \beta| = 0$, where $\{R_\iota\}$ is any directed set of plates $R_\iota \subset R$ exhausting R, β_ι stands for β_{R_ι}, and F is any compact subset of R (consistency relation).

For a proof we recall that $H_2(R)$ is a locally bounded Hilbert space and consider the functional $k_\zeta(u) = (u,K(\cdot,\zeta))$ on $H_2(R)$ for any fixed $\zeta \in R$. It is easily seen that k_ζ is bounded and thus $k_\zeta \in H_2(R)$. It is also readily verified that $\lim_{\zeta\to\zeta_0}\|k_\zeta - k_{\zeta_0}\| = 0$ As a consequence, $H(\cdot,\zeta) = K(\cdot,\zeta) - k_\zeta$ is the beta density on R with pole $\zeta \in R$. By means of the properties of $K(\cdot,\zeta)$ and k_ζ it is not difficult to ascertain that $\beta(z,\zeta) \equiv (H(\cdot,z),H(\cdot,\zeta))$ is continuous on $R\times R$. From $(H(\cdot,\zeta)-H_\iota(\cdot,\zeta),H_\iota(\cdot,z)) = 0$, we obtain on setting $H_\iota(\cdot,\zeta) = H_{R_\iota}(\cdot,\zeta) = 0$ on $R - R_\iota$,

$$\begin{cases} \|H(\cdot,\zeta) - H_\iota(\cdot,\zeta)\|^2 = \|H(\cdot,\zeta)\|^2 - \|H_\iota(\cdot,\zeta)\|^2, \\ |\beta(z,\zeta) - \beta_\iota(z,\zeta)| \le \|H(\cdot,z) - H_\iota(\cdot,z)\|\cdot\|H(\cdot,\zeta) - H_\iota(\cdot,\zeta)\|. \end{cases}$$

Using these relations we deduce $\lim_\iota\|H(\cdot,\zeta) - H_\iota(\cdot,\zeta)\| = 0$ and, in view of the continuity of $\|H(\cdot,\zeta) - H_\iota(\cdot,\zeta)\|^2 = \beta(\zeta,\zeta) - \beta_\iota(\zeta,\zeta)$ on R, obtain the consistency relation. Taking the directed set $\{\Omega\}$ of smooth plates Ω in R as $\{R_\iota\}$ and observing $\Delta^2\beta_\Omega(\cdot,\zeta) = \delta_\zeta$ on Ω we see that

$$(\beta(\cdot,\zeta),\Delta^2\varphi) = \lim_{\Omega\to R}(\beta_\Omega(\cdot,\zeta),\Delta^2\varphi) = \lim_{\Omega\to R}(\Delta^2\beta_\Omega(\cdot,\zeta),\varphi) = \varphi(\zeta)$$

for every $\varphi \in C_0^\infty(R)$, and therefore, $\Delta^2\beta(\cdot,\zeta) = \delta_\zeta$ on R.

3.3. Sharpened consistency relation. An important special case is a plate R for which the biharmonic Green's function of the simply supported plate,

$$\gamma(z,\zeta) = \int_R g(s,z)g(s,\zeta)dpdq,$$

exists. This is the case if and only if $g(\cdot,\zeta) \in L^2(R)$ for some and hence for every $\zeta \in R$. The function γ is continuous on $R\times R$, $\Delta^2\gamma(\cdot,\zeta) = \Delta g(\cdot,\zeta) = \delta_\zeta$ on R, and if a part α of ∂R is an open smooth arc, then $\gamma(\cdot,\zeta) \in C^2(R\cup\alpha - \zeta)$ and $\gamma(\cdot,\zeta) = 0$ on α. In this case, $g(z,\zeta)$ is a fundamental kernel on R and the result in 3.2 applies. Since $g(\cdot,\zeta) - H(\cdot,\zeta) \in H_2(R)$,

$$\begin{cases} \beta(z,\zeta) = \int_R g(s,z)H(s,\zeta)dpdq, \\ \beta(\zeta,\zeta) = \|H(\cdot,\zeta)\|^2 \leq \|g(\cdot,\zeta)\|^2 = \gamma(\zeta,\zeta). \end{cases}$$

In view of $|\beta(z,\zeta)| \leq \gamma(z,z)^{1/2}\beta(\zeta,\zeta)^{1/2}$, $\beta(\cdot,\zeta)$ is continuous on $R \cup \alpha$ and $\beta(\cdot,\zeta) = 0$ on α. We remark that in the case in which γ exists, the following sharpened form of the consistency relation is valid. Suppose $\{R_\iota\}$ is a directed set exhausting R such that ∂R_ι contains an open smooth arc α on ∂R. Then by

$$|\beta(z,\zeta) - \beta_\iota(z,\zeta)| = |\int_R g(s,z)(H(s,\zeta) - H_\iota(s,\zeta))dpdq|$$
$$\leq \gamma(z,z)^{1/2}\|H(\cdot,\zeta) - H_\iota(\cdot,\zeta)\|,$$

$\beta_\iota(z,\zeta)$ converges to $\beta(z,\zeta)$ uniformly on $F_1 \times F_2$, with F_1 any compact subset of $R \cup \alpha$, and F_2 any compact subset of R.

3.4. Infinite strip. Having completed the preparatory part we proceed to our main discussion. We consider, as our basic plate, the infinite strip

$$S = \{z = x + iy \mid -\infty < x < \infty, -1 < y < 1\}.$$

The relative boundary ∂S consists of the lines $y = \pm 1$. We denote by $g(z,\zeta)$ the harmonic Green's function on S. Let $S_m = \{z \in S \mid |x| < m\}$ and denote by $g_m(z,\zeta)$ the harmonic Green's function on S_m, $m = 1,2,\ldots$. Fix an arbitrary $\zeta \in S$, an $n = 1,2,\ldots$, and then an $m = 1,2,\ldots$ such that $\zeta \in S_m$ and $|\mathrm{Re}\, z^{-n}| \in H(S - \bar{S}_m)$. Let c_0 (c_1, resp.) be the supremum (infimum, resp.) of $g(\cdot,\zeta)$ ($|\mathrm{Re}\, z^{-n}|$, resp.) on $S \cap \partial S_m$, and set $c = c_0/c_1$. Comparing boundary values of $g_{m+k}(\cdot,\zeta)$ and $c|\mathrm{Re}\, z^{-n}|$ on $\partial(S_{m+k} - \bar{S}_m)$, we have $g_{m+k}(\cdot,\zeta) \leq c|\mathrm{Re}\, z^{-n}|$ on $S_{m+k} - \bar{S}_m$. On letting $k \to \infty$ we see that $g(\cdot,\zeta) \leq c|\mathrm{Re}\, z^{-n}|$ on $S - \bar{S}_m$, and conclude that

$$\lim_{x \to \pm\infty} g(z,\zeta)/|\mathrm{Re}\, z^{-n}| = 0, \qquad n = 1,2,\ldots,$$

where $x = \mathrm{Re}\, z$ and $z \in S$. In particular, $g(\cdot,\zeta) \in L^2(S)$, and the result in 3.3 applies to S. We denote simply by $H(\cdot,\zeta)$ the beta density on S and by $\beta(z,\zeta)$ the Green's function of the clamped plate S.

We study the class $H_2(S)$ and consider two subspaces $H_2(S)_k$, $k = 1,2$, as follows. First let $H_2(S)_1$ be the subspace of $H_2(S)$ consisting of the functions $u \in H_2(S)$ with $u \in C^\infty(\overline{S})$, $\overline{S} = S \cup \partial S$, and $u(\cdot, \pm 1) \in L^2(-\infty,\infty)$. We maintain that $H_2(S)_1$ is dense in $H_2(S)$ in the L^2 norm,

$$\overline{H_2(S)_1} = H_2(S).$$

To see this, let h be an arbitrary element in $H_2(S)$ and consider $h_\lambda(z) = h(z/\lambda)$ on S with $\lambda \in (1,\infty)$. By the Fubini theorem, since

$$\int_{-1}^{1} \psi(y)dy = \|h\|^2 < \infty, \qquad \psi(y) = \int_{-\infty}^{\infty} h(x,y)^2 dx,$$

we see that $\psi(y) < \infty$ for almost every $y \in (-1,1)$ and a fortiori, $h_\lambda(\cdot,\pm 1) \in L^2(-\infty,\infty)$ for almost every $\lambda \in (1,\infty)$. Thus we can choose a decreasing sequence $\{\lambda_n\}$ converging to 1 such that $h_n(\cdot,\pm 1) \equiv h_{\lambda_n}(\cdot,\pm 1) \in L^2(-\infty,\infty)$ for $n = 1,2,\ldots$. Since $\lim_{n\to\infty}\|h_n - h\| = 0$, as can be easily verified, we conclude that $h \in \overline{H_2(S)_1}$.

3.5. Negligible boundary. We next prove that the ideal boundary $x = \pm\infty$ is negligible for the class $H_2(S)$ in the sense that

$$\{h \in H_2(S)_1 \mid h|\partial S = 0\} = \{0\}.$$

In the notation of the classification theory (e.g., Sario-Nakai [1]) this fact may be expressed as $S \in SO_{H_2}$. For the proof, we choose an arbitrary h in $H_2(S)_1$ with $h|\partial S = 0$ and consider

$$f(x) = \int_{-1}^{1} h(x,y)^2 dy$$

on $(-\infty,\infty)$. Keeping $\Delta h = 0$ in mind, we have

$$\frac{\partial^2}{\partial x^2} h(x,y)^2 = 2\left(\frac{\partial}{\partial x} h(x,y)\right)^2 - 2h(x,y)\frac{\partial^2}{\partial y^2} h(x,y).$$

Since $h(x,\pm 1) = 0$, integration by parts gives

$$\int_{-1}^{1} h(x,y)\frac{\partial^2}{\partial y^2} h(x,y)dy = -\int_{-1}^{1}\left(\frac{\partial}{\partial y} h(x,y)\right)^2 dy.$$

Therefore,

$$\frac{d^2}{dx^2} f(x) = 2\int_{-1}^{1} |\nabla h(x,y)|^2 dy \geq 0,$$

so that $f(x)$ is a nonnegative convex function on $(-\infty,\infty)$. On the other hand, the relation

$$\int_{-\infty}^{\infty} f(x)dx = \|h\|^2 < \infty$$

implies the existence of an increasing (decreasing, resp.) sequence $\{r_n^+\}$ ($\{r_n^-\}$, resp.) converging to $+\infty$ ($-\infty$, resp.) such that $\lim_{n\to\infty} f\{r_n^{\pm}\} = 0$. By the convexity of f,

$$0 \leq \sup_{r_n^- \leq x \leq r_n^+} f(x) = \max(f(r_n^+), f(r_n^-))$$

for every n and hence $f(x) \equiv 0$ on $(-\infty,\infty)$. Therefore, $\|h\| = 0$ and $h \equiv 0$ on S.

<u>3.6</u>. <u>Fundamental Lemma</u>. We now prove a simple lemma which will play a decisive role in our discussion. To state the lemma, it will be convenient to use the notation

$$\begin{cases} [h] = \lim_{|x|\to\infty} \sup [h](x), \\ [h](x) = \sup_{|y|<1} |h(x,y)| + \sup_{|y|<1} |\frac{\partial}{\partial x} h(x,y)| \end{cases}$$

for each $h \in H_2(S)$. We designate by $H_2(S)_2$ the subclass of $H_2(S)_1$ consisting of those $h \in H_2(S)_1$ for which $[h] < \infty$. In view of 3.5, it would seem reasonable to expect that $[h] < \infty$ for all $h \in H_2(S)$ or at least for the majority of h in $H_2(S)$. This expectation is justified in the following form:

<u>FUNDAMENTAL LEMMA</u>. <u>The subspace</u> $H_2(S)_2$ <u>is dense in</u> $H_2(S)_1$ <u>and a fortiori in</u> $H_2(S)$,

$$\overline{H_2(S)_2} = \overline{H_2(S)_1} = H_2(S).$$

The proof will be given in 3.7-3.8.

<u>3.7</u>. <u>Fourier transforms</u>. For any given $h \in H_2(S)$ we have to find a sequence $\{h_n\}$ in $H_2(S)_2$ converging to h in the L^2 norm. By 3.4 we may assume that

$h \in H_2(S)_1$. We choose two sequences $\{\varphi_j\}$ and $\{\psi_j\}$, $j = 1,2,\ldots$, in $C^\infty(-\infty,\infty)$ such that $\varphi_j(x) = h(x,1)$ and $\psi_j(x) = h(x,-1)$ on $|x| \leq j$; $\varphi_j(x) = \psi_j(x) = 0$ on $|x| \geq j + 1$; and

$$\lim_{j\to\infty}\left(\int_{-\infty}^{\infty} (\varphi_j(x) - h(x,1))^2 dx + \int_{-\infty}^{\infty} (\psi_j(x) - h(x,-1))^2 dx\right) = 0.$$

We denote by $\hat{\varphi}_j = \mathfrak{F}\varphi_j$ and $\hat{\psi}_j = \mathfrak{F}\psi_j$ the Fourier transforms of φ_j and ψ_j,

$$\hat{\varphi}_j(p) = (\mathfrak{F}\varphi_j)(p) = \int_{-\infty}^{\infty} e^{-ipx}\varphi_j(x)dx,$$

with $p \in (-\infty,\infty)$. Since φ_j and ψ_j are in the subspace $C_0(-\infty,\infty)$ of the space $\mathcal{S}(-\infty,\infty)$ of rapidly decreasing functions on $(-\infty,\infty)$, $\hat{\varphi}_j$ and $\hat{\psi}_j$ are again in $\mathcal{S}(-\infty,\infty)$.

Consider the function

$$\begin{aligned} u_j(p,y) &= \frac{\hat{\varphi}_j(p)e^{p} - \hat{\psi}_j(p)e^{-p}}{e^{2p} - e^{-2p}}\cdot e^{py} + \frac{\hat{\psi}_j(p)e^{p} - \hat{\varphi}_j(p)e^{-p}}{e^{2p} - e^{-2p}}\cdot e^{-py} \\ &= \frac{e^{p}e^{py} - e^{-p}e^{-py}}{e^{2p} - e^{-2p}}\cdot\hat{\varphi}_j(p) + \frac{e^{p}e^{-py} - e^{-p}e^{py}}{e^{2p} - e^{-2p}}\cdot\hat{\psi}_j(p). \end{aligned}$$

It is easy to see that $u_j \in C^\infty(\overline{S})$, $u_j(\cdot,y) \in \mathcal{S}(-\infty,\infty)$, and

$$\begin{cases} |u_j(p,y)| \leq c(|\hat{\varphi}_j(p)| + |\hat{\psi}_j(p)|), & (p,y) \in S, \\ \lim\limits_{y\to 1} u_j(p,y) = \hat{\varphi}_j(p),\ \lim\limits_{y\to -1} u_j(p,y) = \hat{\psi}_j(p), & p \in (-\infty,\infty), \end{cases}$$

where c is a universal constant. The inverse Fourier transform

$$h_j(x,y) = (\overline{\mathfrak{F}}u_j(\cdot,y))(x) = \int_{-\infty}^{\infty} e^{ixp}u_j(p,y)dp$$

of $u_j(\cdot,y)$ belongs to $H_2(S)_1$ and has boundary values

$$\begin{cases} h_j(x,1) = (\overline{\mathfrak{F}}\hat{\varphi}_j)(x) = (\overline{\mathfrak{F}}\mathfrak{F}\varphi_j)(x) = \varphi_j(x), \\ h_j(x,-1) = (\overline{\mathfrak{F}}\hat{\psi}_j)(x) = (\overline{\mathfrak{F}}\mathfrak{F}\psi_j)(x) = \psi_j(x) \end{cases}$$

on $(-\infty,\infty)$. By the Plancherel theorem,

$$\int_{-\infty}^{\infty} |h_j(x,y) - h_{j+k}(x,y)|^2 dx = \int_{-\infty}^{\infty} |u_j(p,y) - u_{j+k}(p,y)|^2 dp = a_{j,k}(y)$$

and

$$a_{j,k}(y)^{1/2} \le c\left(\int_{-\infty}^{\infty} |\varphi_j(p) - \varphi_{j+k}(p)|^2 dp\right)^{1/2} + c\left(\int_{-\infty}^{\infty} |\psi_j(p) - \psi_{j+k}(p)|^2 dp\right)^{1/2} = b_{j,k}.$$

Therefore, $\|h_j - h_{j+k}\|^2 = \int_{-1}^{1} a_{j,k}(y)dy \le 2b_{j,k}^2$, and by the definitions of $\{\varphi_j\}$ and $\{\psi_j\}$,

$$\lim_{j\to\infty} \|h_j - h_{j+k}\| = 0.$$

In view of the completeness of $H_2(S)$, there exists an $h_\infty \in H_2(S)$ such that $\{h_j\}$ converges to h_∞ in L^2 norm. By the local boundedness of $H_2(S)$ established in IX.§2 and the fact that $h_j(x,\pm 1) - h_{j+k}(x,\pm 1) = 0$ on $|x| \le j$, the convergence of $\{h_j\}$ to h_∞ is also pointwise and uniform on each compact subset of $\overline{S}$. In particular, $h_\infty(x,\pm 1) = h(x,\pm 1)$ on $(-\infty,\infty)$ and $h_\infty \in H_2(S) \cap C^\infty(\overline{S})$. The function $v = h - h_\infty \in H_2(S)$ has vanishing boundary values on ∂S and $v \in H_2(S)_1$. By 3.5, we have $v \equiv 0$ on S and, for $h_j \in H_2(S)_1$,

$$\lim_{j\to\infty} \|h_j - h\| = 0.$$

3.8. Completion of proof. It remains to show that $\{h_j\} \subset H_2(S)_2$, that is, $[h_j] < \infty$ for every $j = 1,2,\ldots$. By 3.7 and the fact that $\hat{\varphi}_j$ and $\hat{\psi}_j$ belong to $\mathcal{S}(-\infty,\infty)$,

$$|h_j(x,y)| \le \int_{-\infty}^{\infty} |u_j(p,y)|dp \le c\int_{-\infty}^{\infty} (|\hat{\varphi}_j(p)| + |\hat{\psi}_j(p)|)dp = c_j < \infty.$$

Similarly,

$$\begin{aligned}
\left|\frac{\partial}{\partial x} h_j(x,y)\right| &= \left|\int_{-\infty}^{\infty} e^{ixp} ipu_j(p,y)dp\right| \\
&\le \int_{-\infty}^{\infty} |pu_j(p,y)|dp \\
&\le c\int_{-\infty}^{\infty} (|p\hat{\varphi}_j(p)| + |p\hat{\psi}_j(p)|)dp = c_j' < \infty,
\end{aligned}$$

since $p\hat{\varphi}_j(p)$ and $p\hat{\psi}_j(p)$ belong to $\mathcal{S}(-\infty,\infty)$ along with $\hat{\varphi}_j$ and $\hat{\psi}_j$. We conclude that $[h_j] \le c_j + c_j' < \infty$. The proof of the Fundamental Lemma is complete.

3.9. Duffin's function. Consider the function

$$D(s,y) = \frac{1}{s^4} + \frac{sy \sinh s \sinh sy - (\sinh s + s \cosh s)\cosh sy}{s^4(s + \cosh s \sinh s)}$$

with $(x,y) \in C \times [-1,1]$. Observe that $s = 0$ is a removable singularity and $D(p,y)$ is a real-valued C^∞ function of $(p,y) \in S$. Take an arbitrary nonnegative function $\rho(x)$ belonging to the class $C_0^\infty(-\infty,\infty)$ and denote by $\hat{\rho}(p)$ the Fourier transform of $\rho(x)$. Since ρ has compact support, $\hat{\rho}$ can be continued analytically to C. In view of $\hat{\rho} \in \mathcal{S}(-\infty,\infty)$, the function

$$w(x,y) = w_\rho(x,y) = \int_{-\infty}^{\infty} e^{ixp} D(p,y)\hat{\rho}(p)dp,$$

to be referred to as Duffin's function with density $\rho(x)$, is well defined on S. We extend ρ to S by $\rho(z) \equiv \rho(x)$, and readily obtain the following properties of w:

$$\begin{cases} w \in C^\infty(S), \\ \Delta^2 w(z) = \rho(z) \quad \text{on} \quad S, \\ w(z) = \dfrac{\partial}{\partial n} w(z) = 0 \quad \text{on} \quad \partial S, \\ [w] = 0. \end{cases}$$

Less obvious is the following result: If $\rho \not\equiv 0$, then

$$\inf_{z\in S} w(z) < 0.$$

The above definitions and properties of $D(s,y)$ and $w(z)$ are due to Duffin [1].

3.10. Nonconstant sign of Duffin's function. For the convenience of the reader we sketch Duffin's proof of $\inf_S w < 0$. In the (p,q)-plane, consider the strip $T = \{|p| < \infty,\ 0 < q < c = 3\pi/4\}$. The function $e^{ixs}D(s,y)\hat{\rho}(s)$, as a function of the complex variable $s = p + iq$, is holomorphic on $\overline{T}$ except for two simple poles $\alpha = a + ib$ with $a,b > 0$, and $-\bar{\alpha} = -a + ib$ on T which are nonzero roots of

$s + \cosh s \sinh s = 0$ on T. We denote by T_n the finite strip $|p| < n$, $0<q<c$, for $n = 1,2,\ldots$. By the residue theorem,

$$\int_{\partial T_n} e^{ixs} D(s,y)\hat{\rho}(s)ds = R,$$

where $n > a$ and R is the $2\pi i$-fold sum of the residues of $e^{ixs}D(s,y)\hat{\rho}(s)$ at α and $-\bar{\alpha}$. Since $\hat{\rho} \in \mathcal{S}(-\infty,\infty)$ and $D(s,y)$ is bounded on $T - T_n$,

$$\lim_{n\to\infty} \int_{T\cap\partial T_n} e^{ixs}D(s,y)\hat{\rho}(s)ds = 0.$$

Therefore,

$$w(z) = R + \int_{\mathrm{Im}\ s=c} e^{ixs}D(s,y)\hat{\rho}(s)ds.$$

Here the last term is dominated by $e^{-cx}\int_{-\infty}^{\infty} |D(p + ic,y)\hat{\rho}(p + ic)|dp$, with the integral bounded for $|y| < 1$. Computing R explicitly, we obtain

$$w(z) = A(y)e^{-bx}\cos(ax + B(y)) + O(e^{-cx}),$$

where $A(y)$ and $B(y)$ are functions of y only, and $A(y) \neq 0$ for some $|y|<1$. In view of $0 < b < c$, we conclude on letting $x \to \infty$ that $\inf_S w < 0$.

3.11. Additional properties. Duffin's function has the following further properties, important from our point of view:

$$\begin{cases} \Delta w \in L^2(S), \\ \Delta w \perp H_2(S). \end{cases}$$

For the proof, observe that $\hat{\rho} \in \mathcal{S}(-\infty,\infty)$ implies the existence of a $\tau \in \mathcal{S}(-\infty,\infty)$ such that $|(p^2D(p,y) - \partial^2D(p,y)/\partial y^2)\hat{\rho}(p)| \leq \tau(p)$ on $(-\infty,\infty)$. By the Plancherel theorem,

$$\int_{-\infty}^{\infty} |\Delta w(x,y)|^2dx = \int_{-\infty}^{\infty} |(p^2D(p,y) - \frac{\partial^2}{\partial y^2} D(p,y))\hat{\rho}(p)|^2dp$$

$$\leq \int_{-\infty}^{\infty} \tau(p)^2dp = k < \infty.$$

Therefore, $\|\Delta w\|^2 = \int_{-1}^{1}\int_{-\infty}^{\infty} |\Delta w(x,y)|^2 dxdy \le k\int_{-1}^{1} dy < \infty$, so that $\Delta w \in L^2(S)$.

To prove that $\Delta w \perp H_2(S)$, we have to show that $(h,\Delta w) = 0$ for every $h \in H_2(S)$. By 3.6, it suffices to establish this for every $h \in H_2(S)_2$. Let $S_n = \{z = x+iy \,\big|\, |x| < n,\ |y| < 1\}$, $n = 1,2,\ldots$. Since h and w are in the class $C^\infty(\bar{S})$, the Green's formula can be applied to h and w on $\bar{S}_n$:

$$\int_{S_n} (h(z)\Delta w(z) - w(z)\Delta h(z))dxdy = -\int_{\partial S_n} (h(z)\frac{\partial}{\partial n}w(z) - w(z)\frac{\partial}{\partial n}h(z))|dz| .$$

By 3.9, we have in the notation of 3.6,

$$|(h,\Delta w)_{S_n}| = |\int_{S_n \cap \partial S_n} (h(z)\frac{\partial}{\partial x}w(z) - w(z)\frac{\partial}{\partial x}h(z))dy|$$
$$\le 2\max([h](n)\cdot[w](n),[h](-n)\cdot[w](-n)).$$

Since h and Δw belong to $L^2(S)$, $|(h,\Delta w)| = \lim_{n\to\infty}|(h,\Delta w)_{S_n}|$ and therefore,

$$|(h,\Delta w)| \le 4[h]\cdot[w].$$

From this and 3.9, we conclude that $(h,\Delta w) = 0$.

3.12. Biharmonic Green's potential. We recall the notation $H(z,\zeta)$ and $\beta(z,\zeta)$ for the beta density and the biharmonic Green's function of the clamped plate S in 3.4. Let ρ be as in 3.9 and denote by S_ρ the support of ρ in S. By 3.3, $|\beta(z,\zeta)|$ is dominated by $\beta(z,z)^{1/2}\gamma(\zeta,\zeta)^{1/2} \le k\beta(z,z)^{1/2}$ on $S \times S_\rho$, with $k = \sup_{S_\rho}\gamma(\zeta,\zeta)^{1/2} < \infty$. Therefore, the biharmonic Green's potential

$$\beta(z;\rho) = \int_S \beta(z,\zeta)\rho(\zeta)d\xi d\eta,$$

$\zeta = \xi + i\eta$, is well defined on S and $|\beta(z;\rho)| \le k\cdot\beta(z,z)^{1/2}\cdot\sup_{S_\rho}\rho\cdot\text{meas}(S_\rho)$. We claim:

$$\begin{cases} \Delta^2\beta(z;\rho) = \rho(z), & z \in S, \\ \Delta\beta(\cdot;\rho) \in L^2(S), \\ \Delta\beta(\cdot;\rho) \perp H_2(S). \end{cases}$$

To establish the first two of these properties of $\beta(z;\rho)$, consider the auxiliary function

$$v(z) = v_\rho(z) = \int_S H(z,\zeta)\rho(\zeta)d\xi d\eta.$$

By 3.3 and the Fubini theorem,

$$\int_S\left(\int_S |H(z,\zeta)|\rho(\zeta)d\xi d\eta\right)^2 dxdy \le \|\rho\|^2 \int_{S_\rho}\left(\int_S H(z,\zeta)^2 dxdy\right)d\xi d\eta$$

$$= \|\rho\|^2 \int_{S_\rho} \beta(\zeta,\zeta)d\xi d\eta \le k^2\|\rho\|^2 \text{meas}(S_\rho) < \infty.$$

Similarly, $(\Delta\varphi, v) = ((\Delta\varphi, H(\cdot,\zeta)),\rho)_\zeta = (\varphi,\rho)$ for any $\varphi \in C_0^\infty(S)$, so that $\Delta v = \rho$ in the sense of distributions, and by $\rho \in C^\infty(\overline{S})$ and $v \in L^2(S)$, in the genuine sense on S:

$$\begin{cases} v \in L^2(S) \cap C^\infty(S), \\ \Delta v(z) = \rho(z). \end{cases}$$

By the definition of $\beta(z;\rho)$, the relation $\beta(z,\zeta) = (g(\cdot,z), H(\cdot,\zeta))$, and the Fubini theorem, we have

$$\beta(z;\rho) = \int_S g(s,z)v(s)dpdq$$

on S. Hence $\Delta\beta(z;\rho) = v(z)$ on S, and the above properties of v imply the first two properties of $\beta(z;\rho)$ we set out to establish. To prove the third, take an arbitrary h in $H_2(S)$ and observe that

$$(h, \Delta\beta(\cdot;\rho)) = (h,v) = ((h,H(\cdot,\zeta)),\rho)_\zeta = 0.$$

<u>3.13</u>. <u>Identity of</u> β <u>and</u> w. A comparison of properties 3.9 and 3.11 of Duffin's function $w = w_\rho$ with properties 3.12 of $\beta(\cdot;\rho)$ suggests that $w \equiv \beta(\cdot;\rho)$ on S. We shall prove that this is indeed the case. Observe that $\Delta(\Delta w - \Delta\beta(\cdot;\rho)) = 0$, that is, $\Delta w - \Delta\beta(\cdot;\rho)$ belongs to $H(S)$ and, in fact, to $H_2(S)$ since both Δw and $\Delta\beta(\cdot;\rho)$ belong to $L^2(S)$. On the other hand, both Δw and $\Delta\beta(\cdot;\rho)$ are orthogonal to $H_2(S)$ and, a fortiori, $\Delta w - \Delta\beta(\cdot;\rho)$ is orthogonal to $H_2(S)$ and at the same time belongs to $H_2(S)$. Therefore,

$$\Delta w(z) \equiv \Delta\beta(z;\rho), \qquad z \in S.$$

Denote by $g_n(z,\zeta)$ the harmonic Green's function on $S_n = \{z \mid |x| < n, |y| < 1\}$, $n = 1,2,\ldots$. Let $h_n \in H(S_n) \cap C(\overline{S}_n)$ such that $h_n|\overline{S}_n \cap \partial S = 0$ and $h_n|S \cap \partial S_n = w$. Note that $|h_n| \leq \max([w](n),[w](-n))$ on ∂S_n and, therefore, on S_n. By 3.9,

$$\lim_{n\to\infty} \sup_{z\in S_n} |h_n(z)| = 0.$$

Since $w(z) - (g_n(\cdot,z),\Delta w)_{S_n}$ is harmonic on S_n with boundary values $w = h_n$ on ∂S_n, we have $w(z) - (g_n(\cdot,z),\Delta w)_{S_n} = h_n$ on S_n and conclude on letting $n \to \infty$ that

$$w(z) = \int_S g(\zeta,z)\Delta w(\zeta)d\xi d\eta$$

on S. Using the expressions in 3.12 for $v(z)$ and $\beta(z;\rho)$, the relation $\Delta w(z) \equiv \Delta\beta(z;\rho)$, and the Fubini theorem, we obtain

$$w(z) = (g(\cdot,z),\Delta\beta(\cdot;\rho)) = (g(\cdot,z),(H(\zeta,\cdot),\rho))_\zeta$$

$$= ((g(\cdot,z),H(\cdot,s)),\rho)_s = (\beta(z,\cdot),\rho) = \beta(z;\rho).$$

We have established the following

MAIN THEOREM. Duffin's function w with the density ρ is a biharmonic Green's potential of the density ρ:

$$w(z) = \int_S \beta(z,\zeta)\rho(\zeta)d\xi d\eta.$$

3.14. Counterexamples, old and new, to Hadamard's conjecture. Consider a plate R with a continuous and consistent biharmonic Green's function $\beta(z,\zeta) = (H(\cdot,z),H(\cdot,\zeta))$ which satisfies the clamping conditions $\beta(\cdot,\zeta) = \partial\beta(\cdot,\zeta)/\partial n = 0$ on ∂R if R is a smooth plate. Let μ and ν be any (signed) Radon measures on R and set

$$(H\mu)(s) = \int_R H(s,z)d\mu(z).$$

The beta mutual energy $\beta[\mu,\nu]$ is given by

$$\beta[\mu,\nu] = \int_{R\times R} \beta(z,\zeta)d\mu(z)d\nu(\zeta) = (H\mu,H\nu).$$

Therefore, the biharmonic Green's function β satisfies the energy principle (strict positive definiteness):

$$\beta[\mu,\mu] \geq 0,$$

and the equality holds if and only if $\mu = 0$. The mere positiveness is clear from the definition. Suppose $\beta[\mu,\mu] = 0$. Then $H\mu \equiv 0$ on R, and the distribution identity $\Delta H\mu = \mu$ implies that $\mu = 0$. As a special case of $\beta[\mu,\mu] \geq 0$, we obtain the relation

$$\beta(z,z) = \beta[\delta_z,\delta_z] = \|H\delta_z\|^2 = \|H(\cdot,z)\|^2 > 0,$$

which, in fact, we have repeatedly used.

The biharmonic Green's function $\beta(z,\zeta)$ certainly takes on positive values on R: $\beta(z,z) > 0$. That $\beta(z,\zeta)$ cannot take on any negative values is Hadamard's conjecture. By $\inf_S w < 0$ and the Main Theorem, the relation $\rho \geq 0$ implies that $\beta_S(z,\zeta)$ takes on negative values on $S \times S$. Thus we have the following counterexample to Hadamard's conjecture:

<u>EXAMPLE</u> (Duffin). <u>The biharmonic Green's function</u> $\beta_S(\cdot,\zeta)$ <u>of the clamped infinite strip</u> $S = \{|x| < \infty, |y| < 1\}$ <u>takes on both positive and negative values for a suitable choice of the pole</u> ζ <u>in</u> S.

Let $\{\Omega_\iota\}$ be a directed set of subregions of S such that $\bigcup_\iota \Omega_\iota = S$. By the consistency relation, $\{\beta_{\Omega_\iota}\}$ converges to β_S uniformly on each compact subset of $S \times S$. Therefore, $\inf \beta_{\Omega_\iota} < 0$ along with β_S if Ω_ι is sufficiently close to S. We have here a good example of the importance and effectiveness of discussing potential theory on noncompact carriers even for the study of compact carriers. As an example, consider in S the ellipse

$$E_n = \{z = x + iy \mid \frac{x^2}{n^2} + y^2 < 1\}$$

whose eccentricity tends to ∞ with n. Since $\{E_n\}$ is increasing and exhausts S, $\{\beta_{E_n}\}$ converges to β_S uniformly on each compact subset and hence $\inf \beta_{E_n} < 0$ for all sufficiently large n. Thus we have a new noncomputational proof of the following

EXAMPLE (Garabedian). The biharmonic Green's function $\beta_E(\cdot,\zeta)$ of a clamped sufficiently eccentric ellipse E takes on both positive and negative values on E for a suitable choice of the pole ζ in S.

Actually, we can produce as many regions as we wish as counterexamples to Hadamard's conjecture by the above method of exhausting Duffin's infinite strip S. We add only one more, the incentive of which was Duffin's [1] suggestion made without proof, that a quadrilateral close to a rectangle be a counterexample. Let $S_n = \{z \mid |x| < n, |y| < 1\}$. Then $\{\beta_{S_n}\}$ converges to β_S as $n \to \infty$ uniformly on each compact subset of S. We thus obtain the following "new" counterexample:

EXAMPLE. The biharmonic Green's function $\beta_R(\cdot,\zeta)$ of a clamped sufficiently elongated rectangle R takes on both positive and negative values on R for a suitable choice of the pole ζ in R.

NOTES TO §3. The expression for $w(z)$ at the end of 3.10 is due to Duffin [1], who thus gave the first example of a solution of nonconstant sign of a biharmonic Poisson equation. The conclusion in 3.14 on the nonconstant sign of $\beta(z,\zeta)$ on Duffin's strip is a slight sharpening of his result, and the example of a rectangle in 3.14 is a proof of his conjecture, for which he gave plausible reasoning. The presentation of §3 follows that in Nakai-Sario [16], where the Fundamental Lemma 3.6 and the Main Theorem 3.13 were established. The essence of §3 is that the disproofs of Hadamard's conjecture by means of Duffin's infinite strip, Garabedian's ellipse, and an elongated rectangle are all obtained in a unified manner and without computations, by making use of beta densities.

BIBLIOGRAPHY

AGMON, S.

[1] Lectures on Elliptic Boundary Value Problems, Van Nostrand, Princeton, 1965, 291 pp.

AHLFORS, L.

[1] Remarks on the classification of open Riemann surfaces, Ann. Acad. Sci. Fenn. A.I. 87 (1951), 1-8.

AHLFORS, L. - ROYDEN, H. L.

[1] A counterexample in the classification of open Riemann surfaces, Ann. Acad. Sci. Fenn. A.I. 120 (1952), 1-5.

AHLFORS, L. - SARIO, L.

[1] Riemann Surfaces, Princeton Math. Series 26, Princeton Univ. Press, 1960, 382 pp.

BELLMAN, R.

[1] On the asymptotic behavior of solutions of $u'' + (1 + f(t))u = 0$, Ann. Mat. Pura Appl. 31 (1950), 83-91.

BERGMAN, S. - SCHIFFER, M.

[1] Kernel Functions and Elliptic Differential Equations in Mathematical Physics, Academic Press, New York, 1953, 432 pp.

BRELOT, M.

[1] Eléments de la théorie classique du potentiel, Centre de Documentation Universitaire, Paris, 1965, 209 pp.

CESARI, L.

[1] Asymptotic Behavior and Stability Problems in Ordinary Differential Equations, Ergebnisse 16, Springer-Verlag, New York - Heidelberg, 1971, 271 pp.

CHEVALLEY, C.

[1] Theory of Lie Groups I, Princeton University Press, Princeton, 1946, 217 pp.

CHUNG, L.

[1] Harmonic and quasiharmonic L^p functions on the Poincaré N-ball, Doctoral dissertation, Univ. of Calif., Los Angeles, Calif., 1974, 67 pp.

[2] Manifolds carrying bounded quasiharmonic but no bounded harmonic functions, Math. Scand. 37 (1975), 264-270.

[3] Asymptotic behavior and degeneracy of biharmonic functions on Riemannian manifolds, Kodai Math. Sem. Rep. 27 (1976), 464-474.

CHUNG, L. - SARIO, L.

[1] Harmonic and quasiharmonic degeneracy of Riemannian manifolds, Tôhoku Math. J. 27 (1975), 487-496.

[2] Harmonic L^p functions and quasiharmonic degeneracy, J. Indian Math. Soc. 39 (1975), 21-28.

CHUNG, L. - SARIO, L. - WANG, C.

[1] Riemannian manifolds with bounded Dirichlet finite polyharmonic functions, Ann. Scuola Norm. Sup. Pisa 27 (1973), 1-6.

[2] Quasiharmonic L^p-functions on Riemannian manifolds, Ann. Scuola Norm. Sup. Pisa, Ser. IV, 2 (1975), 469-478.

[3] Harmonic and polyharmonic degeneracy, Math. Scand. 40 (1977), (to appear).

[4] Biharmonic and quasiharmonic degeneracy, Kōdai Math. Sem. Rep. (to appear).

CONSTANTINESCU, C. - CORNEA, A.

[1] Ideale Ränder Riemannscher Flächen, Ergebnisse 32, Springer-Verlag, Berlin - Göttingen - Heidelberg, 1963, 244 pp.

CORNEA, A.

[1] (CONSTANTINESCU, C. -) Ideale Ränder Riemannscher Flächen, Ergebnisse 32, Springer-Verlag, Berlin - Göttingen - Heidelberg, 1963, 244 pp.

COURANT, R. - HILBERT, D.

[1] Methods of Mathematical Physics, Vol. I, Interscience, New York, 1953, 560 pp.

DUFF, G. F. D.

[1] Partial Differential Equations, Univ. Toronto Press, 1956, 248 pp.

DUFFIN, R. J.

[1] On a question of Hadamard concerning super-biharmonic functions, J. Math. Physics 27 (1949), 253-258.

[2] Some problems of mathematics and science, Bull. Amer. Math. Soc. 80 (1974), 1053-1070.

GARABEDIAN, P. R.

[1] A partial differential equation arising in conformal mapping, Pacific J. Math. 1 (1951), 485-524.

[2] Partial Differential Equations, Wiley, New York - London - Sidney, 1967, 672 pp.

GLASNER, M.

[1] (SARIO, L. - SCHIFFER, M. -) The span and principal functions in Riemannian spaces, J. Analyse Math. 15 (1965), 115-134.

GOLOMB, M. - SHANKS, M.

[1] Elements of Ordinary Differential Equations, McGraw-Hill, New York, 1950, 356 pp.

HADA, D. - SARIO, L. - WANG, C.

[1] N-manifolds carrying bounded but no Dirichlet finite harmonic functions, Nagoya Math. J. 54 (1974), 1-6.

[2] Dirichlet finite biharmonic functions on the Poincaré N-ball, J. Reine Angew. Math. 272 (1975), 92-101.

[3] Bounded biharmonic functions on the Poincaré N-ball, Kōdai Math. Sem. Rep. 26 (1975), 327-342.

HADAMARD, J.

[1] Mémoire sur le problème d'analyse relatif à l'équilibre des plaques élastiques encastrées, Mémoires presentés par divers savants étrangers à l'Académie des Sciences, 33 (1908), 515-629.

HAUPT, O.

[1] Über das asymptotische Verhalten der Lösungen gewisser linearer gewöhnlicher Differentialgleichungen, Math. Z. 48 (1913), 289-292.

HILBERT, D.

[1] (COURANT, R. -) Methods of Mathematical Physics, Vol. I, Interscience, New York, 1953, 560 pp.

HILLE, E.

[1] Behavior of solutions of linear second order differential equations, Ark. Mat. 2 (1952), 25-41.

HÖRMANDER, L.

[1] Linear Partial Differential Operators, Grundlehren 116, Springer-Verlag, New York, 1969, 285 pp.

JOHN, F.

[1] Partial Differential Equations, Appl. Math. Sciences 1, Springer-Verlag, New York - Heidelberg - Berlin, 1971, 221 pp.

KWON, Y. K.

[1] Bounded harmonic but no Dirichlet-finite harmonic, Bull. Amer. Math. Soc. 79 (1973), 491-492.

KWON, Y. K. - SARIO, L. - WALSH, B.

[1] Behavior of biharmonic functions on Wiener's and Royden's compactifications, Ann. Inst. Fourier (Grenoble) 21 (1971), 217-226.

LOEWNER, C.

[1] On generalization of solutions of the biharmonic equation in the plane by conformal mapping, Pacific J. Math. 3 (1953), 417-436.

MIRANDA, C.

[1] Partial Differential Equations of Elliptic Type, Ergebnisse 2, Springer-Verlag, New York - Heidelberg - Berlin, 1970, 370 pp.

MIRSKY, N. - SARIO, L. - WANG, C.

[1] Bounded polyharmonic functions and the dimension of the manifold, J. Math. Kyoto Univ. 13 (1973), 529-535.

[2] Parabolicity and existence of bounded or Dirichlet finite polyharmonic functions, Rend. Ist. Mat. Univ. Trieste 6 (1974), 1-10.

MÜLLER, C.

[1] Spherical Harmonics, Lecture Notes 17, Springer-Verlag, Berlin - Heidelberg - New York, 1966, 46 pp.

MUSKHELISHVILI, N. I.

[1] Some Basic Problems of the Mathematical Theory of Elasticity: Fundamental Equations, Plane Theory of Elasticity, Torsion and Bending, Translated by J. R. M. Rodok, Groningen, P. Noordhoff, 1953.

NAKAI, M.

[1] On a ring isomorphism induced by quasiconformal mappings, Nagoya Math. J. 14 (1959), 201-221.

[2] A function algebra on Riemann surfaces, Nagoya Math. J. 15 (1959), 1-7.

[3] Algebraic criterion on quasiconformal equivalence of Riemann surfaces, Nagoya Math. J. 16 (1960), 157-184.

[4] A measure on the harmonic boundary of a Riemann surface, Nagoya Math. J. 17 (1960), 181-218.

[5] Genus and classification of Riemann surfaces, Osaka Math. J. 14 (1962), 153-180.

[6] On Evans' kernel, Pacific J. Math. 22 (1967), 125-137.

[7] Dirichlet finite biharmonic functions on the plane with distorted metrics, Nagoya Math. J. 51 (1973), 131-135.

[8] A test of Picard's principle for rotation free density, J. Math. Soc. Japan 27 (1975), 412-431.

[9] (SARIO, L. -) Classification Theory of Riemann Surfaces, Grundlehren 164, Springer-Verlag, Berlin - Heidelberg - New York, 1970, 446 pp.

NAKAI, M. - SARIO, L.

[1] Completeness and function-theoretic degeneracy of Riemannian spaces, Proc. Nat. Acad. Sci. 57 (1967), 29-31.

[2] A parabolic Riemannian ball, Proc. 1966 Amer. Math. Soc. Summer Inst., Amer. Math. Soc., Providence, 1968, 341-349.

[3] Biharmonic classification of Riemannian manifolds, Bull. Amer. Math. Soc. 77 (1971), 432-436.

[4] Dirichlet finite biharmonic functions with Dirichlet finite Laplacians, Math. Z. 122 (1971), 203-216.

[5] A property of biharmonic functions with Dirichlet finite Laplacians, Math. Scand. 29 (1971), 307-316.

[6] Quasiharmonic classification of Riemannian manifolds, Proc. Amer. Math. Soc. 31 (1972), 165-169.

[7] Existence of bounded Dirichlet finite biharmonic functions, Ann. Acad. Sci. Fenn. A.I. 505 (1972), 505 (1972), 1-13.

[8] Biharmonic functions on Riemannian manifolds, Continuum Mechanics and Related Problems of Analysis, Nauka, Moscow, 1972, 329-335.

[9] Existence of Dirichlet finite biharmonic functions, Ann. Acad. Sci. Fenn. A.I. 532 (1973), 1-33.

[10] Existence of bounded biharmonic functions, J. Reine Angew. Math. 259 (1973), 147-156.

[11] Parabolic Riemannian planes carrying biharmonic Green's functions of the clamped plate, J. Analyse Math. 30 (1976), 372-389.

[12] A strict inclusion related to biharmonic Green's functions of clamped and simply supported bodies, Ann. Acad. Sci. Fenn. (to appear).

[13] Manifolds with strong harmonic boundaries but without Green's functions of clamped bodies, Ann. Scuola Norm. Sup. Pisa 3 (1976), 665-670.

[14] Harmonic functions on a Riemannian ball, Math. Proc. Cambridge Philos. Soc. 80 (1976), 277-282.

[15] On Hadamard's problem for higher dimensions, J. Reine Angew. Math. (to appear).

[16] Duffin's function and Hadamard's conjecture, Pacific J. Math. (to appear).

[17] Green's functions of the clamped punctured disk, J. Austral. Math. Soc. (to appear).

[18] Existence relations between harmonic and biharmonic Green's functions, Rend. Ist. Mat. Univ. Trieste, (to appear).

[19] A nonexistence test for biharmonic functions of clamped bodies, Math. Scand. 40 (1977), (to appear).

[20] Existence of biharmonic Green's functions, Proc. London Math. Soc. (to appear).

[21] Existence of negative quasiharmonic functions, Jubilee volume dedicated to the 70th anniversary of Academician I. N. Vekua, Academy of Sciences of the USSR, (to appear).

[22] One point clamping and supporting, Rend. Mat. (to appear).

NARASIMHAN, R.

[1] Analysis on Real and Complex Manifolds, Masson, Paris, 1968, 246 pp.

O'MALLA, H.

[1] Dirichlet finite biharmonic functions on the unit disk with distorted metrics, Proc. Amer. Math. Soc. 32 (1972), 521-524.

PARREAU, M.

[1] Sur les moyennes des fonctions harmoniques et analytiques et la classification des surfaces de Riemann, Ann. Inst. Fourier (Grenoble) 3 (1951), (1952), 103-197.

RALSTON, J. - SARIO, L.

[1] A relation between biharmonic Green's functions of simply supported and clamped bodies, Nagoya Math. J. 61 (1976), 59-71.

RANGE, M.

[1] (SARIO, L. - WANG, C. -) Biharmonic projection and decomposition, Ann. Acad. Sci. Fenn. A.I. 494 (1971), 1-14.

DE RHAM, G.

[1] Variétés différentiables, Hermann, Paris, 1960, 196 pp.

RODIN, B. - SARIO, L.

[1] Principal Functions, University Series, Van Nostrand, Princeton, 1968, 347 pp.

ROYDEN, H. L.

[1] Some counterexamples in the classification of open Riemann surfaces, Proc. Amer. Math. Soc. 4 (1953), 363-370.

[2] (AHLFORS, L. -) A counterexample in the classification of open Riemann surfaces, Ann. Acad. Sci. Fenn. A.I. 120 (1952), 1-5.

SARIO, L.

[1] Existence des functions d'allure donnée sur une surface de Riemann arbitraire, C. R. Acad. Sci. Paris 229 (1949), 1293-1295.

[2] Quelques propriétés à la frontière se rattachant à la classification des surfaces de Riemann, C. R. Acad. Sci. Paris 230 (1950), 42-44.

[3] A linear operator method on arbitrary Riemann surface, Trans. Amer. Math. Soc. 72 (1952), 281-295.

[4] An extremal method on arbitrary Riemann surfaces, Trans. Amer. Math. Soc. 73 (1952), 459-470.

[5] Positive harmonic functions, Lectures on Functions of a Complex Variable, Univ. Michigan Press, Ann Arbor, 1955.

[6] Biharmonic and quasiharmonic functions on Riemannian manifolds, Duplicated lecture notes 1968-70, University of California, Los Angeles.

[7] Quasiharmonic degeneracy of Riemannian N-manifolds, Kōdai Math. Sem. Rep. 26 (1974), 53-57.

[8] Completeness and existence of bounded biharmonic functions on a Riemannian manifold, Ann. Inst. Fourier (Grenoble) 24 (1974), 311-317.

[9] Biharmonic measure, Ann. Acad. Sci. Fenn. A.I. 587 (1974), 1-18.

[10] Biharmonic Green's functions and harmonic degeneracy, J. Math. Kyoto Univ. 15 (1975), 351-362.

[11] A criterion for the existence of biharmonic Green's functions, J. Austral. Math. Soc. 21 (1976), 155-165.

[12] (AHLFORS, L. -) Riemann Surfaces, Princeton Math. Series 26, Princeton Univ. Press, 1960, 382 pp.

[13] (CHUNG, L. -) Harmonic and quasiharmonic degeneracy of Riemannian manifolds, Tôhoku Math. J. 27 (1975), 487-496.

[14] (CHUNG, L. -) Harmonic L^p functions and quasiharmonic degeneracy, J. Indian Math. Soc. 29 (1975), 21-28.

[15] (CHUNG, L. -- WANG, C.) Riemannian manifolds with bounded Dirichlet finite polyharmonic functions, Ann. Scuola Norm. Sup. Pisa 27 (1973), 1-6.

[16] (CHUNG, L. -- WANG, C.) Quasiharmonic L^p-functions on Riemannian manifolds, Ann. Scuola Norm. Sup. Pisa, Ser. IV, 2 (1975), 469-478.

[17] (CHUNG, L. -- WANG, C.) Harmonic and polyharmonic degeneracy, Math. Scand. 40 (1977) (to appear).

[18] (CHUNG, L. -- WANG, C.) Biharmonic and quasiharmonic degeneracy, Kōdai Math. Sem. Rep. (to appear).

[19] (HADA, D. -- WANG, C.) N-manifolds carrying bounded but no Dirichlet finite harmonic functions, Nagoya Math. J. 54 (1974), 1-6.

[20] (HADA, D. -- WANG, C.) Dirichlet finite biharmonic functions on the Poincaré N-ball, J. Reine Angew. Math. 272 (1975), 92-101.

[21] (HADA, D. -- WANG, C.) Bounded biharmonic functions on the Poincaré N-ball, Kōdai Math. Sem. Rep. 26 (1975), 327-342.

[22] (KWON, Y. K. -- WALSH, B.) Behavior of biharmonic functions on Wiener's and Royden's compactifications, Ann. Inst. Fourier (Grenoble) 21 (1971), 217-226.

[23] (MIRSKY, N. -- WANG, C.) Bounded polyharmonic functions and the dimension of the manifold, J. Math. Kyoto Univ. 13 (1973), 529-535.

[24] (MIRSKY, N. -- WANG, C.) Parabolicity and existence of bounded or Dirichlet finite polyharmonic functions, Rend. Ist. Mat. Univ. Trieste 6 (1974), 1-9.

[25] (NAKAI, M. -) Completeness and function-theoretic degeneracy of Riemannian spaces, Proc. Nat. Acad. Sci. 57 (1967), 29-31.

[26] (NAKAI, M. -) A parabolic Riemannian ball, Proc. 1966 Amer. Math. Soc. Summer Inst., Amer. Math. Soc., Providence, 1968, 341-349.

[27] (NAKAI, M. -) Biharmonic classification of Riemannian manifolds, Bull. Amer. Math. Soc. 77 (1971), 432-436.

[28] (NAKAI, M. -) Dirichlet finite biharmonic functions with Dirichlet finite Laplacians, Math. Z. 122 (1971), 203-216.

[29] (NAKAI, M. -) A property of biharmonic functions with Dirichlet finite Laplacians, Math. Scand. 29 (1971), 307-316.

[30] (NAKAI, M. -) Quasiharmonic classification of Riemannian manifolds, Proc. Amer. Math. Soc. 31 (1972), 165-169.

[31] (NAKAI, M. -) Existence of bounded Dirichlet finite biharmonic functions, Ann. Acad. Sci. Fenn. A.I. 505 (1972), 1-12.

[32] (NAKAI, M. -) Biharmonic functions on Riemannian manifolds, Continuum Mechanics and Related Problems of Analysis, Nauka, Moscow, 1972, 329-335.

[33] (NAKAI, M. -) Existence of Dirichlet finite biharmonic functions, Ann. Acad. Sci. Fenn. A.I. 532 (1973), 1-33.

[34] (NAKAI, M. -) Existence of bounded biharmonic functions, J. Reine Angew. Math. 259 (1973), 147-156.

[35] (NAKAI, M. -) Parabolic Riemannian planes carrying biharmonic Green's functions of the clamped plate, J. Analyse Math. 30 (1976), 372-389.

[36] (NAKAI, M. -) A strict inclusion related to biharmonic Green's functions of clamped and simply supported bodies, Ann. Acad. Sci. Fenn. (to appear).

[37] (NAKAI, M. -) Manifolds with strong harmonic boundaries but without Green's functions of clamped bodies, Ann. Scuola Norm. Sup. Pisa 3 (1976), 665-670.

[38] (NAKAI, M. -) Harmonic functions on a Riemannian ball, Math. Proc. Cambridge Philos. Soc. 80 (1976), 277-282.

[39] (NAKAI, M. -) On Hadamard's problem for higher dimensions, J. Reine Angew. Math. (to appear).

[40] (NAKAI, M. -) Duffin's function and Hadamard's conjecture, Pacific J. Math. (to appear).

[41] (NAKAI, M. -) Green's functions of the clamped punctured disk, J. Austral. Math. Soc. (to appear).

[42] (NAKAI, M. -) Existence relations between harmonic and biharmonic Green's functions, Rend. Ist. Math. Univ. Trieste (to appear).

[43] (NAKAI, M. -) A nonexistence test for biharmonic functions of clamped bodies, Math. Scand. 40 (1977) (to appear).

[44] (NAKAI, M. -) Existence of biharmonic Green's functions, Proc. London Math. Soc. (to appear).

[45] (NAKAI, M. -) Existence of negative quasiharmonic functions, Jubilee volume dedicated to the 70th anniversary of Academician I. N. Vekua, Academy of Sciences of the USSR, (to appear).

[46] (NAKAI, M. -) One point clamping and supporting, Rend. Math. (to appear).

[47] (RALSTON, J. -) A relation between biharmonic Green's functions of simply supported and clamped bodies, Nagoya Math. J. 61 (1976), 59-71.

[48] (RODIN, B. -) Principal Functions, University Series, Van Nostrand, Princeton, 1968, 347 pp.

[49] (WANG, C. -) Polyharmonic classification of Riemannian manifolds, J. Math. Kyoto Univ. 12 (1972), 129-140.

SARIO, L. - NAKAI, M.

[1] Classification Theory of Riemann Surfaces, Grundlehren 164, Springer-Verlag, Berlin - Heidelberg - New York, 1970, 446 pp.

SARIO, L. - SCHIFFER, M. - GLASNER, M.

[1] The span and principal functions in Riemannian spaces, J. Analyse Math. 15 (1965), 115-134.

SARIO, L. - WANG, C.

[1] The class of (p,q)-biharmonic functions, Pacific J. Math. 41 (1972), 799-808.

[2] Generators of the space of bounded biharmonic functions, Math. Z. 127 (1972), 273-280.

[3] Parabolicity and existence of bounded biharmonic functions, Comm. Math. Helv. 47 (1972), 341-347.

[4] Quasiharmonic functions on the Poincaré N-ball, Rend. Mat. 6 (1973), 1-14.

[5] Existence of Dirichlet finite biharmonic functions on the Poincaré 3-ball, Pacific J. Math. 48 (1973), 267-274.

[6] Radial quasiharmonic functions, Pacific J. Math. 46 (1973), 515-522.

[7] Positive harmonic functions and biharmonic degeneracy, Bull. Amer. Math. Soc. 79 (1973), 182-187.

[8] Harmonic and biharmonic degeneracy, Kōdai Math. Sem. Rep. 25 (1973), 392-396.

[9] Counterexamples in the biharmonic classification of Riemannian 2-manifolds, Pacific J. Math. 50 (1974), 159-162.

[10] Riemannian manifolds of dimension $N \geq 4$ without bounded biharmonic functions, J. London Math. Soc. 7 (1974), 635-644.

[11] Negative quasiharmonic functions, Tôhoku Math. J. 26 (1974), 85-93.

[12] Parabolicity and existence of Dirichlet finite biharmonic functions, J. London Math. Soc. 8 (1974), 145-148.

[13] Harmonic L^p-functions on Riemannian manifolds, Kōdai Math. Sem. Rep. 26 (1975), 204-209.

SARIO, L. - WANG, C. - RANGE, M.

[1] Biharmonic projection and decomposition, Ann. Acad. Sci. Fenn. A.I. 494 (1971), 1-14.

SCHIFFER, M.

[1] The span of multiply connected domains, Duke Math. J. 10 (1943), 209-216.

[2] (BERGMAN, S. -) Kernel Functions and Elliptic Differential Equations in Mathematical Physics, Academic Press, New York, 1953, 432 pp.

[3] (SARIO, L. -- GLASNER, M.) The span and principal functions in Riemannian spaces, J. Analyse Math. 15 (1965), 115-134.

SHANKS, M.

[1] (GOLOMB, M. -) Elements of Ordinary Differential Equations, McGraw-Hill, New York, 1950, 356 pp.

SZEGÖ, G.

[1] Remark on the preceding paper by Charles Loewner, Pacific J. Math. 3 (1953), 437-446.

TIMOSHENKO, S.

[1] Theory of Plates and Shells, McGraw-Hill, New York - London, 1940, 492 pp.

TÔKI, Y.

[1] On the classification of open Riemann surfaces, Osaka Math. J. 4 (1952), 191-201.

[2] On examples in the classification of Riemann surfaces, I, Osaka Math. J. 5 (1953), 267-280.

VIRTANEN, K. I.

[1] Über die Existenz von beschränkten harmonischen Funktionen auf offenen Riemannschen Flächen, Ann. Acad. Sci. Fenn. A.I. 75 (1950), 1-8.

WALSH, B.

[1] (KWON, Y. K. - SARIO, L. -) Behavior of biharmonic functions on Wiener's and Royden's compactifications, Ann. Inst. Fourier (Grenoble) 21 (1971), 217-226.

WANG, C.

[1] Biharmonic Green's functions and quasiharmonic degeneracy, Math. Scand. 35 (1974), 38-42.

[2] Biharmonic Green's functions and biharmonic degeneracy, Math. Scand. 37 (1975 122-128.

[3] (CHUNG, L. - SARIO, L. -) Riemannian manifolds with bounded Dirichlet finit polyharmonic functions, Ann. Scuola Norm. Sup. Pisa 27 (1973), 1-6.

[4] (CHUNG, L. - SARIO, L. -) Quasiharmonic L^p-functions on Riemannian manifold Ann. Scuola Norm. Sup. Pisa 2 (1975), 469-478.

[5] (CHUNG, L. - SARIO, L. -) Harmonic and polyharmonic degeneracy, Math. Scand 40 (1977) (to appear).

[6] (CHUNG, L. - SARIO, L. -) Biharmonic and quasiharmonic degeneracy, Kōdai Math. Sem. Rep. (to appear).

[7] (HADA, D. - SARIO, L. -) N-manifolds carrying bounded but no Dirichlet finite harmonic functions, Nagoya Math. J. 54 (1974), 1-6.

[8] (HADA, D. - SARIO, L. -) Dirichlet finite biharmonic functions on the Poincaré N-ball, J. Reine Angew. Math. 272 (1975), 92-101.

[9] (HADA, D. - SARIO, L. -) Bounded biharmonic functions on the Poincaré N-ball, Kōdai Math. Sem. Rep. 26 (1975), 327-342.

[10] (MIRSKY, N. - SARIO, L. -) Bounded polyharmonic functions and the dimension of the manifold, J. Math. Kyoto Univ. 13 (1973), 529-535.

[11] (MIRSKY, N. - SARIO, L. -) Parabolicity and existence of bounded or Dirichlet finite polyharmonic functions, Rend. Ist. Mat. Univ. Trieste 6 (1974), 1-9.

[12] (SARIO, L. -) The class of (p,q)-biharmonic functions, Pacific J. Math. 41 (1972), 799-808.

[13] (SARIO, L. -) Generators of the space of bounded biharmonic functions, Math. Z. 127 (1972), 273-280.

[14] (SARIO, L. -) Parabolicity and existence of bounded biharmonic functions, Comm. Math. Helv. 47 (1972), 341-347.

[15] (SARIO, L. -) Quasiharmonic functions on the Poincaré N-ball, Rend. Mat. 6 (1973), 1-14.

[16] (SARIO, L. -) Existence of Dirichlet finite biharmonic functions on the Poincaré 3-ball, Pacific J. Math. 48 (1973), 267-274.

[17] (SARIO, L. -) Radial quasiharmonic functions, Pacific J. Math. 46 (1973), 515-522.

[18] (SARIO, L. -) Positive harmonic functions and biharmonic degeneracy, Bull. Amer. Math. Soc. 79 (1973), 182-187.

[19] (SARIO, L. -) Harmonic and biharmonic degeneracy, Kōdai Math. Sem. Rep. 25 (1973), 392-396.

[20] (SARIO, L. -) Counterexamples in the biharmonic classification of Riemannian 2-manifolds, Pacific J. Math. 50 (1974), 159-162.

[21] (SARIO, L. -) Riemannian manifolds of dimension $N > 4$ without bounded biharmonic functions, J. London Math. Soc. 7 (1974), 635-644.

[22] (SARIO, L. -) Negative quasiharmonic functions, Tôhoku Math. J. 26 (1974), 3-93.

[23] (SARIO, L. -) Parabolicity and existence of Dirichlet finite biharmonic functions, J. London Math. Soc. 8 (1974), 145-148.

[24] (SARIO, L. -) Harmonic L^p-functions on Riemannian manifolds, Kōdai Math. Sem. Rep. 26 (1975), 204-209.

[25] (SARIO, L. -- RANGE, M.) Biharmonic projection and decomposition, Ann. Acad. Sci. Fenn. A.I. 494 (1971), 1-14.

WANGE, C. - SARIO, L.

[1] Polyharmonic classification of Riemannian manifolds, J. Math. Kyoto Univ. 12 (1972), 129-140.

AUTHOR INDEX

SUBJECT AND NOTATION INDEX

Underlined numbers refer to definitions